STUDENT SOLUTIONS MANUAL

Shirley Buls

St. Cloud State University

COLLEGE GEOMETRY

A PROBLEM-SOLVING APPROACH WITH APPLICATIONS

SECOND EDITION

Musser ✵ Trimpe ✵ Maurer

Upper Saddle River, NJ 07458

Vice President and Editorial Director, Mathematics: Christine Hoag
Print Supplement Editor: Joanne Wendelken
Senior Managing Editor: Linda Behrens
Associate Managing Editor: Bayani Mendoza de Leon
Project Manager, Production: Traci Douglas
Supplement Cover Manager: Paul Gourhan
Supplement Cover Designer: Victoria Colotta
Senior Operations Supervisor: Diane Peirano

Pearson Prentice Hall
Pearson Education, Inc.
Upper Saddle River, NJ 07458

Printed in the United States of America

ISBN 13: 978-0-13-187971-3

ISBN 10: 0-13-187971-5

Pearson Education Ltd., *London*
Pearson Education Australia Pty. Ltd., *Sydney*
Pearson Education Singapore, Pte. Ltd.
Pearson Education North Asia Ltd., *Hong Kong*
Pearson Education Canada, Inc., *Toronto*
Pearson Educación de Mexico, S.A. de C.V.
Pearson Education—Japan, *Tokyo*
Pearson Education Malaysia, Pte. Ltd.

Table of Contents

1 Problem Solving in Geometry

Section 1.1

Tips:

- ✔ There are often several ways to attack a problem. If your first try does not work, try something else. There may be more than one way to correctly solve a problem.
- ✔ Do not limit yourself by restrictions that are not given in the problem.
- ✔ If you spend 10 minutes on a problem and are not getting anywhere, leave the problem and come back to it later. A break often frees your mind to think of other strategies.

Solutions to odd-numbered textbook problems

1. Step 1: *Understand the Problem.*

   You need to find all of the equilateral triangles of various sizes. Some of them will overlap.

   Step 2: *Devise a Plan.*

   Count the number of triangles of each size and add the numbers together. Use pictures to help you count.

   Step 3: *Carry Out the Plan.*

Side Length	Number of Triangles
3	1
2	3
1	9

   Thus, there are a total of $1 + 3 + 9 = 13$ equilateral triangles.

   Step 4: *Look Back.*

   What would happen if the original large triangle had sides of length 4 or 5?

3. Step 1: *Understand the Problem.*

   Tetrominos consist of 4 squares, connected along a side, like 4 postage stamps might be connected.

   If a tetromino shape, when flipped or turned, looks like another one, then those two tetrominos are the same.

   Step 2 *Devise a Plan.*

   Use graph paper or scratch paper and draw possible arrangements of 4 squares. Rotate or flip, if necessary, to see if two are alike. Try starting with 4 in a row, then 3 in a row, etc.

Section 1.1

3. (continued)

Step 3: *Carry Out the Plan.*

The 5 different tetrominos are shown below.

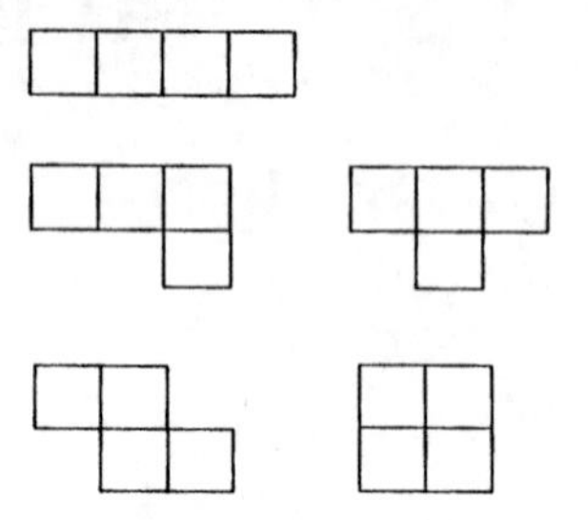

Step 4: *Look Back.*

While it might seem that there are additional solutions, they will look like these when turned or flipped.

How many arrangements of 5 squares might there be? (These are called pentominos-see problem #4.)

5. Step 1: *Understand the Problem.*

You need to fit five *different* pentominos together to make a 5 by 5 square. There should be no gaps or overlaps.

Step 2: *Devise a Plan.*

Cut out of graph paper the 12 pentomino shapes (See problem #3.) and try fitting them together. Use Guess and Test here. Remember that the shapes may be rotated and flipped over as needed.

Step 3: *Carry Out the Plan.*

Two solutions are shown.

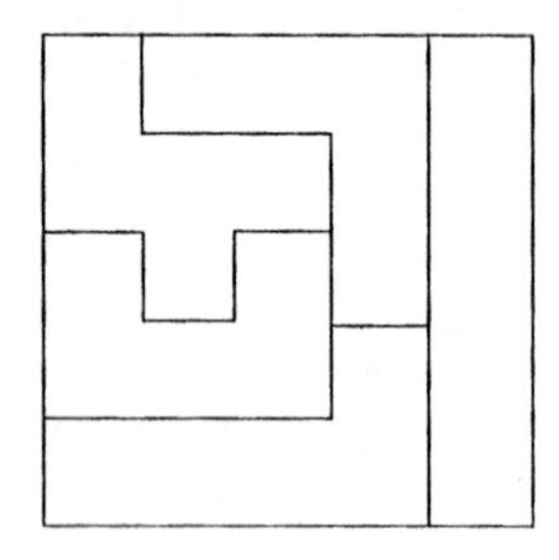

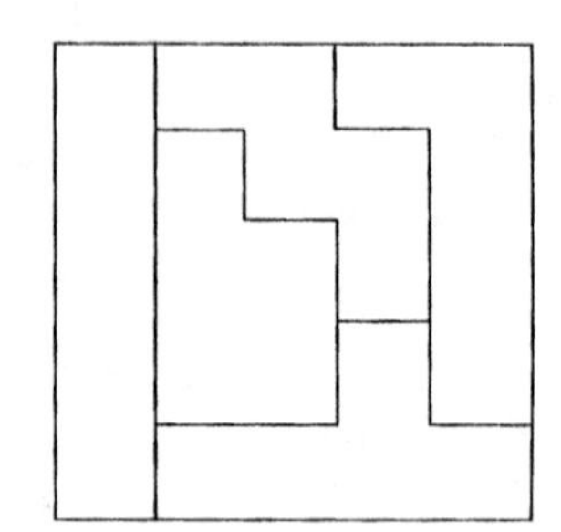

Step 4: *Look Back.*

Are there any other solutions? (See the solution given in your textbook for another example.)

7. Step 1: *Understand the Problem.*

Each shape can be cut out, folded on the edges, and formed into a cube. We must determine what number ends up across from the 1.

Section 1.1

7. (continued)

Step 2: *Devise a Plan.*

Analyze the diagram given to eliminate some possibilities. Cut out the shape and make the cube, if necessary.

Step 3: *Carry Out the Plan.*

(a) Since the squares containing 5, 6 and 3 connect with the 1 square, they cannot be opposite it when folded up. If the 5 square is folded up when 1 is on the bottom, 4 will be on top.

Thus, 4 is opposite 1.

(b) If the shape is folded so that 1 is on the bottom of the cube, 6 will be on a vertical side, and 2 will be on top.

Thus, 2 is opposite 1.

(c) Because the 1 square touches, the 2, 3, and 4 squares, the number opposite 1 must be either 5 or 6. If the shape is folded so that 3 is on the front of the cube as shown,

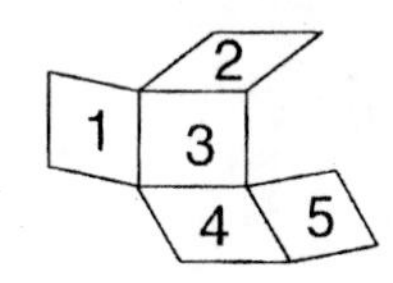

the square with 4 will be on the bottom and the square with 5 will fold up and opposite 1.

Step 4: *Look Back.*

Any square that shares a side or a corner with the 1 square cannot be the square across from it in the cube.

Verify this result and your answers to (a)–(c) by folding the cubes.

9. Step 1: *Understand the Problem.*

The final colored grid must have 4 red, 3 blue, 3 green, 3 white and 3 black squares.

No row, column or diagonal can have two of the same colors in it.

Step 2: *Devise a Plan.*

Draw a blank 4 × 4 grid and fill in squares with letters to represent colors.

R = red　　G = green　　BK = black

BL = blue　　W = white

Section 1.1

9. (continued)

Step 3: *Carry Out the Plan.*

Start with R. Because there are 4 reds, each row and column must have a red in it. One possible arrangement is shown.

R			
			R
	R		
		R	

Note that no two R's lie on the same diagonal

Place the rest of the colors, making sure that no row, column, or diagonal has two of the same colors. One solution is shown.

R	BL	W	G
G	BK	BL	R
W	R	G	BK
BK	W	R	BL

Step 4: *Look Back.*

Verify your answer by checking each color. Make sure that each one appears at most once in each row, column and diagonal.

Are there other solutions? See the solution in the text for a different coloring.

11. Step 1: *Understand the Problem.*

Each cube will be painted on one or more sides. We must find the number of unique cubes, meaning they look different, even when rotated or flipped.

Step 2: *Devise a Plan.*

Sketch cubes and consider all possible ways to paint first one side, then two sides, then three sides, etc.

Step 3: *Carry Out the Plan.*

Nine different cubes are possible:

Section 1.1 11. (continued)

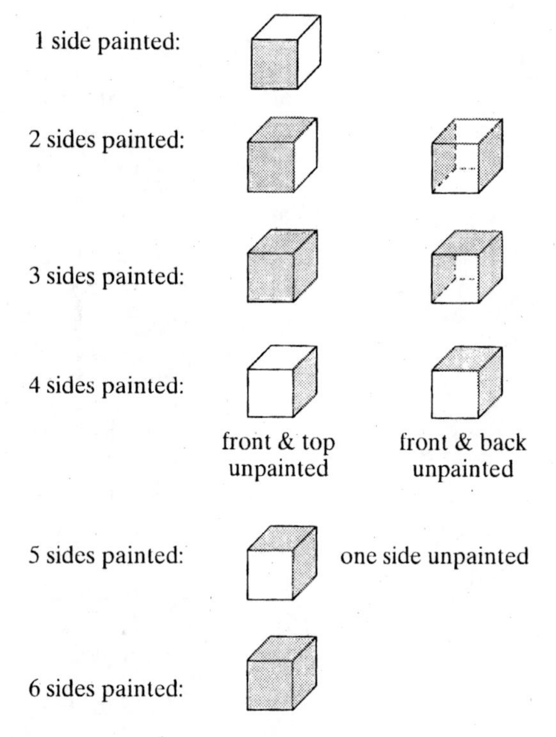

Step 4: *Look Back.*

Verify that no two of your cubes are the same by rotating them or examining them from a different perspective. In some cases, it may be easier to consider what sides are *not* painted.

13. Step 1: *Understand the Problem.*

The 4 cuts must be straight lines and the 11 pizza pieces may not all be the same size.

Step 2: *Devise a Plan.*

Draw a circle and use a ruler to draw straight line cuts, trying to create as many pieces as possible with each successive cut.

Step 3: *Carry Out the Plan.*

One possible solution is shown, with pieces numbered.

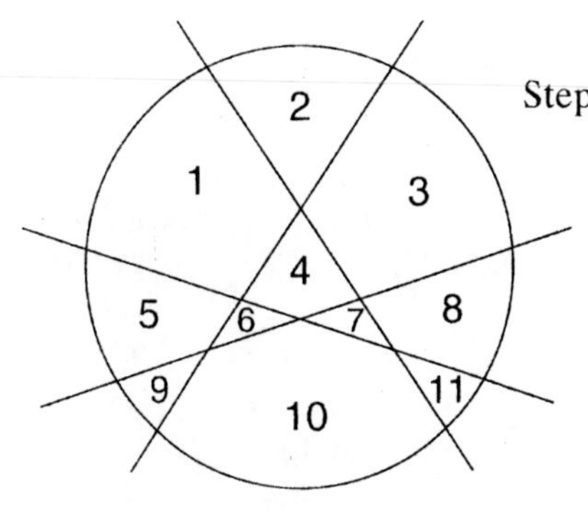

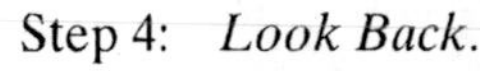

Step 4: *Look Back.*

Count the pieces in your figure to verify your solution. Is there more than one solution?

How many pieces could result if 5 cuts were allowed?

Section 1.1

15. Step 1: *Understand the Problem.*

The posts must be evenly spaced one meter apart in order to fence the 10 m by 10 m garden.

Step 2: *Devise a Plan.*

Draw a diagram with 10 units marked off on each side. Then count the posts.

Step 3: *Carry Out the Plan.*

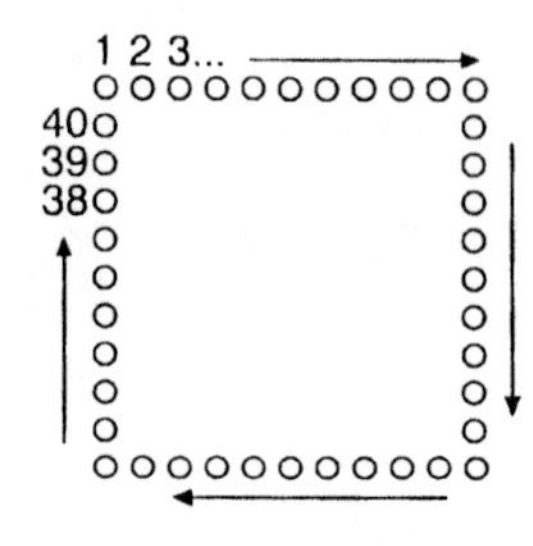

Refer to the diagram shown and count the posts. Forty posts will be needed.

Step 4: *Look Back.*

Check your diagram by counting the spaces on each side to make sure that each side really measures 10 meters. How many posts would you need for a garden that measures 11 meters by 11 meters? 12 meters by 12 meters? n meters by n meters?

17. Step 1: *Understand the Problem.*

Whole toothpicks must be removed, leaving squares. There should be no "leftover" toothpicks and squares may not be the same size.

Step 2: *Devise a Plan.*

Draw a picture of the toothpicks in the starting arrangement and erase toothpicks to leave squares.

Step 3: *Carry Out the Plan.*

(a) Remove toothpicks to leave one large and one small square. ×'s indicate removed toothpicks.

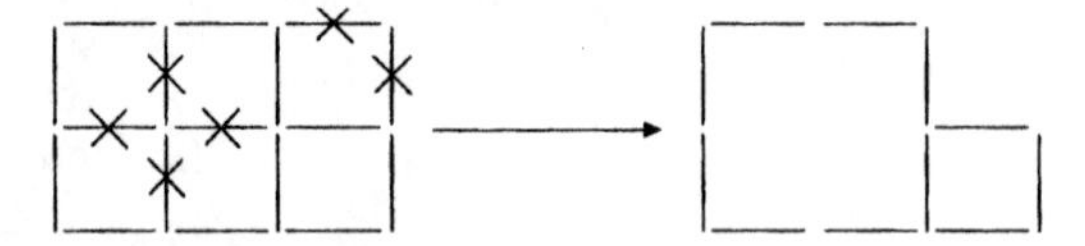

(b) Remove toothpicks to leave 3 squares of the same size.

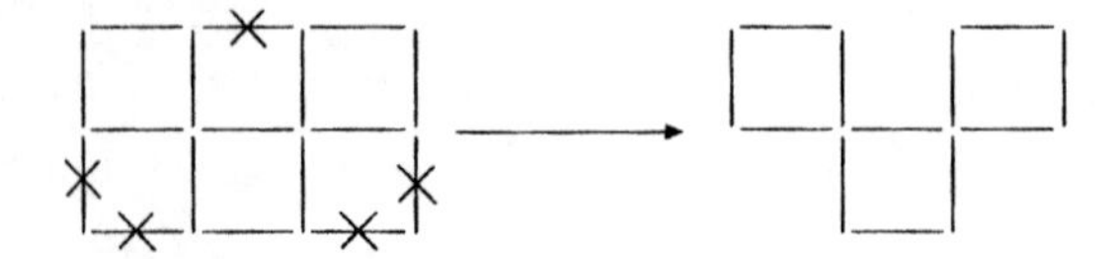

Step 4: *Look Back.*

Are there other solutions? Could there be a different number of squares left when 5 or 6 toothpicks are removed?

19. Step 1: *Understand the Problem.*

Start in any position in the figure and draw it without lifting the pencil from the paper or going over any lines again.

Section 1.1

19. (continued)

Step 2: *Devise a Plan.*

Use Guess and Test, starting in various points in the diagram.

Step 3: *Carry Out the Plan.*

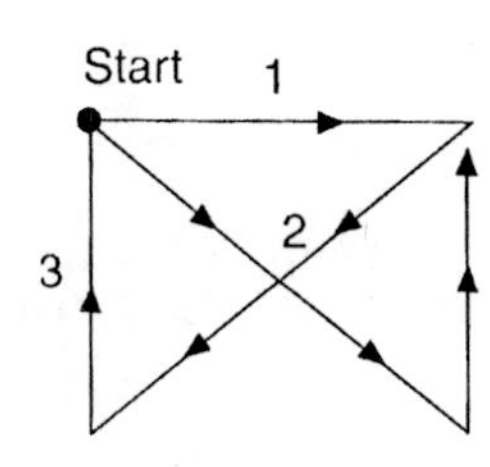

Starting in the upper left corner, we might have the tracing as shown in the diagram.

We end in the upper right corner and would have to retrace a line to exit this corner.

If we started in any other position, we would have the same difficulty. There are 4 corners with 3 line segments leading into them. That means once we enter a corner, we leave it and return to it, with no way to leave. Thus, it is impossible to sketch the figure.

Step 4: *Look Back.*

Might it be possible to sketch the figure if one more line was added to the figure? If one line was deleted?

21. Step 1: *Understand the Problem.*

The figure must be traced without lifting the pencil from the paper and without retracing any lines.

Step 2: *Devise a Plan.*

Start at any location and use Guess and Test to sketch the figure.

Step 3: *Carry Out the Plan.*

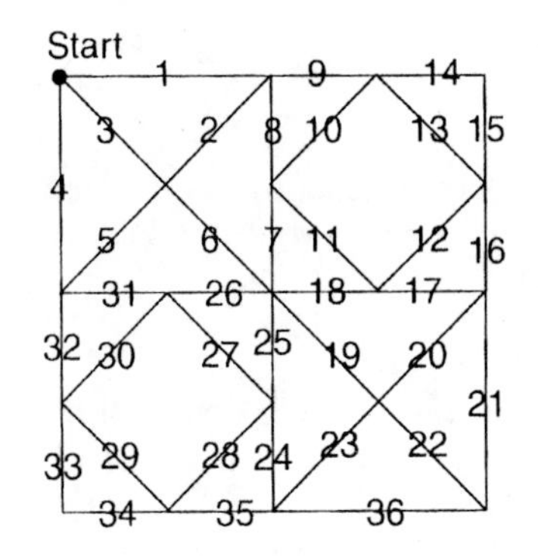

One solution is shown, where numbers indicate the order in which the lines should be traced.

Step 4: *Look Back.*

Are there other solutions? (See the textbook solution for another possibility.)

Does it matter where we begin to draw?

23. Step 1: *Understand the Problem.*

The 72-inch length of wire will be made into a rectangle. We assume there is no leftover wire and no overlap.

Step 2: *Devise a Plan.*

Use a variable to solve the problem.

Let x = width of rectangle in inches and

$3x$ = length of rectangle in inches.

Step 3: *Carry Out the Plan.*

The length of the wire will become the perimeter of the rectangle.

$$x + 3x + x + 3x = 72$$
$$8x = 72$$
$$x = 9$$

So the rectangle is 9 in. wide and 3(9) in. = 27 in. long.

Section 1.1

23. (continued)

Step 4: *Look Back.*

Verify that a rectangle that measures 9 in. by 27 in. has a perimeter of 72 inches.

25. Step 1: *Understand the Problem.*

We must use the perimeters of the square or triangular enclosures to determine the amount of chain link fence Joseph has. We assume he will use every foot of fencing he has.

Step 2: *Devise a Plan.*

Use a variable.

Let x = length (in feet) of a side of the square and

$x + 11$ = length (in feet) of a side of the triangle.

Step 3: *Carry Out the Plan.*

Using the perimeters of the 2 enclosures, we have the following equation

$$4x = 3(x + 11)$$

$$4x = 3x + 33$$

$$x = 33$$

Thus, one side of the square enclosure measures 33 ft. That means the whole enclosure uses up $4(33) = 132$ ft, which is the total amount of chain link fencing.

Step 4: *Look Back.*

To verify the answer, check to see that a 132-ft-long roll of chain link fence could enclose a square pen with a side of 33 ft or a triangular one with a side of 44 ft.

27. Step 1: *Understand the Problem.*

You need two separate towers of the same height, that is, that are the same number of blocks tall. There should be no blocks left over.

Step 2: *Devise a Plan.*

Add the heights of all of the blocks. Divide the sum by 2 to find how high each tower would have to be. Use Guess and Test to try to make towers of that height.

Step 3: *Carry Out the Plan.*

$1 + 2 + 3 + 4 + 5 + 6 + 7 + 8 + 9 + 10 = 55$, and $55 \div 2 = 27.5$. Thus, each tower would have to be 27.5 cm tall. This means it is not possible to make two towers of the same height without splitting a block.

Section 1.1

27. (continued)

Step 4: *Look Back.*

If you allow blocks to be left over, it would be possible to build the towers. Leaving out the 1-cm block, you could construct two towers 27 cm tall by using the 2-, 3-, 5-, 7-, and 10-cm blocks in one tower and the 4-, 6-, 8-, and 9-cm blocks in the other tower.

29. Step 1: *Understand the Problem.*

Fit four of the shapes shown together to form a bigger version of the same shape.

Step 2: *Devise a Plan.*

Trace and cut out 4 copies of the given shape and fit them together to form the desired shape.

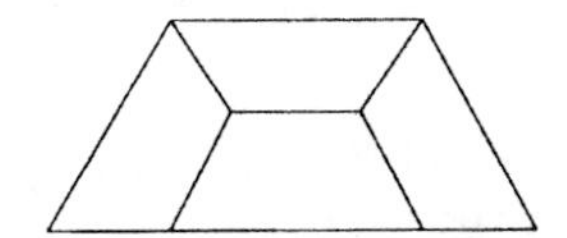

Step 3: *Carry Out the Plan.*

Rotating and inverting some pieces, we get the diagram shown.

Step 4: *Look Back.*

Is there another solution or another approach? Could we have started with the large shape and cut it into four identical pieces?

31. Step 1: *Understand the Problem.*

Using only one cut, divide the given shape into 2 pieces that fit together to make a square.

Step 2: *Devise a Plan.*

Trace the shape, look for parts of it that might fit together and try drawing cut lines to divide it in two.

Step 3: *Carry Out the Plan.*

A solution and cut line are shown.

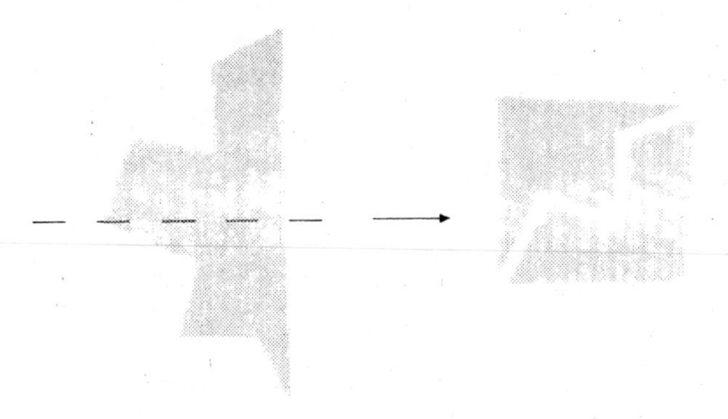

Step 4: *Look Back.*

Are there other ways to make one straight cut in the figure and rearrange the pieces to form a square?

Section 1.1

33. Step 1: *Understand the Problem.*

    The given triangle must be cut into 4 pieces that will be rearranged to form a square. No gaps or overlaps are allowed.

    Step 2: *Devise a Plan.*

    Trace the figure and cut out the pieces of the given shape and fit them together to form a square.

    Step 3: *Carry Out the Plan.*

    Using Guess and Test, rotate and invert some pieces until a square is formed.

    Step 4: *Look Back.*

    Is there another solution or another approach? Can the square be obtained without inverting any pieces?

35. Step 1: *Understand the Problem.*

    A square, triangle, and a rectangle are to each be constructed using all seven tangram pieces. No gaps or overlaps are allowed.

    Step 2: *Devise a Plan.*

    Trace the seven tangram pieces and cut them out of stiff paper, and fit them together to form (a) a square (b) a triangle and (c) a rectangle.

    Step 3: *Carry Out the Plan.*

    Using Guess and Test, rotate and invert some pieces to obtain the following

    a) 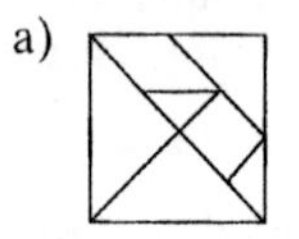 b) c) 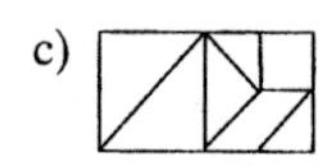

    Step 4: *Look Back.*

    Are there other was to rearrange the pieces to obtain a square, triangle, or rectangle? Can you make a square, triangle, or rectangle out of 2, 3, 4, 5, or 6 tangram pieces?

37. (a) Step 1: *Understand the Problem.*

    The nine toothpicks must form exactly five equilateral triangles. They may overlap, but you may not bend or break the toothpicks. Remember that an equilateral triangle has three equal sides.

    Step 2: *Devise a Plan.*

    Try solving the problem using nine toothpicks and the Guess and Test strategy. Count the triangles you form in each trial.

Section 1.1

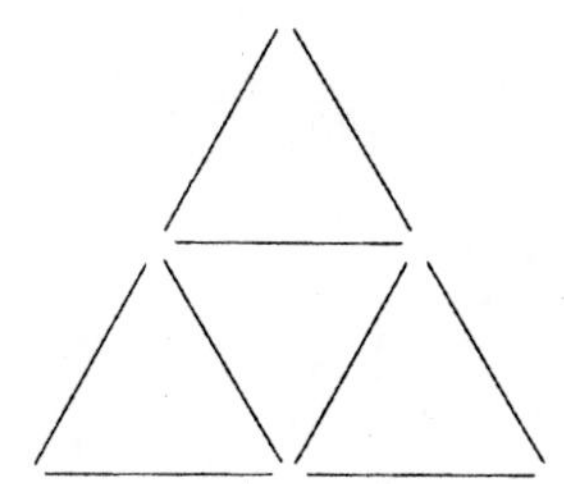

Step 3: *Carry Out the Plan.*

If you try making a large triangle with three toothpicks on a side, you only get one triangle. When you try it with a large triangle that has two toothpicks on a side, you use six toothpicks and have three left over with which to divide the middle of the large triangle.

Count the triangles of various sizes.

1 unit side:	4 triangles
2 unit side:	1 triangle
Total:	5 triangles

Step 4: *Look Back.*

Did you use exactly nine toothpicks? Are there any other sizes of triangles possible?

(b) Step 1: *Understand the Problem.*

You need to use six toothpicks this time. The four equilateral triangles must all be the same size.

Step 2: *Devise a Plan.*

Use toothpicks and the Guess and Test strategy. The toothpicks do not need to all lie on the same flat surface.

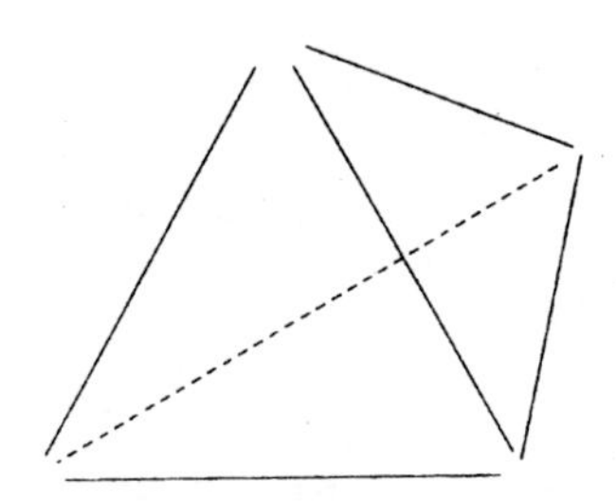

Step 3: *Carry Out the Plan.*

Make a pyramid with a triangular base.

Step 4: *Look Back.*

You need to be careful not to limit yourself by unnecessary conditions, such as the toothpicks all lying flat on the table.

39. Step 1: *Understand the Problem.*

The goal is to travel each street exactly once and pass through each intersection.

Step 2: *Devise a Plan.*

Begin by using Guess and Test to see if a starting and ending intersection can be determined. Once the starting and ending intersections are determined, then we use Guess and Test to find an acceptable route.

Step 3: *Carry Out the Plan.*

Intersections B and C are the only intersections that have an odd number of streets entering or exiting them. Therefore, if the route starts at B, it cannot end at A, D, E, or F as these intersections have an even number of streets entering or exiting them. So the route must end at C. Similarly if the route starts at C it must end at B. The following is one route.

Section 1.1 39. (continued)

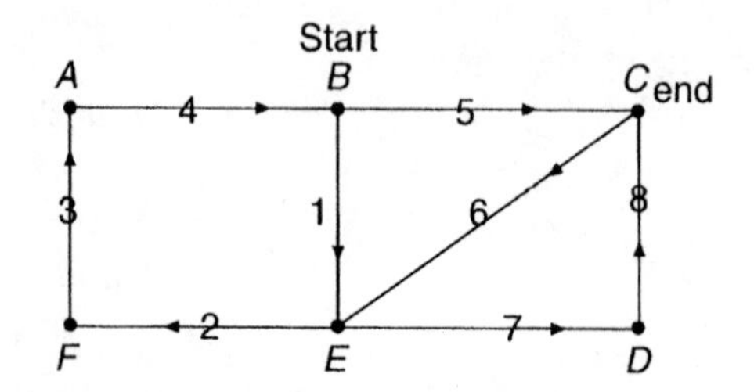

If the driver starts at point A, D, E, or F then, because there are an even number of streets available at those intersections, she must retrace a street.

Step 4: *Look Back.*

Are there any other routes possible? (See the solution given in your textbook for another possible route.)

41. (a) Step 1: *Understand the Problem.*

Each house A, B, and C is to be connected to water, electricity and gas. No pipes or wires are allowed to cross. Also, the pipes and wires are not allowed to connect any of the houses.

Step 2: *Devise a Plan.*

Draw houses A and B and connect these to the utilities. Then add house C and try to make the remaining connections.

Step 3: *Carry Out the Plan.*

Draw houses A and B and connect them to the utilities.

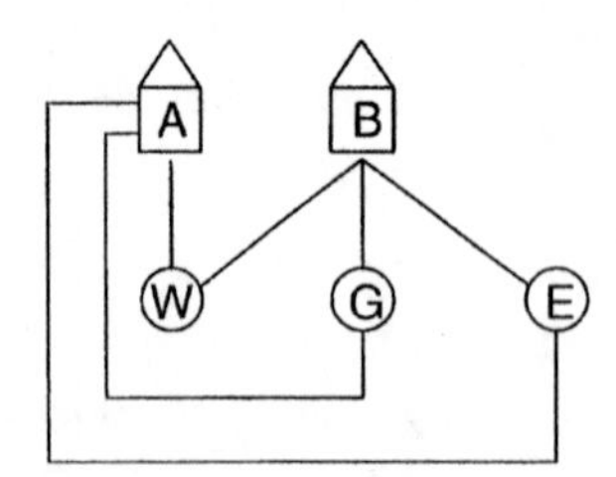

Now add house C. The connection can be made from house C to electricity and house C to water. However, in order for house C to be connected to gas the line must cross another connection.

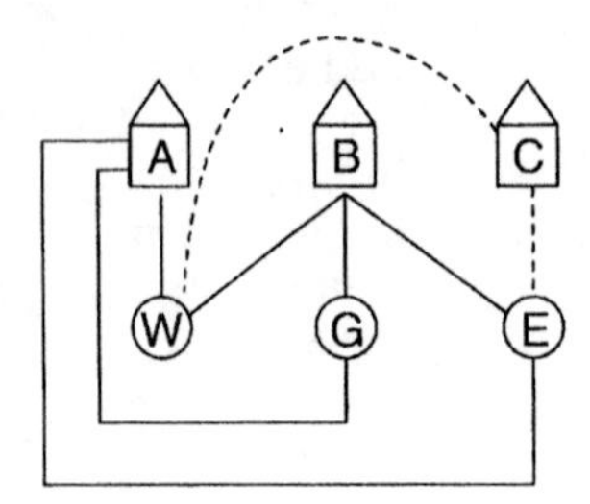

Section 1.1 39. (continued)

Therefore, it is not possible to make all the connections.

Note: This famous problem is not a simple one to prove. The goal here is to give a feel for why the connections cannot all be made without lines crossing.

Step 4: *Look Back.*

Try different ways to make all connections. Convince yourself that it is not possible to make all connections.

(b) Step 1: *Understand the Problem.*

In this case the owner of house B allows a pipe or wire for one of his neighbor's connections to pass through his house. As in part (a), no pipes or wires are allowed to cross.

Step 2: *Devise a Plan.*

Proceed as in part (a) and then see if all connections can be made if house B allows a pipe for one of his neighbor's connections to pass through his house.

Step 3: *Carry Out the Plan.*

Draw houses A, B, and C and connect these to the utilities as in part (a).

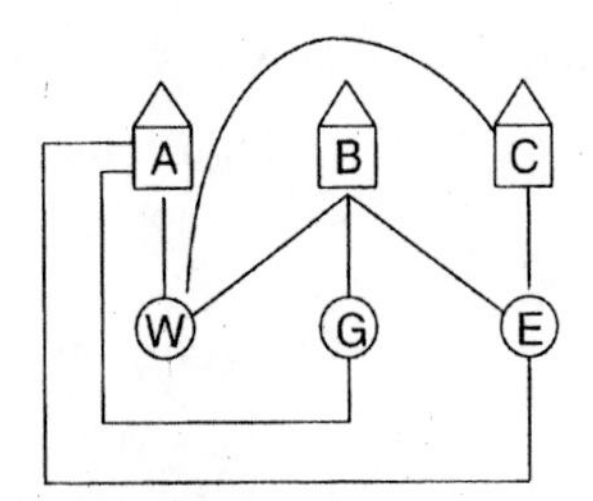

Now, connect house C to house B and then connect to gas. This completes all connections.

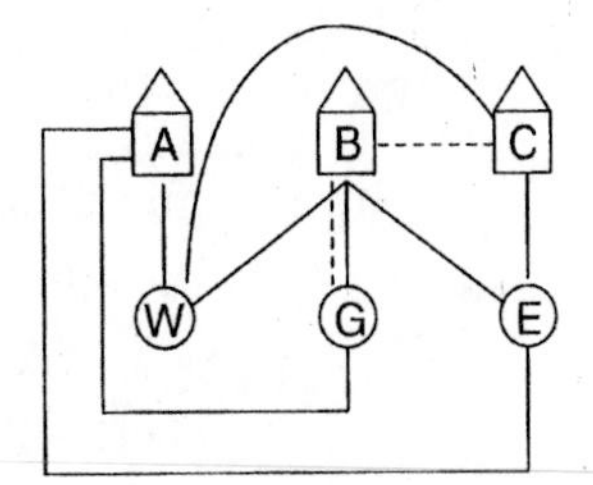

Step 4: *Look Back.*

Are there other ways to make all connections? (See the solution given in your textbook for another way to make all connections.)

43. Step 1: *Understand the Problem.*

The girth is the distance around the box. The girth plus the length must be less than or equal to 130 inches. The end of the box will be square. The box must be 5′10″ long since that is the height of the tree.

Section 1.1

43. (continued)

Step 2: *Devise a Plan.*

Let x be the length of one side of the square end. Then the girth is $4x$. The tree is 5′10″ = 70″. Because of the restriction on the packages, you know that

$$4x + 70 = 130.$$

Step 3: *Carry Out the Plan.*

Solve for x.

$$4x + 70 = 130$$
$$4x = 60$$
$$x = 15$$

The maximum height and width are each 15 inches.

Step 4: *Look Back.*

Verify that girth + length = 130 in.

girth = $4 \cdot 15$ in. = 60 in.

girth + length = 60 in. + 70 in. = 130 in.

The package with these dimensions meets the restriction.

45. Step 1: *Understand the Problem.*

Each map must be colored in the fewest number of colors possible where

- countries that share more than one point are colored differently and
- countries that share only a common point can be colored the same color.

Step 2: *Devise a Plan.*

Begin by coloring two countries that share more than one point different colors. Continue coloring countries that share more than one point with a country that is already colored, one at a time, using a new color only when necessary.

(a) Step 3: *Carry Out the Plan.*

Let R = Red, B = Blue, G = Green, and Y = Yellow. Number each of the six countries one through six as shown.

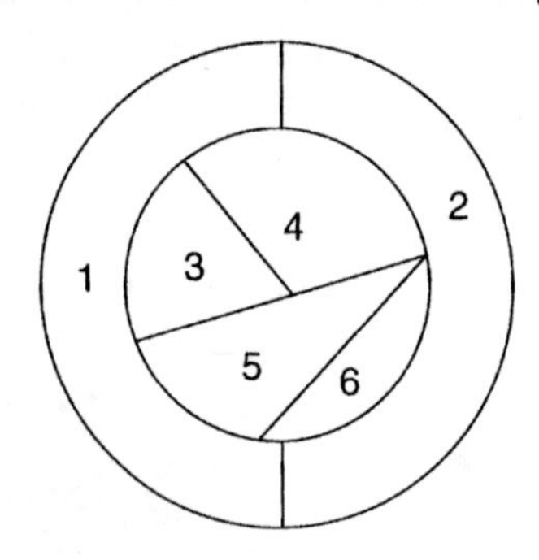

Section 1.1 45. (continued)

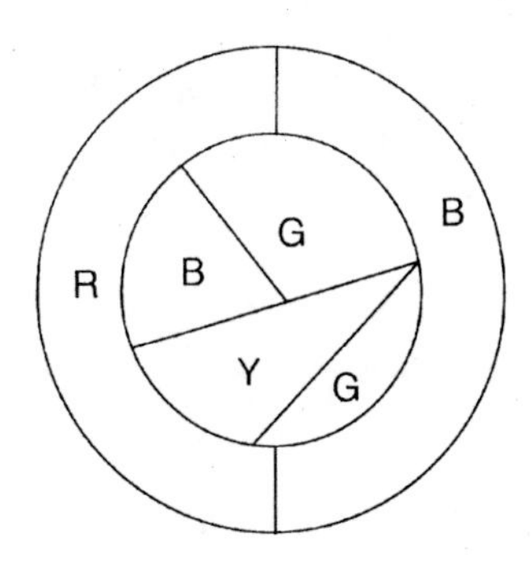

- Countries 1 and 2 share more than one point. Color country 1 red and country 2 blue.
- Country 3 can now be colored blue.
- Country 4 cannot be colored red or blue. Color country 4 green.
- Country 5 shares more than one point with countries 1, 3 and 4, so it cannot be colored red, blue or green. Color country 5 yellow.
- Country 6 shares more than one point with country 1, 2, and 5, so it cannot be colored red, blue, or yellow. Color country 6 green.

Conclusion: Four colors are needed to color this map.

Step 4: *Look Back.*

Begin with two different countries that share more than one point and determine that four colors are still needed.

(b) Step 3: *Carry Out the Plan.*

Let R = Red, B = Blue, and G = Green. Number each of the eight countries one through eight as shown.

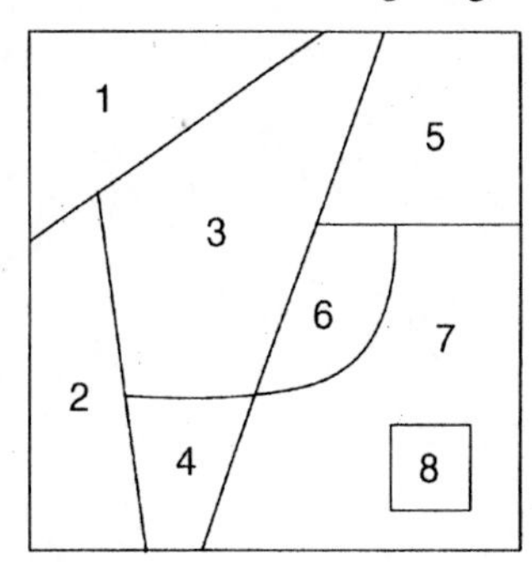

- Countries 1 and 2 share more than one point. Color country 1 red and country 2 blue.
- Country 3 shares more than one point with countries 1 and 2, so it cannot be colored red or blue. Color country 3 green.
- Country 4 cannot be colored blue or green. Color country 4 red.
- Country 5 shares more than one point with country 3, so it cannot be colored green. Color country 5 red.
- Country 6 cannot be colored red or green. Color country 6 blue.
- Country 7 cannot be colored red or blue. Color country 7 green.
- Country 8 cannot be colored green. Color country 8 red.

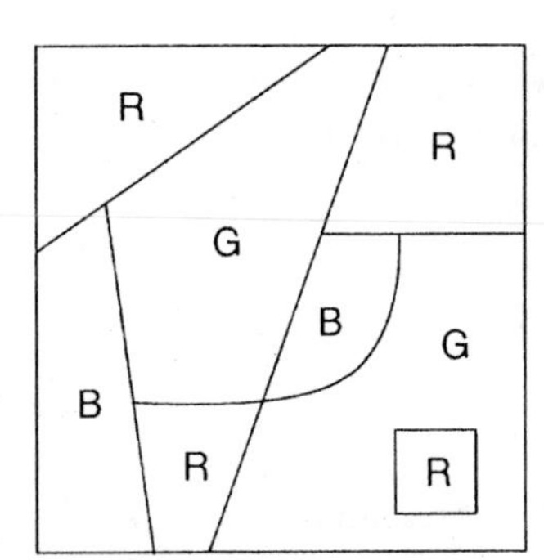

Conclusion: Three colors are needed to color this map.

Section 1.2 45. (continued)

Step 4: *Look Back.*

Begin with two different countries that share more than one point and determine that three colors are still needed. Using only three colors, color the map differently.

(c) Step 3: *Carry Out the Plan.*

Let R = Red and B = Blue. Number each of the eight countries one through eight as shown.

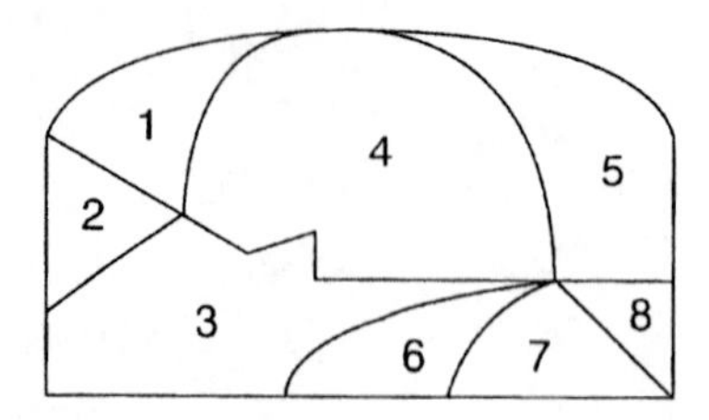

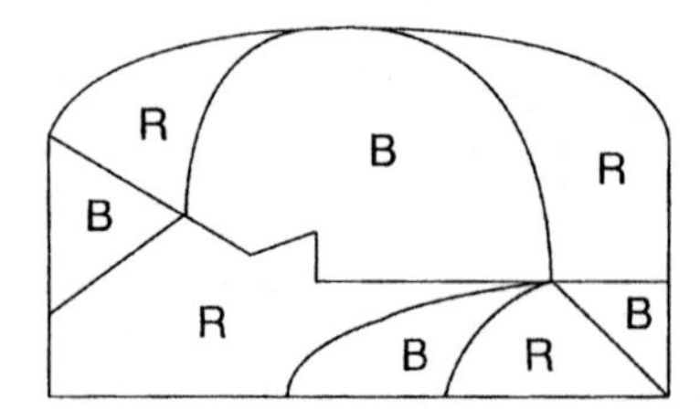

- Countries 1 and 2 share more than one point. Color country 1 red and country 2 blue.
- Country 3 shares more than one point with country 2, so it cannot be colored blue. Color country 3 red.
- Country 4 cannot be colored red. Color country 4 blue.
- Country 5 cannot be colored blue. Color country 5 red.
- Country 6 cannot be colored red. Color country 6 blue.
- Country 7 cannot be colored blue. Color country 7 red.
- Country 8 cannot be colored red. Color country 8 blue.

Conclusion: Two colors are needed to color this map.

Step 4: *Look Back.*

Begin with two different countries that share more than one point and determine that two colors are still needed.

Section 1.2

Tips:

✔ Problems are often solved by a combination of strategies.

✔ Remember, if your first plan does not work, devise another and try it. Often you will go through several different plans before you find a workable one. Keep trying.

Section 1.2 **Solutions to odd-numbered textbook problems**

1. (a)

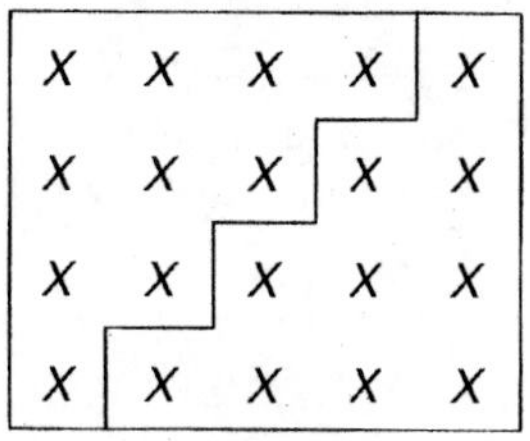

X	X	X	X	X
X	X	X	X	X
X	X	X	X	X
X	X	X	X	X

(b) $1 + 2 + 3 + 4 = \frac{1}{2}(4 \times 5)$

$$10 = 10$$

(c) $1 + 2 = \frac{1}{2}(2 \times 3)$

$1 + 2 + 3 = \frac{1}{2}(3 \times 4)$

$1 + 2 + 3 + 4 = \frac{1}{2}(4 \times 5)$

The highest number in the sum is the first factor of the product. The second factor in the product is one larger than the first factor.

$1 + 2 + ... + 49 + 50 = \frac{1}{2}(50 \times 51) = 1275$

$1 + 2 + ... + 74 + 75 = \frac{1}{2}(75 \times 76) = 2850$

(d) $1 + 2 + ... + (n - 1) + n = \frac{1}{2}[n \times (n + 1)] = \frac{n(n + 1)}{2}$

3. Step 1: *Understand the Problem.*

We cannot break the toothpicks. Toothpicks must be arranged to form rectangles one toothpick in height. The base of each successive rectangle increases by one toothpick.

Step 2: *Devise a Plan.*

Make a few more rectangles, where the base contains four, five and six toothpicks. Then determine if there is pattern between the number of toothpicks and the length of the base of the rectangle.

Step 3: *Carry Out the Plan.*

The next three rectangles require 10, 12, and 14 toothpicks respectively.

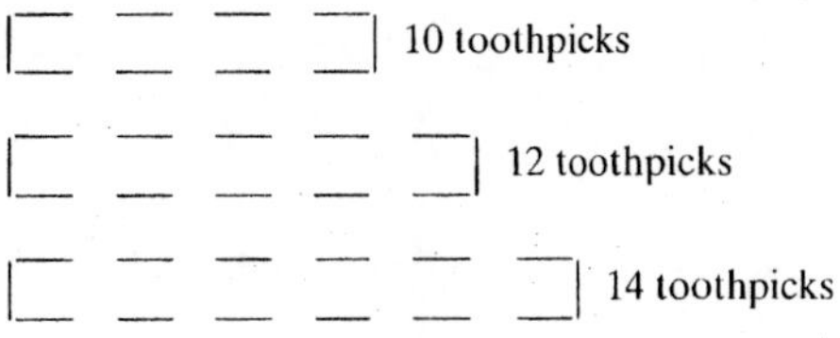

10 toothpicks

12 toothpicks

14 toothpicks

Section 1.2 3. (continued)

Summarize the information in a table.

Number of Toothpicks in Base of Rectangle	Total Number of Toothpicks	Pattern
1	4	$2(1) + 2 = 4$
2	6	$2(2) + 2 = 6$
3	8	$2(3) + 2 = 8$
4	10	$2(4) + 2 = 10$
5	12	$2(5) + 2 = 12$
6	14	$2(6) + 2 = 14$

We see that each successive rectangle requires two additional toothpicks. Therefore the nth rectangle will require $2n + 2$ toothpicks

Step 4: *Look Back.*

Can you explain geometrically why the number of toothpicks needed for the nth rectangle is $2n + 2$?

5. Step 1: *Understand the Problem.*

We cannot break the toothpicks. Toothpicks must be arranged to form a hexagon with squares attached on one of the sides. Each shape has one more square than the previous shape.

Step 2: *Devise a Plan.*

Make a few more figures, where the number of squares in the figure is three, four and five. Then determine if there is pattern between the number of toothpicks and the number of squares attached to the hexagon.

Step 3: *Carry Out the Plan.*

The next three figures in the pattern require 15, 18, and 21 toothpicks respectively.

Figure 4: 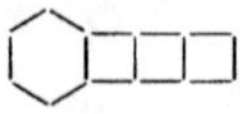 15 toothpicks

Figure 5:  18 toothpicks

Figure 6: 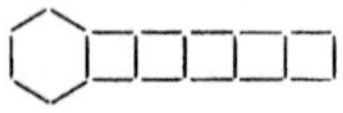 21 toothpicks

Section 1.2

5. (continued)

Summarize the information in a table.

Figure	Number of Squares	Total Number of Toothpicks	Pattern
1	0	6	$3(0) + 6 = 6$
2	1	9	$3(1) + 6 = 9$
3	2	12	$3(2) + 6 = 12$
4	3	15	$3(3) + 6 = 15$
5	4	18	$3(4) + 6 = 18$
6	5	21	$3(5) + 6 = 21$

We see that each successive figure requires three additional toothpicks. Since the first figure contains 0 squares, the nth figure will require $3(n - 1) + 6 = 3n - 3 + 6 = 3n + 3$ toothpicks.

Step 4: *Look Back.*

How would the number of toothpicks for the nth figure change if squares were also attached to the opposite side of the hexagon?

7. Step 1: *Understand the Problem.*

The triangular numbers are whole numbers that can be arranged in a triangular shape. The first four triangular numbers are 1, 3, 6, and 10.

Step 2: *Devise a Plan.*

Arrange dots in the form of a triangle and look for a pattern to determine the number of dots in the nth triangular number. Additional figures may be necessary to determine the pattern.

Step 3: *Carry Out the Plan.*

Arrange dots for the first five triangular numbers. The arrangement given here is slightly different than the arrangement in the text.

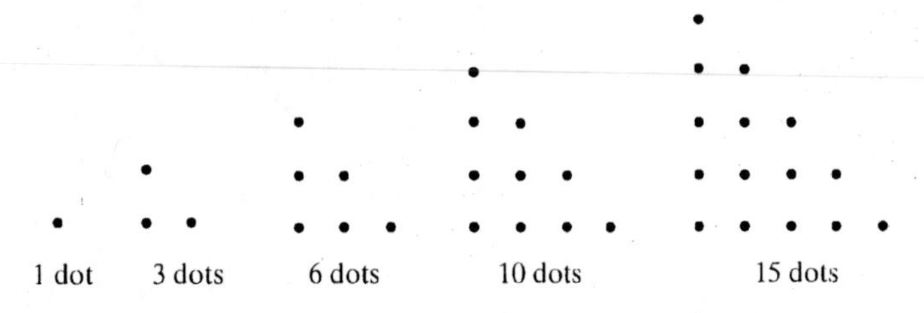

Section 1.2

7. (continued)

Now, copy each triangle, rotate it, and place it to form a rectangle.

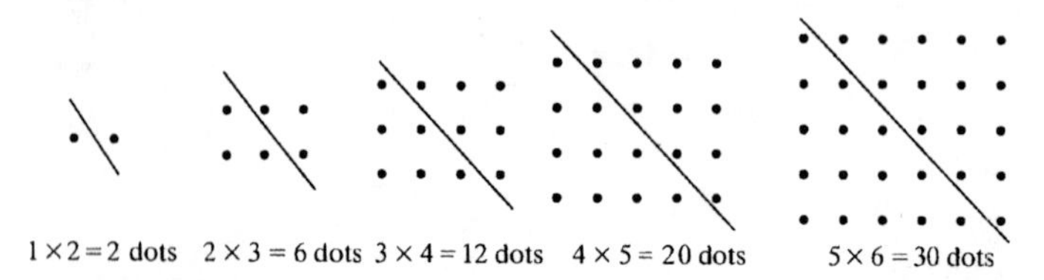

Now to find the nth triangular number, take half of the nth rectangular number where the height of the rectangle is n and the base is $n + 1$. Therefore the nth triangular number is $\frac{n(n+1)}{2}$.

Step 4: *Look Back.*

Verify that the formula found for the nth triangular number works for the first five triangular numbers.

9. Step 1: *Understand the Problem.*

The hexagonal numbers are whole numbers that can be arranged in the shapes as given in the text. The first five hexagonal numbers are 1, 6, 15, 28, and 45.

Step 2: *Devise a Plan.*

Arrange dots in the desired shape and look for a pattern to determine the number of dots in the nth hexagonal number.

Step 3: *Carry Out the Plan.*

Arrange dots for the first five hexagonal numbers. The arrangement given here is slightly different than the arrangement in the text.

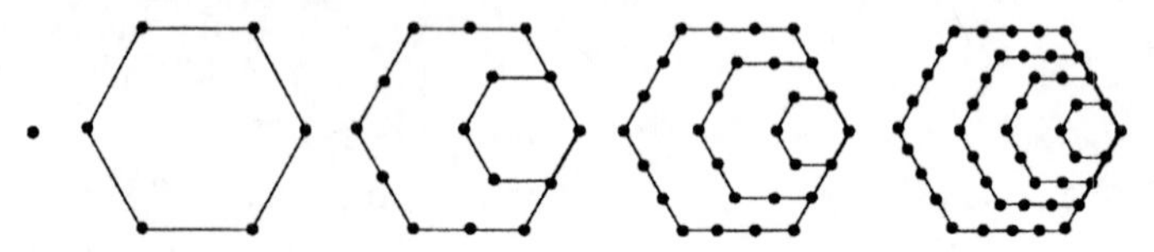

Now, the outside dots of each figure can be rearranged to form the boundary of a rectangle where the base of the first rectangle is 1, the base of the second rectangle is 2, the base of the third rectangle is 3 and so forth. And in general the base of the nth rectangle is n. The heights of the rectangles are 1, 3, 5, 7, and 9 respectively. The height of the nth rectangle is $2n - 1$.

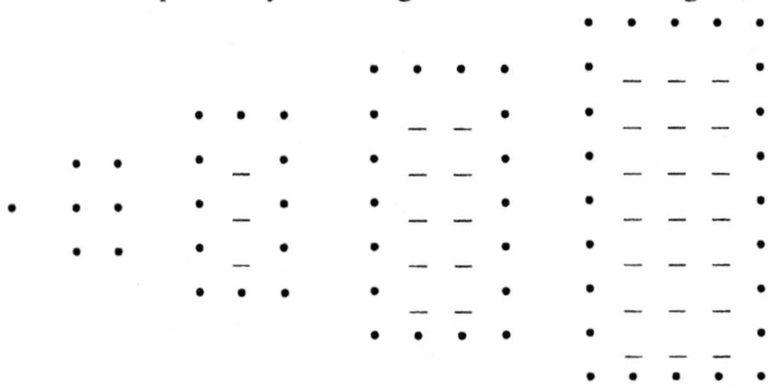

Section 1.2 9. (continued)

The additional dots in each figure will fill the interior of the rectangle.

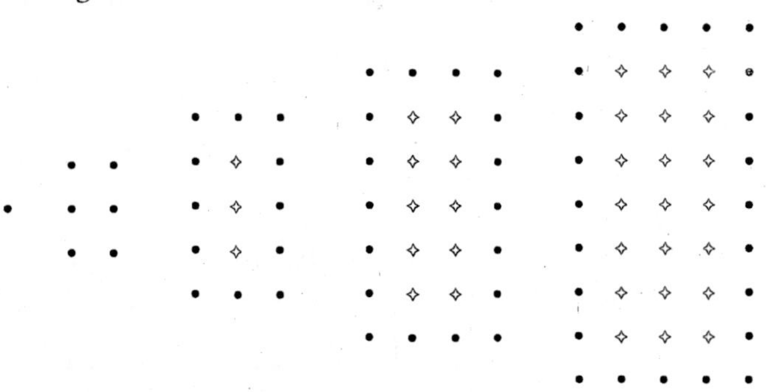

Now to find the nth hexagonal number, find the area of each rectangle. The area of a rectangle is base (n) times height ($2n - 1$). Hence the nth hexagonal number is $n(2n - 1) = 2n^2 - n$.

Step 4: *Look Back.*

Verify that the formula found for the nth hexagonal number works for the first five hexagonal numbers.

11. (a) Figures A and B are the same size, but figure B has a diagonal. Thus (ii) must be the correct choice because it is the same size as X and has a diagonal.

(b) Figures A and B are the same except that the dot has moved to the opposite section. Therefore, (i) must correspond to X.

(c) Figure B is just the same as figure A flipped over. Thus (i) must go with X.

(d) Figure B is the same as figure A, rotated a quarter of a turn clockwise. Therefore, (ii) must correspond to X.

13. (a) The first three figures given begin the pattern. Figure 4 is a result of a reflection about the vertical line of symmetry of the square in figure 1. Figure 5 is a result of a reflection about the horizontal line of symmetry of the square in figure 2. Hence, figure 6 will be the result of a reflection about the vertical line of symmetry of the square in figure 3.

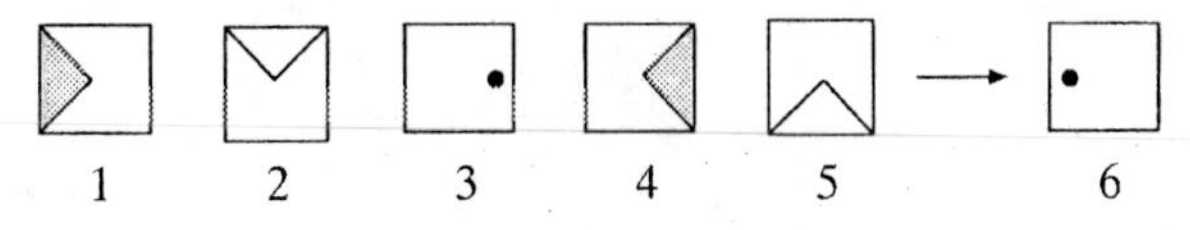

(b) Each figure is obtained by connecting the midpoints of the innermost square of the previous figure.

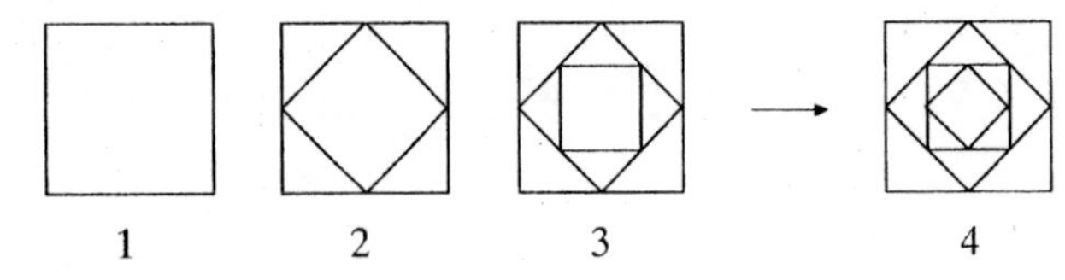

Section 1.2

13. (continued)

(c) Each figure is obtained by rotating the triangle clockwise about the square until a side of the triangle lies on the side of the square.

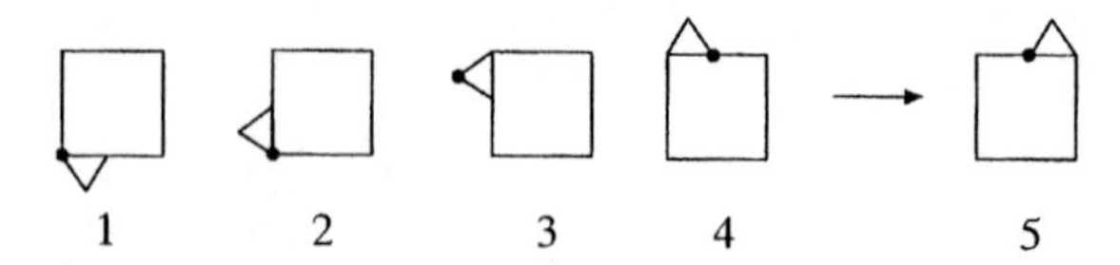

(d) Figure 2 is a result of a reflection about the left side of the square of figure 1. Figure 3 is a result of a reflection about the top side of the square of figure 2. Figure 4 will be the result of a reflection about the right side of the square of figure 3. If the next figure were asked for, it would be the result of a reflection about the bottom side of the square of figure 4.

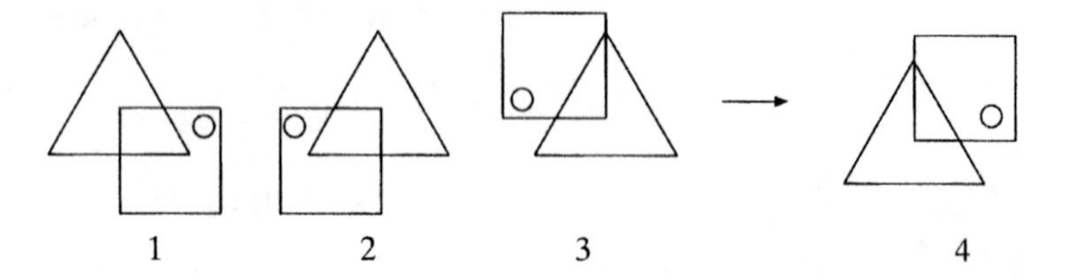

(e) Each figure is obtained by adding a nail on each end of the previous figure so that nail heads are not next to each other. In addition, a horizontal line is drawn alternating from the top of one figure to the bottom of the next figure.

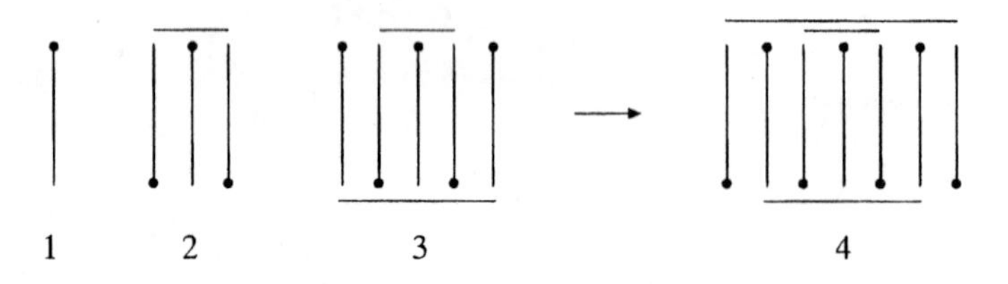

15. (a) The sequence of numbers are perfect squares.

Term	1	2	3	4	5	6	7	...	12	...	n
Value	1	4	9	16	25	36	49	...	144	...	n^2

(b) The value of each term in the sequence is two less than the value of the previous term.

Term	1	2	3	4	5	6	7	...	12	...	n
Value	−2	−4	−6	−8	−10	−12	−14	...	−24	...	$-2n$

Section 1.2

15. (continued)

(c) The value of each term in the sequence is three times the value of the previous term.

Term	1	2	3	4	5	6	7	...	12	...	n
Value	1	3	9	27	81	243	729	...	177,147	...	3^{n-1}
Pattern	3^0	3^1	3^2	3^3	3^4	3^5	3^6	...	3^{11}	...	3^{n-1}

(d) The value of each term in the sequence is two more than the previous term.

Term	1	2	3	4	5	6	7	...	12	...	n
Value	3	5	7	9	11	13	15	...	25	...	$2n + 1$

17. Step 1: *Understand the Problem.*

The people at the party are in couples. Each person shakes hands with every other person *except* for the person that they came to the party with.

Step 2: *Devise a Plan.*

Solve a simpler problem by looking at a party with fewer number of couples than five. This problem has similarities to Example 1.6 in the text. Use what was learned in that example in solving this problem.

Step 3: *Carry Out the Plan.*

Make a table to summarize the number of handshakes.

Number of Couples	Number of People	Number of Handshakes if everyone shakes hands	Number of Handshakes when we eliminate couples shaking each other's hands
1	2	$1 = 1$	$1 - 1 = 0$
2	4	$3 + 2 + 1 = 6$	$6 - 2 = 4$
3	6	$5 + 4 + 3 + 2 + 1 = 15$	$15 - 3 = 12$
4	8	$7 + 6 + 5 + 4 + 3 + 2 + 1 = 28$	$28 - 4 = 24$
5	10	$9 + 8 + 7 + 6 + 5 + 4 + 3 + 2 + 1 = 45$	$45 - 5 = 40$

Conclusion: There are 40 handshakes.

Step 4: *Look Back.*

Compare the similarities and differences between this problem and Example 1.6 in the text. How many handshakes would there be if n couples attend the party? Can you see the hexagonal numbers embedded in this problem?

Section 1.2

19. Step 1: *Understand the Problem.*

If Julie backtracks, her route would be more than 6 blocks, so you can count only paths that do not backtrack.

Step 2: *Devise a Plan.*

You can use a diagram to find the paths. Try drawing them in a consistent pattern so that you do not miss one. Dot paper would be helpful to use.

Step 3: *Carry Out the Plan.*

There are 15 paths of exactly 6 blocks each that Julie can walk from her house to the school.

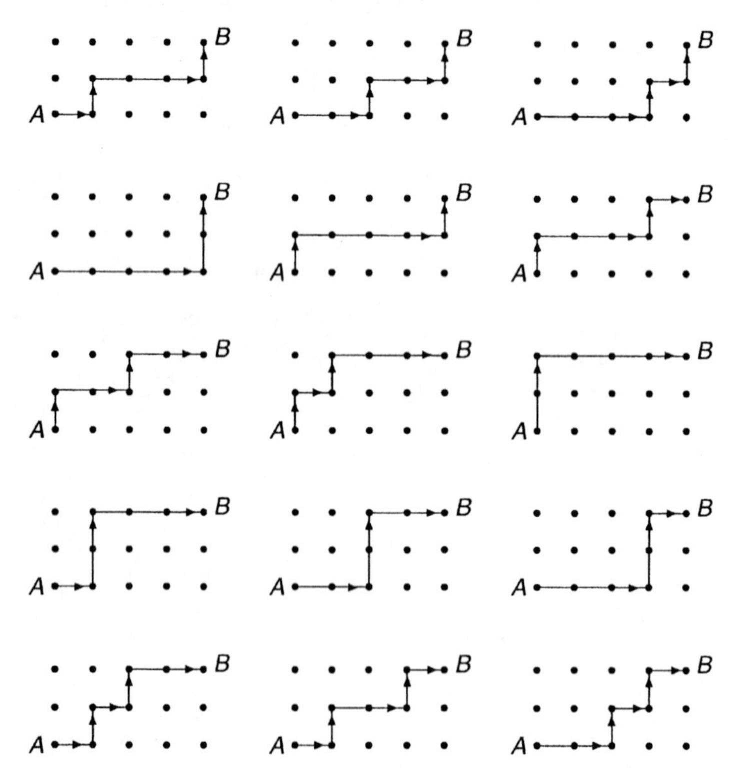

Step 4: *Look Back.*

Are there any duplications or any paths you left out? What system did you use to draw the paths? What if the school were one block closer?

21. The first two terms of the Fibonacci sequence are 1 and 1. Each of the following terms is found by adding the two previous terms.

Fibonacci sequence	1	1	2	3	5	8	13	21	34	55	89	144	233	377	...
Pattern			1 + 1	1 + 2	2 + 3	3 + 5	5 + 8	8 + 13	13 + 21	21 + 34	34 + 55	55 + 89	89 + 144	144 + 233	

Section 1.2

23. Step 1: *Understand the Problem.*

You must count every triangle. Some will be overlapping or contained entirely in another triangle.

Step 2: *Devise a Plan.*

Count the number of each size of triangle and add them up. Try a simpler diagram first, list the results in a table, and look for a pattern.

Step 3: *Carry Out the Plan.*

Start out with a 1×1 square, then a 2×2 square, etc.

Size Square	Size of Triangle								Total
	$1 \times \frac{1}{2}$	1×1	2×1	2×2	$3 \times 1\frac{1}{2}$	3×3	4×2	4×4	
1×1	4	4	0	0	0	0	0	0	8
2×2	16	16	8	4	0	0	0	0	44
3×3	36	36	24	16	8	4	0	0	124
4×4	64	64	48	36	24	16	12	4	268

There are 268 triangles in the 4×4 square.

Step 4: *Look Back.*

Did you count all the triangles? Is there any predictable pattern you can use if you make even larger squares?

25. Step 1: *Understand the Problem.*

With each step, one square is added vertically and two squares are added horizontally.

Step 2: *Devise a Plan.*

Make a table and look for a pattern. If square tiles are available, you may want to use them to model the problem.

Step 3: *Carry Out the Plan.*

Make a table to summarize the number of squares.

Term	Number of squares
1	1
2	$1 + 3 = 4$
3	$4 + 3 = 7$
4	$7 + 3 = 10$
5	$10 + 3 = 13$
6	$13 + 3 = 16$
7	$16 + 3 = 19$
8	$19 + 3 = 21$
9	$21 + 3 = 24$
10	$24 + 3 = 27$
n	$3n - 2$

Conclusion: The 10th term contains 27 squares and the nth term contains $3n - 2$ squares.

Section 1.2

25. (continued)

Step 4: *Look Back.*

Verify that the formula found for the nth set works for the first 10 sets.

27. Step 1: *Understand the Problem.*

With each step, a longer row of squares is added to the bottom.

Step 2: *Devise a Plan.*

Make a table and look for a pattern. If square tiles are available, you may want to use them to model the problem.

Step 3: *Carry Out the Plan.*

Make a table to summarize the number of squares.

Term	Number of squares
1	1
2	$1 + 3 = 4$
3	$1 + 3 + 5 = 9$
4	$1 + 3 + 5 + 7 = 16$
5	$1 + 3 + 5 + 7 + 9 = 25$
6	$1 + 3 + 5 + 7 + 9 + 11 = 36$
7	$1 + 3 + 5 + 7 + 9 + 11 + 13 = 49$
8	$1 + 3 + 5 + 7 + 9 + 11 + 13 + 15 = 64$
9	$1 + 3 + 5 + 7 + 9 + 11 + 13 + 15 + 17 = 81$
10	$1 + 3 + 5 + 7 + 9 + 11 + 13 + 15 + 17 + 19 = 100$
n	$1 + 3 + 5 + \ldots + 2n - 1 = n^2$

Conclusion: The 10th term contains 100 squares and the nth term contains n^2 squares.

Step 4: *Look Back.*

Verify that the formula found for the nth set works for the first 10 sets.

29. Step 1: *Understand the Problem.*

With each step, a cube is added to each of the four bases and the center tower.

Step 2: *Devise a Plan.*

Make a table and look for a pattern. If cubes are available, you may want to use them to model the problem.

Section 1.2

29. (continued)

Step 3: *Carry Out the Plan.*

Make a table to summarize the number of cubes.

Term	Number of cubes	Term	Number of cubes
1	1	7	26 + 5 = 31
2	1 + 5 = 6	8	31 + 5 = 36
3	6 + 5 = 11	9	36 + 5 = 41
4	11 + 5 = 16	10	41 + 5 = 46
5	16 + 5 = 21	n	$5n - 4$
6	21 + 5 = 26		

Conclusion: The 10th term contains 46 cubes and the nth term contains $5n - 4$ cubes.

Step 4: *Look Back.*

Verify that the formula found for the nth set works for the first 10 sets.

31. Step 1: *Understand the Problem.*

With each step you add another, bigger layer of cubes on the bottom.

Step 2: *Devise a Plan.*

Make a table and look for a pattern. If they are available, you may want to use blocks to model the problem.

Step 3: *Carry Out the Plan.*

Height	Number of blocks added by layers
1	1
2	1 + 3 = 4
3	1 + 3 + 6 = 10
4	1 + 3 + 6 + 10 = 20
5	1 + 3 + 6 + 10 + 15 = 35
6	1 + 3 + 6 + 10 + 15 + 21 = 56
7	1 + 3 + 6 + 10 + 15 + 21 + 28 = 84
8	1 + 3 + 6 + 10 + 15 + 21 + 28 + 36 = 120
9	1 + 3 + 6 + 10 + 15 + 21 + 28 + 36 + 45 = 165
10	1 + 3 + 6 + 10 + 15 + 21 + 28 + 36 + 45 + 55 = 220

For $n = 10$ there are 220 blocks. Note that the last layer added has the same number of blocks as the previous layer plus the height of the new stack.

Section 1.2

31. (continued)

Step 4: *Look Back.*

Find the triangular numbers in this problem. Can you find a formula for the nth set of cubes?

33. Step 1: *Understand the Problem.*

The letters of the alphabet are displayed in reverse order. The number of letters in each column increases by 2 at each stage.

Step 2: *Devise a Plan.*

Make a table to display the information.

Step 3: *Carry Out the Plan.*

Part (a) Make a table to summarize the number of letters in the pattern.

Letter	Number of times the letter occurs
Z	1
Y	$1 + 2 = 3$
X	$3 + 2 = 5$
W	$5 + 2 = 7$
V	$7 + 2 = 9$
U	$9 + 2 = 11$
T	$11 + 2 = 13$
S	$13 + 2 = 15$
R	$15 + 2 = 17$
Q	$17 + 2 = 19$
P	$19 + 2 = 21$
O	$21 + 2 = 23$
N	$23 + 2 = 25$
M	$25 + 2 = 27$

Conclusion: There will be 27 letters in the "M" column.

Part (b) The difference between 45 and 27 is 18, and since each letter appears 2 more times than the previous letter, move 9 (18 divided by 2) letters down in the alphabet which results in the "D" column.

Step 4: *Look Back.*

Verify the answer found in part (b) by continuing the table above.

Section 1.2

35. Step 1: *Understand the Problem.*

With each succeeding step a square in the lower right corner is divided into quarters and the upper left quarter is shaded. You need to find out what part of the largest, original square will be shaded if the pattern continues infinitely.

Step 2: *Devise a Plan.*

Each square is $\frac{1}{4}$ of the preceding square. You can write an infinite sum using this fact. To estimate the answer, find the sum for 1 term, 2 terms, 3 terms, etc., and look for a pattern.

Step 3: *Carry Out the Plan.*

$$\text{Sum} = \frac{1}{4} + \frac{1}{4} \times \frac{1}{4} + \frac{1}{4}\left(\frac{1}{4} \times \frac{1}{4}\right) + \frac{1}{4}\left(\frac{1}{4} \times \frac{1}{4} \times \frac{1}{4}\right) + \ldots$$

$$= \frac{1}{4} + \left(\frac{1}{4}\right)^2 + \left(\frac{1}{4}\right)^3 + \left(\frac{1}{4}\right)^4 + \left(\frac{1}{4}\right)^5 + \ldots$$

$$\frac{1}{4} + \left(\frac{1}{4}\right)^2 = \frac{5}{16} = 0.3125$$

$$\frac{1}{4} + \left(\frac{1}{4}\right)^2 + \left(\frac{1}{4}\right)^3 = \frac{5}{16} + \left(\frac{1}{4}\right)^3 = \frac{21}{64} = 0.328125$$

$$\frac{1}{4} + \left(\frac{1}{4}\right)^2 + \left(\frac{1}{4}\right)^3 + \left(\frac{1}{4}\right)^4 = \frac{21}{64} + \frac{1}{256} = \frac{85}{256} = 0.33203125$$

$$\frac{1}{4} + \left(\frac{1}{4}\right)^2 + \left(\frac{1}{4}\right)^3 + \left(\frac{1}{4}\right)^4 + \left(\frac{1}{4}\right)^5 = \frac{85}{256} + \frac{1}{1024} = \frac{341}{1024}$$

$$= 0.3330078125$$

It appears that the sums are approaching $\frac{1}{3}$. Try it with 10 terms.

$$\frac{1}{4} + \left(\frac{1}{4}\right)^2 + \left(\frac{1}{4}\right)^3 + \left(\frac{1}{4}\right)^4 + \left(\frac{1}{4}\right)^5 + \left(\frac{1}{4}\right)^6 + \left(\frac{1}{4}\right)^7 +$$

$$\left(\frac{1}{4}\right)^8 + \left(\frac{1}{4}\right)^9 + \left(\frac{1}{4}\right)^{10} \approx 0.3333330154$$

$\frac{1}{3} = 0.\overline{3}$ seems to be the solution.

Step 4: *Look Back.*

Does this answer seem reasonable?

Section 1.2

37. Make a table to display the information.

(a)

Term	Blue Triangles	White Triangles
1	1	0
2	3	1
3	$9 = 3 \times 3$	$4 = 1 + 3$
4	$27 = 3 \times 3 \times 3$	$13 = 4 + 9$

(b) Each figure contains three times as many blue triangles as the previous figure. The number of white triangles increases by adding the next power of three to the previous number of white triangles.

Term	Blue Triangles	White Triangles
1	1	0
2	3	1
3	$9 = 3 \times 3 = 3^2$	$4 = 1 + 3$
4	$27 = 3 \times 3 \times 3 = 3^3$	$13 = 4 + 3 \times 3$
5	$81 = 3 \times 3 \times 3 \times 3 = 3^4$	$40 = 13 + 3 \times 3 \times 3$

(c) By looking at the table in part (b) we can see a pattern. Therefore, we can expect the nth figure to have 3^{n-1} blue triangles.

39. Step 1: *Understand the Problem.*

A gutter can be formed by using any or all of the gutter lengths given.

Step 2: *Devise a Plan.*

Make a table.

Step 3: *Carry Out the Plan.*

Make a table giving of all possible gutter lengths beginning with one section, then two sections, and finally three sections.

Different Gutter Lengths	Gutter Sections Used	Gutter Length
1	6 ft	6 ft
2	8 ft	8 ft
3	10 ft	10 ft
4	6 ft and 8 ft	14 ft
5	6 ft and 10 ft	16 ft
6	8 ft and 10 ft	18 ft
7	6 ft and 8 ft and 10 ft	24 ft

Section 1.2

39. (continued)

Conclusion: There are 7 different lengths of gutter that can be formed.

Step 4: *Look Back.*

How many different gutter lengths would be possible if a certain type of gutter came in 6-ft, 8-ft, 10-ft, and 12-ft sections?

41. Step 1: *Understand the Problem.*

Each square slab that is added to the decorative concrete floor has side length that is equal to the length of the long side of the rectangle.

Step 2: *Devise a Plan.*

Draw a diagram of the concrete floor as new square slabs are added. make a table of the dimensions of the rectangular floor at each stage and look for a pattern.

Step 3: *Carry Out the Plan.*

A diagram of the stages of the rectangles after each square slab is poured:

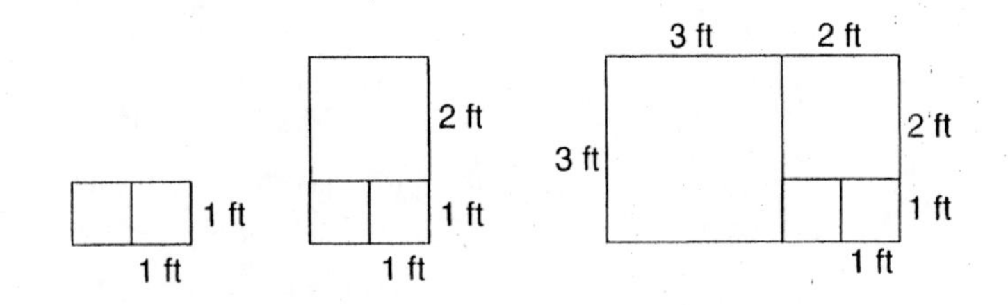

Part (a) The next square slab will be a result of adding a square with side length 5 ft.

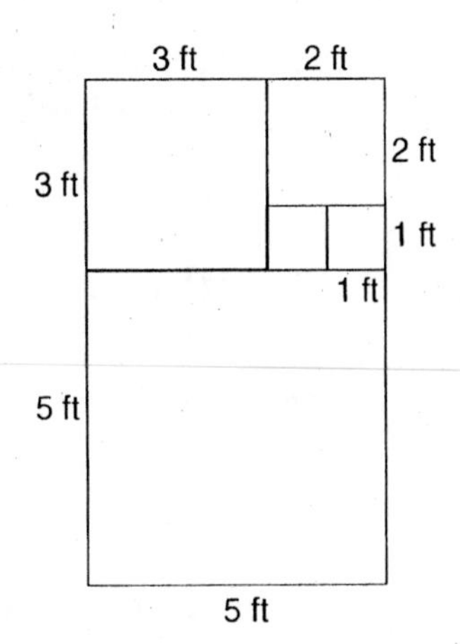

Chapter Review

41. (continued)

Part (b) The following table lists the dimensions of the rectangular floor at each stage.

Number of Square Slabs in Concrete Floor	Dimensions
1	1 by 1
2	1 by 2 $(2 = 1 + 1)$
3	2 by 3 $(3 = 1 + 2)$
4	3 by 5 $(5 = 2 + 3)$
5	5 by 8 $(8 = 3 + 5)$
6	8 by 13 $(13 = 5 + 8)$
7	13 by 21 $(21 = 8 + 13)$

Part (c) Notice that after the first two square slabs are poured, the side length of a new square is the sum of the side lengths of the previous two squares. This gives the sequence of side lengths 1, 1, 2, 3, 5, 8, 13, 21, ...

Part (d) The area of the concrete floor with 7 square slabs is $13 \times 21 = 273$ square feet.

The area of the concrete floor with 8 square slabs is $21 \times 34 = 714$ square feet. $(34 = 13 + 21)$

The area of the concrete floor with 9 square slabs is $34 \times 55 = 1870$ square feet. $(55 = 21 + 34)$

The area of the concrete floor with 10 square slabs is $55 \times 89 = 4895$ square feet. $(89 = 34 + 55)$

Conclusion: There will be 10 square slabs poured when the area of the floor is 4895 square feet.

Step 4: *Look Back.*

What name is given to the sequence of numbers represented by the side lengths of the poured square slabs?

Solutions to Chapter 1 Review

Section 1.1

1. Step 1: *Understand the Problem.*

There are six shapes given in the five different views of the same block. Arrange these six shapes on the flattened block so that when the flattened block is folded it will match the five different views.

Section 1.1

1. (continued)

Step 2: *Devise a Plan.*

Begin by arranging two shapes given on one view of the block so that when the block is folded the shapes are in the correct position. Then choose a block that has one of the already sketched shapes on it and add that information to the flattened block. Continue in this manner until all six shapes are placed on the flattened block. It may help to actually cut out the flattened block and check your results at each stage by folding the block.

Step 3: *Carry Out the Plan.*

Number the different views of the block from 1 to 5.

Place the shapes on the flattened block in the following order.

- Begin with the first block and sketch that information on the flattened block.
- Second, add the moon using the information from block 5.
- Third, add the arrow using the information from block 2.
- Fourth, add the trapezoid using the information from block 3.
- Finally, add the triangle using the information from block 4.

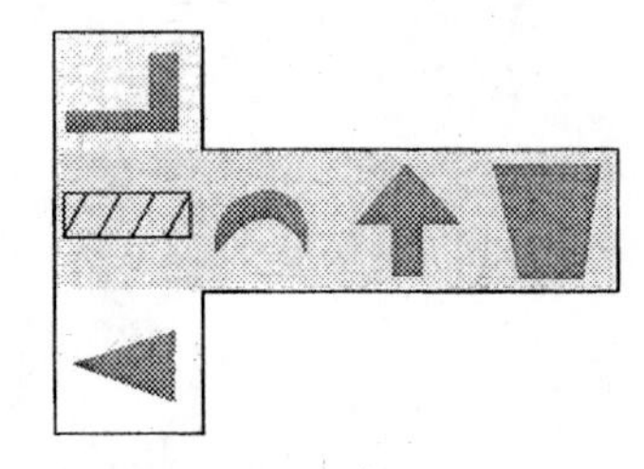

Step 4: *Look Back.*

Fold the flattened block and verify that all five views of the given block are satisfied.

2. Step 1: *Understand the Problem.*

The 4 by 4 grid is to be cut along the lines to form two identical pieces. Two cuts are considered the same if they look alike when the figure is rotated or flipped.

Chapter Review

2. (continued)

Step 2: *Devise a Plan.*

Use Guess and Test to find successful cuts.

Step 3: *Carry Out the Plan.*

Sketch the possible cuts on the grids. Be sure to check and make sure that no duplicate cuts are made. Try to be systematic in your approach to finding the cuts.

Begin by making the cuts where the beginning cut is down the first column. There are three such cuts.

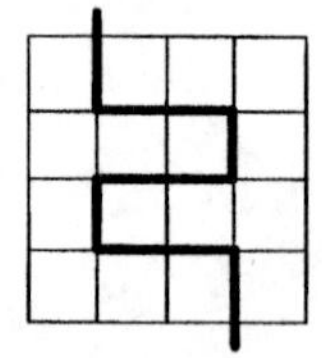
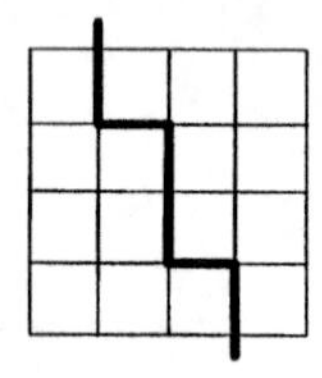
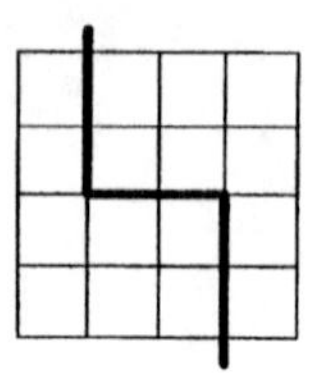

Next make the cuts where the beginning cut is down the second column. There are two such cuts.

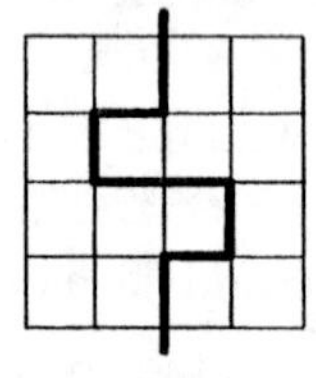
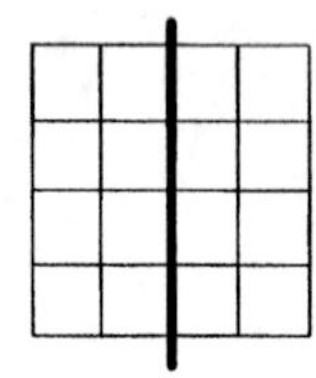

If you make the cuts where the beginning is down the third column, the cuts will be duplicates of the cuts where the beginning is down the first column. There is no need to make the beginning cuts starting at the rows, because if the grid is rotated the beginning cuts would be the same as those starting at the columns.

Conclusion: There are five different ways.

Step 4: *Look Back.*

How many different ways are possible if the grid were 5 by 5?

3. Step 1: *Understand the Problem.*

The triangle that is cut is equilateral. The square has a side length of 10 inches. When finding the distance around the triangle, the dotted line is not included. The length of the two sides of the triangle is one-half the distance of the remaining solid lines in the figure.

Step 2: *Devise a Plan.*

Use a variable.

Let x = the length of the side of the triangle in inches.

Then $2x$ = the distance around the triangle in inches and $30 + (10 - x)$ = the distance around the remaining figure in inches.

Chapter Review

10″ 10″ x x x 10″ 10″−x

3. (continued)

Using the information given in the problem, we can write

$$2x = \frac{30 + (10 - x)}{2}.$$

Step 3: *Carry Out the Plan.*

Now we can solve the equation to find the side length of the triangle.

$$2x = \frac{30 + (10 - x)}{2}$$

$$4x = 40 - x$$

$$5x = 40$$

$$x = 8$$

The side length of the equilateral triangle is 8 inches.

Step 4: *Look Back.*

Verify that 8 inches satisfies the conditions of the problem.

4. Step 1: *Understand the Problem.*

Four straight lines are to be used to connect the nine dots in the figure. Lifting the pencil from the paper is not allowed.

Step 2: *Devise a Plan.*

Use Guess and Test to solve this problem.

Step 3: *Carry Out the Plan.*

Drawing lines that extend from the dots is allowed. The diagram shows a solution to the problem.

Step 4: *Look Back.*

Are there other ways to solve this problem? If the arrangement were a 4 by 4 array of dots, what would be the fewest number of lines possible to connect the sixteen dots?

5. Step 1: *Understand the Problem.*

Four copies of the figure must be arranged to form a larger version of the original figure. Pieces may be rotated and flipped. No gaps or overlaps are allowed.

Step 2: *Devise a Plan.*

Make four copies of the original figure and cut out the shapes and fit them together to form a larger version of the original figure.

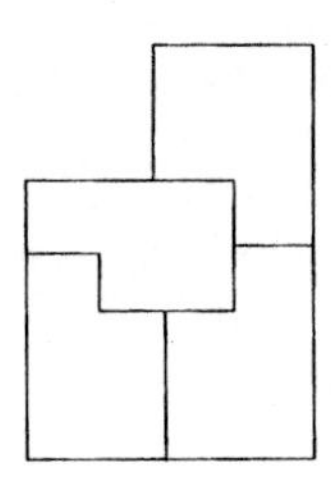

Step 3: *Carry Out the Plan.*

Using Guess and Test, rotate and flip some pieces to obtain the given figure.

5. (continued)

Step 4: *Look Back.*

Is there another solution or another approach? Can the larger version of the original figure be obtained without flipping any of the pieces?

Section 1.2

1. (a) The sequences of numbers are powers of 3. If we write them using the base of 3 we have

$3^0, 3^1, 3^1, 3^2, 3^3,$ _____, _____, _____

Beginning with the 2nd erm the exponents are the Fibonacci numbers. The next three terms in the sequence are

$$3^5 = 243,$$
$$3^8 = 6561, \text{and}$$
$$3^{13} = 1{,}594{,}323.$$

(b) Beginning with the first term, each successive term is 11 greater than the previous term.

3, 14, 25, 36, _____, _____, _____

The next three terms in the sequence are

$$36 + 11 = 47,$$
$$47 + 11 = 58, \text{and}$$
$$58 + 11 = 69.$$

(c) In looking at the three figures given, the octagon is rotated 45° clockwise from one figure to the next. The square is located on the top of the first octagon, the bottom of the second and on the left of the third. One reasonable assumption is that the square will be located on the right of the fourth octagon and then repeat the pattern of top, bottom, left and right. If this is the pattern, then the next three term would be as follows:

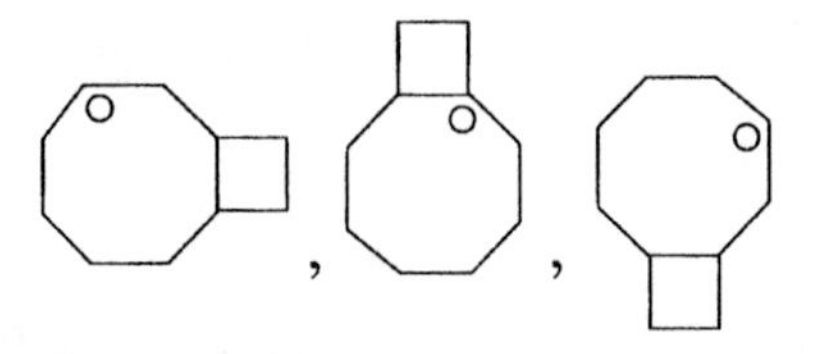

Note: There are other possible reasonable assumptions that can be made regarding the location of the square. One such assumption would be that the square follows the pattern of being located on the top, bottom, left, top, bottom, left etc. If this pattern is followed the three missing terms would be as follows.

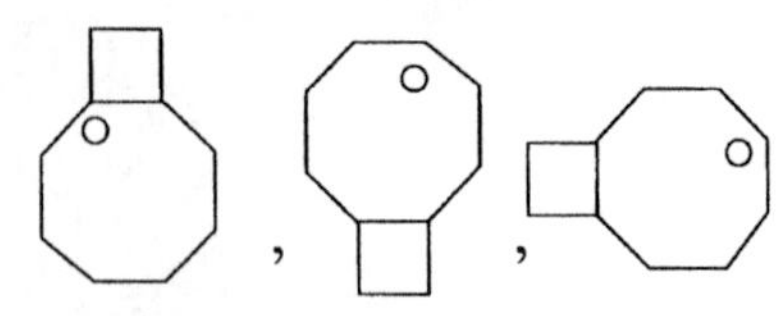

Chapter Review

2. Step 1 *Understand the Problem.*

   Four different tiles are to be placed on a wall side by side. Determine how many ways the tiles can be arranged.

   Step 2: *Devise a Plan.*

   Place the tiles one at a time. Determine how many ways one tile can be placed. Once that tile is placed, determine how many ways the second tile can be placed. Continue until all four tiles are placed.

   Step 3: *Carry Out the Plan.*

   Label the tiles A, B, C, and D. There are four choices for the first tile that is placed on the wall. It could be either Tile A, Tile B, Tile C, or Tile D. Then in placing the second tile we have three choices *for each of the first four choices* of tiles. So in placing the first two tiles, there are 12 (4×3) arrangements. For *each* of these 12 arrangements there are two choices for the third tile, giving us 24 (12×2 or $4 \times 3 \times 2$) arrangements. Finally there is only one tile left. Therefore there are 24 ($4 \times 3 \times 2 \times 1$) different arrangements of tiles.

   The 24 different arrangements are listed as follows:

ABCD	BACD	CABD	DABC
ABDC	BADC	CADB	DACB
ACDB	BCAD	CBAD	DBAC
ACBD	BCDA	CBDA	DBCA
ADBC	BDAC	CDAB	DCAB
ADCB	BDCA	CBDA	DCBA

   Step 4: *Look Back.*

   How many ways could 5 different tiles be arranged side by side on the wall?

3. Step 1: *Understand the Problem.*

   Travel from A to B along the edge of the cube. Determine the number of paths of length 3 units.

   Step 2: *Devise a Plan.*

   Solve a simpler problem by determine the number of paths of length 2 units that are possible when beginning at A and ending exactly one unit away from B. Once this is determined then the number of paths to B will be determined as well. Be sure to travel only on the edges of the cubes. It might be helpful to have a three-dimensional cube to use as a model.

   Step 3: *Carry Out the Plan.*

   There are exactly two ways to travel from point A to each of the three marked vertices. (Note that each of these marked

Chapter Review

3. (continued)

vertices is exactly one unit away from point B.) Thus there are six different ways (2 + 2 + 2) to travel from A to B along the edge of the cube.

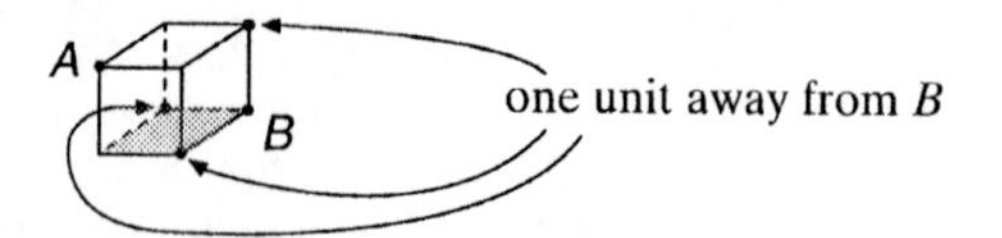

An alternate approach would be to label each of the vertices of the cube and then using Guess and Test, find the paths from A to B.

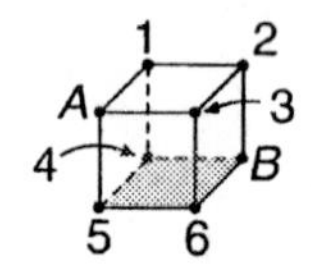

The 6 different paths are listed as follows:

A12B	A14B	A36B
A32B	A54B	A56B

Step 4: *Look Back.*

Verify using a three-dimensional model that each of the paths are 3 units in length.

4. Step 1: *Understand the Problem.*

We cannot break the toothpicks. We must arrange the hexagons in a line side by side. Each figure has one more hexagon than the previous figure. Predict the number of toothpicks required to make the *n*th figure.

Step 2: *Devise a Plan.*

Make a few more figures, where the number of hexagons in the figure is four, five, and six. Then determine if there is pattern between the number of toothpicks and the number of hexagons.

Step 3: *Carry Out the Plan.*

The next three figures in the pattern require 21, 26, and 31 toothpicks respectively.

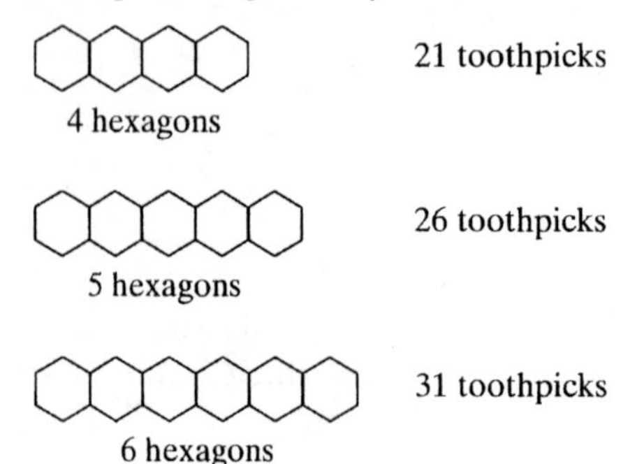

Chapter Review

4. (continued)

Summarize the information in a table.

Number of Hexagons	Total Number of Toothpicks	Pattern
1	6	$5(1) + 1 = 6$
2	11	$5(2) + 1 = 11$
3	16	$5(3) + 1 = 16$
4	21	$5(4) + 1 = 21$
5	26	$5(5) + 1 = 26$
6	31	$5(6) + 1 = 31$

We see that each successive figure requires five additional toothpicks. Therefore the nth figure will require $5(n) + 1 = 5n + 1$ toothpicks.

Step 4: *Look Back.*

Can you explain geometrically why the number of toothpicks needed for the nth figure is $5n + 1$ toothpicks?

5. Step 1: *Understand the Problem.*

Determine the maximum number of points in which three different lines can intersect. Then determine the maximum number of points in which four, five, nine, and n lines can intersect.

Step 2: *Devise a Plan.*

Draw a diagram using three lines, four lines, and five lines. Then look for a pattern in order to determine the maximum number of points that nine lines and n lines intersect.

Step 3: *Carry Out the Plan.*

The following figures show the maximum number of points in which three, four, and five lines intersect.

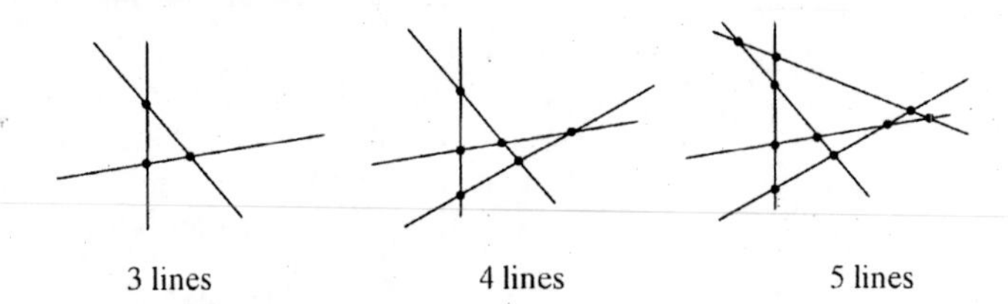

3 lines   4 lines   5 lines

Make a table displaying the information and extend the data in the table. When a new line is added and it intersects each of the previous lines, the maximum number of intersection points occurs. Therefore, the maximum number of intersection points of nine intersecting distinct lines is the sum of the first eight counting numbers. Also, the maximum number of intersection points of n intersecting distinct lines is the sum of the first $n - 1$ counting numbers.

Chapter Test 5. (continued)

Number of Distinct Lines	Maximum Number of Intersection Points
3	$3 = 1 + 2$
4	$6 = 1 + 2 + 3$
5	$10 = 1 + 2 + 3 + 4$
6	
7	
8	
9	$36 = 1 + 2 + 3 + 4 + 5 + 6 + 7 + 8$
n	$\frac{n(n-1)}{2} = 1 + 2 + 3 + 4 + 5 + ... + (n-1)$

Notice in the table that the numbers representing the maximum number of intersection points are the triangular numbers. Therefore, the maximum number of points in which n lines can intersect is the sum of the first $n - 1$ counting numbers or the $(n - 1)$st triangular number; that is

$$\frac{(n-1)n}{2} = \frac{n(n-1)}{2}.$$

Step 4: *Look Back.*

Verify that the formula found works for 3, 4, 5, and 9 distinct lines. Complete the table finding the maximum number of points in which 6, 7, and 8 lines can intersect.

Solutions to Chapter 1 Test

1. Draw a Picture, Guess and Test, Use a Variable, Look for a Pattern, Make a Table, Solve a Simpler Problem.

2. Use a variable is the appropriate strategy because you must find the answer for any number and develop a formula that can be applied to any number.

3. The original pattern is a square spiral. By moving exactly 4 toothpicks you need to make 3 squares which may overlap or be disconnected. Use actual toothpicks or a good picture and solve this by Guess and Test.

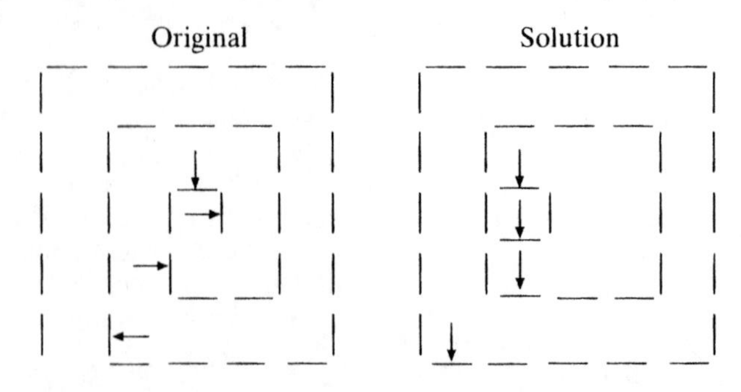

Chapter Test

3. (continued)

Move the indicated toothpicks to the positions shown in the solution. You now have three squares of different sizes.

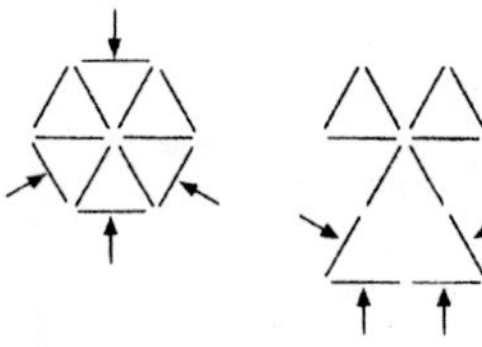

4. You must move exactly 4 toothpicks to change a pattern from 6 equilateral (all sides equal) triangles to 3 equilateral triangles. Use toothpicks and Guess and Test.

   Move the toothpicks indicated by the arrows in the first figure and move them to the positions indicated in the solution to form the large triangle.

5. The triangle has two equal angles and a third angle that is 5° less than 3 times the measure of either of the equal angles. Solve this problem by using a variable and the fact that the measure of the angles of a triangle must add to 180°.

   Let x = measure of one of the equal angles and

   $3x - 5°$ = measure of the third angle.

$$x + x + 3x - 5° = 180°$$
$$5x - 5° = 180°$$
$$5x = 185°$$
$$x = 37°$$

   So $3x - 5° = 3 \cdot 37° - 5° = 106°$.

   The three angles are 37°, 37°, and 106°.

   Check that they add to 180°: $37° + 37° + 106° = 180°$.

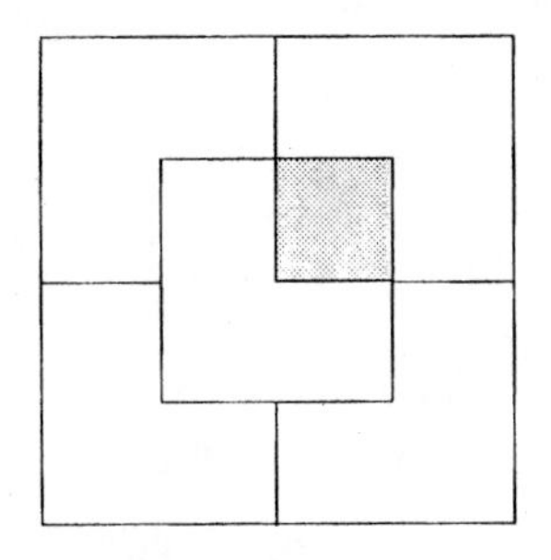

6. You need to subdivide the entire area except the house plot into 5 pieces of the same size and shape. You are not necessarily limited to rectangular or L-shaped lots, although they may work. There are 15 squares so each plot will contain the equivalent of $15 \div 5 = 3$ squares. A good strategy would be to draw a picture and use Guess and Test.

   Experimenting shows that L-shaped plots work.

   Looking back, notice that all plots contain 3 squares and are L-shaped. What would happen if the plots must also all touch the house plot? Are there other shapes of plots that will work?

7. You want the maximum number of pieces possible using the given number of straight cuts on a round cake. The pieces need not be the same size. Start by drawing a picture and using the fewest cuts. Then look for a pattern as you increase the number of cuts.

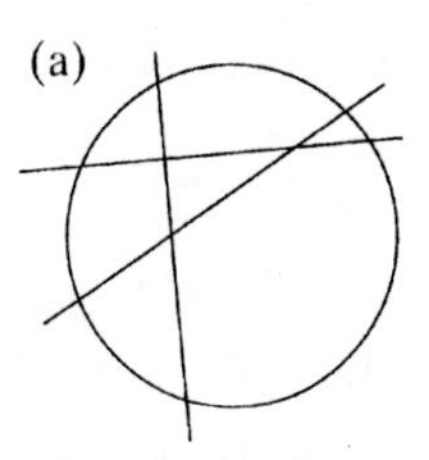

3 cuts = 7 pieces

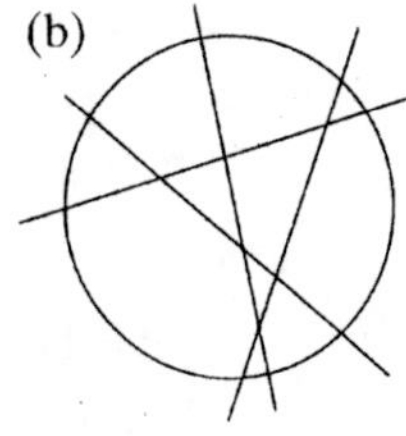

4 cuts = 11 pieces

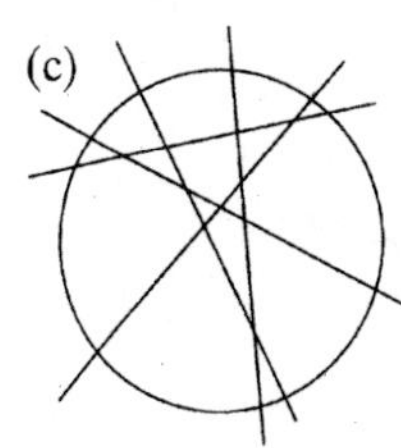

5 cuts = 16 pieces

   Is there a pattern based upon the number of cuts?

Chapter Test

8. The sequence consists of adding a column that is one square taller to both the right and left sides for each step. How many blocks are in the 10th set? What is the general answer for the nth set? Draw a picture, make a table and look for a pattern.

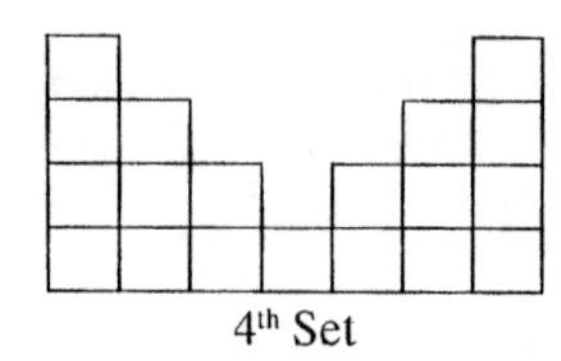

4th Set

Set	Number of blocks
1	1
2	$1 + 4 = 5$
3	$1 + 4 + 6 = 11$
4	$1 + 4 + 6 + 8 = 19$
.	
.	
10	$1 + 4 + 6 + 8 + 10 + 12 + 14 + 16 + 18 + 20 = 109$

There are 109 squares in the 10th set.

All the numbers in the sums except the first are consecutive multiples of 2. Therefore, you could write the sum for the 10th set as

$$2(1 + 2 + 3 + 4 + 5 + 6 + 7 + 8 + 9 + 10) - 1$$

and the fourth set as

$$2(1 + 2 + 3 + 4) - 1$$

so the nth set should be

$$2(1 + 2 + 3 + \ldots + n) - 1.$$

Since $1 + 2 + 3 + \ldots + n = \dfrac{n(n + 1)}{2}$ (Problem #1, Section 1.2),

the nth set will have $2\left(\dfrac{n(n + 1)}{2}\right) - 1 = n^2 + n - 1$ squares.

9. (a) $4, 13, 22, 31, 40, \ldots$

First look at the differences between consecutive terms:

$13 - 4 = 9, \quad 22 - 13 = 9, \quad 31 - 22 = 9, \quad 40 - 31 = 9$

The sequence should continue with a constant difference between successive terms:

$4, 13, 22, 31, 40, 49, 58, 67, 76, 85$

Adding 9 successively, you will find that the 10th term is 85.

To find a general formula you could rewrite the sequence as:

$4, 4 + 9, 4 + 9 + 9, 4 + 9 + 9 + 9, 4 + 9 + 9 + 9 + 9, \ldots$ or $4, 4 + 9, 4 + 2 \cdot 9, 4 + 3 \cdot 9, 4 + 4 \cdot 9, \ldots$

Using this pattern the formula for the nth term is

$4 + (n - 1) \cdot 9 = 4 + 9n - 9 = 9n - 5.$

Check this to make sure it works using $n = 10; 9 \cdot 10 - 5 = 85.$

Chapter Test

9. (continued)

(b) $3, 5, 9, 17, 33$

Look at successive differences.

$5 - 3 = 2, 9 - 5 = 4, 17 - 9 = 8, 33 - 17 = 16$

Notice that all the differences are powers of 2 and each of the numbers in the sequence is 1 more than a power of 2.

$3 = 2^1 + 1, 5 = 2^2 + 1, 9 = 2^3 + 1,$

$17 = 2^4 + 1, 33 = 2^5 + 1$

Thus, the nth term is $2^n + 1$ and the 10th term is $2^{10} + 1 = 1025$.

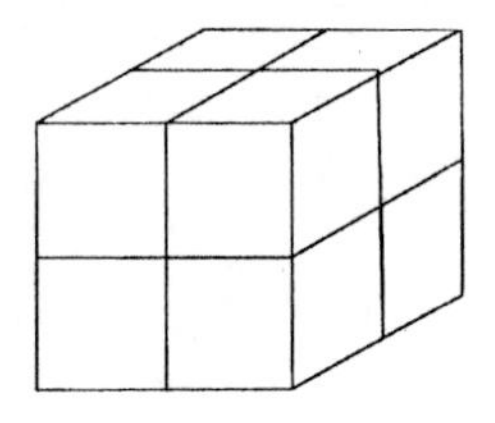

10. It would be easiest to use real blocks and a table here, but a picture will help. Remember you need to find the total number of cubes of *all* sizes, not just small ones.

(a) A 2 by 2 by 2 has one $2 \times 2 \times 2$ cube plus eight $1 \times 1 \times 1$ cubes for a total of nine cubes of any size.

(b) For a 3 by 3 by 3 cube you must count all possible $1 \times 1 \times 1$ cubes plus all possible $2 \times 2 \times 2$ cubes plus the $3 \times 3 \times 3$ cube. If at all possible, try this with actual cubes.

Summarize in a table.

cube size	# of cubes
$1 \times 1 \times 1$	27
$2 \times 2 \times 2$	8
$3 \times 3 \times 3$	1
	36

There are 36 cubes of any size in a 3 by 3 cube.

(c) For a 4 by 4 by 4 cube you need to count all the smaller cubes and the 4 by 4 by 4 cube.

Summarize in a table.

cube size	# of cubes
$1 \times 1 \times 1$	64
$2\ 2 \times 2$	27
$3 \times 3 \times 3$	8
$4 \times 4 \times 4$	1
	100

There are 100 cubes of any size in a 4 by 4 by 4 cube.

Chapter Test

10. (continued)

(d) If you look at the numbers in parts (a), (b) and (c), you may notice that the number of each size of cube is a perfect cube. That is, $1^3 = 1, 2^3 = 8, 3^3 = 27, 4^3 = 64$. For a 2 by 2 by 2 large cube the total number of cubes is $1 + 8 = 1^3 + 2^3$ cubes. For a 3 by 3 by 3 large cube the total number of cubes is $1 + 8 + 27 = 1^3 + 2^3 + 3^3$ cubes. For a 4 by 4 by 4 large cube there are $1^3 + 2^3 + 3^3 + 4^3$ cubes. The number of cubes is the sum of all the perfect cubes of numbers up through the side length of the large cube. In general, the number of cubes of any size in an n by n by n large cube is $1^3 + 2^3 + 3^3 + ... + n^3$.

11. Using Guess and Test by rearranging the pieces yields one solution. Do not forget that you *cannot* turn the pieces over.

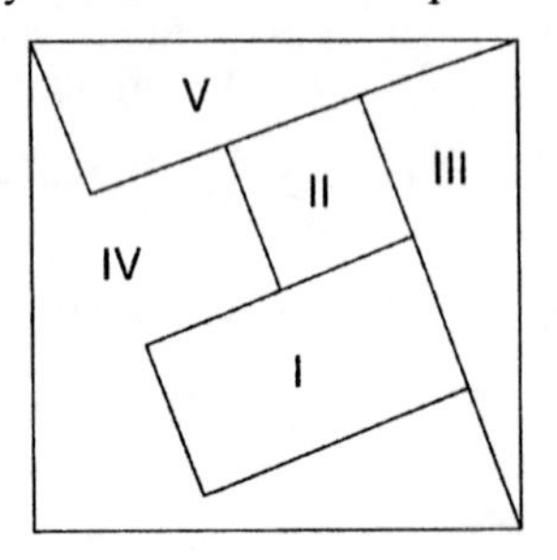

Are there any other possibilities?

2

Geometric Shapes and Measurements

Section 2.1

Tips:

✔ A good way to learn many geometric terms is to use flashcards or make your own glossary which you can add to as you work through the course. You need a good understanding of the terms in this section since they form the basis for the whole system of geometry. Review them often.

✔ Angles, lines, line segments, and rays can be named in more than one way. Answers may vary.

✔ Be careful to read from the correct scale when you use a protractor. Double check by verifying that the measure you obtain is reasonable.

Solutions to odd-numbered textbook problems

1. There are two different names for each line segment: $\overline{AB}$ or $\overline{BA}$, $\overline{AC}$ or $\overline{CA}$, $\overline{AD}$ or $\overline{DA}$, $\overline{AE}$ or $\overline{EA}$, $\overline{BC}$ or $\overline{CB}$, $\overline{BD}$ or $\overline{DB}$, $\overline{BE}$ or $\overline{EB}$, $\overline{CD}$ or $\overline{DC}$, $\overline{CE}$ or $\overline{EC}$, and $\overline{DE}$ or $\overline{ED}$.

   There are 10 different segments.

3. There are 8 different rays in the figure:
   $\overrightarrow{AB}, \overrightarrow{BC}, \overrightarrow{CD}, \overrightarrow{DE}, \overrightarrow{ED}, \overrightarrow{DC}, \overrightarrow{CB}$, and $\overrightarrow{BA}$.

   Note that some rays can be named in more than one way.

   For example, $\overrightarrow{AB}$, $\overrightarrow{AC}$, $\overrightarrow{AD}$, and $\overrightarrow{AE}$ all name the same ray.

5. There are actually 20 different names for $\overleftrightarrow{AE}$:

   $\overleftrightarrow{AB}$ $\overleftrightarrow{AC}$ $\overleftrightarrow{AD}$
   $\overleftrightarrow{BC}$ $\overleftrightarrow{BD}$ $\overleftrightarrow{BE}$
   $\overleftrightarrow{CD}$ $\overleftrightarrow{CE}$ $\overleftrightarrow{DE}$
   $\overleftrightarrow{EA}$ $\overleftrightarrow{BA}$ $\overleftrightarrow{CA}$
   $\overleftrightarrow{DA}$ $\overleftrightarrow{CB}$ $\overleftrightarrow{DB}$
   $\overleftrightarrow{EB}$ $\overleftrightarrow{DC}$ $\overleftrightarrow{EC}$
   $\overleftrightarrow{ED}$

   Any nine of these would be a correct solution.

Section 2.1

7. The distance between points is a nonnegative value so we subtract in an order that keeps the differences positive.

   (a) $PR = 0.56 - (-3.78) = 4.34$

   (b) $RQ = 0.56 - (-1.35) = 1.91$

   (c) $PS = 2.87 - (-3.78) = 6.65$

   (d) $QS = 2.87 - (-1.35) = 4.22$

9. Remember, the letter corresponding to the vertex goes in the middle when you name an angle using three letters.

   (a) $\angle BAC = \angle CAB$, $\angle BAD = \angle DAB$, $\angle CAD = \angle DAC$

   (b) $\angle BAC$ (or $\angle CAB$) and $\angle CAD$ (or $\angle DAC$) are adjacent angles. $\angle BAC$ and $\angle BAD$ are *not* adjacent because they overlap. Adjacent angles must share a side and a common vertex. The common side must lie between $\overrightarrow{AB}$ and $\overrightarrow{AD}$.

11. (a) The are 10 angles less than 180°:

   $\angle EOD$, $\angle DOC$, $\angle COB$, $\angle BOA$, $\angle EOC$, $\angle EOB$, $\angle EOA$, $\angle DOB$, $\angle DOA$, and $\angle COA$.

   (b) Remember that obtuse means between 90° and 180°. There are 4 obtuse angles:

   $\angle EOB$, $\angle EOA$, $\angle DOB$, and $\angle DOA$.

   (c) An acute angle is less than 90°. There are 6 acute angles:

   $\angle EOD$, $\angle EOC$, $\angle DOC$, $\angle COB$, $\angle COA$, and $\angle BOA$.

13. Be sure to place the center mark of the protractor on the vertex of the angle and line up one side of the angle with the "zero line" of the protractor. If your protractor has two sets of numbers, use the set that starts at 0 for the lined-up side (usually the inside set for the left-facing angles). Remember you can rotate your paper or protractor if needed.

   (a) The angle measures 60°, so it is acute.

   (b) The angle measures 114°, so it is obtuse.

   (c) The angle measures 90°, so it is a right angle.

15. Find $\angle AFB$. Because $\angle AFE$ is a straight angle and given $\angle BFE = 120°$ we know that

   $\angle AFB = \angle AFE - \angle BFE$, so $\angle AFB = 180° - 120°$

   $\angle AFB = 60°$.

   Find $\angle CFD$. We know $\angle AFB = 60°$, $\angle BFC = 55°$ and $\angle AFD = 150°$

   and $\angle AFB + \angle BFC + \angle CFD = \angle AFD$ so,

   $\angle CFD = \angle AFD - \angle AFB - \angle BFC$

   $\angle CFD = 150° - 60° - 55°$

   $\angle CFD = 35°$.

Section 2.1

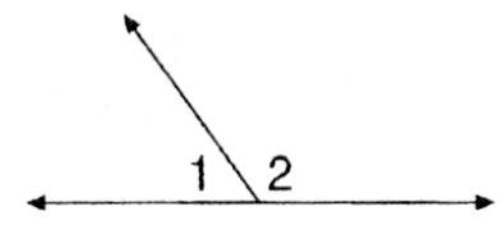

17. Let x = measure of $\angle 2$. Then $\frac{1}{2}x - 9°$ = measure of $\angle 1$.

Since $\angle 1$ and $\angle 2$ together form a straight angle, the sum of their measure is 180°.

$$\angle 1 + \angle 2 = 180°$$

$$\frac{1}{2}x - 9° + x = 180°$$

Multiplying by 2 yields

$$x - 18° + 2x = 360°$$

$$3x = 378°$$

$$x = 126°.$$

So, $\angle 2 = 126°$ and $\angle 1 = \frac{1}{2}x - 9° = 54°$.

19. Let x = measure of $\angle B$. Then $\angle A = 4x$.

Since $\angle$A and $\angle$B are complimentary, they add up to 90°.

$$\angle A + \angle B = 90°$$

$$4x + x = 90°$$

$$5x = 90°$$

$$x = 18°$$

So $\angle B = 18°$ and $\angle A = 4(18) = 72°$.

21. Draw pictures, make a table, and look for patterns. Be careful when you add another line in your drawings that it crosses as many regions as possible and does *not* go through any point of intersection from previously drawn lines. Make your drawings fairly large or there will not be enough room to see all the regions as the number of lines increases.

lines	regions
0	1
1	2 = (1) + 1
2	4 = (1 + 2) + 1
3	7 = (1 + 2 + 3) + 1
4	11 = (1 + 2 + 3 + 4) + 1
5	16 = (1 + 2 + 3 + 4 + 5) + 1
6	22 = (1 + 2 + 3 + 4 + 5 + 6) + 1
7	29 = (1 + 2 + 3 + 4 + 5 + 6 + 7) + 1
8	37 = (1 + 2 + 3 + 4 + 5 + 6 + 7 + 8) + 1
9	46 = (1 + 2 + 3 + 4 + 5 + 6 + 7 + 8 + 9) + 1
10	56 = (1 + 2 + 3 + 4 + 5 + 6 + 7 + 8 + 9 + 10) + 1

Section 2.1

21. (continued)

Notice the pattern in the numbers. From Section 1.2, problem #1, remember that the sum of the first n consecutive numbers is $\frac{n(n+1)}{2}$.

Thus, the number of regions with 10 lines is $\frac{10 \cdot 11}{2} + 1$ or 56 and the number of regions with n lines is $\frac{n(n+1)}{2} + 1$.

23. (a) $3' = \frac{3}{60}^\circ$ or 0.05°, so, $19^\circ 3' = 19.05^\circ$

(b) $36'' = \frac{36}{60}'$ or $0.6'$, so, $6'36'' = 6.6'$

$6.6' = \frac{6.6^\circ}{60} = 0.11^\circ$, so, $12^\circ 6' 36'' = 12.11^\circ$

(c) $56'' = \frac{56^\circ}{60} = 0.93\bar{3}^\circ$, so, $247^\circ 56' \approx 247.933^\circ$

(d) $58'' = \frac{58'}{60} \approx 0.967'$, so, $31'58'' \approx 31.967'$

$31.967' = \frac{31.967^\circ}{60} = 0.533^\circ$, so, $3^\circ 31.967' \approx 3.533^\circ$

25. (a) $0.6^\circ = 0.6(60)' = 36'$, so, $31.6^\circ = 31^\circ 36'$

(b) $0.75^\circ = 0.75(60)' = 45'$, so, $95.75^\circ = 95^\circ 45'$

(c) $0.32^\circ = 0.32(60)' = 19.2'$, so, $241.32^\circ \approx 241^\circ 19'$

(d) $0.48^\circ = 0.48(60)' = 28.8'$, so, $25.48^\circ \approx 25^\circ 29'$

27. (a) $.51^\circ = .51(60)' = 30.6'$. $0.6^\circ = 0.6(60)'' = 36''$

so, $16.51^\circ = 16^\circ 30' 36''$

(b) $0.33^\circ = 0.33(60)' = 19.8'$. $0.8' = 0.8(60)'' = 48''$

so, $0.33^\circ = 0^\circ 19' 48''$

(c) $0.993^\circ = 0.993(60)' = 59.58'$. $0.58' = 0.58(60)'' = 34.8''$

so, $91.933^\circ \approx 91^\circ 59' 35''$

(d) $0.029^\circ = 0.029(60)' = 1.74'$, $0.74' = 0.74(60)'' = 44.4''$

so, $58.029^\circ \approx 58^\circ 1' 44''$

29. The bearing of a line segment always uses the *smallest* angle made with the north-south line either N or S preceding the angle and E or W after it. Lines going up use N, down uses S, left uses W, and right uses E. List N or S first, then the angle, then E or W. For example, a line going up and to the right as shown in the text is described by a bearing of N 20° E.

Section 2.1

To find the bearing of $\overline{BA}$, place B at the orgin instead of A.

(a)

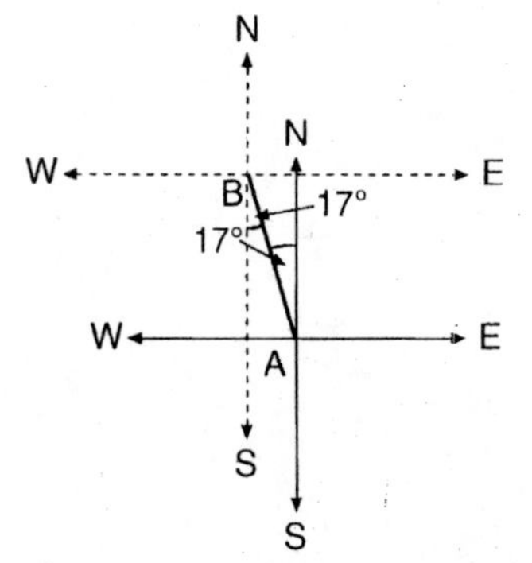

The bearing of $\overline{AB}$ is N17°W. The bearing of $\overline{BA}$ is S17°E.

(b)

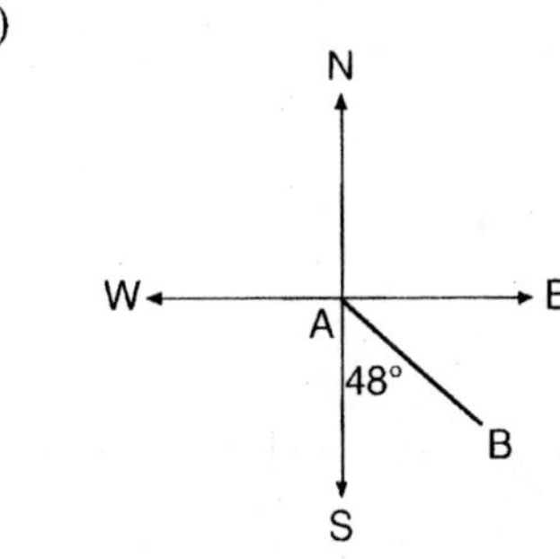

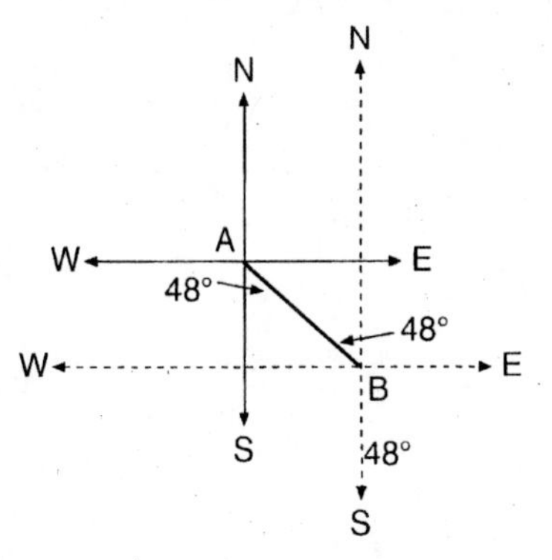

The bearing of $\overline{AB}$ is S48°E. The bearing of $\overline{BA}$ is N48°W.

(c)

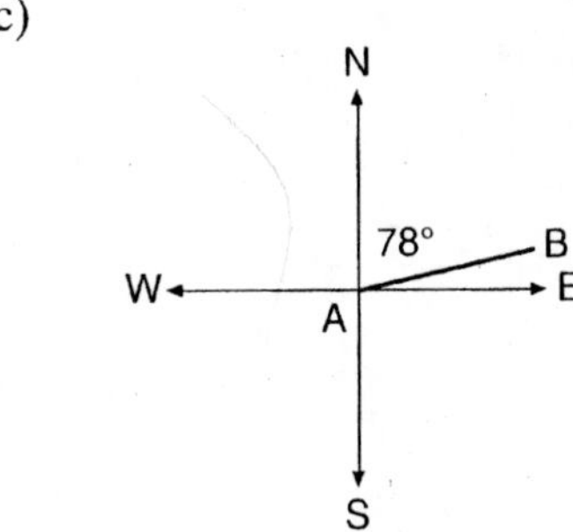

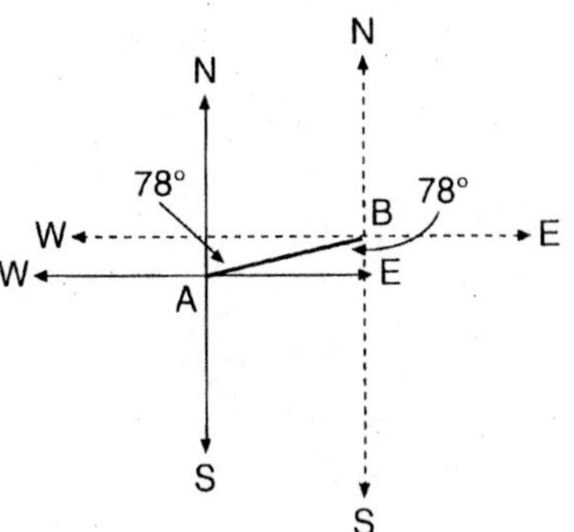

The bearing of $\overline{AB}$ is N78°E. The bearing of $\overline{BA}$ is S78°W.

(d)

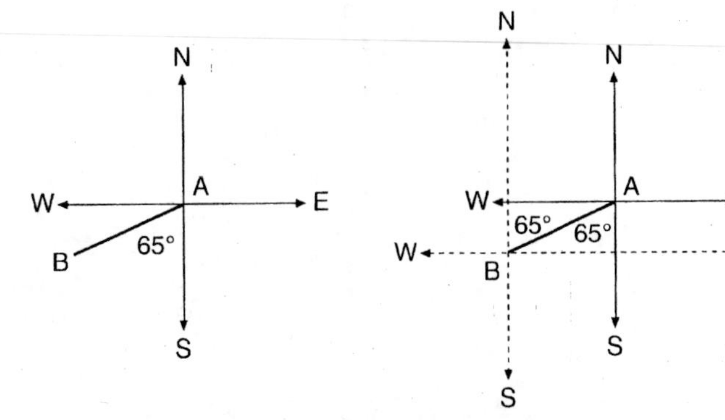

The bearing of $\overline{AB}$ is S65°W. The bearing of $\overline{BA}$ is N65°E.

Section 2.1

31. Draw a line of some measured length x, from A to N. Draw a perpendicular line from N to E, three times the length of x.

    Draw line segment $\overline{AE}$. Use a protractor to find the measure of $\angle NAE$, approximately 72°. The approximate bearing of the plane is N72°E.

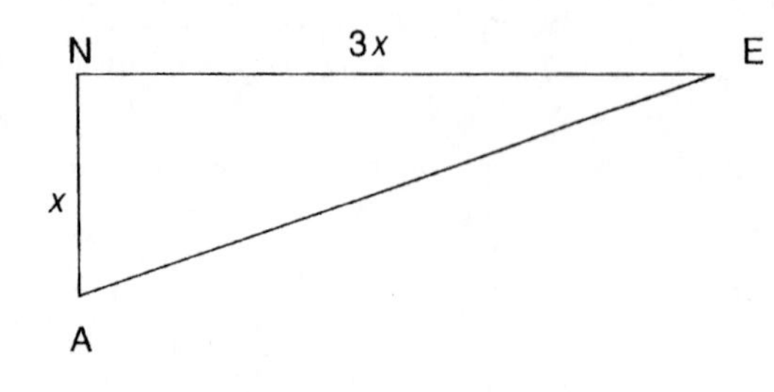

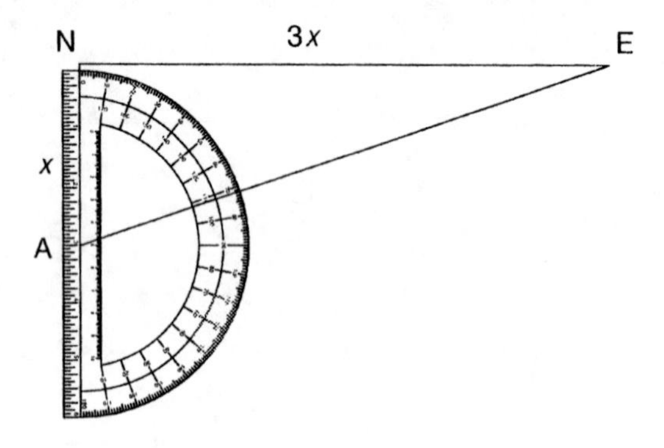

33. (a)

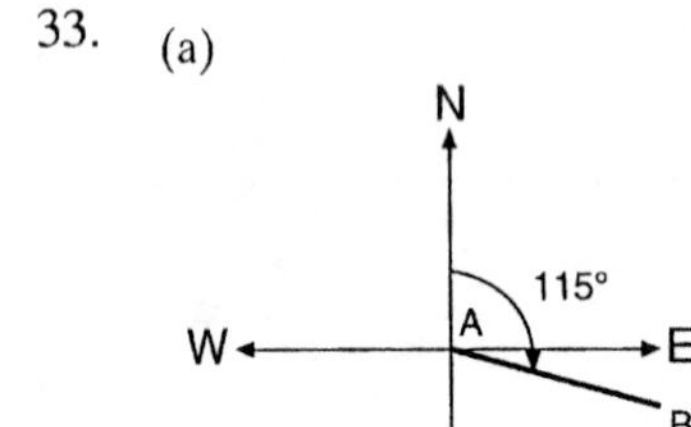

(b)

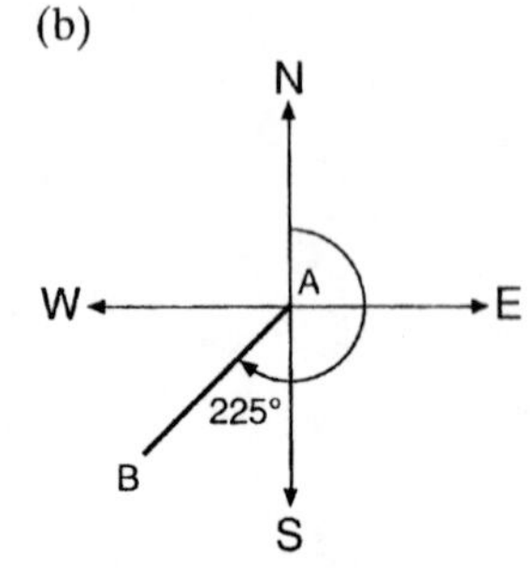

(c)

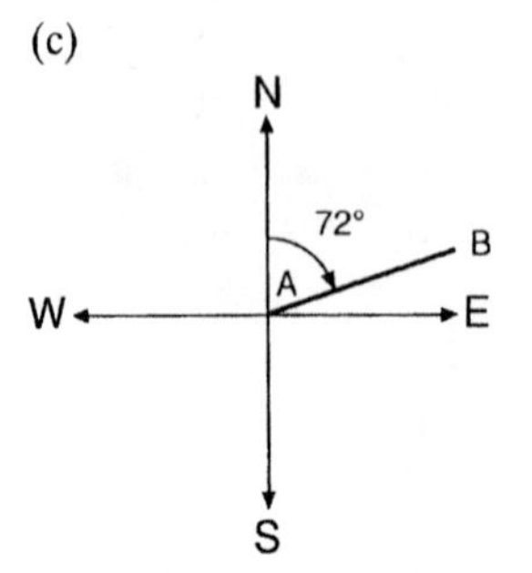

(d)

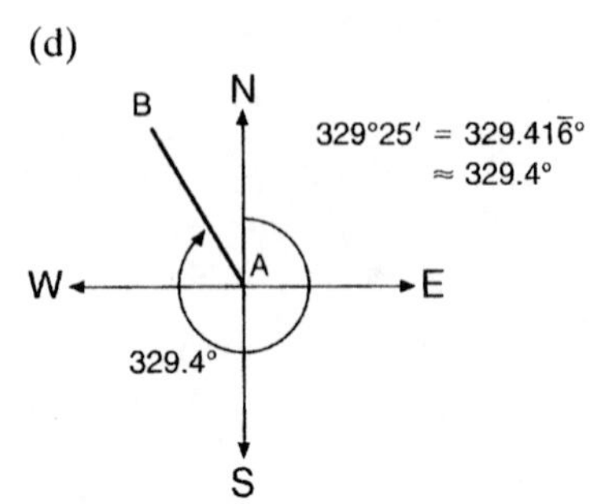

35. (a) N 40° E is measured clockwise from north, so N 40° E corresponds to a 40° azimuth.

    (b) N 40° W is measured counterclockwise from north, so, to reach this angle in a clockwise direction from north we subtract the heading from 360°: 360° − 40° = 320°.

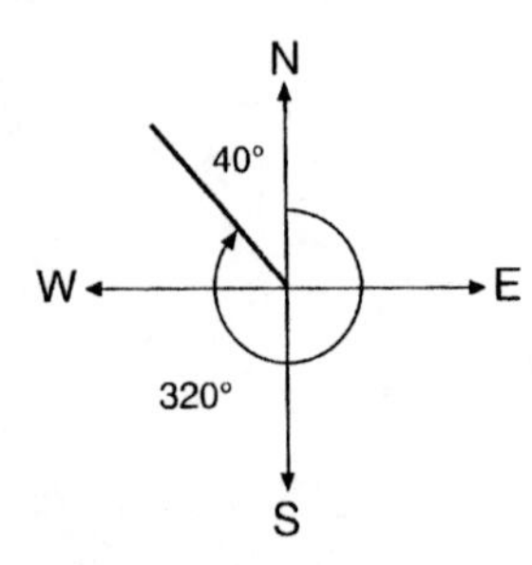

Section 2.1

35. (continued)

(c) A bearing of N 82° 17′ W is measured counterclockwise from the north. The azimuth is $360° - 82°17' = 277°43'$.

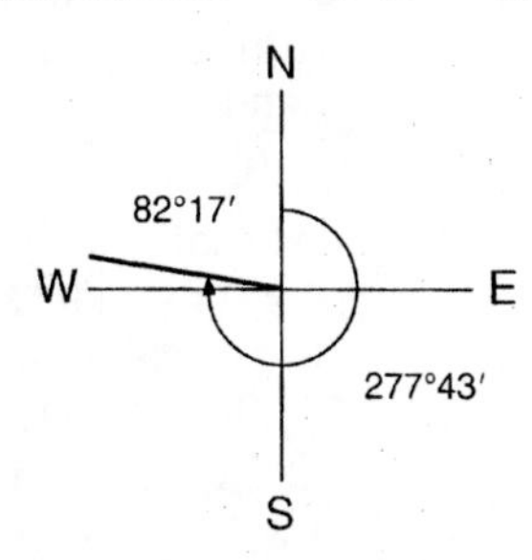

(d) S 34°35′ E is measured counterclockwise from the south. The azimuth is $180° - 24°35' = 155°25'$

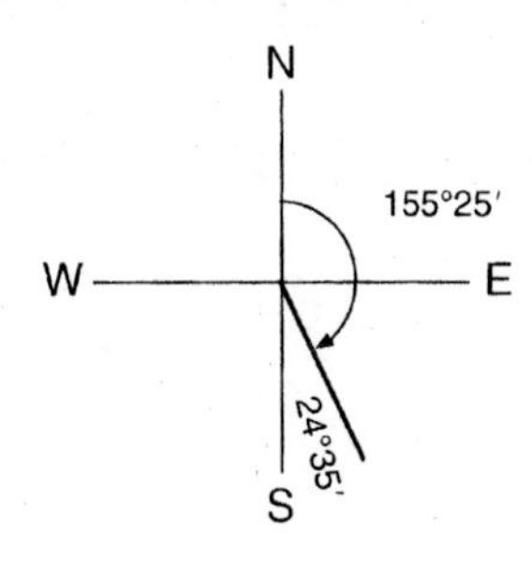

37. $\angle ABC = 36°21' + 72°45'$

$= 108°66'$

$\angle ABC = 109°6'$

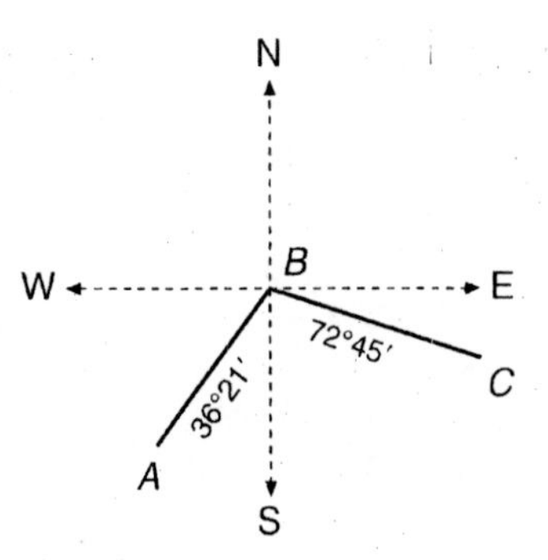

39. (a) Since there are 12 hours on a clock face, in one hour the hour hand rotates $\frac{1}{12}$ of a full 360° circle, or $\frac{1}{12}(360°) = 30°$. Since there are 60 minutes in one hour, the minute hand rotates through one, complete, 360° circle.

(b) In one minute, the minute hand rotates $\frac{1}{60}$ of one full, 360° circle, or $\frac{1}{60}(360°) = 6°$.

In one minute, the hour hand rotates $\frac{1}{60}$ of one hour's rotation. From part (a) the hour hand rotates 30° in one hour. Therefore, in one minute, the hour hand rotates $\frac{1}{60} \cdot 30° = 0.5°$.

Section 2.1

39. (continued)

(c) At 9:10, the hour hand has rotated $9(30°) + 10(0.5°) = 275°$ measured clockwise from the 12. Measuring counterclockwise from the 12, the hour hand is at $360° - 275° = 85°$. The minute hand is at $10(6°) = 60°$ measured clockwise from the 12. The obtuse angle between the hands is $85° + 60° = 145°$.

At 5:56 the hour hand has rotated $5(30°) + 56(0.5°) = 178°$ measured clockwise from the 12. The minute hand is at $56(6°) = 336°$ measured clockwise from the 12. The obtuse angle between the hands is $336° - 178° = 158°$.

(d) When the clock first shows 12:15 A.M., the hour hand has moved $15 \times 0.5° = 7.5°$ and the minute hand has moved $15 \times 6° = 90°$. So the angle between the hour hand and minute hand is $90° - 7.5° = 82.5°$.

When the clock first shows 12:16 A.M., the hour hand has moved $16 \times 0.5° = 8°$ and the minute hand has moved $16 \times 6° = 96°$. So the angle between the hour hand and minute hand is $96° - 8° = 88°$.

When the clock first shows 12:17 A.M., the hour hand has moved $17 \times 0.5° = 8.5°$ and the minute hand has moved $17 \times 6° = 102°$. So the angle between the hour hand and minute hand is $102° - 8.5° = 93.5°$.

Therefore, sometime during the 12:16 A.M. minute the hour hand and minute hand first form a right angle.

41. In order to see the three refracted rays and the three emergent rays it is important to make as large a triangular prism with a 45° angle as possible. Figure 1 shows the white light entering the prism and the red refracted and emergent rays. Figure 2 shows the white light entering the prism and all three refracted and three emergent rays. The difference in the angle of refraction of the three rays is small. It may be difficult to see this difference in the drawing. In fact, it may appear that the rays coincide. The difference in the angle of emergence of the three rays is larger. Therefore, the rays are easier to see.

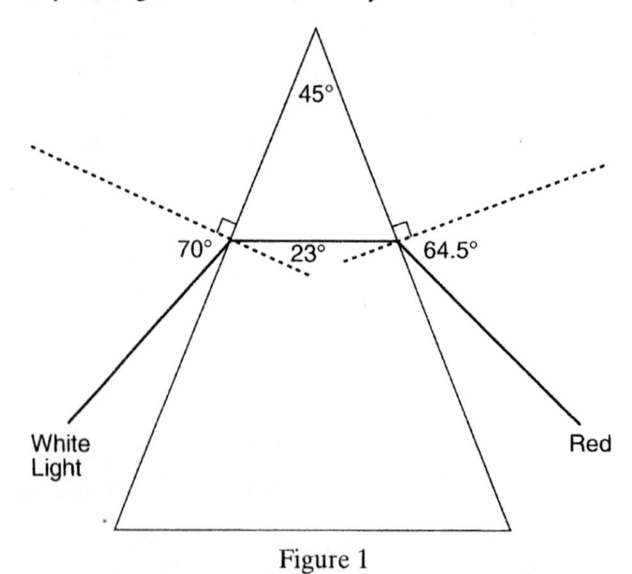

Figure 1

Section 2.2 41. (continued)

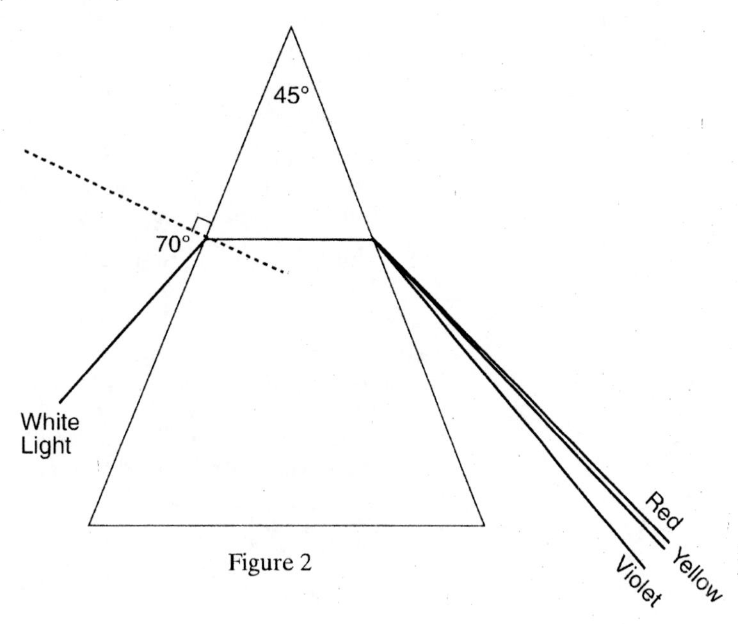

Figure 2

Section 2.2

Tips:

✔ To check for reflection symmetry it often helps to cut out a figure and fold it along possible axes of symmetry to see if the sides match.

✔ To check for rotation symmetry, cut out a copy of the figure and mark matching vertices on the copy and the original. Rotate the copy to see if it matches the original. The angle of rotation is the angle formed by the marked vertices and the center of the rotation.

Solutions to odd-numbered textbook problems

1. (a) Isosceles triangles have *at least* two congruent sides and may have three. Congruent sides are shown by matching numbers of marks on them. ΔDEF, ΔGHI and ΔJKL are isosceles. (Each triangle has other possible names, i.e. $\Delta DEF = \Delta EFD = \Delta FDE = \Delta FED = \Delta EDF = \Delta DFE$. A triangle is often named with its letters in alphabetical order or listed clockwise.)

   (b) An equilateral triangle has three congruent sides. ΔGHI is equilateral.

   (c) A scalene triangle has no congruent sides. ΔABC and ΔMNO are scalene.

   (d) A right triangle has one right (90°) angle. ΔJKL and ΔABC are right triangles.

   (e) An obtuse triangle has one abtuse angle and two acute angles. ΔDEF is obtuse.

Section 2.2

3. A complete solution is given in the answer key at the back of the text.

5. (a) Two sides must be the same length and their sum must be greater than the length of the third side. There are four triangles: 2–9–9, 4–8–8, 6–7–7, and 8–6–6.

   (b) Each side must be a different length and the sum of any two sides must be greater than the length of the third side. There are four triangles: 3–8–9, 4–7–9, 5–6–9 and 5–7–8.

   (c) Since 20 is not divisible by 3, no equilateral triangles can be formed.

7. (a) You must have a strategy for counting triangles in order to keep track of those you have counted. First you might count the triangles for which the third vertex is in the top row, the second row, etc. There are five possibilities in each row and four rows so there will be $4 \cdot 5 = 20$ different triangles with $\overline{AB}$ as one side.

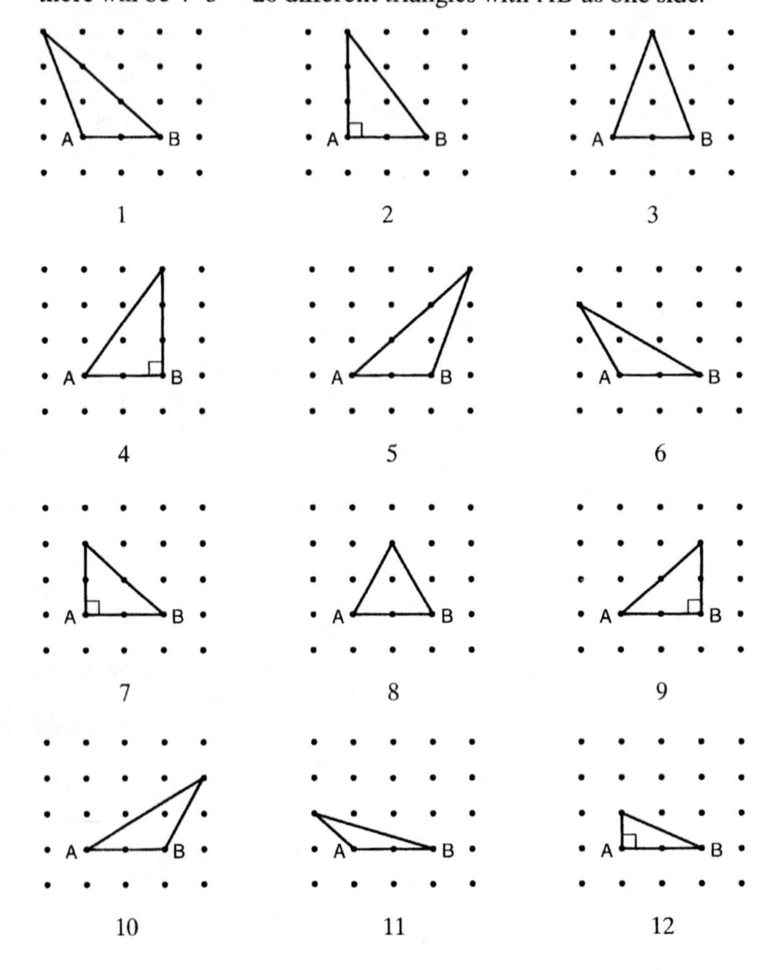

Section 2.2

7. (continued)

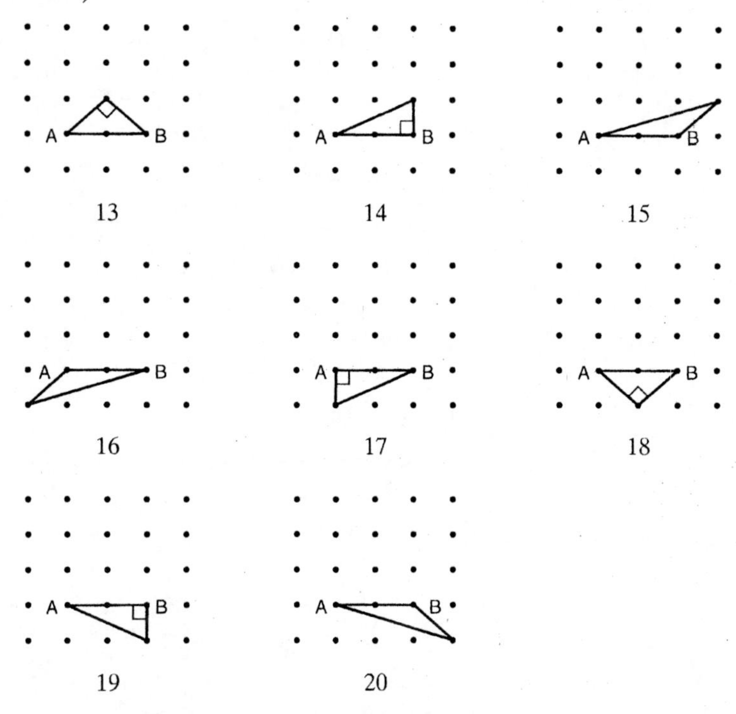

(b) The only isosceles triangles will be ones that contain a vertex in the middle column or a second side of length two. There are six isosceles triangles.

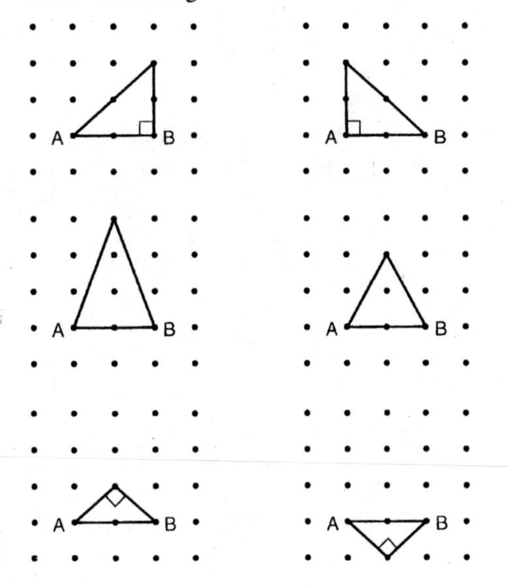

Section 2.2

7. (continued)

(c) Most right triangles will have the third vertex in either the 2nd or the 4th column. There are eight of these. However, right triangles can also be formed by using the middle point just above $\overline{AB}$. This adds two more right triangles for a total of 10.

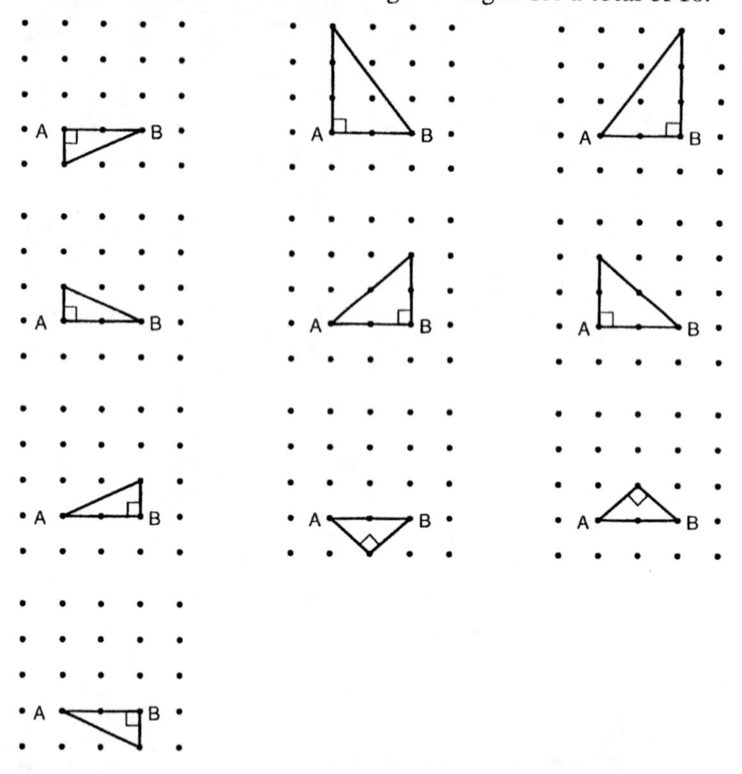

(d) The only acute triangles have the third vertex in the third column. There are four, but two of these are right triangles, so there are only two acute triangles.

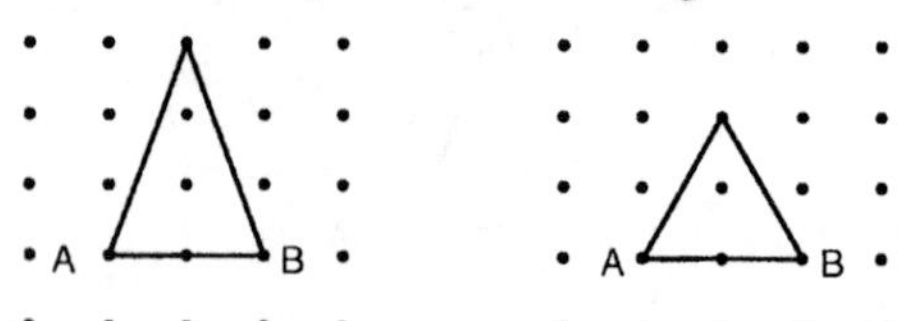

(e) You have a total of 20 triangles and 10 of them are right and 2 of them are acute. Thus, you must have $20 - (10 + 2) = 8$ obtuse triangles. Or, you would simply count the triangles whose third vertex is in the 1st or 5th column.

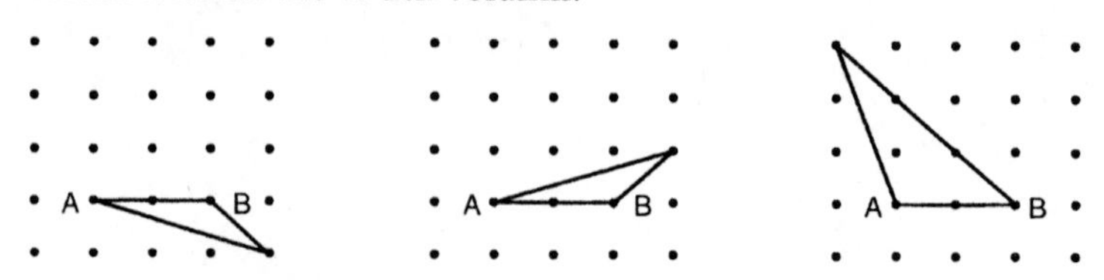

Section 2.2

7. (continued)

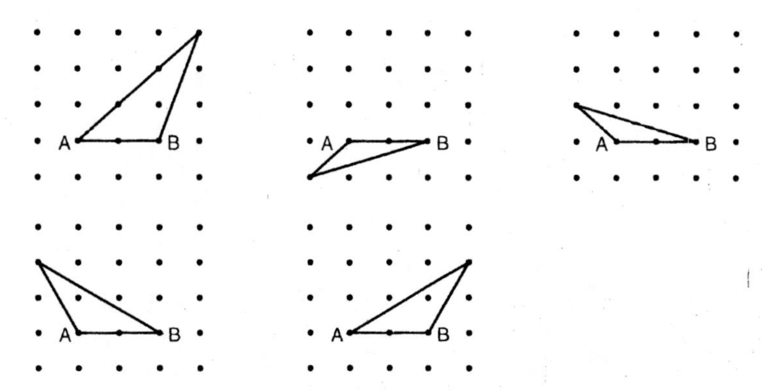

9. There are many answers possible here. The endpoints must be connected without creating any points of intersection or retracing any part of the curve. Here is one possible simple closed curve.

11. The solutions are not unique. One possible set of solutions is given in the answer key at the back of the text.

13. A polygon is a simple closed curve composed of line segments.

(a) This is a polygon.

(b) This is not a polygon since it is not composed entirely of line segments.

(c) Since the figure cannot be drawn without retracing the curve, it is not a polygon.

(d) This is not a simple curve since it has a point of intersection.

15. Hint: On part (b), the lines you draw do not need to originate at a vertex.

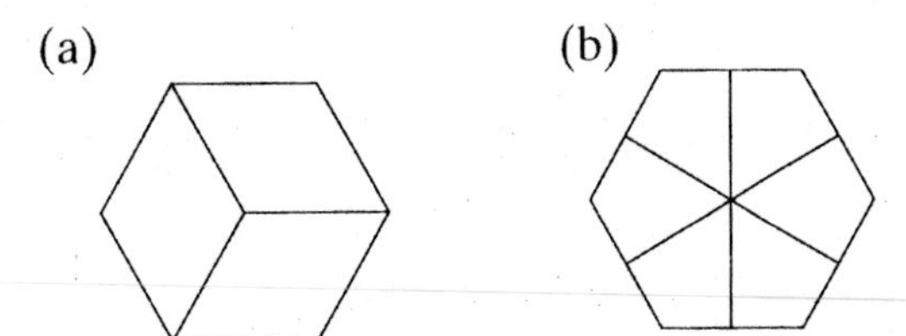

17. (a) A square has four congruent sides and four right angles. *OFHQ* is a square.

(b) A rectangle has four right angles. *ADPK* and *AEGJ* are rectangles which are not squares.

Section 2.2 17. (continued)

(c) A parallelogram has two sets of parallel sides. *MOQK, MNOS,* and *KQHI* are parallelograms.

(d) An isosceles right triangle has one right angle and two congruent sides. ΔFGH, ΔOCD, ΔCEF, ΔABL, ΔOPQ, ΔACK, ΔMCO and ΔIJK are isosceles right triangles.

(e) ΔMNO, ΔMSO and ΔLBC are isosceles triangles with no right angles.

(f) A rhombus has four congruent sides. *MNOS* and *KQHI* are rhombuses but not squares.

(g) A kite has two non-overlapping sets of adjacent congruent sides. *COSM* is a kite that is not also a rhombus.

(h) A scalene triabgle has sides of three different lengths. ΔROQ and ΔLKC are scalene triangles with no right angles.

(i) ΔACL and ΔROP are right scalene triangles.

(j) A trapezoid is a quadrilateral with one pair of parallel sides and one pair of nonparallel sides. When the nonparallel sides are congruent, the trapezoid is called an isosceles trapezoid. *KCOQ*, *CDOM*, *FODE*, *KQHJ*, and *KMOP* are trapezoids that are not isosceles.

(k) *LBCK* is an isosceles trapezoid.

19. (a) Devise a strategy for counting parallelograms in order to be sure that all have been found. One strategy is to determine all possible points C where $\overline{BC}$ is a side of the parallelogram. Remember that the side opposite $\overline{AB}$ must be parallel and congruent to $\overline{AB}$. There are 11 possible locations of point C.

This results in the following parallelograms.

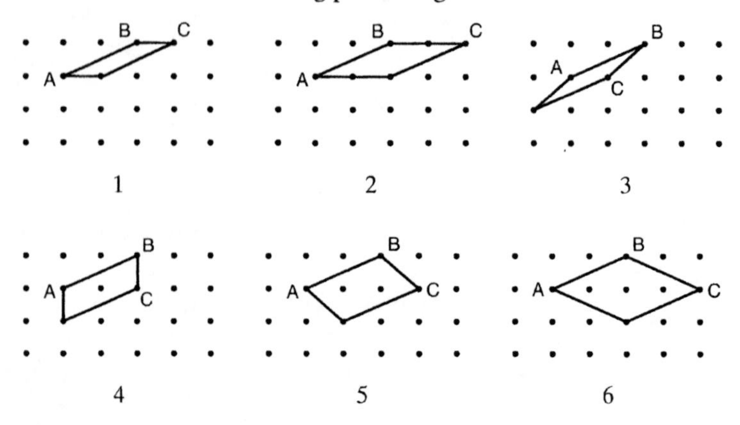

Section 2.2

19. (continued)

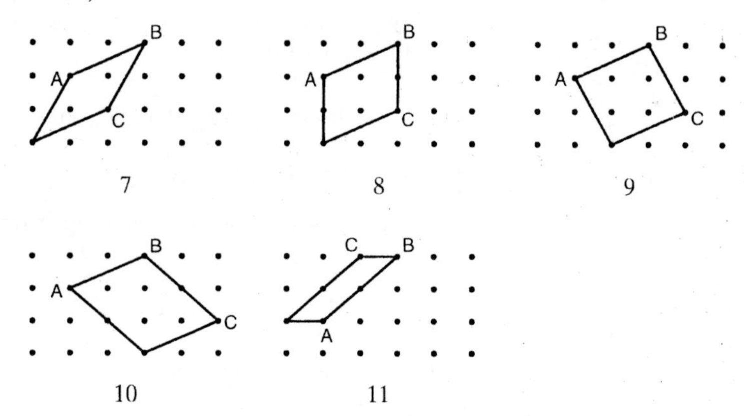

(b) There is one rectangle possible and it is a square.

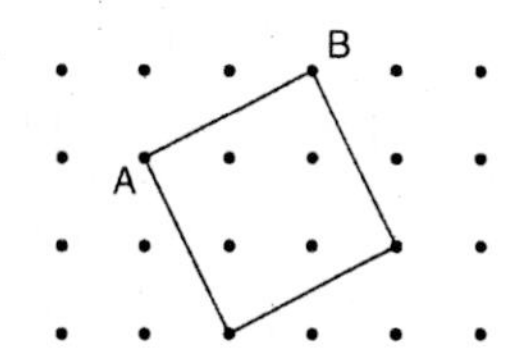

(c) There are three rhombuses and one of them is a square as in Part (b).

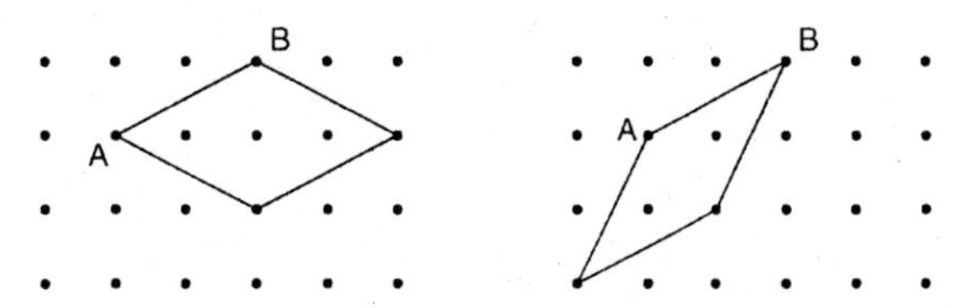

(d) There is one square, as in Part (b).

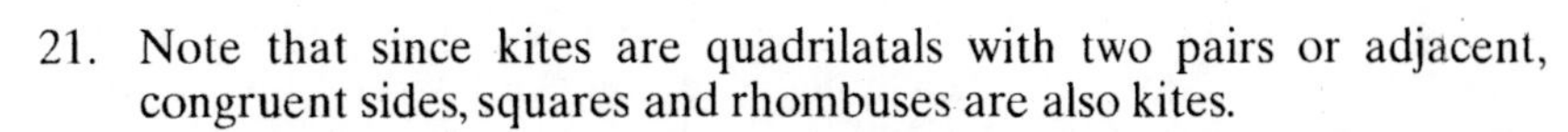

21. Note that since kites are quadrilatals with two pairs or adjacent, congruent sides, squares and rhombuses are also kites.

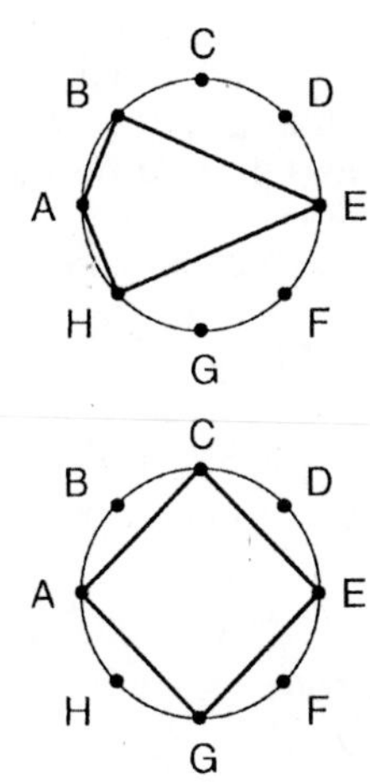

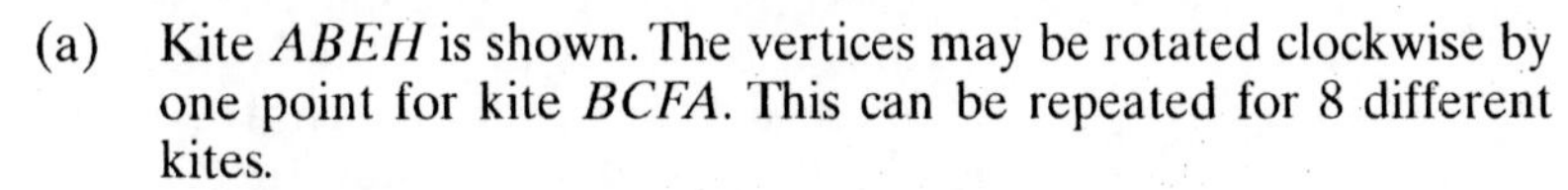

(a) Kite *ABEH* is shown. The vertices may be rotated clockwise by one point for kite *BCFA*. This can be repeated for 8 different kites.

Kite *ACEG*, shown, is a square. The square can be rotated clockwise by one point to form another square kite.

Therefore, there are 10 different kites.

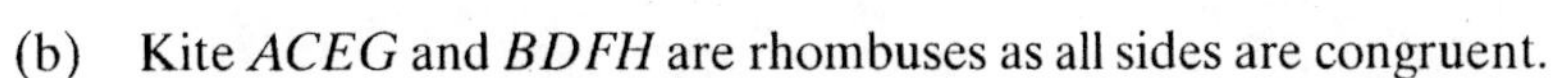

(b) Kite *ACEG* and *BDFH* are rhombuses as all sides are congruent.

(c) Both *ACEG* and *BDFH* are squares. So there are none.

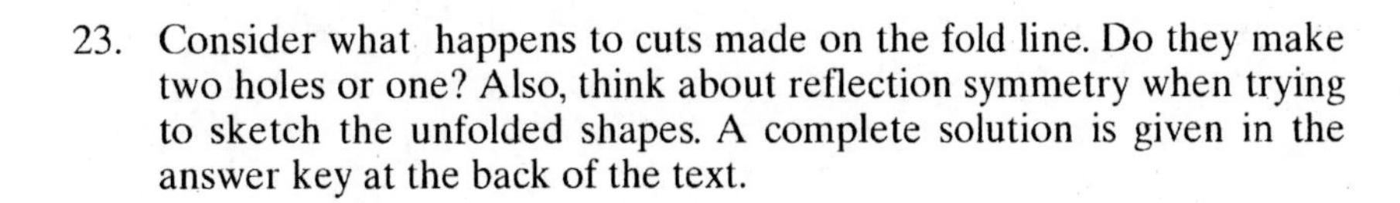

23. Consider what happens to cuts made on the fold line. Do they make two holes or one? Also, think about reflection symmetry when trying to sketch the unfolded shapes. A complete solution is given in the answer key at the back of the text.

Section 2.2

25. Consider an axis of symmetry and draw it in. In each case the axis of symmetry is vertical. A complete solution is given in the answer key at the back of the text.

27. (a) Try connecting the midpoints of opposite sides. This produces two lines of symmetry. Try connecting opposite vertices. Neither of these are lines of symmetry.

    Therefore the rectangle has two axes of symmetry.

    (b) From a point at the center, rotate the rectangle 180°.

    Therefore the rectangle has one rotation of symmetry.

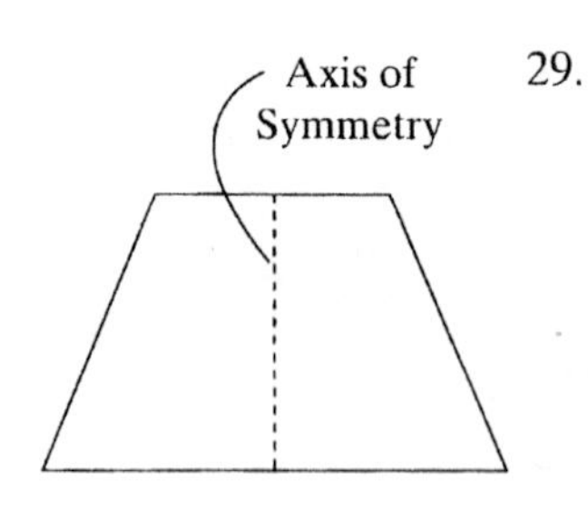

29. (a) Reflection symmetry means you can fold the figure and the sides match. The line on which you fold is the axis of symmetry. One way of testing it is to trace a copy, cut it out and fold it. You will find that there is one axis of symmetry here—the line that joins the midpoints of the top and bottom.

    (b) Take your copy of the trapezoid and put it on top of the original. Rotate the copy until the two trapezoids match. If they match in less than a full (360°) turn, you have a rotation symmetry. There are no rotation symmetries for this trapezoid.

31. (a) A complete solution is shown in the answer key at the back of the text.

    (b) Look for a pattern:

Number of sides	5	6	7	8	9	...	n
Number of lines of symmetry	5	6	7	8	9	...	n

An n-sided polygon has n lines of symmetry.

33. A good strategy would be to cut out a copy of the grid on a thin or clear sheet so you can lay it on the original to check for a match. To check for reflection symmetry, fold your copy to find out if the sides match. To check for rotation symmetry, rotate it and place it on the original to see if they match. The angle through which it is rotated is the angle of rotation symmetry.

    (a) This pattern has reflection symmetry with one axis of symmetry (a vertical line down the middle). It has no rotation symmetry.

    (b) This pattern has reflection symmetry with each of the two diagonals being an axis of symmetry. It also has rotation symmetry of 180°.

    (c) This pattern has reflection symmetry with a vertical axis of symmetry down the middle. It does not have rotation symmetry.

    (d) This pattern has only reflection symmetry. The axis of symmetry is the diagonal from the upper left to the lower right.

Section 2.2

33. (continued)

(e) This pattern has reflection symmetry. Both the vertical and horizontal axes of symmetry are right down the middle. It also has rotation symmetry of 180°.

(f) This pattern has only reflection symmetry with a vertical axis of symmetry through the middle of the pattern.

35. (a) Four axes of reflection.

Three rotational symmetries of 90°, 180° and 270°.

(b) Two axes of reflection.

One rotation symmetry of 180°.

(c) Two axes of reflection.

One rotation symmetry of 180°.

(d) Four axes of reflection.

Three rotational symmetries of 90°, 180° and 270°.

(e) Four axes of reflection.

Three rotational symmetries of 90°, 180° and 270°.

(f) Two axes of reflection.

One rotation symmetry of 180°.

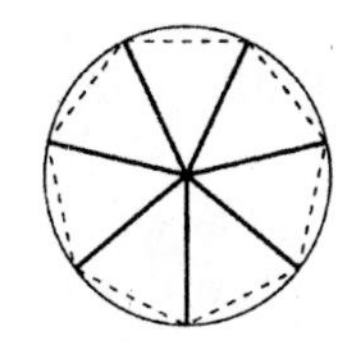

37. (a) Sketch a wheel with 7 axes of reflection symmetry. This can be done inscribing a regular heptagon in a circle. The spokes are then drawn from the center of the circle to the vertices of the heptagon.

There are seven spokes and 6 rotational symmetries.

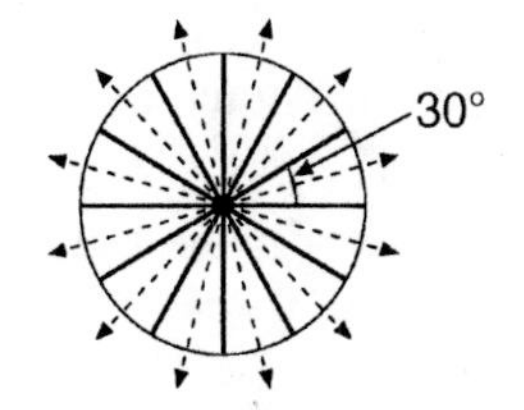

(b) If the measure of the angle between two adjacent spokes is 30°, then there are $\frac{360°}{30°} = 12$ spokes.

The wheel has 12 axes of symmetry. Six of them coincide with the spokes on the wheel. The other six are represented by the dotted lines.

Section 2.3

Tips:

- ✔ Keep in mind that the sum of the angle measures in any triangle is 180° and the sum of the angle measures in any polygon can be found by dividing it into triangles.
- ✔ Remember that in a tessellation, the measures of the vertex angles surrounding a point must be exactly 360°.
- ✔ To create a tessellation with a certain polygon, use tracing paper to copy the polygon, then try rotating it or re-positioning it around a fixed point.

Section 2.3 Solutions to odd-numbered textbook problems

1. Make xerox copies of ΔABC and carefully cut them out with scissors. Use a flat surface to arrange the pieces into a tessellation.

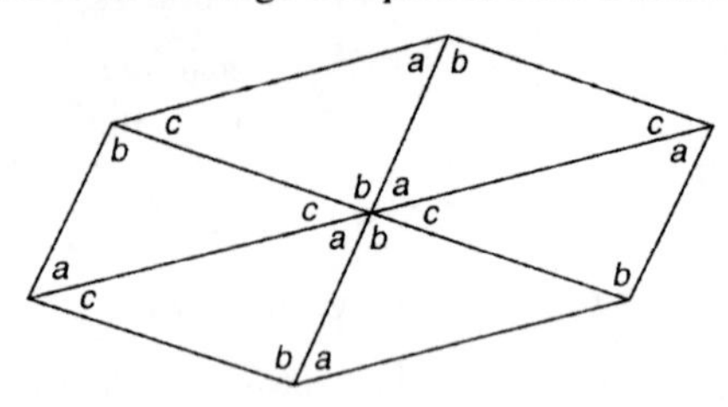

3. Remember, a tessellation or tiling uses copies of the shape to cover a surface with no gaps or overlaps. There need to be two of each angle of the triangle around every point (because having one of each angle totals 180°, so two of each will total 360°, covering the whole region around the point).

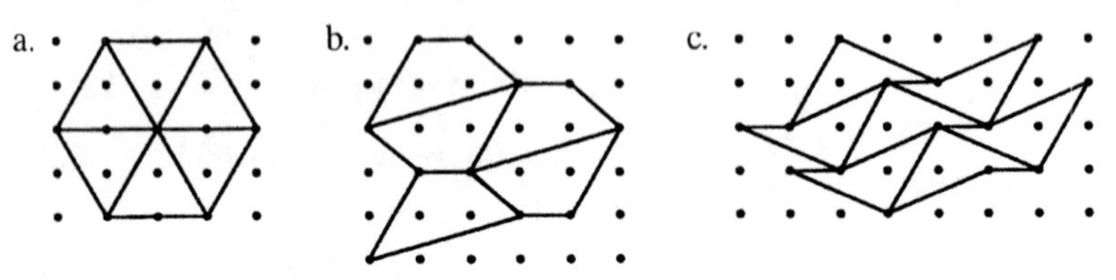

5. Remember, if your protractor has two sets of numbers, read the smaller one if the angle is an acute angle and read the larger number if the angle is obtuse.

   (a) $\angle BAC \approx 54°$, $\angle ABC \approx 90°$, $\angle ACB \approx 36°$. Your answers may differ from these slightly. The angle sum should be 180° (or close to it because of small errors in measurement).

   (b) You should have $\angle HDE \approx 100°$, $\angle DEF \approx 126°$, $\angle EFG \approx 89°$, $\angle FGH \approx 115°$, and $\angle GHD \approx 110°$. The angle sum should be 540°. Answers may vary slightly.

7. To find the measure of the third angle, add the two given angle measures and subtract from 180°.

   (a) $180° - (90° + 60°) = 30°$

   (b) $180° - (120° + 40°) = 20°$

   (c) $180° - (85° + 33°) = 62°$

   (d) $180° - (79° + 67°) = 34°$

   Looking back, would it be possible to have angles of 130° and 70° in a triangle? No, because they add to more than 180° before you even get to the third angle.

9, To start with, fill in any angle measure you can determine, even if it is not one you want. The angle measures may be found in any convenient order. For reference, points have been labeled in the picture here. As you work, keep track of what you have and what you still need to find.

Section 2.3

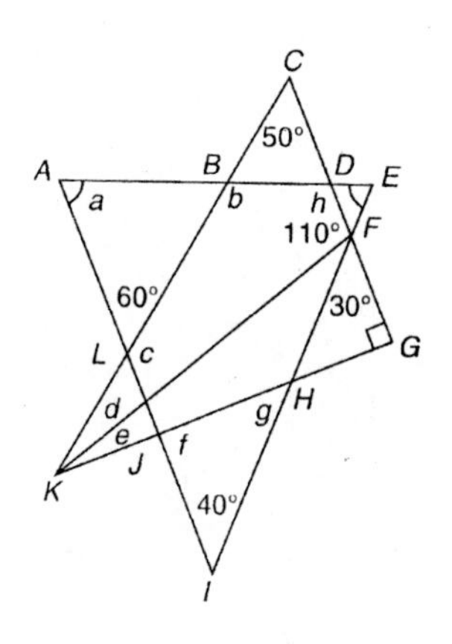

9. (continued)

Find $\angle e$.

$110° + \angle KFH + 30° = 180°$, so $\angle KFH = 40°$

$\angle KFH + 30° + 90° + \angle e = 180°$ because they make up ΔKFG, so

$$\angle e = 180° - 120° - \angle KFH$$
$$= 60° - 40°$$
$$\angle e = 20°.$$

Find $\angle a$.

$\angle DEF = \angle a$ and $\angle DEF + \angle a + 40° = 180°$ (from ΔAEI)

Substituting $\angle a$ for $\angle DEF$, we get

$$2\angle a + 40° = 180°$$
$$2\angle a = 140°$$
$$\angle a = 70°.$$

Find $\angle d$.

$\angle d + \angle e$, 50°, and 90° are the angles ofΔCKG so

$\angle d + \angle e + 50° + 90° = 180°$.

Since $\angle e = 20°$, $\angle d + 20° + 50° = 90° = 180°$, and $\angle d = 20°$.

Find $\angle c$.

$\angle c$ and the 60° angle make a straight angle, so

$\angle c = 180° - 60° = 120°$.

Find $\angle b$.

$\angle a + \angle ABL + 60° = 180°$

Substituting 70° for $\angle a$ gives

$70° + \angle ABL + 60° = 180°$.

So $\angle ABL = 50°$.

$\angle b$ and $\angle ABL$ form a straight angle so you have

$\angle b + \angle ABL = 180°$. Since $\angle ABL = 50°$

$\angle b + 50° = 180°$

$\angle b = 130°$.

Find $\angle g$.

Since KFH is a triangle,

$$\angle e + \angle KHF + \angle KFH = 180° \text{ so}$$
$$20° + \angle KHF + 40° = 180°$$
$$\angle KHF = 120°.$$

Since $\angle g$ and $\angle KFH$ make a straight angle,

$$\angle g + \angle KHF = 180°$$
$$\angle g + 120° = 180°$$
$$\angle g = 60°.$$

Section 2.3

9. (continued)

Find $\angle f$.

Since HIJ is a triangle, $\angle f + \angle g + \angle 40° = 180°$

$$\angle f + 60° + 40° = 180°.$$

$$\text{So } \angle f = 80°.$$

Find $\angle h$.

Since $\angle KBDG$ is a quadrilateral, its angles add to 360°.

$$\angle d + \angle e + \angle b + \angle h + 90° = 360°$$

$$20° + 20° + 130° + \angle h + 90° = 360°$$

$$\angle h = 100°$$

In summary,

$\angle a = 70°$ $\quad$ $\angle e = 20°$

$\angle b = 130°$ $\quad$ $\angle f = 80°$

$\angle c = 120°$ $\quad$ $\angle g = 60°$

$\angle d = 20°$ $\quad$ $\angle h = 100°$.

11. You know the sum of the angles in a 4-sided polygon is $(4-2) \cdot 180° = 360°$.

(a) Let x = the measure of the unknown angle.

$$100° + 90° + 72° + x = 360°$$

$$262° + x = 360°$$

$$x = 98°$$

(b) Let x = the measure of the unknown angles.

$$x + x + 55° + 55° = 360°$$

$$2x + 110° = 360°$$

$$2x = 250°$$

$$x = 125°$$

Each of the unknown angles measures 125°.

(c) Let x = the measure of the unknonwn angle.

$$90° + 90° + 60° + x = 360°$$

$$240° + x = 360°$$

$$x = 120°$$

(d) $x - 10° + x - 15° + x + 5° + x = 360°$

$$4x - 20° = 360°$$

$$4x = 380°$$

$$x = 95°$$

And it follows that $x + 5 = 100°$, $x - 15° = 80°$ and $x - 10° = 85°$.

13. Using thorem 2.2 (angle measure in a polygon), where you know the angle sum is 1620°. The number of sides is represented by n.

$$(n - 2)\,180° = 1620°$$

$$n - 2 = 9$$

$$n = 11$$

Section 2.3

15. Using Theorem 2.2 (angle measure in a polygon), where you know the number of sides (n).

$$(29 - 2)\,180° = 27 \cdot 180° = 4,860°$$

17. Using Theorem 2.3 (vertex angle measure in a regular polygon), where you know the number of sides (n).

(a) $\dfrac{(12 - 2)180°}{12} = \dfrac{10 \cdot 180°}{12} = 150°$

(b) $\dfrac{(16 - 2)180°}{16} = \dfrac{14 \cdot 180°}{16} = 157.5°$

(c) $\dfrac{(10 - 2)180°}{10} = \dfrac{8 \cdot 180°}{10} = 144°$

(d) $\dfrac{(20 - 2)180°}{20} = \dfrac{18 \cdot 180°}{20} = 162°$

(e) $\dfrac{(18 - 2)180°}{18} = \dfrac{16 \cdot 180°}{18} = 160°$

(f) $\dfrac{(36 - 2)180°}{36} = \dfrac{34 \cdot 180°}{36} = 170°$

19. Using Theorem 2.3 (vertex angle measure in a regular polygon), where you know the measure of the vertex angle. The number of sides is represented by n.

(a)
$$\begin{aligned}\frac{(n-2)180°}{n} &= 172.5°\\ (n-2)180° &= 172.5°n\\ 180°n - 360° &= 172.5°n\\ -360° &= -7.5°n\\ \frac{-360°}{-7.5°} &= n\\ 48 &= n\end{aligned}$$

(b)
$$\begin{aligned}\frac{(n-2)180°}{n} &= 175°\\ (n-2)180° &= 175°n\\ 180°n - 360° &= 175°n\\ -360° &= -5°n\\ \frac{-360°}{-5°} &= n\\ 72 &= n\end{aligned}$$

(c)
$$\begin{aligned}\frac{(n-2)180°}{n} &= 171°\\ (n-2)180° &= 171°n\\ 180°n - 360° &= 171°n\\ -360° &= -9°n\\ \frac{-360°}{-9°} &= n\\ 40 &= n\end{aligned}$$

(d)
$$\begin{aligned}\frac{(n-2)180°}{n} &= 176.4°\\ (n-2)180° &= 176.4°n\\ 180°n - 360° &= 176.4°n\\ -360° &= -3.6°n\\ \frac{-360°}{-3.6°} &= n\\ 100 &= n\end{aligned}$$

Section 2.3

21. To have congruent sides, there must be the same number of dots on the circle between consecutive vertices of the polygon.

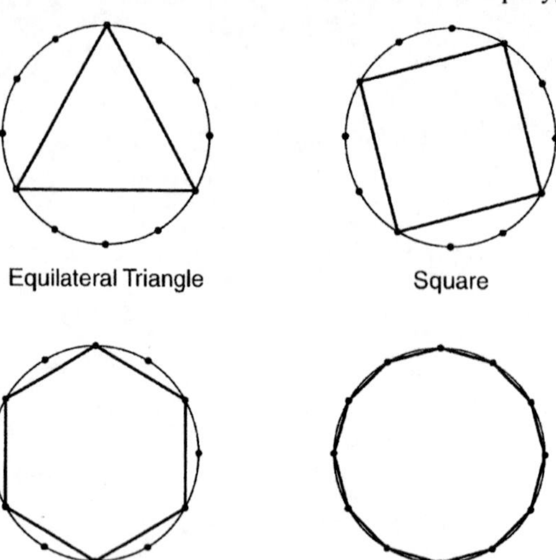

23. If a polygon has n congruent angles and is not regular, then not all sides are the same length. Make some sides longer than others. Some examples have been given in your text. Others are possible.

25. A regular heptagon is a 7-sided polygon. The measure of each vertex angle is $\dfrac{(7-2)180°}{7} \approx 128.6°$.

    The sum of the vertex angles in a regular tessellation is 360°, which is not divisible by 128.6°. Thus, a regular tessellation cannot be composed of regular heptagons.

27. The solutions are given for triangles and quadrilaterals. Start with 5 sides.

    5 sides

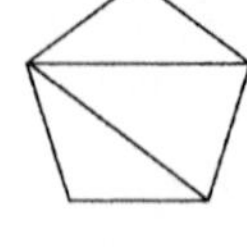

    Draw the diagonals all from the same vertex. There are 3 triangles formed so the sum of the angle measures is $3 \cdot 180° = 540°$ and thus one angle of the regular pentagon measures $540°/5 = 108°$.

    6 sides

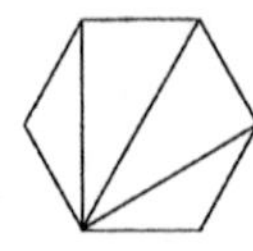

    There are four triangles here. The sum of the angle measures is $4 \cdot 180° = 720°$ and thus one angle of the regular hexagon measures $720°/6 = 120°$.

    8 sides

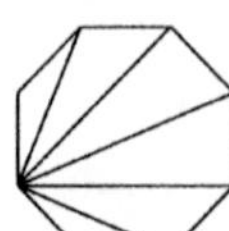
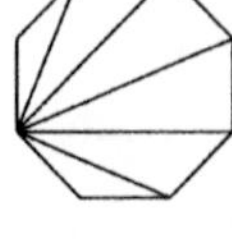

    There are now six triangles. The sum of the angle measures is $6 \cdot 180° = 1080°$ and thus one angle of the regular octagon measures $1080°/8 = 135°$.

    10 sides

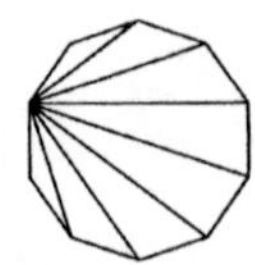

    There are eight triangles. The sum of the angle measures is $8 \cdot 180° = 1440°$ and thus one angle of the regular decagon measures $1440°/10 = 144°$.

    For each n-gon the number of triangles formed is 2 less than the number of sides or $n - 2$. The sum of the angle measures is $(n - 2) \cdot 180°$. Since

Section 2.3

27. (continued)

there are n angles and all angles in a regular n-gon are congruent, you can find one angle measure by dividing the sum of the angle measures by n, i.e., $\frac{(n-2)\cdot 180^\circ}{n}$.

29. Using ΔABC, you know that

$$20^\circ 35'50'' + 90^\circ + \angle ACB = 180^\circ$$
$$\angle ACB = 180^\circ - 90^\circ - 20^\circ 35'50''$$
$$= 90^\circ - 20^\circ 35'50''.$$

To do this subtraction you need to take 1° from 90° and make it into 60′, then take 1′ and make it into 60″.

That is, $90^\circ = 89^\circ\ 60' =$

$$\begin{array}{rrr} 89^\circ & 59' & 60'' \\ -\ 20^\circ & 35' & 50'' \\ \hline 69^\circ & 24' & 10'' \end{array} = \angle ACB.$$

31. (a)

$$\begin{array}{rr} 141^\circ & 32' \\ 78^\circ & 17' \\ 63^\circ & 25' \\ 97^\circ & 10' \\ +\ 159^\circ & 36' \\ \hline 538^\circ & 120' \end{array}$$

Since $120' = 2^\circ$

$538^\circ\ 120' = 540^\circ$.

Using the angle sum formula $(n-2)\ 180^\circ$ where $n = 5$:

$(5-2)\ 180^\circ = 540^\circ$. There is no error.

(b)

$$\begin{array}{rr} 109^\circ & 28' \\ 92^\circ & 15' \\ 136^\circ & 46' \\ 112^\circ & 53' \\ 145^\circ & 35' \\ +\ 121^\circ & 8' \\ \hline 715^\circ & 185' \end{array}$$

Since $185' = 3^\circ\ 5'$, $715^\circ\ 185' = 718^\circ\ 5'$.

Using the angle sum formula $(n-2)\ 180^\circ$ where $n = 6$, $(6-2)\ 180^\circ = 720^\circ$. To find the error you subtract $718^\circ 5'$ from 720°. To do this subtraction take 1° from 720° and make it into 60′.

$$\begin{array}{rr} 719^\circ & 60' \\ -\ 718^\circ & 5' \\ \hline 1^\circ & 55' \end{array}$$

So, the error is $1^\circ\ 55'$.

33. There is just one combination of 90° and 60° angles that has a sum of 360°; 90° + 90° + 60° + 60° + 60°, not necessarily in that order. You know, then, that each point in the tessellation where vertices meet must include the vertices of two squares and three equilateral triangles. Try making some paper copies and arranging them in various ways. Two ways to lay the floor are given.

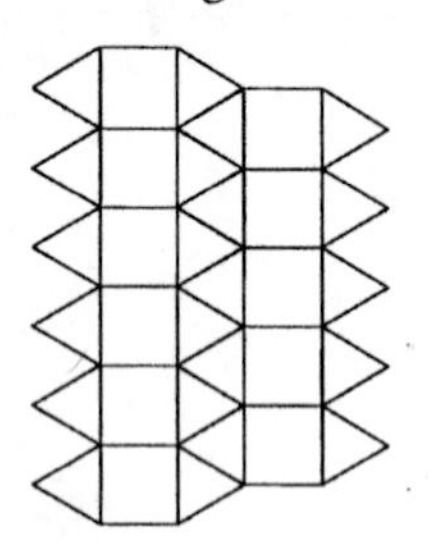

Section 2.4

Tips:

- ✔ All polyhedra must have (flat) polygons as faces.
- ✔ To have a prism you must have *parallel congruent* polygons for bases and parallelograms for lateral faces.
- ✔ Any prism, pyramid, cylinder or cone that seems to "lean" to the side when standing on its base is oblique, while those that do not "lean" are right.
- ✔ You name a prism or pyramid by specifying its base shape and by using oblique or right as in "oblique pentagonal prism" or "right square pyramid."

Solutions to odd-numbered textbook problems

1. Figures a and b are polyhedra. A polyhedron is composed of polygonal regions, and a polygon is composed of line segments. Figure c has circular regions, which are not polygonal, so it is not a polyhedron.

3. Top, front, and side views are two-dimensional shapes of three-dimensional figures. It helps to examine only one face of each cube that is visible from a particular view.

(a)

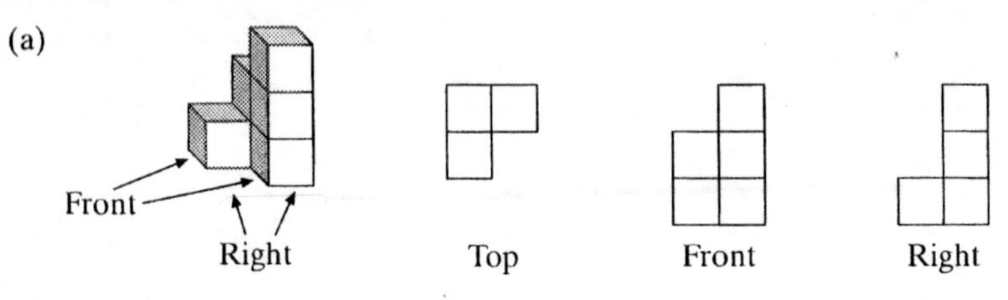

Section 2.4

3. (continued)

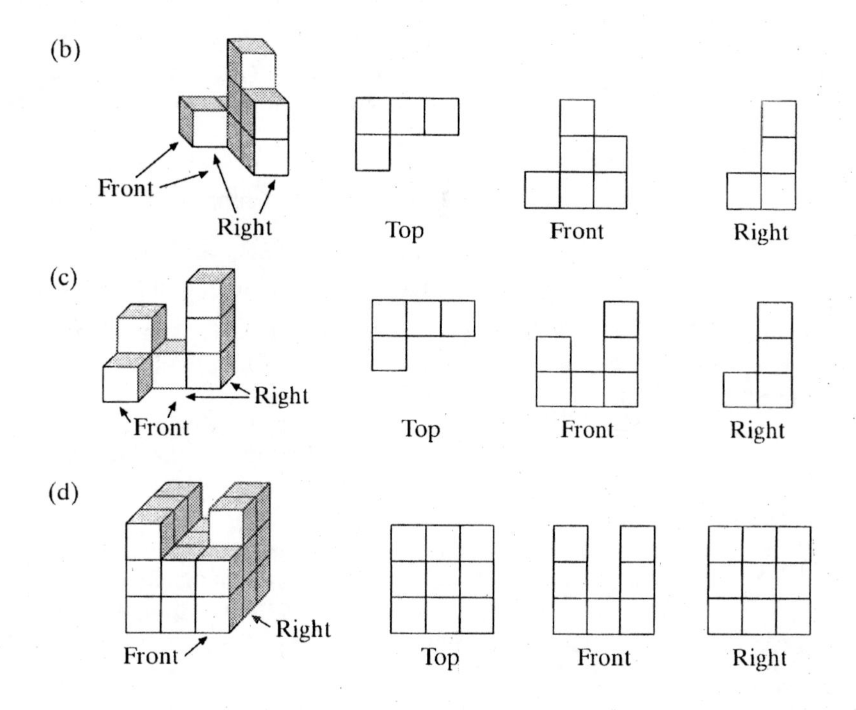

5. A good strategy here would be to cut out each of the 12 pentomino shapes and try to fold each one into an open-topped cube. The ones that will form a cube are shown below with the bottom of the cube marked with an x.

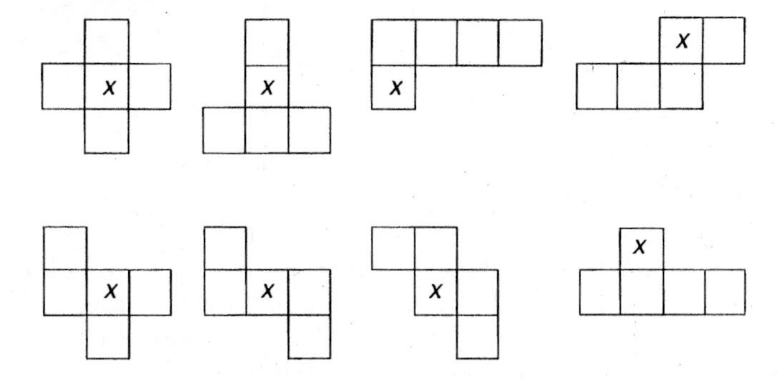

7. (a) The bases are the two parallel faces, *ABCDE* and *FGHIJ*.

   (b) The lateral faces are sides, *ABGF, BCHG, CDIH, EDIJ* and *AEJF*.

   (c) The faces you would not see are those with some dotted edges. Remember that the bases are also faces. Hidden faces are *FGHIJ, ABGF, BCHG* and *CDIH*.

   (d) The bases are pentagons and the prism appears to be upright so it is a right pentagonal prism.

Section 2.4

9. (a) Figure a. has lateral faces that are parallelograms (not rectangles). This is an oblique rectangular prism.

   (b) The base is an octagon and the lateral faces are rectangles. This is a right octagonal prism. If the bases are regular octagons then the polyhedron is a right octagonal prism.

11. The bases are regular hexagons and the lateral faces are congruent squares. The resulting polyhedron is a right regular hexagonal prism.

13. (a) The pyramid "leans" and the base is a regular hexagon, so the figure is an oblique regular hexagonal pyramid.

    (b) This pyramid does not appear to "lean" and the base is a regular polygon (square), so the figure is a right square pyramid.

15. Try making copies of the given nets. Cut them out and fold along the lines. Nets a. and b. both make tetrahedrons. Figure c. makes a right square pyramid without a base.

17. The base is a regular pentagon and the lateral faces are equilateral triangles. (Equilateral triangles are also isosceles triangles.) This net will make a right regular pentagonal pyramid. Check your answer by cutting out the net and folding it up.

19. An octahedron has eight congruent equilateral triangular faces. It is essentially two right square pyramids stuck together at the base. There is more than one way to make a net for it. Here is one possibility:

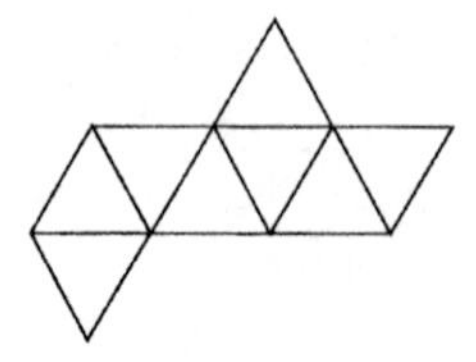

To check your solution, cut out the net and fold it up to form the octahedron.

21. Picture a point of rotation at the center of the "A" on the cube at the left. Rotate the cube 90° counterclockwise about the point. Think about the resulting position of sides "B" and "F." Cube i has an incorrect positioning of "B" and cube ii has an incorrect positioning of "F." Cube iii is the only cube that could be different view of the cube on the left.

23. To complete the row of entries for the n-gon remember that n represents the number of sides in the polygonal base. Look for a pattern. A 3-sided base (triangle) produces $3 + 2$ or 5 faces. As for each entry in the F column, start with the number of sides in each polygon and add 2. Similarly, each entry in the V column starts with the number of sides of the polygon base and multiplies by 2. The number of edges, E, is 3 times the number of sides in the polygon base.

Section 2.4

23. (continued)

Base	F	V	$F + V$	E
Triangle	5	6	11	9
Quadrilateral	6	8	14	12
Hexagon	8	12	20	18
n-gon	$n + 2$	$2n$	$3n + 2$	$3n$

To show that Euler's formula holds for each figure, substitute the table values into the formula $F + V = E + 2$ to see if the result is a true statement. Yes, Euler's formula holds for these figures.

To show that the formula holds for an n-gon base, substitute $n + 2$ for F, $2n$ for V and $3n$ for E to get: $n + 2 + 2n = 3n + 2$ and $3n + 2 = 3n + 2$. Thus, Euler's formula holds for an n-gon base.

25. Euler's formula is $F + V = E + 2$ where F is the number of faces, V is the number of vertices and E is the number of edges of a polyhedron. Be sure to count hidden edges and vertices, too. A complete table is given in the answer key in the text.

27. To determine the number of vertices, substitute $E = 18$ and $F = 8$ into Euler's formula and solve for V.

$$F + V = E + 2$$
$$8 + V = 18 + 2$$
$$V = 12$$

Thus, the polyhedron has 12 vertices.

One possible polyhedron is a right hexagonal prism.

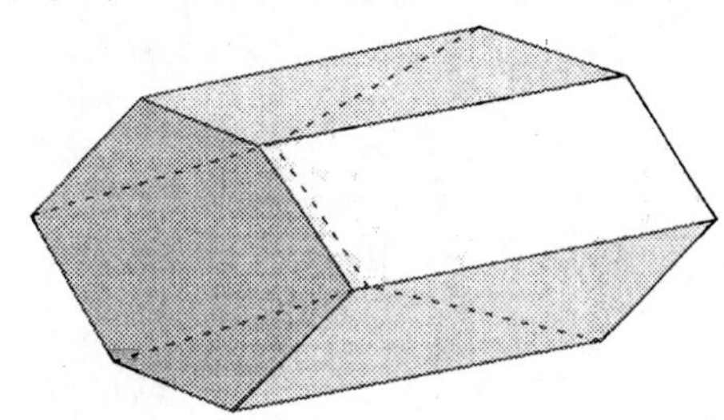

29. (a) This cone "leans" so it is an oblique circular cone.
    (b) This "leaning" cylinder is an oblique circular cylinder.

31. Modeling clay may help you to visualize the cross sections.
    (a) The cross-section is the shape of the top you see when the cut-off section is lifted. You will have a circular cross-section when a horizontal cut is made through this cylinder.
    (b) When a vertical cut is made, the two sections look like the diagram shown.
    The cross-section is a rectangle.

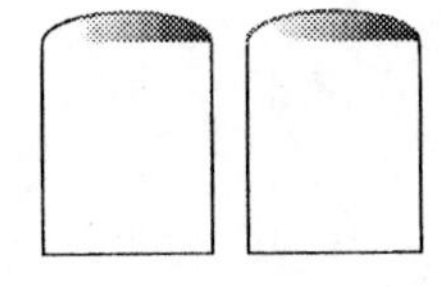

Section 2.4

33. (a) Each plane of symmetry cuts midway through two pairs of opposing faces.

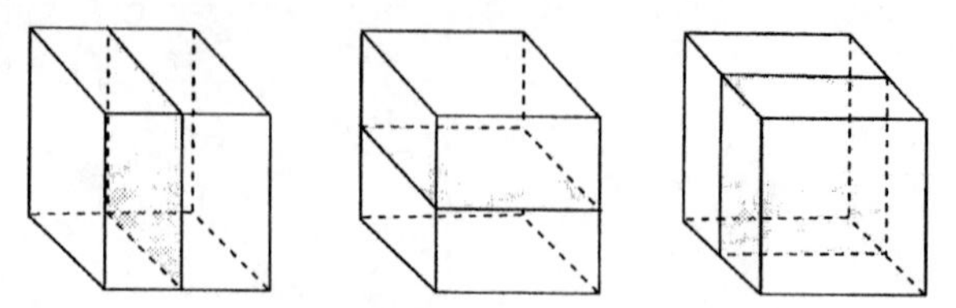

(b) There are six different pairs of opposite edges.

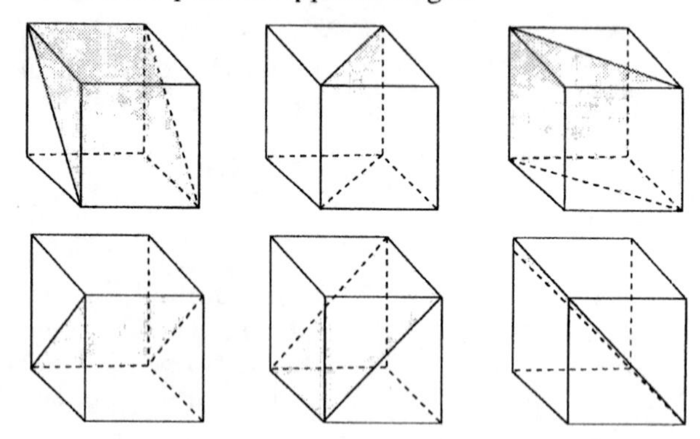

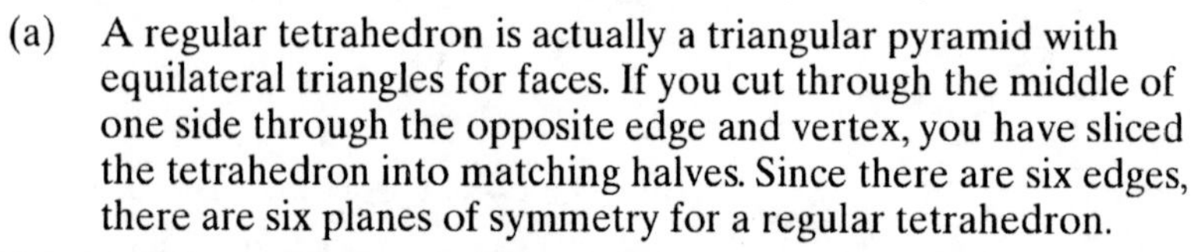

(c) There are a total of nine planes of symmetry for a cube.

35. See problem #33 for a definition of plane of symmetry.

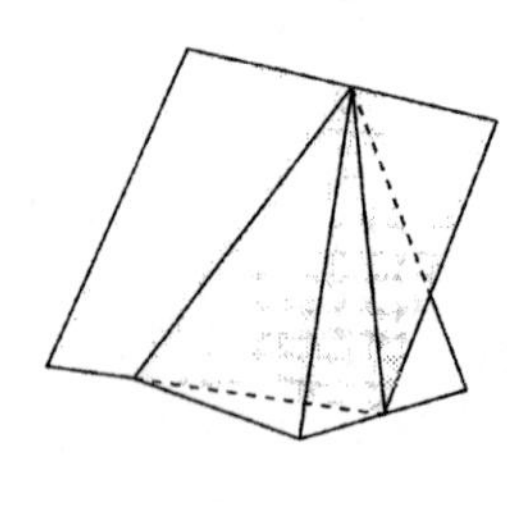

(a) A regular tetrahedron is actually a triangular pyramid with equilateral triangles for faces. If you cut through the middle of one side through the opposite edge and vertex, you have sliced the tetrahedron into matching halves. Since there are six edges, there are six planes of symmetry for a regular tetrahedron.

(b) By slicing opposing triangular faces of a right square pyramid in half, you get two planes of symmetry. There are two more planes cutting through opposite edges. That makes four planes of symmetry for a right square pyramid.

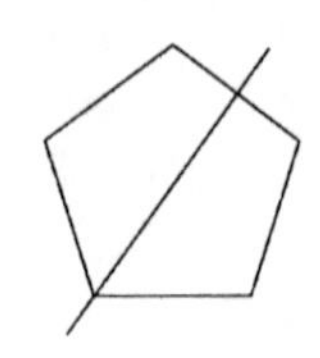

(c) A plane which cuts vertically through the middle of one face of a right regular pentagonal prism and out through the opposite edge will be a plane of symmetry. The picture shows how the plane would cut the base of the prism. There are five of these planes of symmetry. In addition, you can slice a prism in the middle horizontally, making $5 + 1 = 6$ planes of symmetry.

(d) Any vertical plane that passes through the center of a base of the cylinder is a plane of symmetry. Thus, there are infinitely many planes of symmetry for a right circular cylinder.

37. (a) The faces of a regular tetrahedron are congruent equilateral triangles. Thus the length of the cut is the same across each of the faces. Since there are three faces, the shape of the intersection is an equilateral triangle.

(b) This time the slice cuts through four faces. The result is an intersection that has four sides of equal length. The intersection is a square.

39. (a) The top and bottom of the soda can are congruent circular regions on parallel planes. The lateral surface of the can is perpendicular to the circular bases. The shape is a right circular cyclinder.

    (b) The tissue box has retangular opposite faces that are identical and on parallel planes. The other four sides are perpendicular and rectangular. The shape is a right rectangular pyramid.

    (c) If you assume the base of this figure to be square and the lateral faces to be congruent, this is a right square pyramid.

    (d) Join two right square pyramids at their bases. The result is eight congruent, equilateral triangle faces. This is a regular octahedron.

41. The base of the tipi has 12 edges. The lateral faces are scalene triangles, and they join at an apex. This tipi is an oblique dodecagonal (12-gon) pyramid. It has 24 edges, 13 vertices and 13 faces.

    Substitute into Euler's formula.

$$F + V = E + 2$$
$$13 + 13 = 24 + 2$$
$$26 = 26$$

Hence, Euler's formula holds for Native American tipis.

Section 2.5

Tips:

✔ Unit ratios are formed using two equal quantites.

✔ It is important to learn to use unit ratios for conversions even when you can do it other ways. Practicing the dimensional analysis method of conversion for simple conversions will enable you to apply it to more difficult conversions later.

Solutions to odd-numbered textbook problems

1. Start out by putting the original measurement over 1. Then create a unit ratio by placing the unit you want to eliminate in the denominator and the unit to which you want to convert in the numerator.

    (a) $\frac{\text{inches}}{1} \times \frac{1 \text{ ft}}{12 \text{ inches}}$. The unit ratio is $\frac{1 \text{ ft}}{12 \text{ in.}}$.

    (b) $\frac{\text{miles}}{1} \times \frac{5280 \text{ ft}}{1 \text{ mile}}$. The unit ratio is $\frac{5280 \text{ ft}}{1 \text{ mi}}$.

    (c) $\frac{\text{pounds}}{1} \times \frac{16 \text{ ounces}}{1 \text{ pound}}$. The unit ratio is $\frac{16 \text{ oz}}{1 \text{ lb}}$.

    (d) $\frac{\text{quarts}}{1} \times \frac{1 \text{ gallon}}{4 \text{ quarts}}$. The unit ratio is $\frac{1 \text{ gal}}{44 \text{ qt}}$.

Section 2.5

3. Start out by putting the original measurement over 1. Then create a unit ratio by placing the unit you want to eliminate in the denominator and the unit to which you want to convert in the numerator.

   (a) $\frac{17\text{ hours}}{1} \times \frac{60\text{ minutes}}{1\text{ hour}}$. The unit ratio is $\frac{60\text{ min}}{1\text{ hr}}$.

   (b) $\frac{360\text{ seconds}}{1} \times \frac{1\text{ minute}}{60\text{ seconds}}$. The unit ratio is $\frac{1\text{ min}}{60\text{ sec}}$.

   (c) $\frac{720\text{ inches}}{1} \times \frac{1\text{ yard}}{36\text{ inches}}$. The unit ratio is $\frac{1\text{ yd}}{36\text{ in.}}$.

   (d) $\frac{1440\text{ work-hours}}{1} \times \frac{1\text{ work-day}}{24\text{ work-hours}}$. The unit ratio is $\frac{1\text{ work-day}}{24\text{ work-hours}}$.

5. Since you aren't given a unit ratio that shows a relationship between ounces and pecks, this conversion requires more than one step, thus, more than one conversion ratio. Start by placing the original measurement over 1. Set up the unit ratios so that unwanted units cancel.

$$\frac{\text{ounces}}{1} \times \frac{1\text{ cup}}{8\text{ ounes}} \times \frac{1\text{ quart}}{4\text{ cups}} \times \frac{1\text{ gallon}}{4\text{ quarts}} \times \frac{1\text{ peck}}{2\text{ gallons}}$$

7. When you multiply the unit ratios, the unwanted units should "cancel", leaving you with the desired units.

   (a) $\frac{4.5\ \cancel{\text{lb}}}{1} \times \frac{16\text{ oz}}{1\ \cancel{\text{lb}}} = 72\text{ oz}$

   (b) $\frac{3744\ \cancel{\text{min}}}{1} \times \frac{1^\circ}{60\ \cancel{\text{min}}} = 62.4^\circ$

   (c) Since you probably do not know the relationship between miles and inches, do this problem in two steps. Convert miles to feet and then from feet to inches.

   $$\frac{25\ \cancel{\text{mi}}}{1} \times \frac{5280\ \cancel{\text{ft}}}{1\ \cancel{\text{mi}}} \times \frac{12\text{ in.}}{1\ \cancel{\text{ft}}} = 1{,}584{,}000\text{ in.}$$

   (d) $\frac{10500\ \cancel{\text{mL}}}{1} \times \frac{1\ \cancel{\text{L}}}{1000\ \cancel{\text{mL}}} \times \frac{1\text{ kL}}{1000\ \cancel{\text{L}}} = 0.0105\text{ kL}$

9. (a) $\frac{6\ \cancel{\text{in.}}}{1} \times \frac{2.54\text{ cm}}{1\ \cancel{\text{in.}}} = 15.24\text{ cm}$

   (b) $\frac{100\ \cancel{\text{yd}}}{1} \times \frac{36\ \cancel{\text{in.}}}{1\ \cancel{\text{yd}}} \times \frac{2.54\ \cancel{\text{cm}}}{1\ \cancel{\text{in.}}} \times \frac{1\text{ m}}{100\ \cancel{\text{cm}}} = 91.44\text{ m}$

   (c) $\frac{420\ \cancel{\text{ft}}}{1} \times \frac{12\ \cancel{\text{in.}}}{1\ \cancel{\text{ft}}} \times \frac{2.54\ \cancel{\text{cm}}}{1\ \cancel{\text{in.}}} \times \frac{1\text{ m}}{100\ \cancel{\text{cm}}} = 128.016\text{ m}$

   (d) $\frac{1\ \cancel{\text{km}}}{1} \times \frac{100\cancel{\text{m}}}{1\ \cancel{\text{km}}} \times \frac{100\text{ cm}}{1\ \cancel{\text{m}}} = 100000\text{ cm}$

   $$\frac{100000\ \cancel{\text{cm}}}{1} \times \frac{1\ \cancel{\text{in.}}}{2.54\ \cancel{\text{cm}}} \times \frac{1\ \cancel{\text{ft}}}{12\ \cancel{\text{in.}}} \times \frac{1\text{ mi}}{5280\ \cancel{\text{ft}}} \approx 0.621\text{ mi}$$

Section 2.5

11. (a) $\dfrac{45\ \cancel{\text{mi}}}{1\cancel{\text{hr}}} \times \dfrac{5280\text{ ft}}{1\ \cancel{\text{mi}}} \times \dfrac{1\ \cancel{\text{hr}}}{60\ \cancel{\text{min}}} \times \dfrac{1\ \cancel{\text{min}}}{60\text{ sec}} = \dfrac{45(5280)(1)(1)}{1(1)(60)(60)}\dfrac{\text{ft}}{\text{sec}}$

$= 66\,\dfrac{\text{ft}}{\text{sec}}$

(b) $\dfrac{200\text{ dollars}}{1\ \cancel{\text{day}}} \times \dfrac{1\ \cancel{\text{day}}}{24\text{ hours}} = \dfrac{200(1)}{1(24)}\dfrac{\text{dollars}}{\text{hour}} = 8.\bar{3}\,\dfrac{\text{dollars}}{\text{hour}} = 8\dfrac{1}{3}\,\dfrac{\text{dollars}}{\text{hr}}$

(c) $\dfrac{0.3\ \cancel{\text{in.}}}{1\ \cancel{\text{year}}} \times \dfrac{1\text{ ft}}{12\ \cancel{\text{in.}}} \times \dfrac{100\ \cancel{\text{years}}}{1\text{ century}} = \dfrac{0.3(1)(100)}{1(12)(1)}\dfrac{\text{ft}}{\text{century}}$

$= 2.5\,\dfrac{\text{ft}}{\text{century}}$

(d) $\dfrac{16\ \cancel{\text{km}}}{1\ \cancel{\text{L}}} \times \dfrac{1000\text{ m}}{1\ \cancel{\text{km}}} \times \dfrac{1\ \cancel{\text{L}}}{1000\text{ mL}} = \dfrac{16(1000)(1)\text{ m}}{1(1)(1000)\text{ mL}} = 16\,\dfrac{\text{m}}{\text{mL}}$

13. (a) $\dfrac{100\ \cancel{\text{ft}}}{1\text{ sec}} \times \dfrac{12\ \cancel{\text{in.}}}{1\ \cancel{\text{ft}}} \times \dfrac{2.54\ \cancel{\text{cm}}}{1\ \cancel{\text{in.}}} \times \dfrac{1\text{ m}}{100\ \cancel{\text{cm}}} = \dfrac{100(12)(2.54)(1)}{1(1)(1)(100)}\dfrac{\text{m}}{\text{sec}}$

$= 30.48\,\dfrac{\text{m}}{\text{sec}}$

(b) $\dfrac{0.4\ \cancel{\text{kg}}}{1\ \cancel{\text{m}}} \times \dfrac{2.205\text{ lb}}{1\ \cancel{\text{kg}}} \times \dfrac{1\ \cancel{\text{m}}}{100\cancel{\text{cm}}} \times \dfrac{2.54\ \cancel{\text{cm}}}{1\ \cancel{\text{in.}}} \times \dfrac{12\ \cancel{\text{in.}}}{1\ \cancel{\text{ft}}} \times \dfrac{3\ \cancel{\text{ft}}}{1\text{ yd}}$

$= \dfrac{0.4(2.205)(1)(2.54)(12)(3)}{1(1)(100)(1)(1)(1)}\dfrac{\text{lb}}{\text{yd}} \approx 0.807\,\dfrac{\text{lb}}{\text{yd}}$

15. When we are given a rate such as 65 miles per hour it can be written as $\dfrac{65\text{ mi}}{1\text{ hr}}$. Use unit ratios to change to the desired units. To cancel miles, we need miles in the denominator of a ratio. The text gives us the approximate conversion ratio of $\dfrac{1\text{ km}}{0.6214\text{ mi}}$ (See Example 2.13).

So we have $\dfrac{65\ \cancel{\text{mi}}}{1\text{ hr}} \times \dfrac{1\text{ km}}{0.6214\ \cancel{\text{mi}}} \approx 105\,\dfrac{\text{km}}{\text{hr}}$.

An alternate approach would be the following:

$\dfrac{65\ \cancel{\text{mi}}}{1\text{ hr}} \times \dfrac{5280\ \cancel{\text{ft}}}{1\ \cancel{\text{mi}}} \times \dfrac{12\ \cancel{\text{in.}}}{1\ \cancel{\text{ft}}} \times \dfrac{2.54\ \cancel{\text{cm}}}{1\ \cancel{\text{in.}}} \times \dfrac{1\ \cancel{\text{m}}}{100\ \cancel{\text{cm}}} \times \dfrac{1\text{ km}}{1000\ \cancel{\text{m}}}$

$= \dfrac{(65)(5280)(12)(2.54)(1)(1)}{(1)(1)(1)(1)(100)(100)}\dfrac{\text{km}}{\text{hr}} \approx 105\,\dfrac{\text{km}}{\text{hr}}$.

17. The speed she runs can be written as the rate $\dfrac{1\text{ mile}}{4\text{ min }17.6\text{ sec}}$.

First, simpifly the denominator by converting 4 minutes to seconds.

$\dfrac{4\ \cancel{\text{min}}}{1} \times \dfrac{60\text{ sec}}{1\ \cancel{\text{min}}} = 240\text{ sec}$

Section 2.5

17. (continued)

Her rate of speed can be written $\frac{1\text{ mile}}{257.6\text{ sec}}$. Now, to convert to $\frac{\text{km}}{\text{hr}}$:

$$\frac{1\ \cancel{\text{mi}}}{257.6\ \cancel{\text{sec}}} \times \frac{1\text{ km}}{0.6214\ \cancel{\text{mi}}} \times \frac{60\ \cancel{\text{sec}}}{1\ \cancel{\text{min}}} \times \frac{60\ \cancel{\text{min}}}{1\text{ hr}}$$

$$= \frac{1(1)(60)(60)}{257.6(0.6214)(1)(1)}\frac{\text{km}}{\text{hr}} \approx 22.49\frac{\text{km}}{\text{hr}}$$

19. This rate of growth can be written $\frac{0.12\text{ cm}}{1\text{ week}}$. Now convert to inches per year.

$$\frac{0.12\ \cancel{\text{cm}}}{1\ \cancel{\text{week}}} \times \frac{1\text{ in.}}{2.54\ \cancel{\text{cm}}} \times \frac{52\ \cancel{\text{weeks}}}{1\text{ year}} = \frac{0.12(1)(52)}{1(2.54)(1)}\frac{\text{in.}}{\text{year}} \approx 2.5\frac{\text{in.}}{\text{year}}$$

21. The rate of oil consumption can be written $\frac{20{,}000{,}000\text{ barrels}}{1\text{ day}}$. You can use "one barrel of crude oil contains 42 gallons" to build a conversion ratio.

$$\frac{20{,}000{,}000\ \cancel{\text{barrels}}}{1\text{ day}} \times \frac{42\ \cancel{\text{gal}}}{1\ \cancel{\text{barrel}}} \times \frac{4\ \cancel{\text{qt}}}{1\ \cancel{\text{gal}}} \times \frac{1\text{ L}}{1.057\ \cancel{\text{qt}}}$$

$$= \frac{20{,}000{,}000(42)(4)(1)}{1(1)(1)(1.057)}\frac{\text{L}}{\text{day}} \approx 3{,}178{,}807{,}947\frac{\text{L}}{\text{day}}$$

23. The rate of hair growth is $\frac{\frac{1}{2}\text{ in.}}{4\text{ weeks}}$ or, using a decimal for $\frac{1}{2}$, we have $\frac{0.5\text{ in.}}{4\text{ weeks}}$. The required rate is mph which, written as a ratio is $\frac{\text{mi}}{\text{hr}}$. The conversion follows.

$$\frac{0.5\ \cancel{\text{in.}}}{4\ \cancel{\text{weeks}}} \times \frac{1\ \cancel{\text{ft}}}{12\ \cancel{\text{in.}}} \times \frac{1\text{ mi}}{5280\ \cancel{\text{ft}}} \times \frac{1\ \cancel{\text{week}}}{7\ \cancel{\text{days}}} \times \frac{1\ \cancel{\text{day}}}{24\text{ hr}}$$

$$\frac{0.5(1)(1)(1)(1)}{4(12)(5280)(7)(24)}\frac{\text{mi}}{\text{hr}} \approx 1.17 \times 10^{-8}\text{ mph}$$

25. Rates can be compared by changing both to the same units.

change $186{,}282\frac{\text{mi}}{\text{sec}}$ to $\frac{\text{km}}{\text{hr}}$.

$$\frac{186{,}282\ \cancel{\text{mi}}}{1\ \cancel{\text{sec}}} \times \frac{60\ \cancel{\text{sec}}}{1\ \cancel{\text{min}}} \times \frac{60\ \cancel{\text{min}}}{1\text{ hr}} \times \frac{1\text{ km}}{0.6214\ \cancel{\text{mi}}}$$

$$= \frac{(186{,}282)(60)(60)(1)}{(1)(1)(1)(0.6214)}\frac{\text{km}}{\text{hr}} \approx 1{,}079{,}000{,}000\frac{\text{km}}{\text{hr}}.$$

Section 2.5

25. (continued)

So the ratio of the speed of light to the speed of sound is approximate

$$\frac{1{,}079{,}000{,}000\,\frac{\text{km}}{\text{hr}}}{1000\,\frac{\text{km}}{\text{hr}}} = \frac{1{,}079{,}000}{1}.$$

Thus light travels about 1,079,000 times faster than sound. (Answers will vary depending on the rounding.)

27. (a) The starting rate is $\frac{186{,}282\text{ mi}}{1\text{ sec}}$ and you want to convert to $\frac{\text{mi}}{1\text{ yr}}$.

$$\frac{186{,}282\text{ mi}}{1\ \cancel{\text{sec}}} \times \frac{60\ \cancel{\text{sec}}}{1\ \cancel{\text{min}}} \times \frac{60\ \cancel{\text{min}}}{1\ \cancel{\text{hr}}} \times \frac{24\ \cancel{\text{hr}}}{1\ \cancel{\text{day}}} \times \frac{365\ \cancel{\text{day}}}{1\text{ yr}}$$

$$= \frac{186{,}282(60)(60)(24)(365)}{1(1)(1)(1)(1)}\frac{\text{mi}}{\text{yr}} \approx 5.88 \times 10^{12}\text{ mi/yr}$$

(b) Seventy-six light years is the distance light travels in 76 years. The starting ratio can be written $\frac{76\text{ years}}{1}$. From part (a) you know the unit ratio for one light year is $\frac{5.88 \times 10^{12}\text{ mi}}{1\text{ yr}}$, so the conversion follows.

$$\frac{76\ \cancel{\text{yr}}}{1} \times \frac{5.88 \times 10^{12}\text{ mi}}{1\ \cancel{\text{yr}}} \approx 4.47 \times 10^{14}\text{ mi}$$

(c) The starting measurement is 480,000,000 miles.

$$\frac{480{,}000{,}000\ \cancel{\text{mi}}}{1} \times \frac{1\ \cancel{\text{yr}}}{5.88 \times 10^{12}\ \cancel{\text{mi}}} \times \frac{365\ \cancel{\text{days}}}{1\ \cancel{\text{yr}}} \times \frac{24\text{ hrs}}{1\ \cancel{\text{day}}}$$

$$= \frac{480{,}000{,}000(1)(365)(24)}{1(5.88 \times 10^{12})(1)(1)}\text{ hrs} \approx 0.716\text{ hour}$$

29. (a) One smoot is 5 feet 7 inches. To convert to inches, we need only convert 5 feet to inches, then add 7 in. $\frac{5\ \cancel{\text{ft}}}{1} \times \frac{12\text{ in.}}{1\ \cancel{\text{ft}}} = 60$ in. So, one smoot is 60 in. + 7 in. = 67 in. As a ratio, $\frac{1\text{ smoot}}{67\text{ in.}}$, or $\frac{67\text{ in.}}{1\text{ smoot}}$.

(b) To convert smoots to meters we find the number of meters in one smoot. Unit ratio $\frac{67\text{ in.}}{1\text{ smoot}}$ we convert to meters per smoot.

$$\frac{67\ \cancel{\text{in.}}}{1\text{ smoot}} \times \frac{2.54\ \cancel{\text{cm}}}{1\ \cancel{\text{in.}}} \times \frac{1\text{ m}}{100\ \cancel{\text{cm}}} = \frac{67(2.54)(1)\ \ \text{m}}{(1)(1)(100)\,\text{smoot}}$$

$$= \frac{1.7018\text{ m}}{1\text{ smoot}} \text{ or } \frac{1\text{ smoot}}{1.7018\text{ m}}$$

Additional Problems

29. (continued)

(c) The starting measure is 364.4 smoots.

$$\frac{364.4 \text{ smoots}}{1} \times \frac{1.7018 \text{ m}}{1 \text{ smoot}} = \frac{364.4(1.7018)}{(1)(1)} \text{ m} \approx 620 \text{ m}$$

(d) $$\frac{120 \text{ yd}}{1} \times \frac{3 \text{ ft}}{1 \text{ yd}} \times \frac{12 \text{ in.}}{1 \text{ ft}} \times \frac{1 \text{ smoot}}{67 \text{ in.}} = \frac{120(3)(12)(1)}{1(1)(1)(67)} \text{ smoot}$$

$\approx$ 64.5 smoots

Additional Problems

1. Let x = original height of beanstalk.

   Then, after day 1 the height is $x \cdot \frac{3}{2}$,

   after day 2 the height is $x \cdot \frac{3}{2} \cdot \frac{4}{3}$, and

   after day 3 the height is $x \cdot \frac{3}{2} \cdot \frac{4}{3} \cdot \frac{5}{4}$.

   After n days the height would be given by

   $$x\left(\frac{3}{2}\right)\left(\frac{4}{3}\right)\left(\frac{5}{4}\right)\cdots\left(\frac{n+1}{n}\right) = x\left(\frac{n+1}{2}\right).$$

   We want $\frac{n+1}{2} \geq 100$

   so $n + 1 \geq 200$

   and $n \geq 199$.

   Therefore, the fewest number of days is 199.

2. Pairing the first ray on the left with the remaining rays produces 99 angles. Pairing the second ray from the left with the remaining rays to the right produces 98 angles. Continuing in this way, we obtain $99 + 98 + 97 + \cdots + 1$ angles.

   The formula for the sum of the first n counting numbers is $\frac{n(n+1)}{2}$.

   Therefore, the number of different (nonzero) angles formed is the sum of the first 99 counting numbers. That is, there are $\frac{99(100)}{2} = 4950$ angles.

Chapter Review

Solutions to Chapter 2 Review

Section 2.1

1. A line is straight and infinite in length. The infinite length is shown with arrowheads. A line segment has two endpoints and a ray has one endpoint.

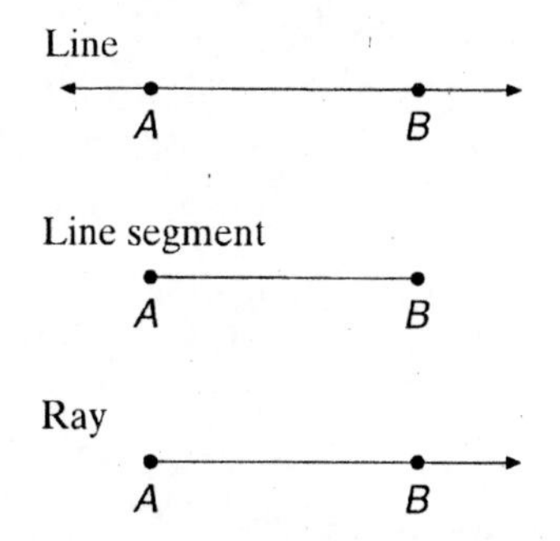

2. Remember to center the protractor at the vertex of the angle. Be sure to use the correct scale by starting at 0° when reading the angle. See the text for definitions of types of angles.

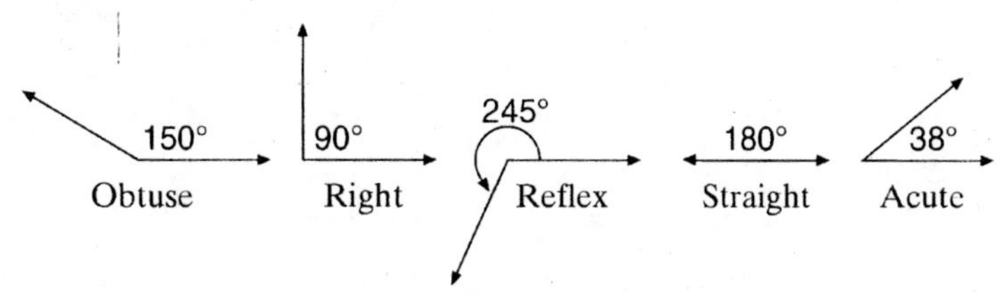

3. The coordinates are $A = 1.7$, $B = 0.4$, and $C = -1.2$. The distance between two points is the non-negative difference between their coordinates.

   $AB = 1.7 - 0.4 = 1.3$, $CB = 0.4 - (-1.2) = 1.6$, and
   $AC = 1.7 - (-1.2) = 2.9$.

4. Complementary angles have measures that add to 90° and congruent angles have the same measure. Let x = the measure of each angle, so, $x + x = 90°$, $2x = 90°$, and $x = 45°$. Thus each of the angles measure 45°.

5. (a) $\angle CFD$ is adjacent to $\angle EFD$ and has an angle measure between 0° and 90°.

   (b) $\angle CFD$ and $\angle DFE$ are adjacent angles and the sum of these angle measures is 90°.

   (c) Since $\angle CFD$ is adjacent to $\angle DFE$, and they are complementary angles. $62° + \angle DFE = 90°$, $\angle DFE = 90° - 62°$, and $\angle DFE = 28°$.

   (d) $\angle BFC = 90°$ and $\angle CFE = 90°$, so $\angle CFE$ is congruent to $\angle BFC$.

5. (continued)

(e) $\angle BFD$ is supplementary to $\angle BFA$ since $\angle BFA + \angle BFD = 180°$. $\angle BFD$ is also an obtuse angle.

$\angle AFE$ is supplementary to $\angle BFA$ since $\angle BFA + \angle AFE = 180°$. $\angle AFE$ is also an obtuse angle.

(f) You know that the measure of $\angle EFA$ is the sum of adjacent angles $\angle EFB$ and $\angle BFA$. $\angle EFB$ is straight, and so,

$$\angle EFB + \angle BFA = 214°$$
$$180° + \angle BFA = 214°$$
$$\angle BFA = 34°.$$

(g) Remember that a straight angle forms a straight line. Therefore $\angle BFE$ and $\angle AFD$ are straight angles.

Section 2.2

1. (a) There are 5 rotation symmetries at 60° intervals: 60°, 120°, 180°, 240°, 300°.

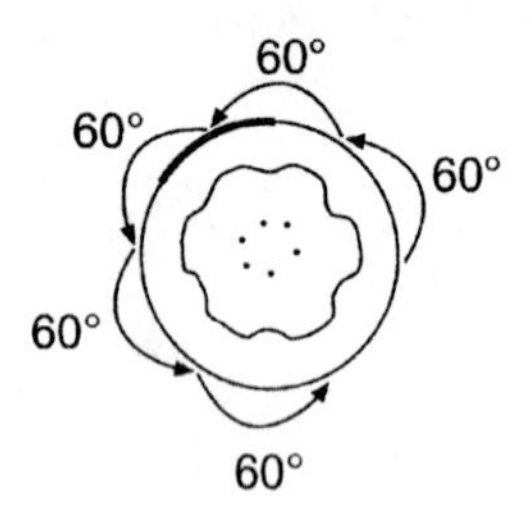

The figure has 6 axes of symmetry.

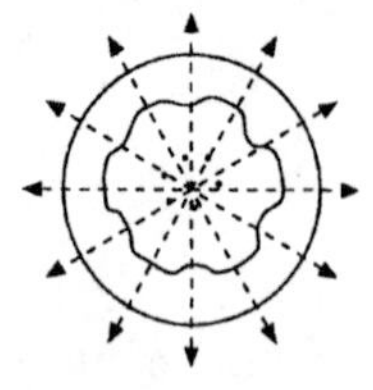

(b) There is no rotation symmetry since the figure must be rotated a full 360° to result in an image exactly like the original. The figure has one axis of symmetry.

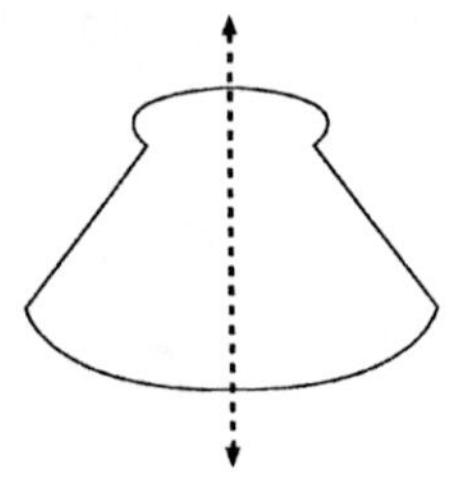

Chapter Review

1. (continued)

   (c) There is one rotation symmetry of 180° about the center of the flag. There are two axes of symmetry as shown.

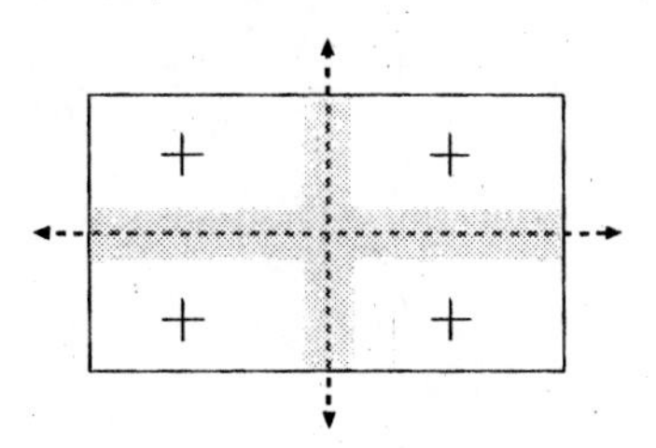

2. (a) An isosceles trapezoid has 1 axis of symmetry.

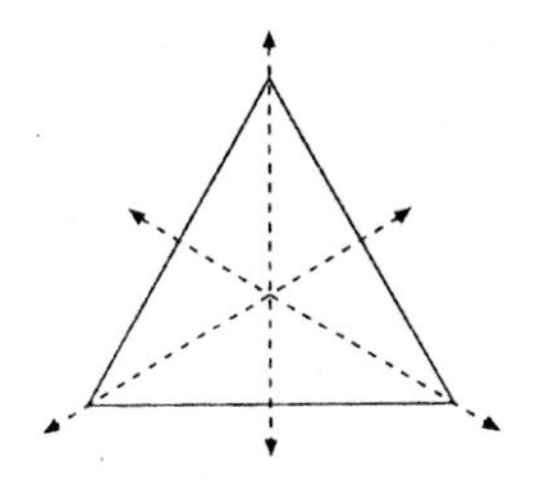

   (b) An equilateral triangle has 3 axes of symmetry, one through each vertex and the midpoint of the opposite side.

   (c) A scalene triangle has no axis of symmetry.

   (d) A rectangle has two axes of symmetry.

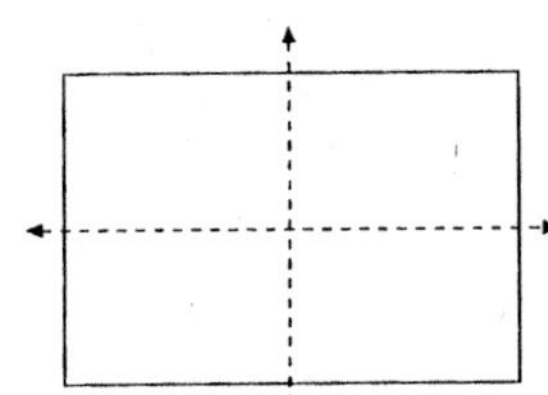

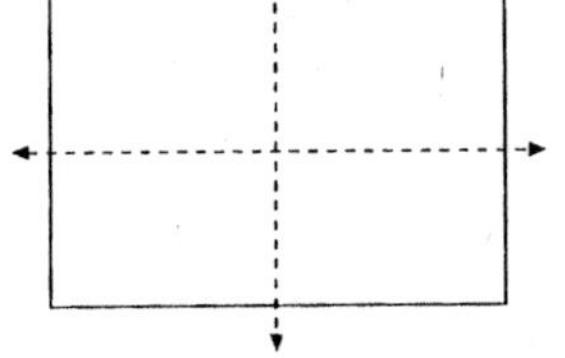

3. (a) A square has 3 rotation symmetries about the center of the square.

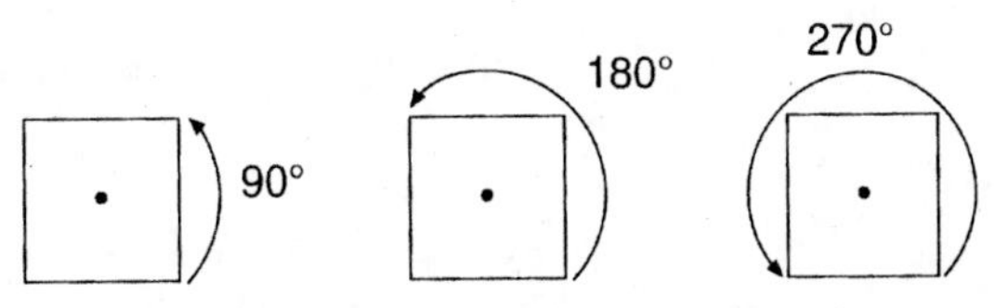

   (b) A non square rhombus has one rotation symmetry.

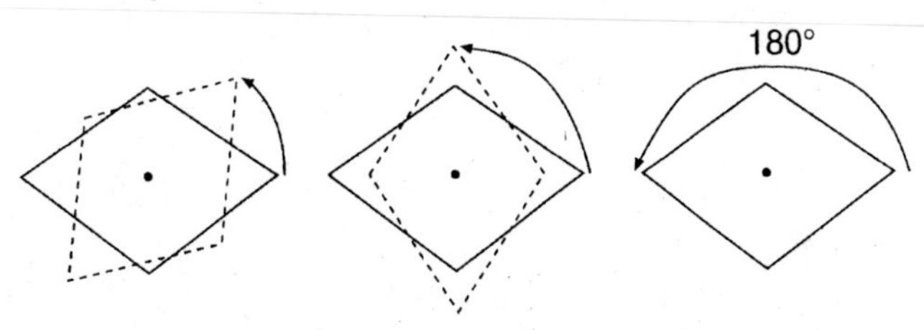

   (c) An isosceles triangle has no rotational symmetry unless it is equilateral. An equilateral triangle has two rotation symmetries. They are 120° and 240°.

Chapter Review

4. (a) A trapezoid has exactly one pair of parallel sides, so a trapezoid is not a parallelogram. A kite is not a parallelogram unless it is a rhombus or a square. There are other non-regular quadrilaterals that are not parallelograms.

   (b) A parallelogram has two pair of opposite parallel sides. Remember that squares and rhombuses are also parallelograms.

   (c) Remember that rectangles have "square," or 90° corners. Thus, all four angles must be right angles.

   (d) All four sides must be congruent. The angles do not need to be right angles.

   (e) The four sides must be congruent and the four angles must be right angles.

5. (a) A triangle with an obtuse angle between two congruent sides.

   (b) All three angles of an acute triangle must be less than 90°. A right triangle has one 90° angle. It is impossible for a triangle to be both acute and right.

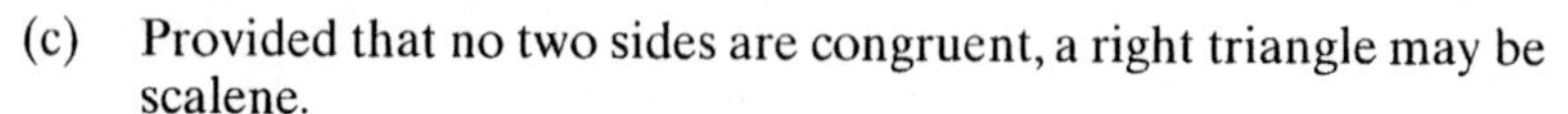

   (c) Provided that no two sides are congruent, a right triangle may be scalene.

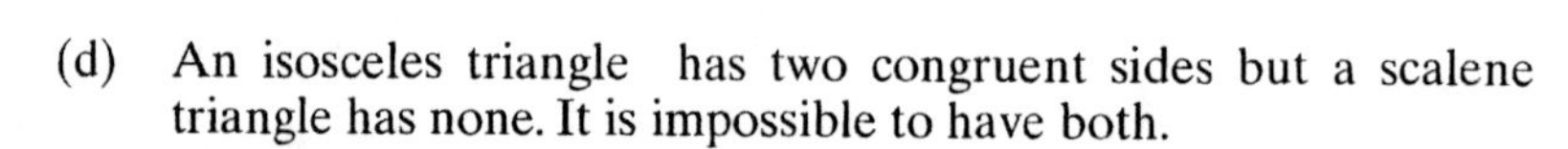

   (d) An isosceles triangle has two congruent sides but a scalene triangle has none. It is impossible to have both.

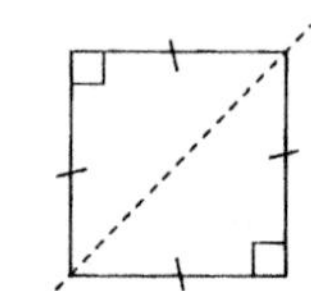

6. An axis of reflection drawn through two vertices of a square divides the square into two isosceles right triangles.

Section 2.3

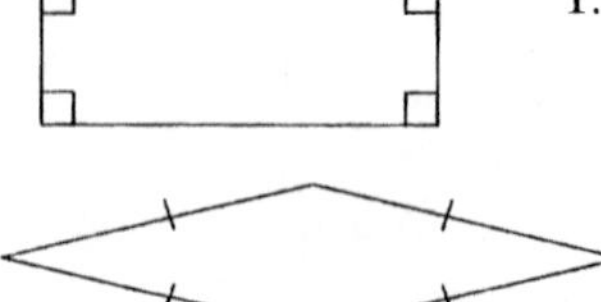

1. (a) The sum of the angle measures in a quadrilateral is 360°. Since $360° \div 4 = 90°$, each of the four angles must be 90°. Thus, the figure must be a rectangle, but not a square.

   (b) Remember, the sides must be congruent. But the angles can not all be the same. The shape must be a rhombus, but not a square.

2. (a) The sum of the measures of the vertex angles in a polygon with n sides is $(n - 2)\,180°$.

$$(n - 2)\,180° = 2520°$$

$$180°\,n - 360° = 2520°$$

$$180°\,n = 2880$$

$$n = 16$$

So, the polygon has 16 sides.

Chapter Review

2. (continued)

(b) The measure of a vertex angle of a regular n-gon is $\frac{(n-2)180^\circ}{n}$.

Substituting $n = 16$ we obtain

$$\frac{(16-2)180^\circ}{16} = \frac{(14)180^\circ}{16} = 157.5^\circ.$$

3. The vertex angle measure of the regular octagon is:

$$\frac{(8-2)180^\circ}{8} = \frac{(6)180^\circ}{8} = 135^\circ.$$

The vertex angle measure of the regular hexagon is:

$$\frac{(6-2)180^\circ}{6} = \frac{(4)180^\circ}{6} = 120^\circ.$$

The vertex angle measure of a square is 90°. Note that $\angle x$ and the vertex angle of the square are adjacent, and their sum is equal to the measure of the vertex angle of the regular hexagon. Solve the equation:

$$\angle x + 90^\circ = 120^\circ$$

$$\angle x = 30^\circ.$$

Next, note that the sum of the measures of $\angle y$, $\angle x$ and 90° must equal the measure of the vertex angle of the regular octagon. Solve the equation:

$$\angle y + \angle x + 90^\circ = 135^\circ$$

$$\angle y + 30^\circ + 90^\circ = 135^\circ$$

$$\angle y + 120^\circ = 135^\circ$$

$$\angle y = 15^\circ.$$

4. Since the sum of the measures of the angles is a triangle is 180°, you can solve the equation:

$$x + x + 3 + 2x = 180$$

$$4x + 3 = 180$$

$$4x = 177$$

$$x = 44.25.$$

So, the value of x is 44.25°.

5. A diagram of the solution is in the answer key at the back of the text.

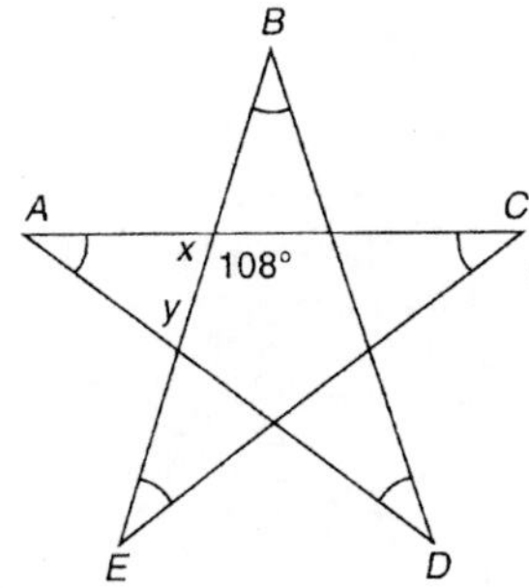

6. The vertex angle measure of a regular pentagon is

$$\frac{(5-2)180^\circ}{5} = \frac{(3)180^\circ}{5} = 108^\circ.$$

In the diagram the angle, labeled x, is adjacent and supplementary to the 108° vertex angle. So, you know that the sum of $\angle x$ and 108° equals 180°. Solve the equation:

$$\angle x + 108^\circ = 180^\circ$$

$$\angle x = 72^\circ.$$

Chapter Review

6. (continued)

Similarly, you know that the sum of the angle labeled y is also adjacent and supplementary to 108°.

It follows that $\angle y + 108° = 180°$ and $\angle y = 72°$. The sum of the angle measure in a triangle is 180°, so:

$$\angle A + \angle x + \angle y = 180°$$
$$\angle A + 72° + 72° = 180°$$
$$\angle A = 36°.$$

Since angles A, B, C, D, and E are congruent, the sum of their measures is: $36° + 36° + 36° + 36° + 36°$ or $5\,(36°) = 180°$.

7. A complete explanation is given in the answer key at the back of the text.

Section 2.4

1. (a) Prisms and pyramids have polygonal bases. The base of this figure is circular, and the figure "leans." The figure is an oblique circular cone.
   (b) The figure has two opposite, triangular bases. This is a triangular prism.
   (c) This figure has a pentagonal base and lateral faces that meet at one point, an apex. The figure "leans," so it is oblique. It is an oblique pentagonal pyramid.
   (d) the figure has two opposite, petagonal bases. This is a right, pentagonal prism.

2. (a) A round smooth (without flat faces) ball is a sphere.
   (b) Since opposite, identical, rectangular faces are in parallel planes, the box-shaped figure is a right rectangular prism.
   (c) Since two identical circular bases are in parallel planes, the figure is a right circular cylinder.
   (d) This regular polyhedron is a cube, also known as a regular hexahedron.

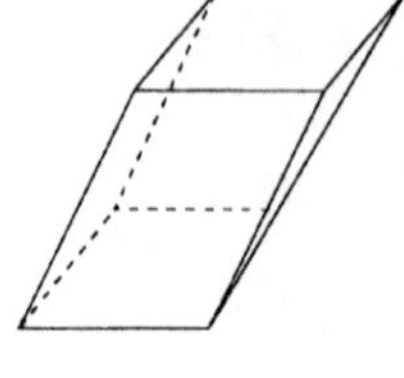

3. Try drawing a picture. Note the figure cannot be a right rectangular prism since it has two parallelogram faces. The figure "leans" so it is an oblique square prism.

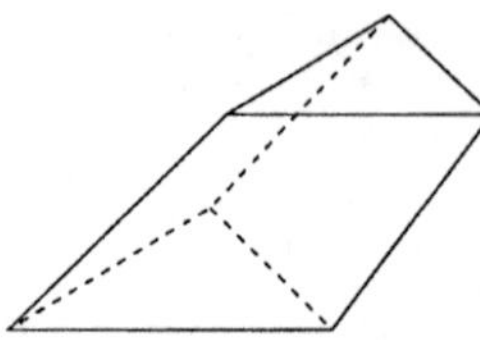

4. Draw and equilateral triangular prism that "leans." The two opposite bases are equilateral triangles. The three lateral faces are parallelograms.

Chapter Review

5. The base is a regular hexagon and each lateral face is an isosceles triangle.

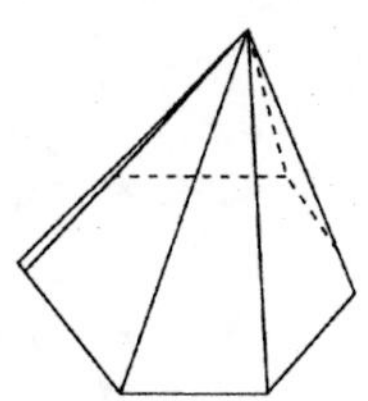

6. You can use Euler's formula, $F + V = E + 2$, where $E = 31$ and $V = 2F$.

So,
$$\begin{aligned} F + V &= E + 2 \\ F + 2F &= 31 + 2 \\ 3F &= 33 \\ F &= 11. \end{aligned}$$

And $V = 2F$, so $V = 2(11) = 22$. Thus, the polyhedron has 11 faces and 22 vertices.

Section 2.5

1. $$\frac{39.48 \cancel{\text{wingnuts}}}{1} \times \frac{1 \cancel{\text{blog}}}{2.8 \cancel{\text{wingnuts}}} \times \frac{1 \text{ marlock}}{0.5 \cancel{\text{blogs}}}$$
$$= \frac{39.48(1)(1)}{1(2.8)(0.5)} \text{marlocks}$$
$$= 28.2 \text{ marlocks}$$

2. $$\frac{0.68 \cancel{\text{gal}}}{1} \times \frac{4 \cancel{\text{qt}}}{1 \cancel{\text{gal}}} \times \frac{2 \cancel{\text{pt}}}{1 \cancel{\text{qt}}} \times \frac{2 \cancel{\text{c}}}{1 \cancel{\text{pt}}} \times \frac{16 \text{ T}}{1 \cancel{\text{c}}}$$
$$= \frac{0.68(4)(2)(2)(16)}{1(1)(1)(1)(1)} \text{T} = 174.08 \text{ T}$$

3. $$\frac{430 \cancel{\text{mi}}}{1} \times \frac{1 \text{ km}}{0.6214 \cancel{\text{mi}}} = \frac{430(1) \text{ km}}{1(0.6214)} \approx 692 \text{ km}$$

4. Eight 8-ounce glasses contains $8 \cdot 8$ oz $= 64$ oz of water. Convert 64 oz per day to liters per year.
$$\frac{64 \cancel{\text{oz}}}{1 \cancel{\text{day}}} \times \frac{1 \cancel{\text{c}}}{8 \cancel{\text{oz}}} \times \frac{1 \cancel{\text{pt}}}{2 \cancel{\text{c}}} \times \frac{1 \cancel{\text{qt}}}{2 \cancel{\text{pt}}} \times \frac{1 \text{ L}}{1.057 \cancel{\text{qt}}} \times \frac{365 \cancel{\text{days}}}{1 \text{ year}}$$
$$= \frac{64(1)(1)(1)(1)(365) \text{ L}}{1(8)(2)(2)(1.057)(1) \text{ year}} = 691 \text{ L/yr}$$

5. $$\frac{100 \cancel{\text{km}}}{1 \cancel{\text{hr}}} \times \frac{1000 \text{ m}}{1 \cancel{\text{km}}} \times \frac{1 \cancel{\text{hr}}}{60 \cancel{\text{min}}} \times \frac{1 \cancel{\text{min}}}{60 \text{ sec}}$$
$$= \frac{100(1000)(1)(1) \text{ m}}{1(1)(60)(60) \text{ sec}} \approx 27.8 \text{ m/sec}$$

Chapter Test

Solutions to Chapter 2 Test

1. F If the three points are collinear, there are infinitely many planes which contain the line. The statement is true for three *non-collinear* points.

2. T Two adjacent angles, share one side and do not overlap.

3. F This statement is true except when the isosceles triangle is also equilateral.

4. F The statement is true if the hexagon is *regular*.

5. F The diameter is twice the radius.

6. T All angles in an equiangular triangle are 60°.

7. T The fixed point is the center and the fixed distance is the radius.

8. T A rhombus is a quadrilateral with four congruent sides.

9. T You can trace the triangle so that the starting and ending points are the same.

10. F The faces of a polyhedron must be polygons.

11. The angles must all be equal. In other words, you must have four congruent sides and four right angles.

12. For convenience the vertices have been labeled.

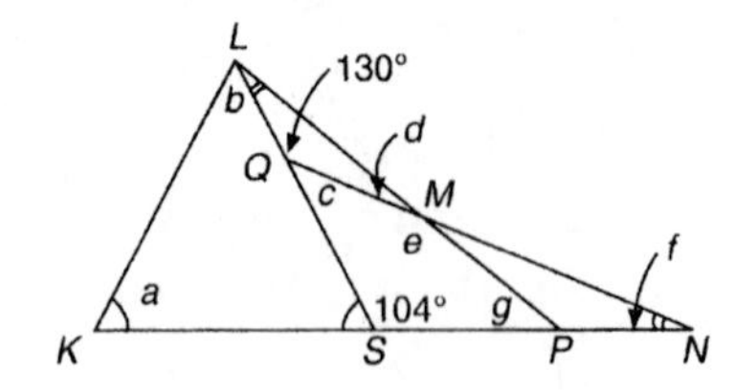

Find $\angle a$.

$$\angle KSL = 180° - 104° = 76°$$

Since $\angle KSL \cong \angle a$ we have $\angle a = 76°$.

Find $\angle b$.

$$\angle KSL + \angle a + \angle b = 180°$$

$$76° + 76° + \angle b = 180°$$

$$\angle b = 28°$$

Find $\angle c$.

$$\angle c = 180° - 130° = 50°$$

Chapter Test

Find $\angle f$.

$$\text{From } \Delta SQN, \angle c + 104° + \angle f = 180°$$
$$50° + 104° + \angle f = 180°$$
$$\angle f = 26°.$$

Find $\angle d$.

$$\angle QLM = \angle f = 26°$$
$$\text{From } \Delta LQM, \angle QLM + 130° + \angle d = 180°$$
$$26° + 130° + \angle d = 180°$$
$$\angle d = 24°.$$

Find $\angle e$.

$$\angle d + \angle e = 180°$$
$$24° + \angle e = 180°$$
$$\angle e = 156°$$

Find $\angle g$.

From quadrilateral $SQMP$

$$104° + \angle c + \angle e + \angle g = 360°$$
$$104° + 50° + 156° + \angle g = 360°$$
$$\angle g = 50°.$$

Summarizing, $\angle a = 76°$, $\angle b = 28°$, $\angle c = 50°$, $\angle d = 24°$, $\angle e = 156°$, $\angle f = 26°$, and $\angle g = 50°$.

13. The formula $\frac{(n-2)180°}{n}$ gives the measure of a vertex angle in a regular n-gon. Use this formula with $n = 24$ sides.

$$\frac{(24-2)180°}{24} = 165°$$

14. A complete solution is given in the answer key at the back of the text.

15. (a) Supplementary pairs of angles (with a sum of 180°): $\angle 1$ and $\angle 2$, $\angle 3$ and $\angle 4$, $\angle 1$ and $\angle 3$, $\angle 2$ and $\angle 4$, $\angle 6$ and $\angle 9$.
    (b) Complementary pairs of angles (with a sum of 90°): $\angle 5$ and $\angle 8$, $\angle 7$ and $\angle 10$, $\angle 5$ and $\angle 7$, $\angle 8$ and $\angle 10$.
    (c) Right angles: $\angle 6$, $\angle 9$.
    (d) Adjacent angles (sharing a side): $\angle 1$ and $\angle 2$, $\angle 2$ and $\angle 4$, $\angle 4$ and $\angle 3$, $\angle 3$ and $\angle 1$, $\angle 5$ and $\angle 6$, $\angle 6$ and $\angle 7$, $\angle 7$ and $\angle 10$, $\angle 10$ and $\angle 9$, $\angle 8$ and $\angle 5$, $\angle 9$ and $\angle 8$.
    (e) Acute angles (measuring less than 90°): $\angle 1$, $\angle 4$, $\angle 5$, $\angle 7$, $\angle 8$, $\angle 10$.
    (f) Obtuse angles (measuring more than 90°): $\angle 2$, $\angle 3$.

Chapter Test

16. (a) There are six points in this star. Connecting each pair of opposite points gives three different lines of symmetry. Similarly, joining the opposite indentations between points gives three more, making six lines of symmetry for reflection.

    Rotating one point to the next causes the figures to coincide. Since you can rotate each point to five other points, there are five rotation symmetries.

    (b) There are two lines of symmetry for reflection symmetry—one vertical line through the middle of the figure and one horizontal line through the middle of the figure.

    This figure only has one rotation symmetry of 180° (when you turn it so the top points become the bottom points).

    (c) For a circle, any diameter is a line of symmetry, and any rotation causes the circles to coincide. Thus, there are infinitely many reflection symmetries and infinitely many rotation symmetries.

17. A tessellation must be a repeatable pattern that will completely cover a plane with no overlapping or gaps. Remember that any quadrilateral can be used to tessellate the plane. Each vertex angle appears once at each point. Here is one solution.

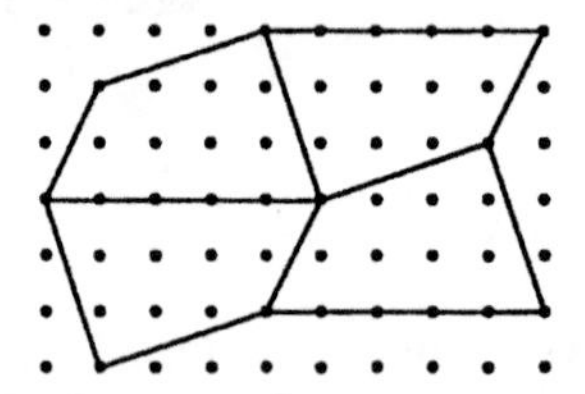

18. (a) You can make a triangle with $\overline{AB}$ as a side using as the third vertex any other point not on the same line. There are 24 points in other rows and thus, 24 different triangles.

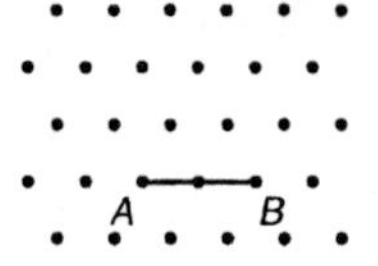

    (b) Using $\overline{AB}$ as the base of the isosceles triangle there is only one triangle possible.

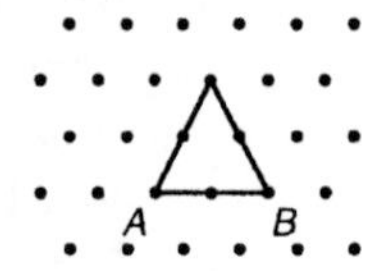

    However, using $\overline{AB}$ as one of the congruent sides, you will find two more triangles for a total of three isosceles triangles.

Chapter Test

17. (continued)

(c) An obtuse triangle with an obtuse angle at A can be formed using any point to the left of A that is not in the same row as A. There are eight of these. Similarly an obtuse angle can be formed at B by making the third vertex of the triangle any point to the right of B, but not in the same row as B, making seven more. That makes a total of 15 obtuse triangles.

(d) To form a right angle you need a square corner. By making the third vertex a dot directly above or below A or B you will make a right triangle. There are two of these. Using the dots just above and below but between A and B makes four more right triangles, each having $\overline{AB}$ as its hypotenuse. Thus, there are six right triangles possible.

19. One millisecond $= \frac{1}{1000}$ sec $= 0.001$ sec

$$\frac{3.25\ \cancel{hr}}{1} \times \frac{60\ \cancel{min}}{1\ \cancel{hr}} \times \frac{60\ \cancel{sec}}{1\ \cancel{min}} \times \frac{1\ \text{millisec}}{0.001\ \cancel{sec}} = 11{,}700{,}000 \text{ milliseconds}$$

20. Be careful not to limit yourself by conditions that are not stated in the problem.

(a) One example is shown.

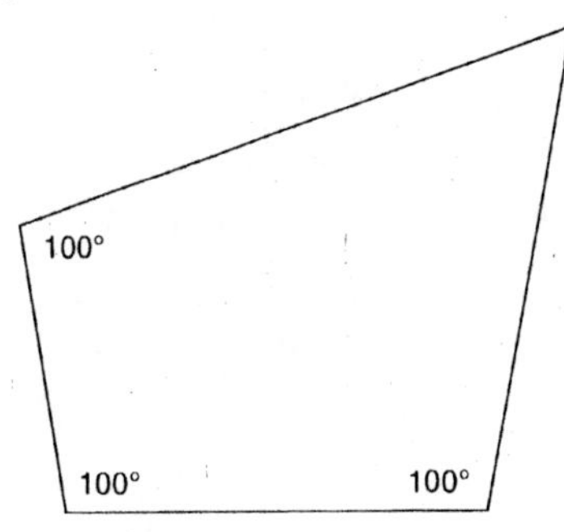

(b) One example is shown.

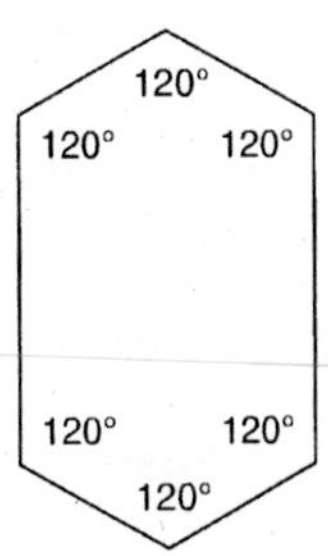

Chapter Test

21. Use the fact that 1 $ft^3 \approx 7.48$ gal, 1 gal = 4 qt, 1 liter ≈ 1.057 qt to change the units to volume first.

(Note: The ounces here are dry weight, not fluid ounces.)

$$\frac{6 \text{ oz}}{1 \cancel{\text{ft}^3}} \times \frac{1 \cancel{\text{ft}^3}}{7.48 \cancel{\text{gal}}} \times \frac{1 \cancel{\text{gal}}}{4 \cancel{\text{qt}}} \times \frac{1.0.57 \cancel{\text{qt}}}{1 \text{ L}} \approx 0.212 \frac{\text{oz}}{\text{L}}$$

Now using 1 kg ≈ 2.2 lb, 1000 g = 1 kg and 1 lb = 16 oz,

$$\frac{0.212 \cancel{\text{oz}}}{1 \text{ L}} \times \frac{1 \cancel{\text{lb}}}{16 \cancel{\text{oz}}} \times \frac{1 \cancel{\text{kg}}}{2.2 \cancel{\text{lb}}} \times \frac{1000 \text{ g}}{1 \cancel{\text{kg}}} \approx 6.02 \frac{\text{g}}{\text{L}}.$$

22. The sum of the interior angles of any pentagon is $(5-2)(180°) = 540°$. Check to see if the sum of these angle measures is 540°.

$$\begin{array}{r} 121°45' \\ 118°25' \\ 87°32' \\ 98°34' \\ +\ 113°54' \\ \hline 537°190' = 540°10' \end{array}$$

No the traverse does not close because of the 10′ error.

23. Find the number of revolutions on the inner track:

$$\frac{45 \cancel{\text{sec}}}{1} \times \frac{1 \cancel{\text{min}}}{60 \cancel{\text{sec}}} \times \frac{500 \text{ rev}}{1 \cancel{\text{mph}}} = \frac{45(1)(500) \text{ rev}}{1(60)(1)}$$

= 375 revolutions.

Find the number of revolutions on the outer track: First, change 20 seconds to minutes.

$$\frac{20 \cancel{\text{sec}}}{1} \times \frac{1 \text{ min}}{60 \cancel{\text{sec}}} = \frac{20(1)}{1(60)} = 0.\overline{3} \text{ min}$$

So, 2 minutes 20 seconds is $2.\overline{3}$ min. Now, to find the number of revolutions:

$$\frac{2.\overline{3} \cancel{\text{min}}}{1} \times \frac{180 \text{ rev}}{1 \cancel{\text{min}}} = \frac{2.\overline{3}(180)}{1(1)} \text{ rev}$$

= 420 revolutions on the outer track.

The total number of revolution on both tracks is 375 + 420 = 795 revolutions.

24. This shape is a right trapezoidal prism. It is lying on one of its lateral faces (sides).

Perimeter, Area, and Volume

Section 3.1

Tips

- ✔ Perimeter is always measured in simple units of length such as feet or meters, never square or cubic units.
- ✔ Area is always measured in square units such as ft^2 or m^2.
- ✔ Any side may be used as the base of a triangle.
- ✔ The height of a triangle is the perpendicular distance from a vertex to the opposite side (which is then considered the base). The height is not usually the length of a side.

Solutions to odd-numbered textbook problems

1. Note the marks indicating congruence.

   (a) $P = 7 + 7 + 4 + 4 = 22$ in.

   (b) $P = 11.2 + 8.7 + 24.6 + 8.7 = 53.2$ m

3. Note that sides are congruent in regular polygons.

   (a) 5 (4 cm) = 20 cm

   (b) 6 (10 in.) = 60 in.

   (c) To find the perimeter (P) of a regular polygon, multiply the number of sides (n) times the side length (s).

   The formula is $P = ns$.

5. 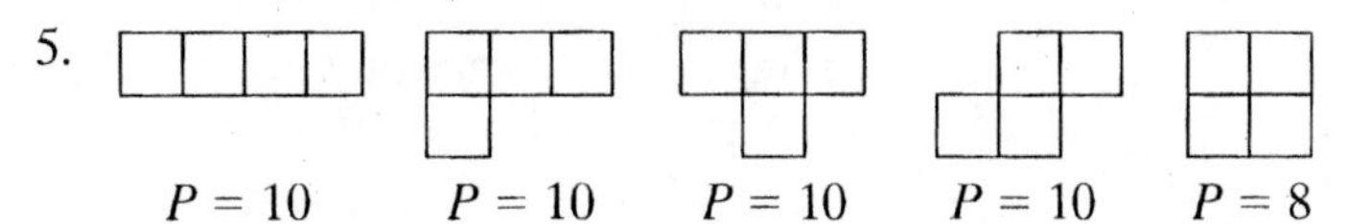

   $P = 10$ $P = 10$ $P = 10$ $P = 10$ $P = 8$

   (a) There are 5 tetrominos, 4 of which have the maximum perimeter of 10.

   (b) The 2 by 2 square has the minimum perimeter of 8.

7. The perimeter is the length of the outside edge (in toothpicks).

   (a) Make a table and look for a pattern.

Number of triangles	1	2	3	4	5 . . .
perimeter	3	4	5	6	7

   There are always 2 more toothpicks in the perimeter than the number of triangles. For n triangles the perimeter is $n + 2$.

Section 3.1

7. (continued)

(b)

Number of squares	1	2	3	4...
perimeter	4	6	8	10

There are always n toothpicks on the top, n toothpicks on the bottom, and 2 toothpicks on the ends for a perimeter of $2n + 2$ toothpicks for n squares.

(c)

Number of pentagons	1	2	3	4...
perimeter	5	8	11	14

For n pentagons there are n toothpicks on the bottom, $2n$ toothpicks on the top, and 2 toothpicks on the ends for a perimeter of $3n + 2$ toothpicks.

(d)

Number of hexagons	1	2	3...
perimeter	6	10	14

There are $2n$ toothpicks on the top, $2n$ toothpicks on the bottom and 2 toothpicks on the ends, making the perimeter $4n + 2$ toothpicks.

(e)

Number of octagons	1	2	3...
perimeter	8	14	20

There are $3n$ toothpicks on the top, $3n$ toothpicks on the bottom and 2 toothpicks on the ends for a perimeter of $6n + 2$ toothpicks.

(f) How about when there are n m-sided polygons?

Summarize what you have already.

Number of sides (m)	3	4	5	6	8
perimeter	$n + 2$	$2n + 2$	$3n + 2$	$4n + 2$	$6n + 2$

Notice that the coefficient of n is always 2 less than the number of sides of the polygon. When you have n m-sided polygons, the perimeter is $(m - 2)n + 2$.

9. Depending on the dimension given in the problem, diameter (d) or radius (r), you may use either formula, $C = \pi d$ or $C = 2\pi r$.

(a) $r = 4$ in. so, $C = 2\pi(4)$

≈ 25.13 in.

(b) $d = 10.8$ m so, $C = \pi\,(10.8)$

≈ 33.93 m

11. The area of a square with side length s is $A = s^2$. To determine the length s, solve the equation $25 = s^2$, by taking a square root, $s = 5$ cm. The length of the wire is the perimeter of the square. Using $P = 4s$, when $s = 5$ cm:

Section 3.1

11. (continued)

$P = 4(5)$ or 20 cm. So, the perimeter of the square, and length of wire, is 20 cm. To make two identical circles, the wire must be cut into two equal lengths. $20 \div 2 = 10$ cm.

Each circle is created from a piece of wire 10 cm long, which gives each circle a circumference of 10 cm. Use the formula $C = 2\pi r$ where $C = 10$, and solve for r. $10 = 2\pi r$ and $r = \frac{5}{\pi} \approx 1.59$. So, the radius of the circles is approximately 1.59 cm.

13. (a) There are 6 whole square units plus 2 half-squares for a total of 7 square units.

(b) There are 9 whole square units. The bottom right corner uses half of a rectangle containing 3 square units. Half of 3 is $1\frac{1}{2}$ square units. In the point on top are two halves, making 1 square unit. That makes a total of $9 + 1\frac{1}{2} + 1 = 11\frac{1}{2}$ square units.

(c) There are 8 whole square units and 8 halves for a total of $8 + 8 \cdot \frac{1}{2} = 12$ square units.

15. (a) To find perimeter, $P = 2l + 2w$ where $l = 10$ in. and $w = 4$ in.

$$\begin{aligned} P &= 2(10) + 2(4) \\ &= 28 \text{ in.} \end{aligned}$$

To find area, $A = lw$ where $l = 10$ in. and $w = 4$ in.

$$\begin{aligned} A &= 10(4) \\ &= 40 \text{ in}^2 \end{aligned}$$

(b) To find perimeter, $P = 4s$ where $s = 5.7$ cm.

$$\begin{aligned} P &= 4(5.7) \\ &= 22.8 \text{ cm} \end{aligned}$$

To find area, $A = s^2$ where $s = 5.7$ cm.

$$\begin{aligned} A &= 5.7^2 \\ &= 32.49 \text{ cm}^2 \end{aligned}$$

$l = 2w - 6$

w

17. Use a variable and draw a picture.

$$\begin{aligned} 2l + 2w &= P \\ 2(2w - 6) + 2w &= 2 \\ 4w - 12 + 2w &= 72 \\ 6w &= 72 + 12 \\ 6w &= 84 \\ w &= 14 \text{ in.} \\ l = 2w - 6 &= 22 \text{ in.} \end{aligned}$$

Section 3.1

19. The formula for perimeter of a rectangle is $P = 2l + 2w$ or $P = 2(l + w)$.
   - Multiply length and width by 2.

     $P = 2(2l) + 2(2w)$

     $P = 4l + 4w$ or $P = 4(l + w)$

     The perimeter is 2 times as large as the original.
   - Multiply length and width by 3.

     $P = 2(3l) + 2(3w)$

     $P = 6l + 6w$ or $P = 6(l + w)$

     The perimeter is 3 times as large as the original.
   - Multiply length and width by 10.

     $P = 2(10l) + 2(10w)$

     $= 20l + 20w$ or $P = 20(l + w)$

     The perimeter is 10 times as large as the original.
   - Now, multiply length and width by n.

     $P = 2(nl) + 2(nw)$

     $= 2nl + 2nw$ or $P = n[2(l + w)]$

     The perimeter is n times as large as the original.

21. $P = 2 \cdot 14 + 2 \cdot 6 = 40$ m

    $A = 6 \cdot 14 = 84$ m^2

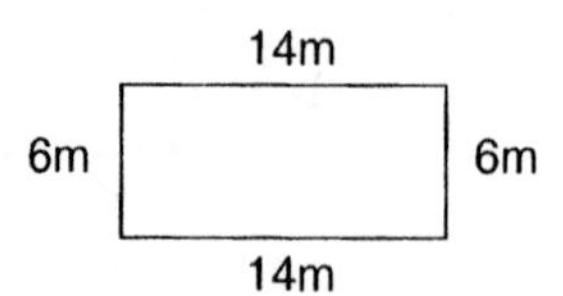

$P = 2 \cdot 8 + 2 \cdot 12 = 40$ m

$A = 8 \cdot 12 = 96$ m^2

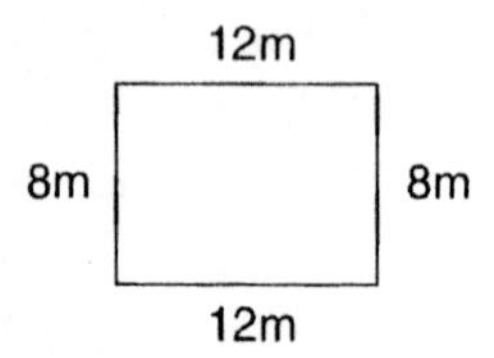

Therefore, two rectangles with the same perimeter do not have to have the same area. Other counter examples are given in the answer key at the back of the text.

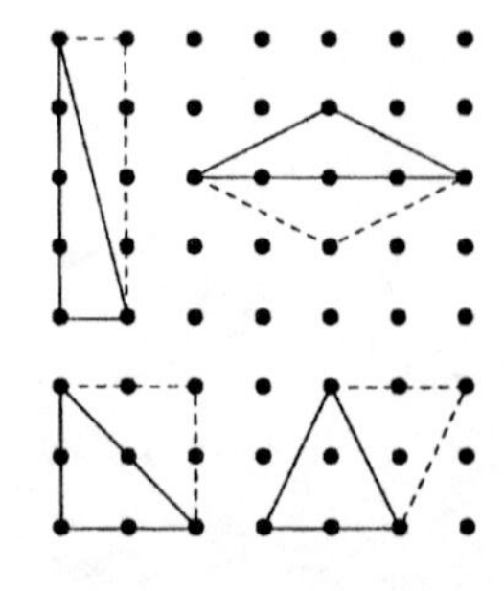

23. Many answers are possible. One approach is to sketch a parallelogram with an area of 4 square units. Connect two opposite vertices with a straight line. The result is two triangles each with area of 2 square units.

Section 3.1

25. (a) $P = 6 + 8 + 10 = 24$ m

$$A = \frac{1}{2}(8 \cdot 6) = 24 \text{ m}^2$$

(b) $P = 3.25 + 5.75 + 6.60 = 15.60$ cm

$$A = \frac{1}{2}(5.75)(3.25) = 9.34375 \approx 9.34 \text{ cm}^2$$

(c) $P = 5 + 5\sqrt{2} + 5 = 10 + 5\sqrt{2}$ in. ≈ 17.07 in.

$$A = \frac{1}{2}(5)(5) = 12.5 \text{ in}^2$$

27. The formula for area of a rectangle is $A = lw$. So, the area of $ABCD$ is $A = AD \cdot CD$.

The formula for area of a triangle is $A = \frac{1}{2}bh$. So, the area of ΔAED is

$A = \frac{1}{2} \cdot AD \cdot ED$. Since $AD = 2CD$, we can substitute 2 CD into the equation for ED, resulting in:

$A = \frac{1}{2} \cdot AD \cdot 2\,CD$ and simplifying,

$$A = \frac{1}{2} \cdot 2 \cdot AD \cdot CD$$

$$A = AD \cdot CD.$$

Thus, the area of $ABCD$ is equal to the area of ΔAED.

29. A table with the measurements is given in the answer key at the back of the text. Remember to measure the height perpendicular to each side. Measurements may vary depending on the quality of your measuring device, how careful you are, and the quality of the picture in your text. Don't expect your measurements to match exactly with those given in the book. Also, don't expect your area calculations to be identical, but they should be approximately the same.

31. (a) $$\frac{50 \cancel{\text{in}^2}}{1} \cdot \frac{1 \cdot 1 \text{ ft}^2}{12 \cdot 12 \cancel{\text{in}^2}} = 0.347\overline{2} \text{ ft} \approx 0.35 \text{ ft}^2$$

(b) $$\frac{2000 \cancel{\text{m}^2}}{1} \cdot \frac{100 \cdot 100 \text{ cm}^2}{1 \cdot 1 \cancel{\text{m}^2}} = 20{,}000{,}000 \text{ cm}^2$$

(c) $$\frac{3.5 \cancel{\text{yd}^2}}{1} \cdot \frac{36 \cdot 36 \cancel{\text{in}^2}}{1 \cdot 1 \cancel{\text{yd}^2}} \cdot \frac{(2.54)(2.54) \cancel{\text{cm}^2}}{1 \cdot 1 \cancel{\text{in}^2}} \cdot \frac{1 \cdot 1 \text{ m}^2}{100 \cdot 100 \cancel{\text{cm}^2}}$$

$$= \frac{3.5(36)(36)(2.54)(2.54)(1)(1)}{1(1)(1)(1)(1)(100)(100)} \text{ m}^2$$

$$\approx 2.93 \text{ m}^2$$

Section 3.1

33. (a) Convert 8″ to feet.

$$\frac{8 \text{ in.}}{1} \cdot \frac{1 \text{ ft}}{12 \text{ in.}} = 0.\overline{6} \text{ ft}$$

So, $4'8'' = 4.\overline{6}$ ft.

Convert 3″ to feet.

$$\frac{3 \text{ in.}}{1} \cdot \frac{1 \text{ ft}}{12 \text{ in.}} = 0.25 \text{ ft}$$

So, $8'3'' = 8.25$ ft.

The area of the rectangle is

$$\begin{aligned} A &= 4.\overline{6} \text{ ft} \cdot 8.25 \text{ ft} \\ &= 38.5 \text{ ft}^2. \end{aligned}$$

(b) Convert 38.5 ft² to m².

$$\frac{38.5 \text{ ft}^2}{1} \cdot \frac{12 \cdot 12 \text{ in}^2}{1 \cdot 1 \text{ ft}^2} \cdot \frac{(2.54)(2.54) \text{ cm}^2}{1 \cdot 1 \text{ in}^2} \cdot \frac{1 \cdot 1 \text{ m}^2}{100 \cdot 100 \text{ cm}^2}$$

$$= \frac{38.5(12)(12)(2.54)(2.54)(1)(1)}{1(1)(1)(1)(1)(100)(100)} \text{ m}^2$$

$$\approx 3.58 \text{ m}^2$$

35. Before adding 30 feet to the length, the cable is 7926 mi long. In feet, this is $\frac{7926 \text{ mi}}{1} \cdot \frac{5280 \text{ ft}}{1 \text{ mi}} = 41{,}849{,}280$ ft. The cable length represents the circumference of the earth. Find the radius of the earth using $C = 2\pi r$ where $C = 41{,}849{,}280$ ft.

So, $41{,}849{,}280 = 2\pi r$ or $r \approx 6{,}660{,}519.78$ ft.

Thus, the earth had a radius of approximately 6,660,519.78 ft.

After adding 30 feet to the length of the cable, the cable is approximately 41,849,310 ft. Find the radius when $C = 41{,}849{,}310$ ft.

So $41{,}849{,}310 = 2\pi r$ or $r \approx 6{,}660{,}524.55$ ft.

Then the distance between the surface of the earth and the cable is approximately $6{,}660{,}524.55 - 6{,}660{,}519.78 = 4.77$ ft.

You could crawl under the cable but not walk under it, unless you were less than 4.77 ft tall.

37. (a) $s = \frac{5 + 12 + 13}{2} = 15$ cm

$$A = \sqrt{15(15 - 5)(15 - 12)(15 - 13)}$$

$$A = \sqrt{900}$$

$$A = 30 \text{ cm}^2$$

Section 3.1

37. (continued)

(b) $s = \dfrac{4 + 5 + 6}{2} = 7.5$ m

$A = \sqrt{7.5(7.5 - 4)(7.5 - 5)(7.5 - 6)}$

$A = \sqrt{98.4375}$

$\approx 9.92 \text{ m}^2$

(c) $s = \dfrac{4 + 5 + 8}{2} = 8.5$ ft

$A = \sqrt{8.5(8.5 - 4)(8.5 - 5)(8.5 - 8)}$

$A = \sqrt{66.9375}$

$\approx 8.18 \text{ ft}^2$

39. (a) Using b_1 for the base, the area of ΔABD is $A = \frac{1}{2}b_1h$.

(b) Using b_2 for the base, the area of ΔCBD is $A = \frac{1}{2}b_2h$.

(c) The sum of the area of ΔABD and ΔCBD is $A = \frac{1}{2}b_1h + \frac{1}{2}b_2h$.

(d) Factor out the common factors $\frac{1}{2}$ and h from the right side to obtain $A = \frac{1}{2}h(b_1 + b_2)$.

From the diagram, $b_1 + b_2$ is side b, the base of ΔABC. Substituting b for $b_1 + b_2$ gives us

$A = \frac{1}{2}hb$ or $A = \frac{1}{2}bh$.

(e) Using Theorem 3.3, the area of ΔABC is $A = \frac{1}{2}bh$.

This agrees with the answer to part (d).

41. Suppose the circumference of a basketball was 29.5 in., the smallest allowable. Use the circumference formula $C = \pi d$ where $C = 29.5$ in. and solve for diameter d.

$29.5 = \pi d$

$\dfrac{29.5}{\pi} = d$

$9.39 \text{ in.} \approx d$

The greatest distance across the two basketballs when side by side is approximately 2(9.39)= 18.78 in.

The balls would not fit through the ring side by side because the diameter of the ring is less than 18.78 in.

43. (a) The perimeter of the room is:

$P = 2(17'9'') + 2(24'8'') + 2(2'6'')$

$= 34'18'' + 48'16'' + 4'12''$

$= 89'10''$.

Section 3.1 43. (continued)

The total allowance for doors is 7 ft + 2(3 ft) = 13 ft.
The total amount of quarter round required is
89′10″ − 13′ = 76′10″ or approximately 77ft.

(b) The room can be divided into three rectangles as shown in the diagram at the left.

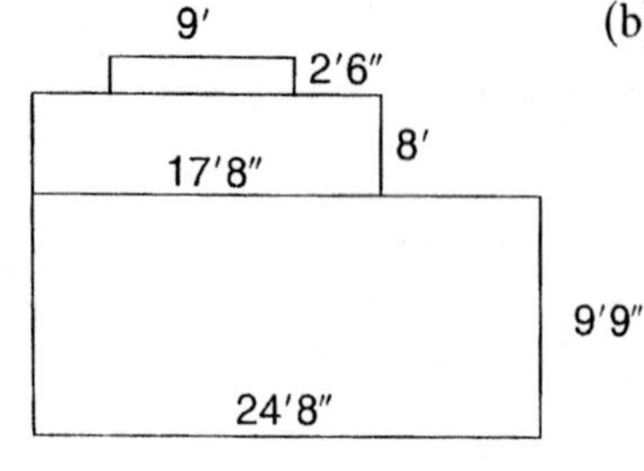

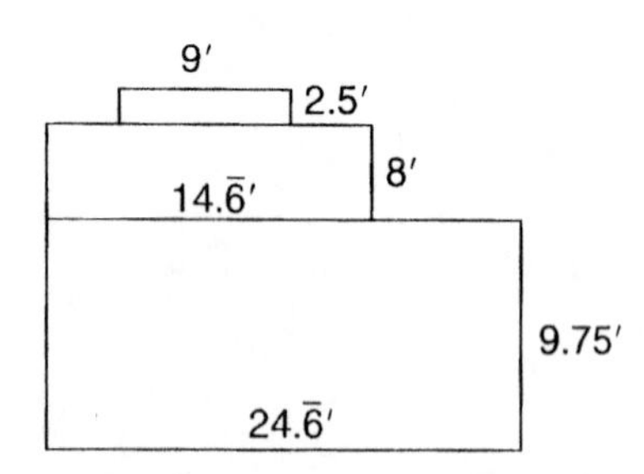

Converting dimensions from feet and inches to feet gives us the diagram at the left.

Find the sum of the areas or the three rectangles.

$\text{Area}_{\text{Total}} = 9(2.5) + 14.\overline{6}(8) + 24.\overline{6}(9.75) = 380.\overline{3}\ \text{ft}^2$

Convert $380.\overline{3}\ \text{ft}^2$ to square yards.

$$\frac{380.\overline{3}\ \cancel{\text{ft}^2}}{1} \cdot \frac{1 \cdot 1\ \text{yd}^2}{3 \cdot 3\ \cancel{\text{ft}^2}} \approx 42\ \text{yd}^2$$

45. Subtracting 10% of the 10 ft molding, 90% will be used for the two frames. Hence there is (0.90)(10) = 9 ft of molding used for the two frames.

Let s = length of side of smaller frame in feet.

Then $4s$ = amount of molding used for the smaller frame.

$3s$ = length of side of larger frame in feet.

So $4(3s)$ = amount of molding used for larger frame.

We have the equation $4s + 4(3s) = 9$

$$16s = 9$$

$$s = \frac{9}{16}\ \text{ft.}$$

The length of the side of the smaller frame is $\frac{9}{16}$ ft and the length of the side of the larger frame is $3\left(\frac{9}{16}\right) = \frac{27}{16}$ ft.

Section 3.1

45. (continued)

So, the dimensions of the smaller frame are

$$\frac{9}{16}\text{ ft by }\frac{9}{16}\text{ ft or 0.5625 ft by 0.5625 ft.}$$

And the dimensions of the larger frame are

$$\frac{27}{16}\text{ ft by }\frac{27}{16}\text{ ft or 1.6875 ft by 1.6875 ft.}$$

47. 6.5 billion people would reach $6.5 \cdot 2 = 13$ billion yards.

The circumference of the earth is $C = 2\pi r$ where $r = 3960$ mi.

So, $C = 2\pi(3960) = 7920\pi$ miles.

Convert 13 billion yards to miles.

$$\frac{13{,}000{,}000{,}000\ \cancel{\text{yd}}}{1}\cdot\frac{3\ \cancel{\text{ft}}}{1\ \cancel{\text{yd}}}\cdot\frac{1\text{ mi}}{5280\ \cancel{\text{ft}}}$$

$$=\frac{13{,}000{,}000{,}000\ (3)(1)}{1\ (1)(5280)}\text{ mi} = \frac{81{,}250{,}000}{11}\text{ mi}$$

Divide the reach of the people by the circumference of the earth.

$$\frac{81{,}250{,}000}{11} \div 7920\pi \approx 296.9$$

Thus, the line of people would wrap around the earth about 297 times.

49. If the diameter of the tire is 29.78 inches, then the circumference is $C = \pi d = \pi(29.78) = 29.78\pi$ inches. So the car travels 29.78π inches for every revolution of the wheels.

If the diameter of the tire is 30.78 inches, then the car will travel $C = \pi d = \pi(30.78) = 30.78\pi$ inches for each revolution.

With the larger tires, the car travels $30.78\pi - 29.78\pi = \pi$ inches farther with each revolution of the tire. As a percent, this is

$$\frac{\pi}{29.78\pi} \approx 0.03358 \quad\text{or}\quad 3.358\%\text{ farther.}$$

So the car is traveling about 3.358% faster than the speedometer shows, and the actual speed when it shows 65 mph is about

$1.03358(65) \approx 67.2$ mph.

51. The circumference of the tire is $C = 22\pi$ in. To find the number of revolutions in 1 minute, use dimensional analysis.

$$\frac{30\ \cancel{\text{mi}}}{1\ \cancel{\text{hr}}}\times\frac{5280\ \cancel{\text{ft}}}{1\ \cancel{\text{mi}}}\times\frac{12\ \cancel{\text{in.}}}{1\text{ ft}}\times\frac{1\text{ revoution}}{22\pi\ \cancel{\text{in.}}} = \frac{27516\text{ rev}}{\text{hr}}$$

$$\frac{27516\text{ rev}}{1\ \cancel{\text{hr}}}\times\frac{1\ \cancel{\text{hr}}}{60\text{ min}} \approx 458.4\text{ rev/min}$$

Section 3.2

Tips:

✔ The height of a trapezoid or a parallelogram is the perpendicular distance between the bases.

✔ To find the area of a regular polygon you need the perpendicular distance from the center to one side, not the distance across the polygon.

✔ Sometimes you must divide a composite or irregular figure into several areas which you add or subtract to find the total area of the figure.

Solutions to odd-numbered textbook problems

1. (a) The figure shown is a parallelogram so

$$A = bh$$
$$A = 7 \cdot 4$$
$$A = 28 \text{ cm}^2.$$

The 5 cm side length is not used to calculate the area.

(b) This quadrilateral is also a parallelogram. Since the base and the height must be perpendicular to each other, you can use 5 inches for the base and 4 inches for the height.

$$A = bh$$
$$A = 5 \cdot 4$$
$$A = 20 \text{ in}^2$$

3. (a) $A = bh$ and when $A = 40, h = 6$.

$$b \cdot 6 = 40$$
$$b = \frac{40}{6}$$
$$b = 6.\overline{6}$$
$$b \approx 6.67$$

(b) When $b = 3\sqrt{2}$ and $h = 2\sqrt{3}$

$$A = 3\sqrt{2} \cdot 2\sqrt{3}$$
$$A = 6\sqrt{6}$$
$$A \approx 14.70.$$

5. The table is given in the answer key at the back of the text. Answers may vary slightly due to small differences in measuring and rounding.

The areas are close since a square millimeter is quite small.

Section 3.2

7. Using the formula for area of a trapezoid (Theorem 3.4), we have

$$A = \frac{1}{2}h\,(b_1 + b_2) = \frac{1}{2}(14.6)(18 + 11.5) = 215.35 \text{ mm}^2.$$

9. Think of this composite figure as a rectangle with a trapezoid piece subtracted.

18″ 10″ − 6″ 4″ 10″

$$l \cdot w - \frac{1}{2}h\,(b_1 + b_2) = 10(18) - \frac{1}{2}(4)(6 + 10)$$
$$= 180 - 32 = 148 \text{ in}^2$$

11. If $A = 69 \text{ in}^2$, $b_1 = 9$ in., and $b_2 = 14$ in., find h.

$$A = \frac{1}{2}h\,(b_1 + b_2)$$
$$69 = \frac{1}{2}h\,(9 + 14)$$
$$69 = 11.5h$$
$$6 \text{ in.} = h$$

13. First, find the area of the hexagon.

$$A = \frac{1}{2}Ph = \frac{1}{2}(6 \cdot 3.4)(2.94) = 29.988 \text{ mm}^2$$

Next, find the area of the unshaded triangle. Note the height of the triangle is twice the apothem of the hexagon $h = 2(2.94) = 5.88$ mm.

$$A = \frac{1}{2}bh = \frac{1}{2}(3.4)(5.88) = 9.996 \text{ mm}^2$$

Finally, subtract the area of the triangle from the area of the hexagon to get the area of the shaded region.

$A = 29.988 - 9.996 = 19.992 \approx 19.99 \text{ mm}^2$

15. (a) The figure is a circle.

You are given the diameter, but need the radius to find the area.

$$r = \frac{1}{2}d = \frac{1}{2}(6\sqrt{2}) = 3\sqrt{2}$$

$$A = \pi r^2$$
$$= \pi(3\sqrt{2})^2$$
$$= \pi \cdot 9 \cdot 2$$
$$= 18\pi \text{ cm}^2 \approx 56.55 \text{ cm}^2$$

Section 3.2

15. (continued)

(b) The figure is a semicircle with radius $r = 5$ in.

$$A = \frac{1}{2}\pi r^2$$
$$= \frac{1}{2}\pi(5)^2$$
$$= 12.5\pi$$
$$\approx 39.27 \text{ in}^2$$

17. (a) Find the area of the three semicircles, each with radius $r = 8$ in.

$$A = 3 \cdot \frac{1}{2}\pi r^2$$
$$= 3 \cdot \frac{1}{2}\pi(8)^2$$
$$= 96\pi$$
$$\approx 301.59 \text{ in}^2$$

Find the area of ΔABC using Heron's formula (See Section 3.1 problem #37).

$$A = \sqrt{s(s-a)(s-b)(s-c)}, \text{ where}$$
$$s = \frac{a+b+c}{2} = \frac{16+16+16}{2} = 24.$$
$$A = \sqrt{24(24-16)(24-16)(24-16)}$$
$$= 64\sqrt{3}$$
$$\approx 110.85 \text{ in}^2$$
$$A_{total} = 96\pi + 64\sqrt{3}$$
$$\approx 412.44 \text{ in}^2$$

(b) Find the area of the square with side length 12.4 cm.

$$A = s^2$$
$$= 12.4^2$$
$$\approx 153.76 \text{ cm}^2$$

Find the area of the semicircle with radius $r = \frac{12.4}{2} = 6.2$ cm.

$$A = \frac{1}{2}\pi r^2$$
$$= \frac{1}{2}\pi(6.2)^2$$
$$= 19.22\pi$$
$$\approx 60.38 \text{ cm}^2$$

Subtract the area of the semicircle from the area of the square.

$$A_{total} = 153.76 - 19.22\pi$$
$$\approx 93.38 \text{ cm}^2$$

Section 3.2

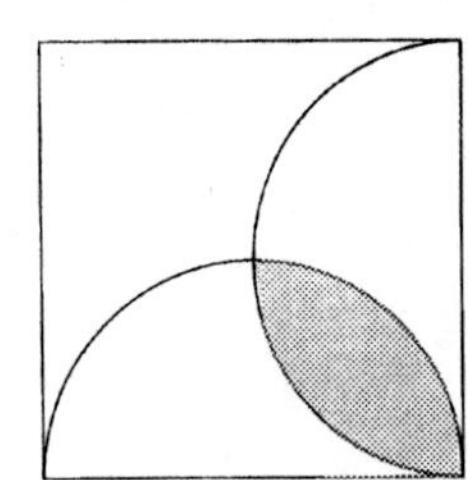

19. Each petal is formed by the intersection of two semicircles (See figure).

The shaded area is the area of 4 semicircles minus the area of the square. If you add the area of the 4 semicircles together, each shaded area is counted twice. Subtracting the area of the square subtracts the extra petal area as well as the unshaded areas.

$$A_{\text{shaded}} = 4A_{\text{semicircle}} - A_{\text{square}}$$

$$A_{\text{shaded}} = 4 \cdot \frac{1}{2} \cdot \pi(4)^2 - 8^2$$

$$= 32\pi - 64 \approx 36.53 \text{ in}^2$$

21. Find the side length of the square with perimeter of 20 inches.

$$P = 4s$$

$$20 = 4s$$

$$5 = s$$

So the square has side length 5 inches.

$$A_{\text{square}} = s^2$$

$$= 5^2$$

$$A_{\text{square}} = 25 \text{ in}^2$$

Find the radius of the circle with circumference of 20 inches.

$$C = 2\pi r$$

$$20 = 2\pi r$$

$$\frac{20}{2\pi} = r$$

$$\frac{10}{\pi} = r$$

$$A_{\text{circle}} = \pi r^2$$

$$= \pi\left(\frac{10}{\pi}\right)^2$$

$$= \frac{100}{\pi}$$

$$A_{\text{circle}} \approx 31.83 \text{ in}^2$$

Thus, the area of the circle is larger.

23. The area of a circle is $A = \pi r^2$.

If radius is doubled:

$$A = \pi(2r)^2$$

$$= \pi \cdot 4r^2$$

$$= 4\pi r^2.$$

The resulting area is 2^2 or 4 times as large.

Section 3.2

23. (continued)

If radius is multiplied by 5:

$$\begin{aligned} A &= \pi(5r)^2 \\ &= \pi \cdot 25r^2 \\ &= 25\pi r^2. \end{aligned}$$

The resulting area is $5^2 = 25$ times as large.

If radius is multiplied by n:

$$\begin{aligned} A &= \pi(nr)^2 \\ &= \pi \cdot n^2r^2 \\ &= n^2\pi r^2. \end{aligned}$$

The resulting area is n^2 times as large.

25. $ABCD$ is a parallelogram as shown.

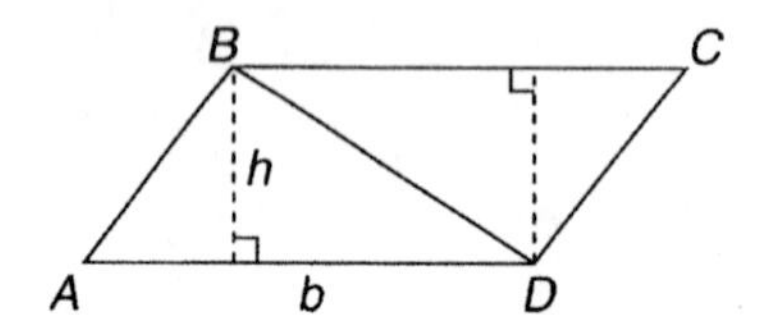

(a) $A_{\Delta ABD} = \frac{1}{2}bh$

(b) ΔABD is congruent to ΔCDB.

$A_{\Delta ABD} = A_{\Delta CDB} = \frac{1}{2}bh$

(c) $A_{ABCD} = A_{\Delta ABD} + A_{\Delta CDB}$

(d) $A_{ABCD} = A_{\Delta ABD} + A_{\Delta CDB} = \frac{1}{2}bh + \frac{1}{2}bh = bh$

Yes, the result agrees with Theorem 3.4.

27. One possible solution is to subdivide the figure in $\Delta 1$, $\Delta 2$, $\Delta 3$, and $\Delta 4$ (shown) and use Heron's formula to find the area of each triangle. Then find the sum of the areas.

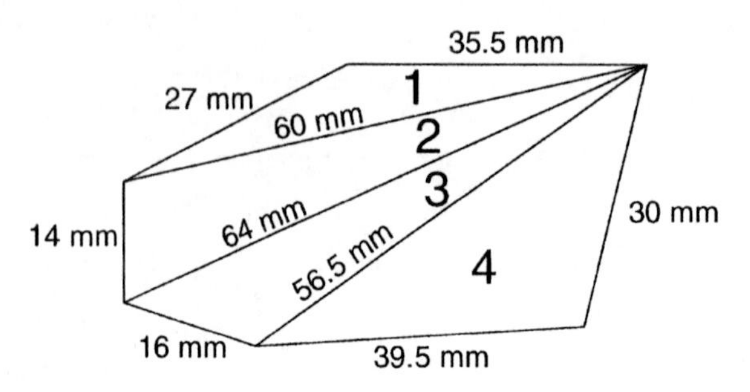

Section 3.2

$$\Delta 1: \quad s_1 = \frac{27 + 35.5 + 60}{2} = 61.25 \text{ mm}$$

$$A_1 = \sqrt{61.25(61.25 - 27)(61.25 - 35.5)(61.25 - 60)}$$

$$A_1 \approx 259.8525 \text{ mm}^2$$

$$\Delta 2: \quad s_2 = \frac{14 + 64 + 60}{2} = 69$$

$$A_2 = \sqrt{69(69 - 14)(69 - 64)(69 - 60)}$$

$$A_2 \approx 413.2493 \text{ mm}^2$$

$$\Delta 3: \quad s_3 = \frac{64 + 16 + 56.5}{2} = 68.25$$

$$A_3 = \sqrt{68.25(68.25 - 64)(68.25 - 16)(68.25 - 56.5)}$$

$$A_3 \approx 421.9955 \text{ mm}^2$$

$$\Delta 4: \quad s_4 = \frac{56.5 + 39.5 + 30}{2} = 63$$

$$A_4 = \sqrt{63(63 - 56.5)(63 - 39.5)(63 - 30)}$$

$$A_4 \approx 563.5310 \text{ mm}^2$$

$$A_1 + A_2 + A_3 + A_4 \approx 259.8525 + 413.2493 + 421.9955 + 563.5310$$

$$\approx 1658.6 \text{ mm}^2$$

29. Assuming that pizzas of every size have the same thickness, we can compare cost per square inch of area for each size.

Small pizza: $A_{\text{small}} = \pi \cdot 5^2 = 25\pi \text{ in}^2$

$$\text{Cost} = \frac{\$10.99}{25\pi \text{ in}^2} \approx 0.14 \ \$/\text{in}^2$$

Medium pizza: $A_{\text{medium}} = \pi \cdot 6^2 = 36\pi \text{ in}^2$

$$\text{Cost} = \frac{\$12.75}{36\pi \text{ in}^2} \approx 0.11 \ \$/\text{in}^2$$

Large pizza: $A_{\text{large}} = \pi \cdot 8^2 = 64\pi \text{ in}^2$

$$\text{Cost} = \frac{\$15.25}{64\pi \text{ in}^2} \approx 0.08 \ \$/\text{in}^2$$

The large pizza is the best buy since the cost is the least per square inch.

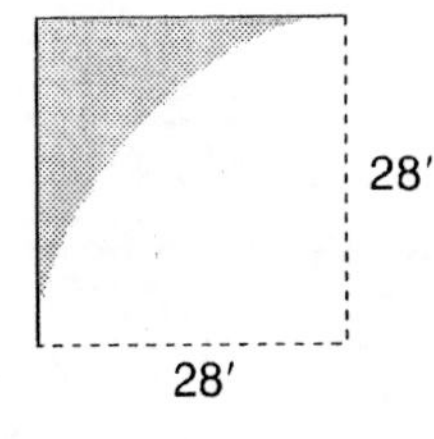

31. Each rounded corner cuts off an area equal to the shaded region in the figure.

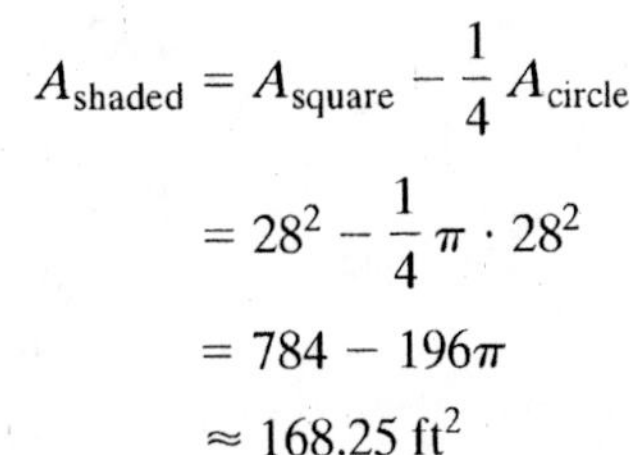

$$A_{\text{shaded}} = A_{\text{square}} - \frac{1}{4} A_{\text{circle}}$$

$$= 28^2 - \frac{1}{4}\pi \cdot 28^2$$

$$= 784 - 196\pi$$

$$\approx 168.25 \text{ ft}^2$$

Section 3.2

31. (continued)

Find the total area by subtracting four shaded corner areas from the area of the rectangle.

$$\begin{aligned} A_{\text{total}} &= A_{\text{rectangle}} - 4 \cdot A_{\text{shaded}} \\ &= 85(200) - 4(784 - 196\pi) \\ &= 17{,}000 - 3136 + 784\pi \\ &= 13{,}864 + 784\pi \\ &\approx 16{,}327 \text{ ft}^2 \end{aligned}$$

33. (a) Find the radius of the circle. $r = 8 \cdot 165 = 1320$ ft.

Now $A = \pi r^2$

$$\begin{aligned} &= \pi(1320)^2 \\ &= 1{,}742{,}400\pi \text{ ft}^2. \end{aligned}$$

So $\dfrac{1{,}742{,}400\pi \cancel{\text{ft}^2}}{1} \times \dfrac{1 \text{ acre}}{43560 \cancel{\text{ft}^2}} = 40\pi \text{ ft}^2$

$$\approx 125.66 \text{ ft}^2.$$

(b) The length of one side of the square is the diameter of the circle $= 2 \cdot 1320 = 2640$ ft.

$$\begin{aligned} A_{\text{square}} - A_{\text{circle}} &= (2640)^2 - 1{,}742{,}400\pi \\ &= (6{,}969{,}600 - 1{,}742{,}400\pi) \text{ ft}^2 \end{aligned}$$

So, the number of acres of wheat is

$$\frac{(6{,}969{,}600 - 1{,}742{,}400\pi) \cancel{\text{ft}^2}}{1} \times \frac{1 \text{ acre}}{43560 \cancel{\text{ft}^2}} \approx 34.34 \text{ acres.}$$

35. Find the area of the plate and subtract the areas of two $\frac{3}{4}$ in. diameter holes and two $\frac{1}{2}$ in. diameter holes. Note: radius is $\frac{1}{2}$ the diameter, so the radius of the $\frac{3}{4}$ in. holes is $\frac{1}{2}\left(\frac{3}{4}\right) = \frac{3}{8}$ in. and the radius of the $\frac{1}{2}$ in. holes is $\frac{1}{2}\left(\frac{1}{2}\right) = \frac{1}{4}$ in.

$$\begin{aligned} A_{\text{total}} &= A_{\text{plate}} - 2A_{\frac{3}{4}\text{ hole}} - 2A_{\frac{1}{2}\text{ hole}} \\ &= 10.2(8.4) - 2\pi\left(\frac{3}{8}\right)^2 - 2\pi\left(\frac{1}{4}\right)^2 \\ &= 85.68 - \frac{5\pi}{32} \\ &\approx 85.19 \text{ in}^2 \end{aligned}$$

Section 3.2

35. (continued)

Convert in² to ft².

$$\frac{85.19\ \cancel{\text{in}^2}}{1} \times \frac{(1)(1)\ \text{ft}^2}{(12)(12)\ \cancel{\text{in}^2}} = 0.592\ \text{ft}^2$$

To find the resulting weight, multiply area times $7.5\frac{\text{lb}}{\text{ft}^2}$.

$$\frac{0.592\ \cancel{\text{ft}^2}}{1} \times \frac{7.5\ \text{lb}}{1\ \cancel{\text{ft}^2}} = 4.4\ \text{lbs}$$

37. The diagram shows the stretch of road to be paved with dimensions given in feet. Bike lanes are included. The diagram is not to scale.

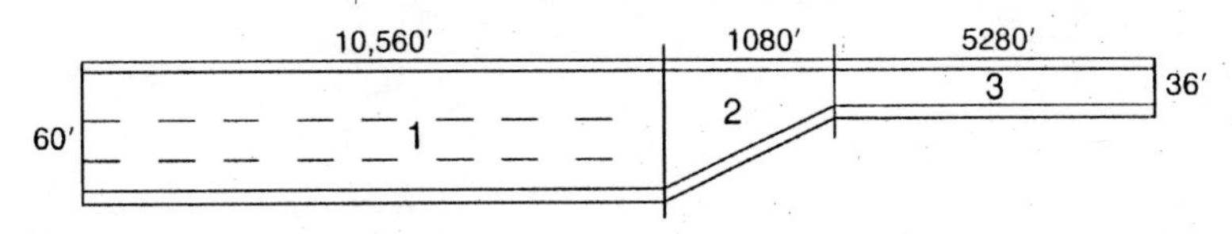

Area 1 is a rectangle.

$$A = l \cdot w$$
$$A_1 = 10560(60)$$
$$= 633{,}600\ \text{ft}^2$$

Area 2 is a trapezoid.

$$A = \frac{1}{2}h\,(b_1 + b_2)$$
$$A_2 = \frac{1}{2}(1080)(60 + 36)$$
$$= 51{,}840\ \text{ft}^2$$

Area 3 is a rectangle.

$$A = l \cdot w$$
$$A_3 = 5280(36)$$
$$= 190{,}080\ \text{ft}^2$$
$$A_{\text{Total}} = A_1 + A_2 + A_3$$
$$= 633{,}600 + 51{,}840 + 190{,}080$$
$$= 875{,}520\ \text{ft}^2$$

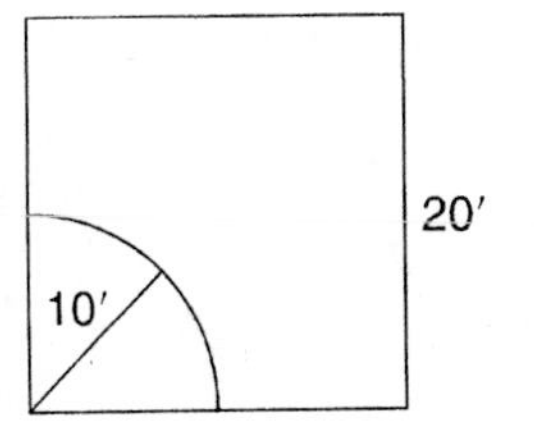

39. (a) To find the percentage of the grass the goat can reach, first find the area inside the pen and the area which the goat can reach.

$$A_{\text{pen}} = 20 \cdot 20 = 400\ \text{ft}^2$$

$$A_{\text{goat}} = \frac{1}{4} \cdot \pi(10)^2 = 25\pi\ \text{ft}^2$$

$$\frac{25\pi}{400} \approx 0.196 = 19.6\%$$

The goat has access to approximately 19.6% of the grass in the pen.

Section 3.2

39. (continued)

(b) For the goat to reach half the grass, the area of its quarter circle must be half of 400 ft^2, or 200 ft^2.

$$200 = \frac{1}{4}\pi r^2$$

$$800 = \pi r^2$$

$$\frac{800}{\pi} = r^2$$

$$16.0 \text{ ft} \approx r$$

41. Find the area of the shaded region.

$$A_{\text{shaded}} = 4 \cdot \frac{1}{4}\pi r^2$$
$$= \pi(40)^2$$
$$= 1600\pi \text{ m}^2$$

Find the area of the square field.

$$A_{\text{field}} = s^2$$
$$= 100^2$$
$$= 10{,}000 \text{ m}^2$$

Find the area of the unshaded region.

$$A_{\text{unshaded}} = A_{\text{field}} - A_{\text{shaded}}$$
$$= 10{,}000 - 1600\pi$$
$$\approx 4973.45 \text{ m}^2$$

The probability of missing the briar patches is

$$P = \frac{A_{\text{unshaded}}}{A_{\text{field}}} = \frac{10{,}000 - 1600\pi}{10{,}000} \approx 0.497.$$

43. (a) Area of the entire dart board is

$$A_{\text{total}} = \pi r^2$$
$$= \pi(8)^2$$
$$= 64\pi$$
$$\approx 201.06 \text{ in}^2.$$

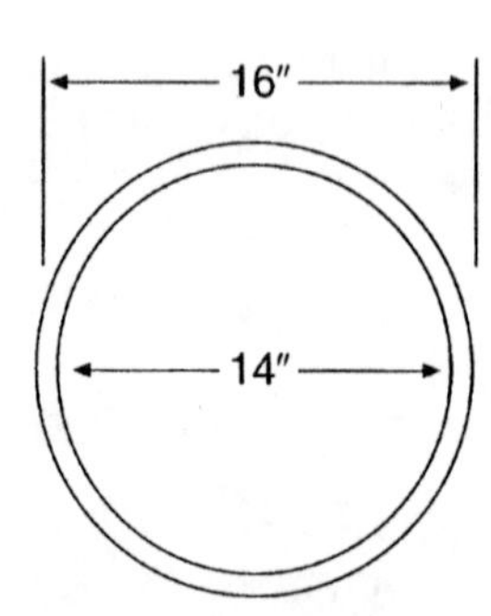

(b) Area of the no-score ring is total area subtract the area of a 14″ diameter circle.

$$A_{\text{total}} - A_{14''} = 64\pi - \pi r^2$$
$$= 64\pi - 49\pi$$
$$= 15\pi$$
$$\approx 47.12 \text{ in}^2$$

Section 3.3

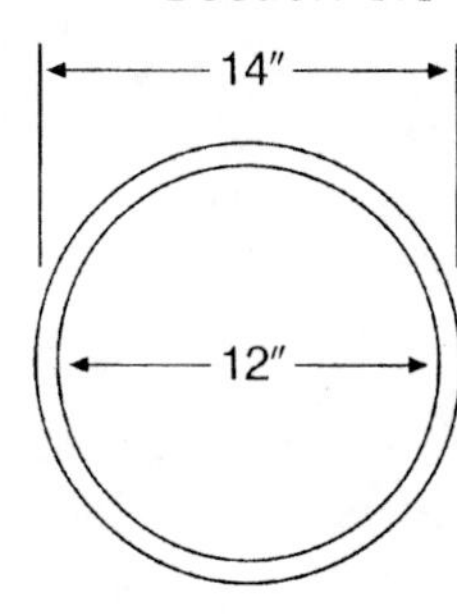

43. (continued)

(c) The outer diameter of the double-score ring is 14″ and the inner diameter is 13″.

Subtract the area of the inner circle from the outer circle.

$$A_{\text{outer}} - A_{\text{inner}} = \pi(7)^2 - \pi(6.5)^2$$
$$= 6.75\pi$$
$$\approx 21.21 \text{ in}^2$$

(d) The probability of a dart randomly landing in the no-score ring is the area of the no-score ring divided by the total area of the dart board.

$$P = \frac{15\pi}{64\pi} = \frac{15}{64} \approx 0.23$$

(e) The probability of a dart randomly hitting the double-score ring is the area of the double-score ring divided by the total area of the dart board.

$$P = \frac{6.75\pi}{64\pi} = \frac{6.75}{64} \approx 0.11$$

Section 3.3

Tips:

- ✔ You can use the Pythagorean Theorem for *right* triangles only.
- ✔ The longest side of a right triangle is always the hypotenuse c. The other two sides are a and b. It does not matter which side you call a or which side you call b.
- ✔ A key feature of a 30°–60° right triangle is that the length of the side opposite the 30° angle is always half the length of the hypotenuse.
- ✔ A main feature of the 45°–45° right triangle is that it is isosceles, i.e., the legs are equal in length.
- ✔ Whenever possible use exact values when finding the solution. This minimizes round off errors.

Solutions to odd-numbered textbook problems

1. (a) Remember, the length of the long side (or hypotenuse) is always represented by c in the Pythagorean Theorem.

$$a^2 + b^2 = c^2$$
$$16^2 + x^2 = 34^2$$
$$256 + x^2 = 1156$$
$$x^2 = 900$$
$$x = 30 \text{ in.}$$

Section 3.3

1. (continued)

(b)

$$a^2 + b^2 = c^2$$
$$4^2 + (7.5)^2 = x^2$$
$$16 + 56.25 = x^2$$
$$72.25 = x^2$$
$$8.5 \text{ cm} = x$$

(c)

$$a^2 + b^2 = c^2$$
$$9^2 + x^2 = 12^2$$
$$81 + x^2 = 144$$
$$x^2 = 63$$
$$x = \sqrt{63} \approx 7.94 \text{ m}$$

3. (a) In the triangle with legs of 7.4 cm and 8.1 cm, x is the hypotenuse. Use the Pythagorean Theorem.

$$x^2 = 7.4^2 + 8.1^2$$
$$= 54.76 + 65.61$$
$$= 120.37$$
$$x = \sqrt{120.37} \approx 11.0 \text{ cm}$$

Side y is a leg of the triangle with hypotenuse x, so,

$$4^2 + y^2 = 120.37$$
$$y^2 = 120.37 - 4^2$$
$$= 120.37 - 16$$
$$= 104.37$$
$$y = \sqrt{104.37} \approx 10.2 \text{ cm}$$

(b) Side x is the hypotenuse of the triangle with legs of length 12 in. and 20 in. Use the Pythagorean Theorem.

$$x^2 = 12^2 + 20^2$$
$$= 144 + 400$$
$$= 544$$
$$x = \sqrt{544} \approx 23.3 \text{ in.}$$

Side y is the hypotenuse of the triangle with the two legs labeled x, which is $\sqrt{544}$ in.

$$y^2 = x^2 + x^2$$
$$= 544 + 544$$
$$= 1088$$
$$y = \sqrt{1088} \approx 33.0 \text{ in.}$$

Section 3.3

3\. (continued)

Side z is a leg of the triangle with hypotenuse of 40 in. and leg y, which is $\sqrt{1088}$ in.

$$z^2 + 1088 = 40^2$$
$$z^2 = 40^2 - 1088$$
$$= 1600 - 1088$$
$$z = \sqrt{512} \approx 22.6 \text{ in.}$$

5\. (a) The legs of the right triangle are perpendicular, so we can call side a the base and side b the height. By the triangle area formula, $A = \frac{1}{2}ab$.

$$10.60 = \frac{1}{2}(2.30) \cdot b$$
$$10.60 = 1.15b$$
$$\frac{10.60}{1.15} = b$$
$$\frac{212}{23} = b$$
$$9.22 \approx b$$

Use the Pythagorean Theorem.

$$2.30^2 + \left(\frac{212}{23}\right)^2 = c^2$$
$$\sqrt{2.30^2 + \left(\frac{212}{23}\right)^2} = c$$
$$9.50 \approx c$$

(b) Use the triangle area formula.

$$A = \frac{1}{2}ab$$
$$30.5 = \frac{1}{2}a(6.1)$$
$$30.5 = 3.05a$$
$$10 = a$$

Use the Pythagorean Theorem.

$$10^2 + 6.1^2 = c^2$$
$$100 + 37.21 = c^2$$
$$\sqrt{137.21} = c$$
$$11.71 \approx c$$

Find the perimeter.

$$P = 10 + 6.1 + \sqrt{137.21}$$
$$P \approx 27.81$$

Section 3.3

5. (continued)

(c) Use the Pythagorean Theorem.

$$(\sqrt{7})^2 + (\sqrt{7})^2 = c^2$$
$$14 = c^2$$
$$\sqrt{14} = c$$
$$3.74 \approx c$$

Find the perimeter.

$$P = \sqrt{7} + \sqrt{7} + \sqrt{14}$$
$$P \approx 9.03$$

Use the triangle area formula.

$$A = \frac{1}{2}(\sqrt{7})(\sqrt{7})$$
$$A = \frac{1}{2} \cdot 7$$
$$A = 3.5$$

(d) Because we have a right triangle where the length of a leg is half the hypotenuse, we know this is a 30°–60° right triangle.

Thus, side a is $\frac{\sqrt{3}}{2}x$.

Use the triangle area formula.

$$\frac{\sqrt{3}}{2} = \frac{1}{2}\left(\frac{\sqrt{3}}{2}x\right)\left(\frac{x}{2}\right)$$
$$\frac{\sqrt{3}}{2} = \frac{\sqrt{3}}{8}x^2$$
$$4\sqrt{3} = \sqrt{3}x^2$$
$$4 = x^2$$
$$2 = x$$

Since x is 2,

$$c = 2$$
$$b = \frac{2}{2} = 1$$
$$a = \frac{\sqrt{3}}{2} \cdot 2$$
$$a = \sqrt{3} \approx 1.73.$$

Find the perimeter.

$$P = \sqrt{3} + 1 + 2$$
$$P \approx 4.73$$

Section 3.3

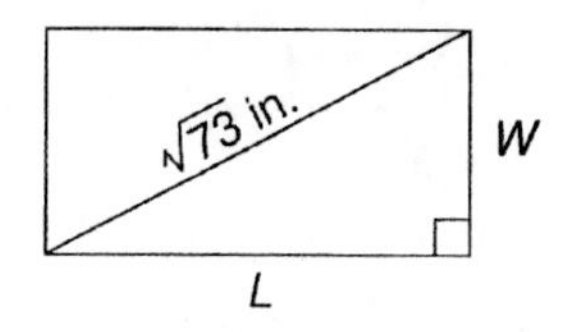

7. A diagram will help (See figure).

   You know that $A = 24 \text{ in}^2 = LW$. Because the diagonal forms a right triangle with the length and width,

$$L^2 + W^2 = (\sqrt{73})^2$$
$$L^2 + W^2 = 73$$

   Solving $LW = 24$ for W, you get $W = \dfrac{24}{L}$.

   Substituting,
$$L^2 + W^2 = 73$$
$$L^2 + \left(\frac{24}{L}\right)^2 = 73$$
$$L^2 + \frac{576}{L^2} = 73.$$

   Multiplying by L^2 yields
$$L^4 + 576 = 73L^2$$
$$L^4 - 73L^2 + 576 = 0.$$

   Factoring to solve for L,
$$(L^2 - 64)(L^2 - 9) = 0$$
$$L^2 = 64 \text{ or } L^2 = 9.$$

   Hence $L = 8$ or $L = -8$, or $L = 3$ or $L = -3$. We can eliminate $L = -8$ and $L = -3$ as they do not make sense in the context of the problem.

$$\text{So } L = 8 \text{ or } L = 3.$$

   If $L = 8$ inches, then $W = \dfrac{24}{8} = 3$ inches.

   If $L = 3$ inches, then $W = \dfrac{24}{3} = 8$ inches, so the rectangle measures 8 inches by 3 inches.

9. Using the Pythagorean Theorem, if $a^2 + b^2 \neq c^2$, then ΔABC is not right triangle

   (a) Let $a = 7, b = 24$, and $c = 25$ which is the long side of the triangle.

   $a^2 + b^2 = 7^2 + 24^2 = 625, c^2 = 25^2 = 625$

   Since $7^2 + 24^2 = 25^2$, a triangle with sides of length 7, 24, and 25 is a right triangle

   (b) Let $a = 12, b = 24$, and $c = 26$ which is the long side of the triangle.

   $a^2 + b^2 = 12^2 + 24^2 = 720, c^2 = 26^2 = 676$

   Because $12^2 + 24^2 \neq 26^2$, a triangle with side lengths 12, 24, and 26 is not a right triangle.

Section 3.3

9. (continued)

(c) Let $a = 28, b = 21$, and $c = 35$ which is the long side of the triangle.

$a^2 + b^2 = 28^2 + 21^2 = 1225, c^2 = 35^2 = 1225$

Since $28^2 + 21^2 = 35^2$, a triangle with sides of length 28, 21, and 35 is a right triangle.

11. (a) The triangle will always be a right triangle when you have $a^2 + b^2 = c^2$.

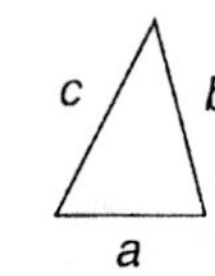

(b) The triangle will be acute when $a^2 + b^2 > c^2$ because sides a and b will have to form an angle smaller than 90° in order to allow c to be small enough that $c^2 < a^2 + b^2$. (See figure)

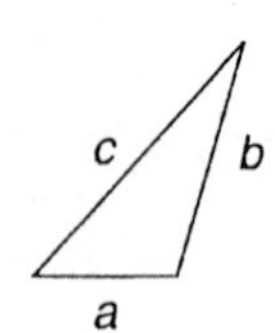

(c) The triangle will be obtuse when $a^2 + b^2 < c^2$ because sides a and b will have to form an angle larger than 90° to allow c to be large enough that $c^2 > a^2 + b^2$. (See figure)

13. (a) You have been given the side opposite the 60° angle. If s is the length of the hypotenuse, you know that

$$\frac{\sqrt{3}}{2}s = 7.$$

Thus, the length of the hypotenuse is s and

$$s = 7 \cdot \frac{2}{\sqrt{3}} = \frac{14}{\sqrt{3}} = \frac{14\sqrt{3}}{3} \text{ cm} \approx 8.08 \text{ cm}.$$

The length of the side opposite the 30° angle $= \frac{1}{2}s = \frac{7\sqrt{3}}{3}$ cm $\approx$ 4.04 cm.

(b) You have been given the length of the hypotenuse. Use the hypotenuse to find the length of each leg, s.

$$s\sqrt{2} = 12$$

$$s = \frac{12}{\sqrt{2}} = \frac{12\sqrt{2}}{2} = 6\sqrt{2} \text{ in.} \approx 8.49 \text{ in.}$$

15. The height, h, is the leg of a 30°–60° right triangle. Since h is opposite the 60° angle,

$$h = \frac{\sqrt{3}}{2}(17.4)$$

$$h = 8.7\sqrt{3} \approx 15.07 \text{ cm}.$$

The short leg of the 30°–60° right triangle is $\frac{17.4}{2} = 8.7$ cm. Thus, the measure of $\overline{AD}$ is $8.7 + 31.8 = 40.5$ cm.

Section 3.3

15. (continued)

Use the trapezoid area formula.

$$A = \frac{1}{2}h\,(b_1 + b_2)$$
$$= \frac{1}{2}(8.7\sqrt{3})(24.5 + 40.5)$$
$$\approx 489.7 \text{ cm}^2$$

17. It helps to draw a picture.

h is opposite the 30° angle in a right triangle, so

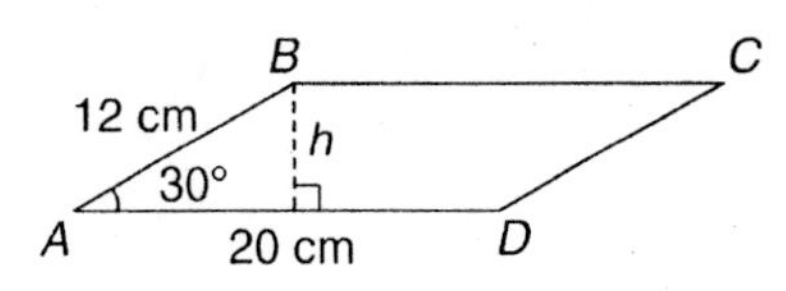

$$h = \frac{1}{2} \cdot \text{hypotenuse}$$
$$h = \frac{1}{2} \cdot 12 = 6 \text{ cm.}$$

The area of the parallelogram is $A = bh = 20 \cdot 6 = 120 \text{ cm}^2$.

19. Draw a picture of the rhombus. The height h forms a leg of a 45°–45° right triangle, and the hypotenuse is 9.4 m, so

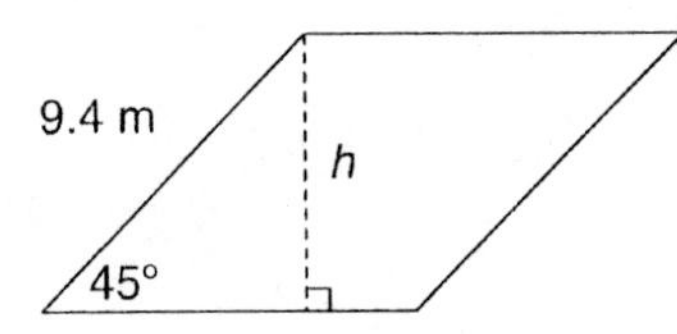

$$9.4 = h\sqrt{2}$$
$$\frac{9.4}{\sqrt{2}} = h$$
$$4.7\sqrt{2} \text{ m} = h.$$

The area of the rhombus is

$$A = bh$$
$$A = 9.4(4.7\sqrt{2})$$
$$A = 44.18\sqrt{2}$$
$$A \approx 62.48 \text{ m}^2.$$

21. The vertex angle measure in a regular hexagon is $\frac{(6-2)180°}{2} = 120°$.

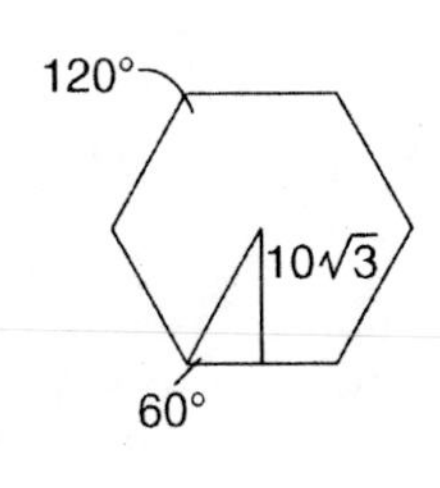

Draw a picture of the hexagon. A line drawn from the center to the vertex cuts the vertex angle in half, forming a 30°–60° right triangle using Theorem 3.9 on this triangle, as shown.

30°
s
10√3
60°
$\frac{s}{2}$

$$10\sqrt{3} = \frac{\sqrt{3}}{2}s$$
$$20\sqrt{3} = \sqrt{3}\,s$$
$$20 = s \quad \text{and} \quad \frac{s}{2} = 10 \text{ cm}$$

Section 3.3

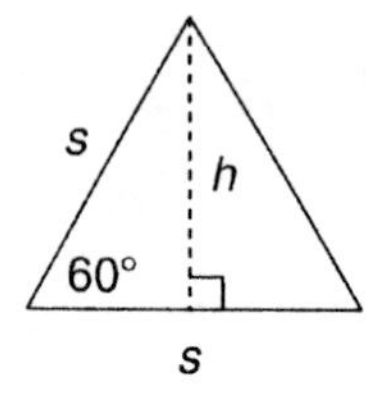

21. (continued)

The area of this triangle is

$$\frac{1}{2}\cdot 10(10\sqrt{3}) \approx 50\sqrt{3}\ \text{cm}^2.$$

There are 12 such triangles in the hexagon, so the area of the regular hexagon is $12(50\sqrt{3}) = 600\sqrt{3} \approx 1039.2\ \text{cm}^2$.

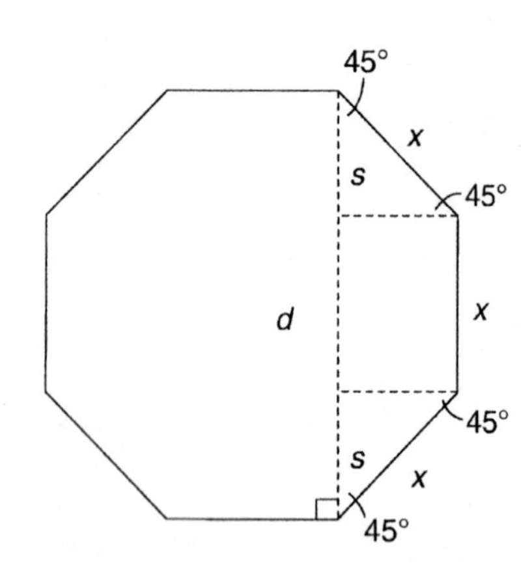

23. From Theorem 3.9,

$$h = \frac{\sqrt{3}}{2}s \quad \text{so} \quad \frac{2h}{\sqrt{3}} = s.$$

Use the triangle area formula.

$$A = \frac{1}{2}\left(\frac{2h}{\sqrt{3}}\right)\cdot h$$

$$A = \frac{h^2}{\sqrt{3}} = \frac{h^2\sqrt{3}}{3}$$

25. The vertex angle measure in a regular octagon is

$$\frac{(8-2)\cdot 180°}{2} = 135°.$$

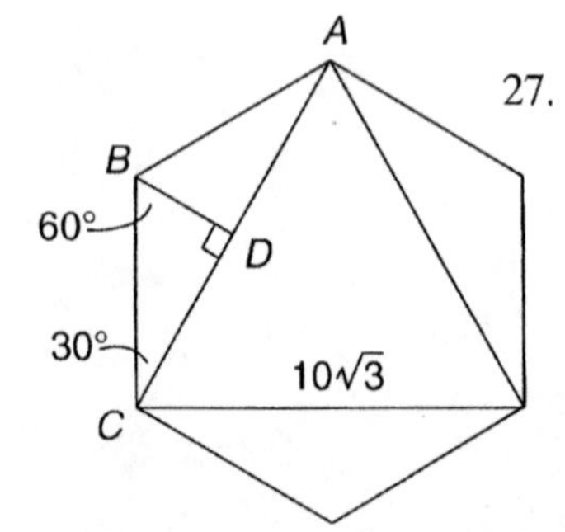

The distance, d, forms two $45° - 45° - 90°$ triangles and one rectangle as shown. Use Theorem 3.10.

$$x = s\sqrt{2}$$

$$\frac{x}{\sqrt{2}} = s$$

$$\frac{x\sqrt{2}}{2} = s$$

So, the distance, d, is the following sum.

$$\begin{aligned} d &= s + x + s \\ &= x + 2s \\ &= x + 2\left(\frac{x\sqrt{2}}{2}\right) \\ &= x + x\sqrt{2} \end{aligned}$$

27. The vertex angle measure in a regular hexagon is $\dfrac{(6-2)\cdot 180°}{2} = 120°$.

So, $\angle ABC = 120°$.

ΔABC is isosceles. We draw $\overline{BD}$ perpendicular to $\overline{AC}$. So, $\overline{BD}$ bisects (cuts in half) $\overline{AC}$. Hence $DC = 5\sqrt{3}$ cm.

$\overline{BD}$ also bisects $\angle B$ to form $\angle DBC$ which is 60°.

Thus ΔBCD is a $30° - 60°–90°$ triangle where leg $DC = 5\sqrt{3}$.

Section 3.3

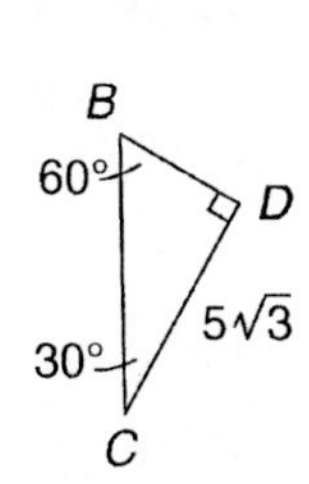

21. (continued)

Use Theorem 3.9.

$$5\sqrt{3} = \frac{\sqrt{3}}{2} \cdot BC$$
$$10\sqrt{3} = \sqrt{3} \cdot BC$$
$$10 \text{ cm} = BC$$

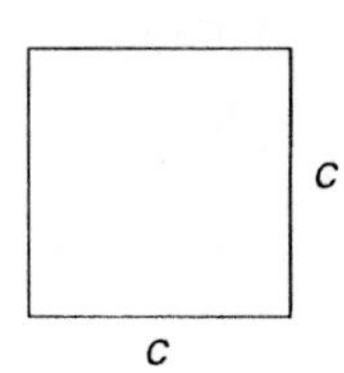

29. The area of the large square is

$$\text{Area} = c^2.$$

The area of a right triangle is

$$\text{Area} = \frac{1}{2}ab.$$

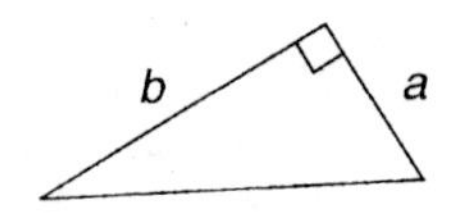

There are four of these triangles for an area of

$$\text{Area} = 4\left(\frac{1}{2}ab\right)$$
$$= 2ab.$$

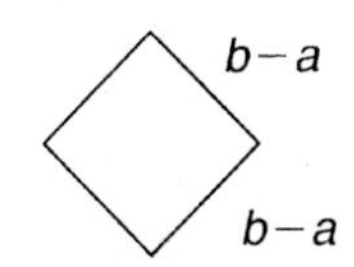

The area of the small square is

$$\text{Area} = (b - a)^2$$
$$= b^2 - 2ab + a^2.$$

The area of the large square is equal to the sum of the areas of the four right triangles and the small square.

$$c^2 = 2ab + b^2 - 2ab + a^2$$
$$c^2 = a^2 + b^2$$

31. Using the Pythagorean Theorem, let $x =$ the side length of the square in inches.

$$x^2 + x^2 = 6.5^2$$
$$2x^2 = 42.25$$
$$x^2 = 21.125$$
$$x = \sqrt{21.125}$$
$$x \approx 4.6 \text{ in.}$$

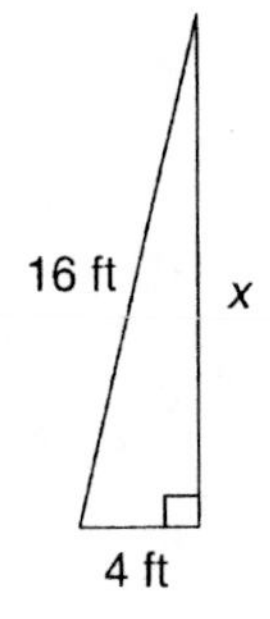

33. Using the Pythagorean Theorem, let $x =$ height of the house from the ground to the top of the ladder in feet.

$$x^2 + 4^2 = 16^2$$
$$x^2 + 16 = 256$$
$$x^2 = 240$$
$$x = \sqrt{240}$$
$$= 4\sqrt{15} \approx 15.5 \text{ ft}$$

Section 3.3

35. Let x = the length of the escalator in feet. The triangle formed is a 30°–60°–90° triangle. Use Theorem 3.9.

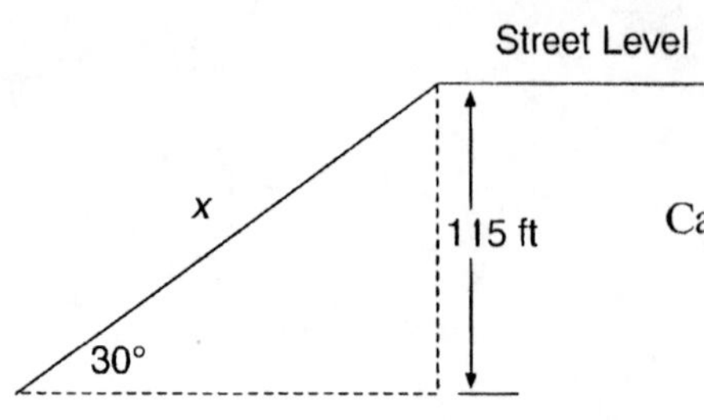

$$\frac{x}{2} = 115$$

$$x = 230 \text{ ft}$$

Calculate the rate in miles per hour.

$$\frac{230 \cancel{\text{ft}}}{3 \cancel{\text{min}}} \cdot \frac{60 \cancel{\text{min}}}{1 \text{ hr}} \cdot \frac{1 \text{ mi}}{5280 \cancel{\text{ft}}} = \frac{230(60)(1)\text{mi}}{3(1)(5280)\text{ hr}} \approx 0.87 \frac{\text{mi}}{\text{hr}}$$

37. Let A = altitude of plane in feet. The triangle formed is a 30°–60°–90° triangle. Use Theorem 3.9.

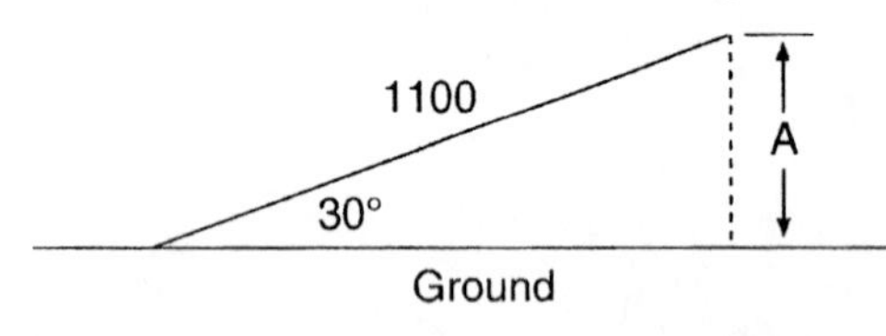

$$A = \frac{1100}{2}$$

$$A = 550 \text{ ft}$$

39. Assuming the yard is rectangular, the hose must reach the opposite corner of the yard. Let h = the length of the hose in feet.

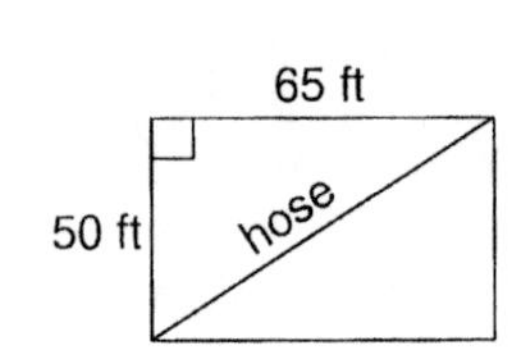

$$h^2 = 50^2 + 65^2$$
$$= 6725$$
$$h = \sqrt{6725}$$
$$= \sqrt{25 \cdot 269}$$
$$= 5\sqrt{269}$$
$$h \approx 82 \text{ ft}$$

41. Use the Pythagorean Theorem.

$$PR^2 = 83^2 + 170^2$$
$$PR^2 = 35{,}789$$
$$PR = \sqrt{35{,}789}$$
$$PR \approx 189 \text{ ft}$$

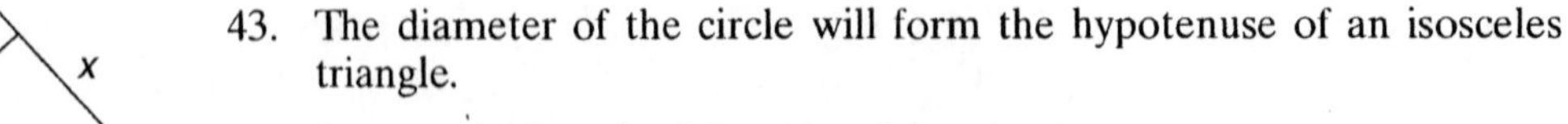

43. The diameter of the circle will form the hypotenuse of an isosceles triangle.

x x

3.16 cm

Let x = the length of the side of the plug in cm.

$$x^2 + x^2 = 3.16^2$$
$$2x^2 = 3.16^2$$
$$x^2 = \frac{3.16^2}{2}$$
$$x^2 = 4.9928$$
$$x = \sqrt{4.9928}$$
$$x \approx 2.2 \text{ cm}$$

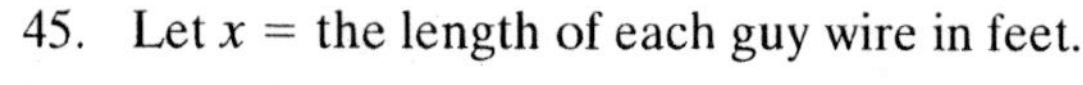

Section 3.3

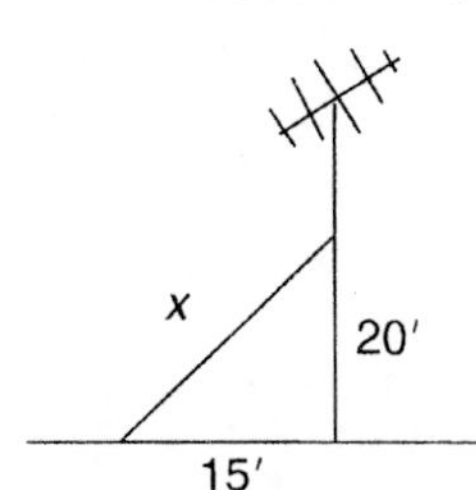

45. Let x = the length of each guy wire in feet.

$$x^2 = 15^2 + 20^2$$
$$= 625$$
$$x = 25 \text{ ft}$$

The wire used for the four guy wires plus fastening is $4 \cdot 25 + 20 = 120$ ft.

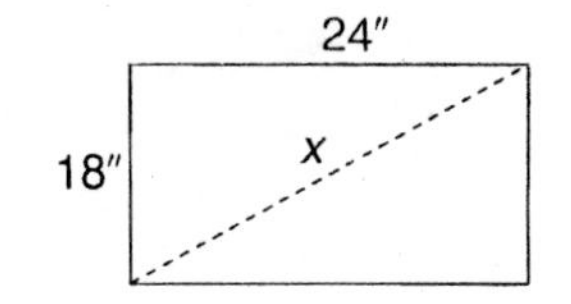

47. The umbrella should be placed with one end in a *bottom* corner and the other end in the *top* corner diagonally across from it. The umbrella will lie along the hypotenuse of a right triangle formed by the height of the box and the diagonal of the bottom of the box. To see if the umbrella will fit you must first find the length of the bottom diagonal.

$$a^2 + b^2 = x^2$$
$$18^2 + 24^2 = x^2$$
$$900 = x^2$$
$$30 \text{ in.} = x$$

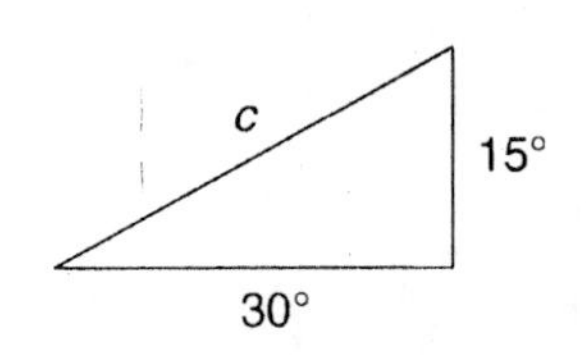

Using this diagonal now as the base for the right triangle formed by the umbrella and the height of the box, you can find the hypotenuse of the new triangle to see if it is as long as the umbrella.

$$30^2 + 15^2 = c^2$$
$$1125 = c^2$$
$$15\sqrt{5} = c$$
$$33.54 \text{ in.} \approx c$$

Yes, the 32-inch umbrella will fit in the packing box.

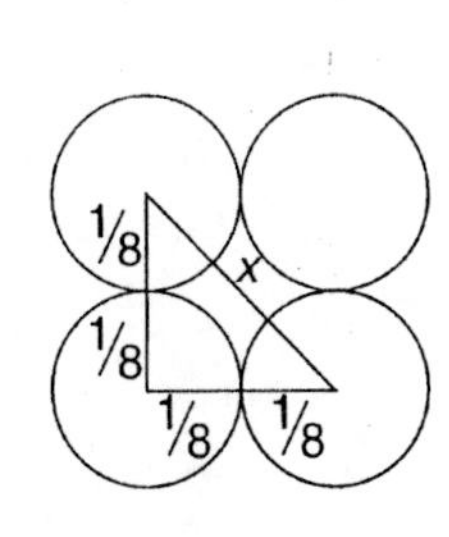

49. The radius of each wire is $\frac{1}{8}$ in., so the center-to-center distance between two adjacent wires is $\frac{1}{4}$ in. Let x be the distance in inches between the centers of two nonadjacent wires. Then x is the hypotenuse of a right triangle with legs each of length $\frac{1}{4}$ in. Solve for x.

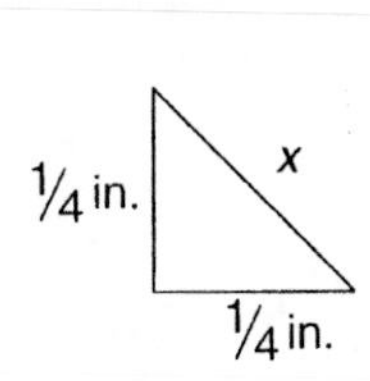

$$x^2 = \left(\frac{1}{4}\right)^2 + \left(\frac{1}{4}\right)^2$$
$$= \frac{1}{16} + \frac{1}{16}$$
$$= \frac{1}{8}$$
$$x = \sqrt{\frac{1}{8}} \text{ in.}$$

49. (continued)

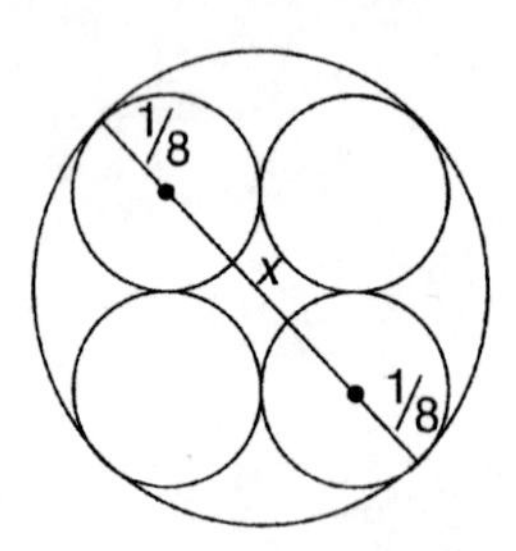

The diameter of the conduit is

$$d = \frac{1}{8} + x + \frac{1}{8}$$
$$= \frac{1}{8} + \sqrt{\frac{1}{8}} + \frac{1}{8}$$
$$= \frac{1}{4} + \sqrt{\frac{1}{8}}$$
$$d \approx 0.60 \text{ in.}$$

51. Let x = the diagonal of the square in inches.

x
6 in.
6 in.

$$x^2 = 6^2 + 6^2$$
$$= \sqrt{72}$$
$$x = 6\sqrt{2}$$
$$x \approx 8.4853 \text{ in.}$$

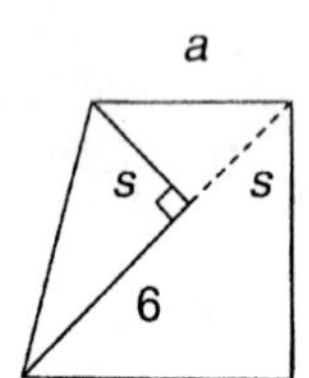

The second fold forms an isosceles triangle with equal sides, s, as shown. The length of side s is:

$$s = (6\sqrt{2} - 6) \text{ in.}$$

Use the Pythagorean Theorem to find the length of side a.

$$a^2 = s^2 + s^2$$
$$= 2(s)^2$$
$$= 2(6\sqrt{2} - 6)^2$$
$$a = 12 - 6\sqrt{2}$$
$$a \approx 3.5147 \text{ in.}$$

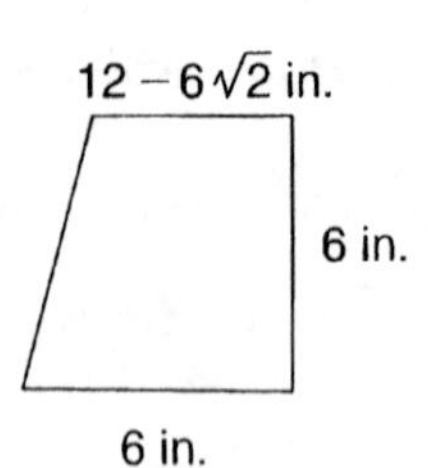

Find the area of the trapezoid.

$$A = \frac{1}{2}(a + b)h$$
$$A = \frac{1}{2}(12 - 6\sqrt{2} + 6)6$$
$$= \frac{1}{2}(18 - 6\sqrt{2})6$$
$$= 3(18 - 6\sqrt{2})$$
$$= 54 - 18\sqrt{2}$$
$$A \approx 28.54 \text{ in}^2$$

Section 3.4

Section 3.4

Tips:

✔ Surface area is expressed in square units of measure.

✔ In the equation for surface area of a prism, A is the area of the base and P is the perimeter of the base. h is the height of the prism. That is, h is the perpendicular distance between the bases.

✔ For pyramids and cones, l is the slant height and h is the perpendicular distance from the apex to the base.

✔ For a pyramid, you measure l as the height of a triangular face, not the length of an edge.

Solutions to odd-numbered textbook problems

1. (a) The base of this rectangular prism is the 10-in. by 15-in. rectangle on the bottom.

$$SA = 2A + Ph$$

$$A = 10 \cdot 15 = 150 \text{ in}^2$$

$$P = 2 \cdot 10 + 2 \cdot 15 = 50 \text{ in.}$$

$$h = 10 \text{ in.}$$

$$SA = 2(150) + (50)(10)$$

$$SA = 800 \text{ in}^2$$

(b) This is a right trapezoidal prism. Again you have the height of the base (4 cm) used to find the area of the base and the height of the prism (10 cm) used to find the surface area of the entire prism.

$$A = \frac{1}{2} \cdot 4(8 + 2) = 20 \text{ cm}^2$$

$$P = 8 + 2 + 5 + 5 = 20 \text{ cm}$$

$$SA = 2A + Ph$$

$$= 2(20) + 20 \cdot 10$$

$$= 240 \text{ cm}^2$$

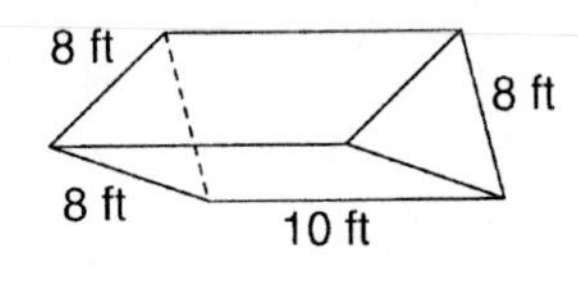

3. (a) This is a right triangular prism. The equilateral triangular base has a height of

$$h^2 = 8^2 - 4^2$$

$$h = \sqrt{48} = 4\sqrt{3} \text{ ft.}$$

The area of one base is

$$A = \frac{1}{2}(8)(4\sqrt{3}) = 16\sqrt{3} \text{ ft}^2.$$

Section 3.4

3. (continued)

The perimeter of the base is

$$P = 3 \cdot 8 = 24 \text{ ft.}$$

Now find surface area.

$$\begin{aligned} SA &= 2A + Ph \\ &= 2(16\sqrt{3}) + 24(10) \\ &= 32\sqrt{3} + 240 \\ SA &\approx 295 \text{ ft}^2 \end{aligned}$$

(b) The base is a trapezoid.

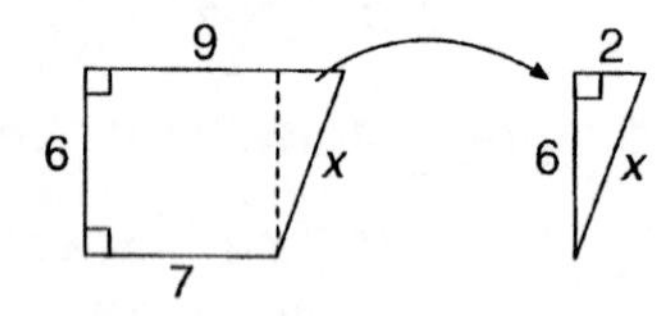

Find dimension x.

$$x^2 = 6^2 + 2^2$$

$$x = \sqrt{40} = 2\sqrt{10} \text{ cm}$$

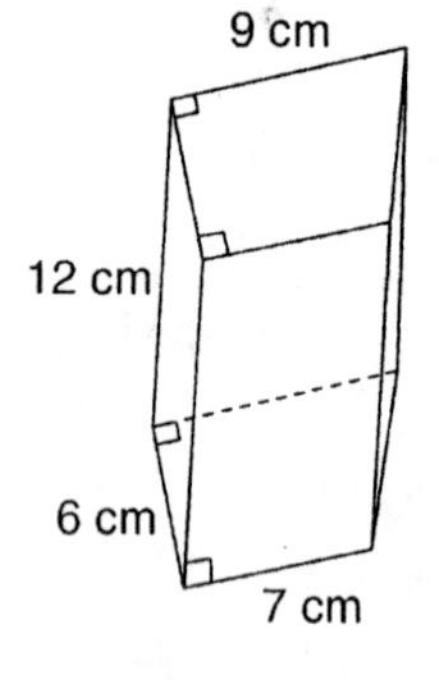

The base has area of

$$A = \frac{1}{2} \cdot 6(9 + 7) = 48 \text{ cm}^2.$$

The perimeter has the base of

$$\begin{aligned} P &= 9 + 6 + 7 + 2\sqrt{10} \\ &= 22 + 2\sqrt{10} \\ &\approx 28.32 \text{ cm.} \end{aligned}$$

The height of the prism is $h = 12$ cm.

$$\begin{aligned} SA &= 2A + Ph \\ &= 2(48) + 22 + 2\sqrt{10} \\ &= 96 + 264 + 24\sqrt{10} \\ &= 360 + 24\sqrt{10} \\ SA &\approx 436 \text{ cm}^2 \end{aligned}$$

5. Area of the top $= LW = 96 \text{ in}^2$

Area of the side $= Lh = 72 \text{ in}^2$

Area of the end $= Wh = 48 \text{ in}^2$

Solve $LW = 96$ for W.

$$W = \frac{96}{L}$$

Section 3.4

5. (continued)

Solve $Lh = 72$ for h.

$$h = \frac{72}{L}$$

Substitute for W and h in $Wh = 48$.

$$\frac{96}{L} \cdot \frac{72}{L} = 48$$

$$\frac{6912}{L^2} = 48$$

$$6912 = 48L^2$$

$$144 = L^2$$

$$12 \text{ in.} = L$$

$$W = \frac{96}{L} = \frac{96}{12} = 8 \text{ in.}$$

$$h = \frac{72}{L} = \frac{72}{12} = 6 \text{ in.}$$

The box is 12 inches by 8 inches by 6 inches high.

7. (a) This right regular pyramid has a square base.

$$A = 10 \cdot 10 = 100 \text{ ft}^2$$

$$P = 4 \cdot 10 = 40 \text{ ft}$$

$$l = 13 \text{ ft}$$

$$SA = 100 + \frac{1}{2}(40)(13) = 360 \text{ ft}^2$$

(b) This right regular pyramid has a square base.

$$A = 14 \cdot 14 = 196 \text{ cm}^2$$

$$P = 4 \cdot 14 = 56 \text{ cm}$$

$$l = 25 \text{ cm}$$

$$SA = 196 + \frac{1}{2}(56)(25) = 896 \text{ cm}^2$$

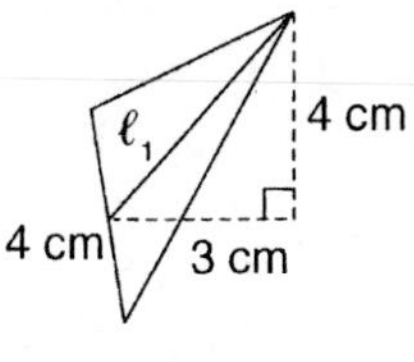

9. Since this right pyramid is not regular, we must find the areas of the triangular sides. To do that, we first need the slant height.

$$l_1^2 = 4^2 + 3^2$$

$$l_1 = 5 \text{ cm}$$

The area of the triangular side with a 4 cm base is

$$S_1 = \frac{1}{2} \cdot 4(5) = 10 \text{ cm}^2.$$

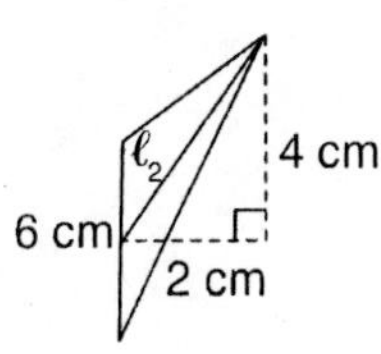

Now find slant height l_2.

$$l_2^2 = 2^2 + 4^2$$

$$l_2 = \sqrt{20} = 2\sqrt{5} \text{ cm}$$

Section 3.4

9. (continued)

The area of the triangular side with a 6 cm base is

$$S_2 = \frac{1}{2} \cdot 6(2\sqrt{5}) = 6\sqrt{5}\ \text{cm}^2.$$

Find the area of the rectangular base of the pyramid.

$$A = 4 \cdot 6 = 24\ \text{cm}^2$$

Now find the total surface area.

$$\begin{aligned} SA &= 2\,S_1 + 2\,S_2 + A \\ &= 2(10) + 2(6\sqrt{5}) + 24 \\ &= 44 + 12\sqrt{5} \\ SA &\approx 71\ \text{cm}^2 \end{aligned}$$

11. The hexagonal base can be divided into six equilateral triangles. A line of symmetry drawn through a vertex forms two 30°–60° right triangles. The line of symmetry cuts the base of the triangle in half. Use the 30°–60° right triangle relationship to find h.

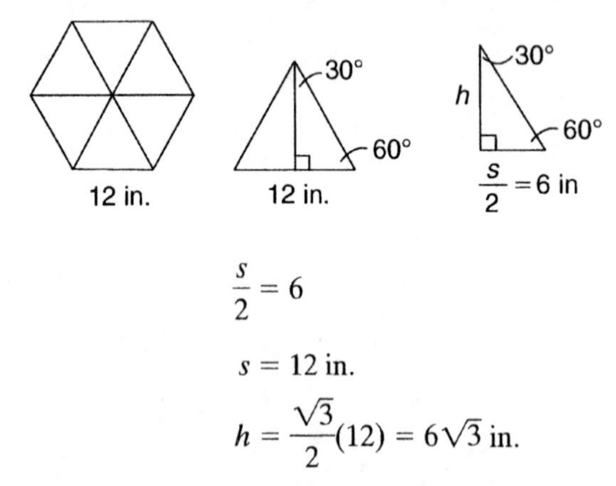

$$\frac{s}{2} = 6$$

$$s = 12\ \text{in.}$$

$$h = \frac{\sqrt{3}}{2}(12) = 6\sqrt{3}\ \text{in.}$$

Find the area of the hexagonal base where $P = 6 \cdot 12 = 72$ in. and $h = 6\sqrt{3}$ in.

$$A = \frac{1}{2}Ph$$

$$\begin{aligned} A &= \frac{1}{2}(72)(6\sqrt{3}) \\ &= 216\sqrt{3}\ \text{in}^2 \end{aligned}$$

Find the slant height of the pyramid.

ℓ

14 cm

$6\sqrt{3}$ in.

$$l = 14^2 + (6\sqrt{3})^2$$

$$l = \sqrt{304}$$

$$l = 4\sqrt{19}\ \text{in.}$$

Section 3.4

11. (continued)

Find total surface area.

$$SA = A + \frac{1}{2}Pl$$

$$SA = 216\sqrt{3} + \frac{1}{2}(72)(4\sqrt{19})$$

$$= 216\sqrt{3} + 144\sqrt{19}$$

$$SA \approx 1002 \text{ in}^2$$

13. (a) $SA = 2\pi r^2 + 2\pi rh$

$SA = 2\pi(7.6)^2 + 2\pi(7.6)(16.3)$

$SA \approx 1141 \text{ cm}^2$

(b) $SA = 2\pi(3.3)^2 + 2\pi(3.3)(10)$

$SA \approx 276 \text{ cm}^2$

15. $112\pi = 2\pi r^2 + 2\pi r(10)$

$112\pi = 2\pi(r^2 + 10r)$

$56 = r^2 + 10r$

$0 = r^2 + 10r - 56$ We could solve using the quadratic formula or by factoring.

$$0 = (r + 14)(r - 4)$$

$$r = \cancel{-14}, 4$$

We eliminate $r = -14$ as it does not make sense to have a radius with negative value.

So, $d = 2r = 8$ units.

17. (a) $SA = \pi r(r + l)$

$SA = \pi(12)(12 + 20)$

$SA \approx 1206 \text{ in}^2$

(b) Use the Pythagorean Theorem to find the slant height.

$$l^2 = 12^2 + 15^2$$

$$= 369$$

$$l = \sqrt{369}$$

$$l = 3\sqrt{41}$$

$SA = \pi(15)(15 + 3\sqrt{41})$

$= 225\pi + 45\pi\sqrt{41}$

$SA \approx 1612 \text{ m}^2$

Section 3.4

19. (a)
$$SA = 4\pi r^2$$
$$= 4\pi(6)^2$$
$$= 144\pi$$
$$SA \approx 452 \text{ in}^2$$

(b)
$$r = \frac{1}{2}d$$
$$r = \frac{1}{2}(24) = 12 \text{ m}$$
$$SA = 4\pi(12)^2$$
$$= 576\pi$$
$$SA \approx 1810 \text{ m}^2$$

(c)
$$C = 2\pi r$$
$$C = 7\pi$$
so, $2\pi r = 7\pi$
$$r = \frac{7}{2}.$$
$$SA = 4\pi\left(\frac{7}{2}\right)^2$$
$$= 49\pi$$
$$SA \approx 154 \text{ cm}^2$$

21.
$$SA = 4\pi r^2$$
$$215.8 = 4\pi r^2$$
$$\frac{215.8}{4\pi} = r^2$$
$$\frac{53.95}{\pi} = r^2$$
$$\sqrt{\frac{53.95}{\pi}} = r$$
$$4.144 \text{ cm} \approx r$$

So $r \approx 41.4$ mm.

Hence $d = 2r \approx 2(41.4)$
$= 82.8$ mm.

Thus, $d \approx 83$ mm.

23. If $r = \frac{1}{2}d = 4$ ft

then $SA = 4\pi(4)^2$
$SA = 64\pi$ ft.

Section 3.4

23. (continued)

Doubling the surface area, $2(64\pi) = 128\pi$.

Now let x = increase in radius in feet.

Then $4 + x$ is the new radius.

$$\begin{aligned} 128\pi &= 4\pi(4 + x)^2 \\ 32 &= (4 + x)^2 \\ \sqrt{32} &= 4 + x \\ \sqrt{32} - 4 &= x \\ 1.66 \text{ ft} &\approx x \end{aligned}$$

The radius should be increased by approximately 1.66 ft

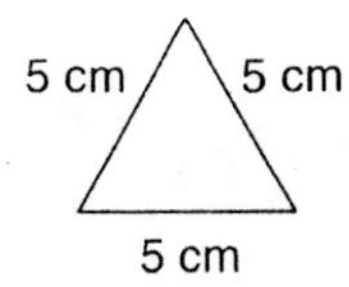

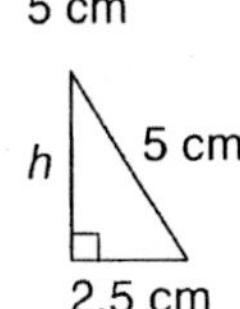

25. A regular tetrahedron has four equilateral triangle faces. To find the area of one face we divide the triangle into two 30°–60° right triangles.

Using the 30°–60° right triangle relationship: $h = \frac{\sqrt{3}}{2} \cdot 5 = \frac{5\sqrt{3}}{2}$ cm.

Find the area of one face.

$$\begin{aligned} A &= \left(\frac{1}{2}\right)(2.5)\left(\frac{5\sqrt{3}}{2}\right) \\ &= 6.25\sqrt{3} \end{aligned}$$

So,

$$\begin{aligned} SA &= 4(6.25\sqrt{3}) \\ &= 25\sqrt{3} \\ &\approx 43 \text{ cm}^2. \end{aligned}$$

27. (a) Find the areas of each face.

Top face:

$$A_T = 2 \cdot 4 + 5 \cdot 4$$
$$A_T = 2 \text{ cm}^2$$

Front face:

$$A_F = 2 \cdot 4 + 4 \cdot 4$$
$$A_F = 24 \text{ cm}^2$$

Back face:

$$A_b = 6 \cdot 4$$
$$A_b = 24 \text{ cm}^2$$

Bottom face:

$$A_{bot} = 7 \cdot 4$$
$$A_{bot} = 28 \text{ cm}^2$$

Section 3.4 27. (continued)

Left face:

$$A_l = 7 \cdot 4 + 2 \cdot 2$$
$$A_l = 32 \text{ cm}^2$$

Right face:

$$A_r = 7 \cdot 4 + 2 \cdot 2$$
$$= 32 \text{ cm}^2$$

Thus, $SA = A_t = A_f + A_b + A_{bot} + A_l + A_r$

$$= 28 + 24 + 24 + 28 + 32 + 32$$
$$SA = 168 \text{ cm}^2.$$

(b) The radius of the circular base is

$$r = \frac{1}{2}d$$
$$r = 4.5 \text{ in.}$$

Find the surface area of the cylinder and subtract the area of one circular end.

$$A_{cyl} = 2\pi r^2 + 2\pi rh - \pi r^2$$
$$= \pi r^2 + 2\pi rh$$
$$= \pi(4.5)^2 + 2\pi(4.5)(8)$$
$$= 20.25\pi + 72\pi$$
$$= 92.25\pi$$

To find the area of the cone we first need to find the slant height.

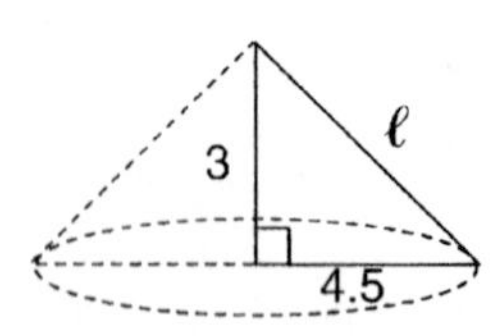

$$l^2 = 3^2 + 4.5^2$$
$$= 9 + 20.25$$
$$= 29.25$$
$$l = \sqrt{29.25}$$
$$l \approx 5.41 \text{ in.}$$

Find the lateral surface area of the cone and subtract the area of one circular end.

$$A_{\text{lateral } SA \text{ of cone}} = \pi rl$$
$$= \pi(4.5)\sqrt{29.25}$$

Find total surface area.

$$SA = A_{\text{cyl}} + A_{\text{lateral } SA \text{ of cone}}$$
$$= 92.25\pi + 4.5\sqrt{29.25}\pi$$
$$\approx 366 \text{ in}^2$$

Section 3.4

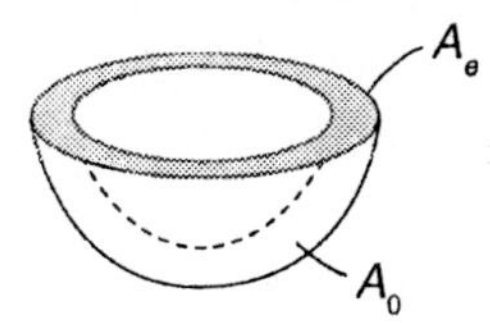

29. There are three surfaces to consider here: the inner hemisphere, outer hemisphere, and the edge of the hemisphere. Find the area of the inner hemisphere.

$$r = \frac{1}{2}d$$
$$r = \frac{1}{2}(1.86)$$
$$r = 0.93 \text{ cm}$$

$$\begin{aligned} A_i &= \frac{1}{2} \cdot 4\pi(0.93)^2 \\ &= 1.7298\pi \\ A_i &\approx 5.43 \text{ cm}^2 \end{aligned}$$

Find the area of the outer hemisphere.

$$\begin{aligned} A_o &= \frac{1}{2} \cdot 4\pi(1.25)^2 \\ &= 3.125\pi \\ A_o &\approx 9.82 \text{ cm}^2 \end{aligned}$$

Find the area of the edge of the hemisphere.

$$\begin{aligned} A_e &= \pi r_1^2 - \pi r_2^2 \quad \text{where } r_1 = 1.25, r_2 = 0.93 \\ &= \pi(1.25)^2 - \pi(0.93)^2 \\ &= \pi(1.5625 - 0.8649) \\ &= 0.6976\pi \\ A_e &\approx 2.19 \text{ cm}^2 \end{aligned}$$

Find the surface area.

$$\begin{aligned} SA &= A_i + A_o + A_e \\ SA &= 1.7298\pi + 3.125\pi + 0.6976\pi \\ &= 5.5524\pi \\ SA &\approx 17.44 \text{ cm}^2 \end{aligned}$$

31. (a)

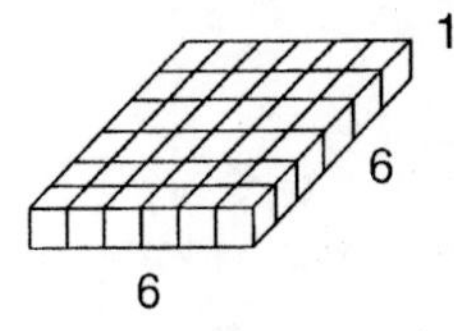

$$\begin{aligned} SA &= 2 \cdot (6 \cdot 6) + 2 \cdot (1 \cdot 6) + 2 \cdot (1 \cdot 6) \\ &= 72 + 12 + 12 \\ SA &= 96 \text{ units}^2 \end{aligned}$$

(b)

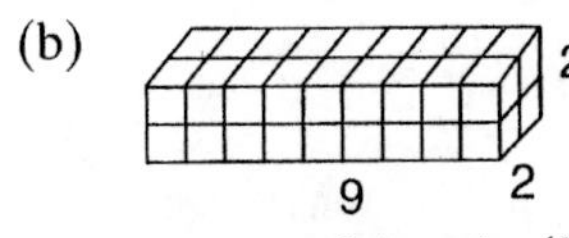

$$\begin{aligned} SA &= 4 \cdot (2 \cdot 9) \cdot 2 \cdot (2 \cdot 2) \\ &= 72 + 8 \\ SA &= 80 \text{ units}^2 \end{aligned}$$

Section 3.4

21. (continued)

(c)

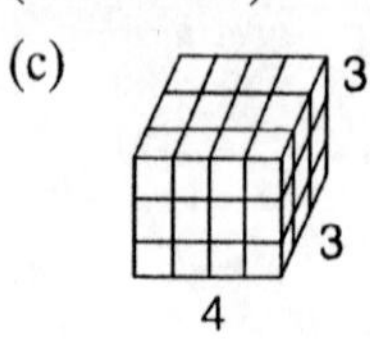

$$SA = 4 \cdot (3 \cdot 4) + 2 \cdot (3 \cdot 3)$$
$$= 48 + 18$$
$$SA = 66 \text{ units}^2$$

The smallest possible total surface area is 66 square units.

(d)

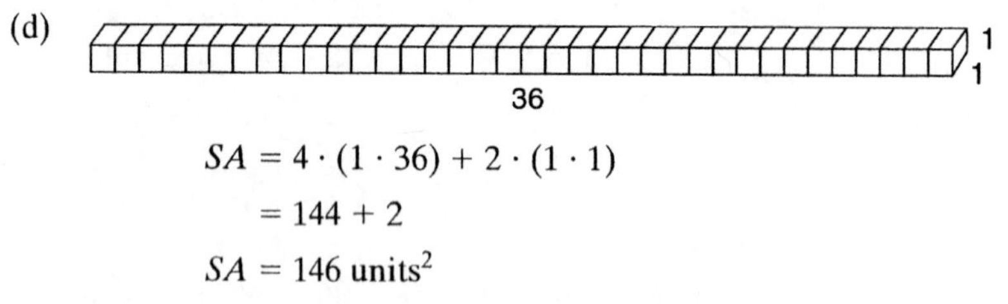

$$SA = 4 \cdot (1 \cdot 36) + 2 \cdot (1 \cdot 1)$$
$$= 144 + 2$$
$$SA = 146 \text{ units}^2$$

The largest possible total surface area is 146 square units.

33. One approach is to convert measurements to feet

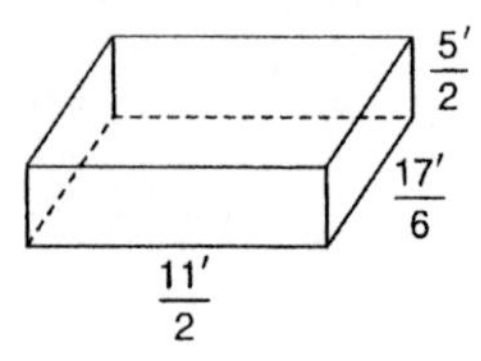

$$5'6'' = 5\frac{1}{2} \text{ ft} = \frac{11}{2} \text{ ft}$$

$$2'6'' = 2\frac{5}{6} \text{ ft} = \frac{17}{6} \text{ ft}$$

$$1'8'' = 1\frac{2}{3} \text{ ft} = \frac{5}{3} \text{ ft}$$

Now find surface area.

$$SA = 2\left(\frac{11}{2}\right)\left(\frac{5}{3}\right) + 2\left(\frac{17}{6}\right)\left(\frac{5}{3}\right) + \frac{11}{2}\left(\frac{17}{6}\right)$$
$$= \frac{55}{3} + \frac{85}{9} + \frac{187}{12}$$
$$= \frac{660 + 340 + 561}{36}$$
$$= \frac{1561}{36}$$
$$= 43\frac{13}{36} \approx 43.4 \text{ ft}^2$$

Another approach is to convert measurements to inches.

$5'6'' = 66''$, $2'10'' = 34''$, $1'8'' = 20''$

$$SA = 2(66)(20) + 2(34)(20) + 66(34)$$
$$= 6244 \text{ in}^2$$

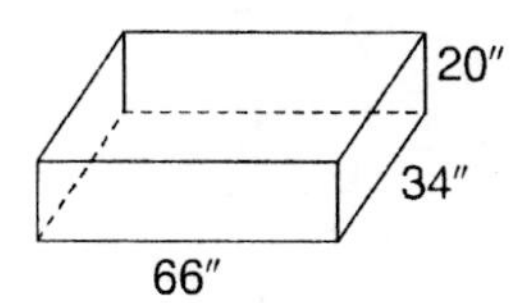

33. (continued)

Now convert in^2 to ft^2.

$$\frac{6244\ \cancel{\text{in}^2}}{1} \times \frac{1\ \text{ft}^2}{144\ \cancel{\text{in}^2}} \approx 43.44\ \text{ft}^2$$

35. The greenhouse is a half-cylinder with radius, $r = \frac{1}{2}(24) = 12$ ft and height, $h = 100$ ft.

$$\begin{aligned} SA &= \frac{1}{2}(2\pi r^2 + 2\pi rh) \\ &= \frac{1}{2}(2\pi \cdot 12^2 + 2\pi \cdot 12 \cdot 100) \\ &= 1344\pi \\ SA &\approx 4222.3\ \text{ft}^2 \end{aligned}$$

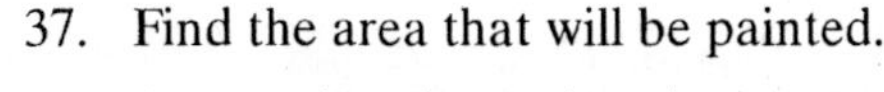

37. Find the area that will be painted.

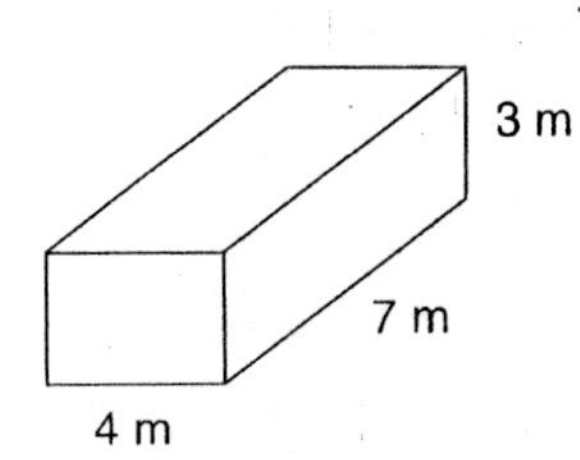

$SA = 2\,(3 \cdot 7) + 2\,(3 \cdot 4) + 4 \cdot 7$

$SA = 42 + 24 + 28$

$SA = 94\ \text{m}^2$

Determine the number of liters of paint required.

$$\frac{SA}{\text{coverage per liter}} = \frac{94}{20} = 4.7\ \text{liters}$$

39. The surface area will be the surface area of a right hexagonal prism plus the lateral surface area of the cylindrical hole minus the circular ends of the hole.

A = area of hexagon

P = perimeter of hexagon

h = height of prism

C = circumference of hole

r = radius of hole

The area of the hexagon is the sum of the areas of the six equilateral triangles with base of 0.375 in. and height of 0.325 in.

$$\begin{aligned} A &= 6\left[\frac{1}{2}(0.375)(0.325)\right] \\ &= 0.365625\ \text{in}^2 \end{aligned}$$

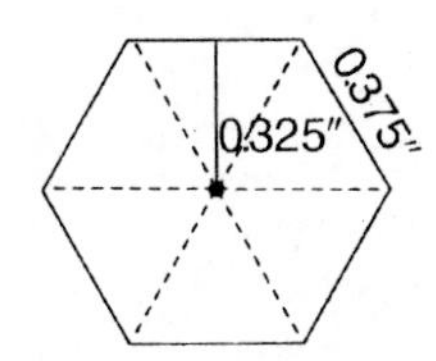

The perimeter of the base of the prism is

$$\begin{aligned} P &= 6(0.375) \\ &= 2.25\ \text{in.} \end{aligned}$$

Section 3.4

39. (continued)

Since the height of the prism is $h = 0.375$ in., the surface area of the right hexagonal prism is

$$\begin{aligned} SA_{\text{prism}} &= 2A + Ph \\ &= 2(0.365625) + 2.25(0.375) \\ &= 1.575 \text{ in}^2. \end{aligned}$$

The lateral surface area of the cylindrical hole is $C \times h$, where $C = 2\pi r$. That is

$$\begin{aligned} SA_{\text{cyl}} &= 2\pi rh \\ &= 2\pi(0.25)(0.375) \\ &= 0.1875\pi \text{ in}^2. \end{aligned}$$

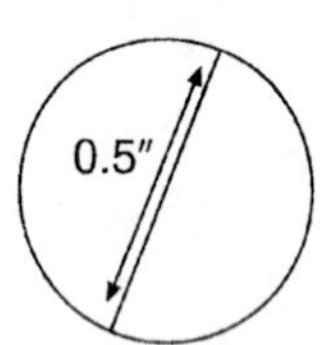

The surface area of the holes is

$$\begin{aligned} SA_{\text{holes}} &= 2(\pi r^2) \\ &= 2[\pi(0.25)^2] \\ &= 0.125\pi \text{ in}^2. \end{aligned}$$

Therefore the total surface area of the machined piece is $SA = SA_{\text{prism}} + SA_{\text{cyl}} - SA_{\text{holes}}$ or

$$\begin{aligned} SA &= 1.575 + 0.1875\pi - 0.125\pi \\ &= 1.575 + 0.0625\pi \\ &\approx 1.8 \text{ in}^2. \end{aligned}$$

41. The figure consists of a cylinder without ends and two hemispheres. The tank has radius, $r = \frac{1}{2} \cdot 5 = 2.5$ ft and height, $h = 14$ ft.

$$\begin{aligned} A_{\text{cyl}} &= 2\pi r^2 + 2\pi rh - 2\pi r^2 \\ &= 2\pi rh \\ &= 2\pi(2.5)(14) \\ &= 70\pi \end{aligned}$$

The two hemispheres combined form a sphere with radius, $r = 2.5$ ft.

$$\begin{aligned} A_s &= 4\pi(2.5)^2 \\ &= 25\pi \\ A_s &\approx 78.54 \text{ ft}^2 \end{aligned}$$

Now find the surface area.

$$\begin{aligned} SA &= A_{\text{cyl}} + A_s \\ SA &= 70\pi + 25\pi \\ SA &= 95\pi \end{aligned}$$

Find the liters of paint required.

$$\frac{SA}{\text{coverage}} = \frac{95\pi}{200} = \frac{19\pi}{40}$$

41. (continued)

The cost of the paint used is

$$\frac{19\pi}{40}(21.95) \approx \$32.76.$$

43. Since there are three colors on the pole, $\frac{1}{3}$ of the surface area is red.

The radius, $r = 10$ cm and the height, $h = 1$ m $= 100$ cm. Find surface area excluding the ends.

$$\begin{aligned} SA &= 2\pi r^2 + 2\pi rh - 2\pi r^2 \\ &= 2\pi rh \\ &= 2\pi(10)(100) \\ &= 2000\pi \end{aligned}$$

Area covered by the red stripe is

$$\begin{aligned} \frac{1}{3}SA &= \frac{1}{3}(2000\pi) \\ &\approx 2094 \text{ cm}^2. \end{aligned}$$

Section 3.5

Tip:

✔ All volumes should be expressed in cubic units (such as cm^3 or ft^3) or in units that are strictly volume (such as liters or gallons).

Solutions to odd-numbered textbook problems

1. (a) $V = 3 + 4 + 2 + 3 + 1 + 1 = 14$ cubic units

To find the surface area, look at different views.

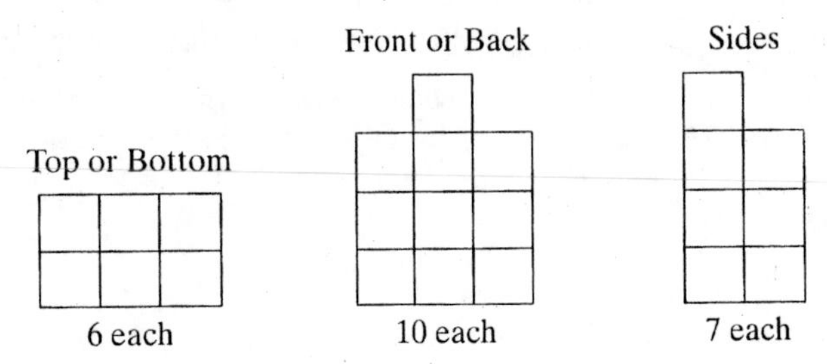

$SA = 2 \cdot 6 + 2 \cdot 10 + 2 \cdot 7 = 46$ square units

Section 3.5

1. (continued)

(b) $V = 1 + 4 + 2 = 7$ cubic units

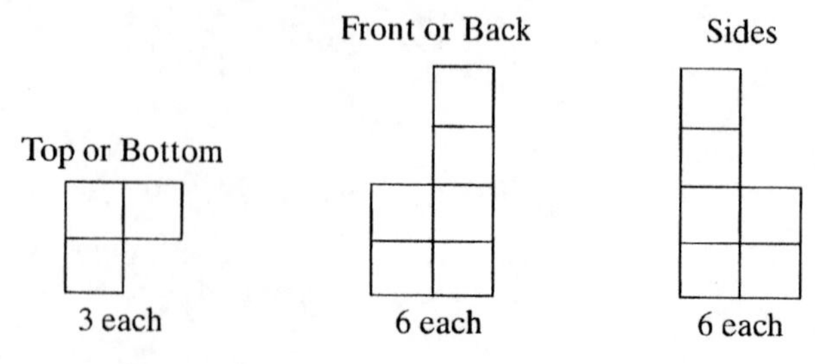

$SA = 2 \cdot 3 + 2 \cdot 6 + 2 \cdot 6 = 30$ square units

(c) $V = 3 + 3 + 3 + 3 + 2 + 1 + 3 + 2 + 1 = 21$ cubic units

Top or Bottom — 9 each

Front or Back — 9 each

Sides — 9 each

$SA = 2 \cdot 9 + 2 \cdot 9 + 2 \cdot 9 = 54$ square units

3. (a) $V = lwh$

$V = (10)(15)(10)$

$V = 1500 \text{ in}^3$

(b) The base of the prism is a trapezoid with area $A = \frac{1}{2}(a + b)h$.

Find the volume of the prism.

$$V = Ah$$

$$V = \frac{1}{2}(8 + 2)(4)(10)$$

$$V = 200 \text{ cm}^3$$

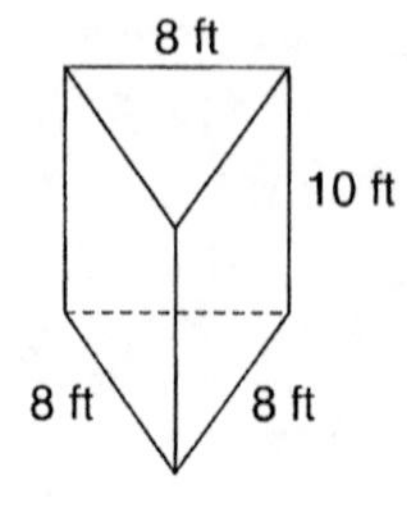

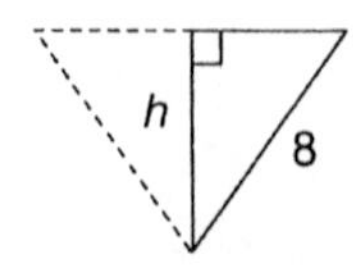

5. (a) Volume is $V = Ah$ where the height of the prism is $h = 10$ ft.

The base of the prism is an equilateral triangle. We use the 30°–60°–right triangle relationship to find the height.

$$h = \frac{\sqrt{3}}{2}(8) = 4\sqrt{3} \text{ ft}$$

Find the area of the triangular base.

$$A = \frac{1}{2}(8)(4\sqrt{3})$$

$$A = 16\sqrt{3} \text{ ft}^2$$

Section 3.5

5. (continued)

Find the volume of the prism.

$$V = 16\sqrt{3}(10)$$
$$V = 160\sqrt{3}$$
$$V \approx 277 \text{ ft}^2$$

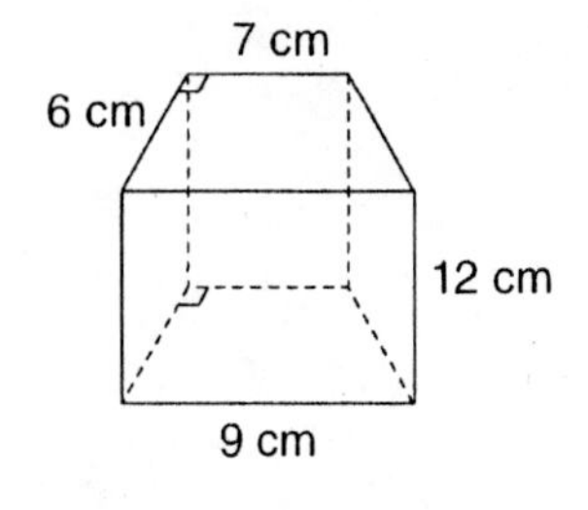

(b) $V = Ah$

$h = 12$ cm

$$A = \frac{1}{2}(7 + 9)(6)$$
$$= 48 \text{ cm}^2$$
$$V = (48)(12)$$
$$= 576 \text{ cm}^3$$

7. (a) $V = \frac{1}{3}Ah$ where the base is a right triangle with area, $A = \frac{1}{2}bh$.

$$V = \frac{1}{3}\left[\frac{1}{2}(6)(3)\right] \cdot 5$$
$$V = 15 \text{ ft}^3$$

(b) $V = \frac{1}{3}Ah$

$$V = \frac{1}{3}(38)(9)$$
$$V = 114 \text{ in}^3$$

9.

$$V = \frac{1}{3}Ah$$
$$V = \frac{1}{3}(23.5)(18.7)(9.9)$$
$$V = 1450.185 \text{ cm}^3$$

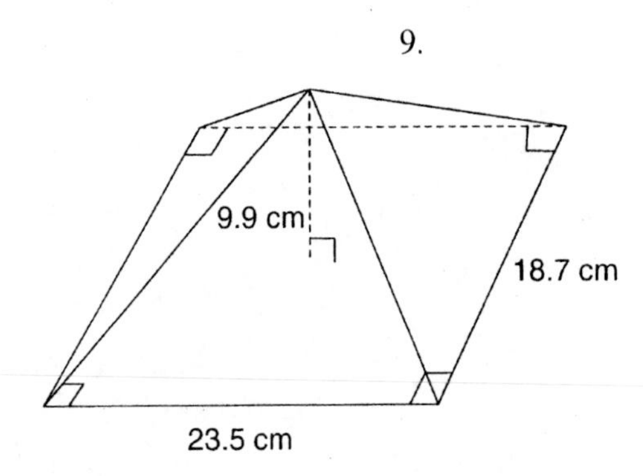

11. (a) $V = \pi r^2 h$

$$V = \pi(7.6)^2(16 \cdot 3)$$
$$V = 941.488\pi$$
$$V \approx 2958 \text{ cm}^3$$

Section 3.5

11. (continued)

(b) $V = \pi r^2 h$ where $r = \frac{1}{2}(6.6) = 3.3$ cm

$V = \pi(3.3^2)(10)$

$= 108.9\pi$

$V \approx 342 \text{ cm}^3$

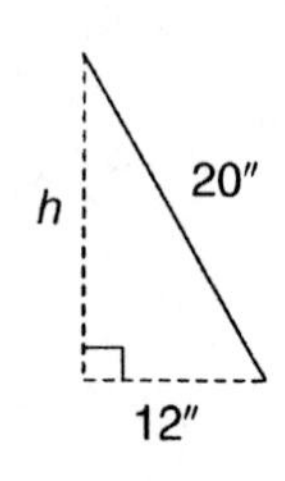

13. (a) $V = \frac{1}{3}\pi r^2 h$

$V = \frac{1}{3}\pi(15^2)(12)$

$= 900\pi$

$V \approx 2827 \text{ m}^3$

(b) $V = \frac{1}{3}\pi r^2 h$ where $h = \sqrt{20^2 - 12^2} = 16$ in.

$V = \frac{1}{3}\pi(12^2)(16)$

$= 768\pi$

$V \approx 2413 \text{ in}^3$

15. (a) $V = \frac{4}{3}\pi r^3$

$V = \frac{4}{3}\pi(6^3)$

$= 288\pi$

$V \approx 905 \text{ in}^3$

(b) $V = \frac{4}{3}\pi r^3$ where $r = \frac{1}{2}(24) = 12$ m

$V = \frac{4}{3}\pi(12^3)$

$= 2304\pi$

$V \approx 7238 \text{ m}^3$

(c) $C = 2\pi r = 18\pi$ cm

$2\pi r = 18\pi$

$r = 9$ cm

$V = \frac{4}{3}\pi \cdot 9^3$

$= 972\pi$

$V \approx 3054 \text{ cm}^3$

Section 3.5

17. (a) $\frac{400 \text{ in}^3}{1} \cdot \frac{1 \text{ ft}^3}{12^3 \text{ in}^3} \approx 0.23 \text{ ft}^3$

(b) $\frac{1.2 \text{ m}^3}{1} \cdot \frac{100^3 \text{ cm}^3}{1 \text{ m}^3} = 1{,}200{,}000 \text{ cm}^3$

(c) $\frac{0.4 \text{ ft}^3}{1} \cdot \frac{12^3 \text{ in}^3}{1 \text{ ft}^3} \cdot \frac{2.54^3 \text{ cm}^3}{1 \text{ in}^3} \cdot \frac{10^3 \text{ mm}^3}{1 \text{ cm}^3} \approx 11{,}326{,}738.64 \text{ mm}^3$

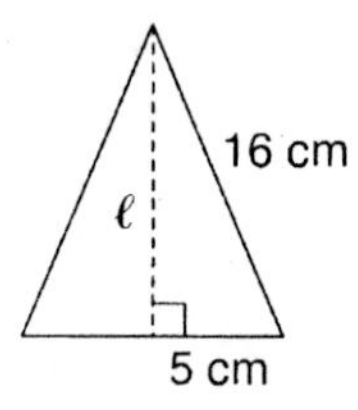

19. The figure shown in the text is a net for a square pyramid. To find the surface area of the pyramid, you need to find l, the altitude of a triangular face (See figure).

$$l^2 + 5^2 = 16^2$$
$$l^2 + 25 = 256$$
$$l^2 = 231$$
$$l = \sqrt{231} \approx 15.2 \text{ cm}$$

Then $$A = 10 \cdot 10 = 100,$$

and $$SA = A + \frac{1}{2}Pl$$
$$= 100 + \frac{1}{2} \cdot 4 \cdot 10\,(\sqrt{231})$$
$$= 100 + 20\sqrt{231}$$
$$SA \approx 404 \text{ cm}^2.$$

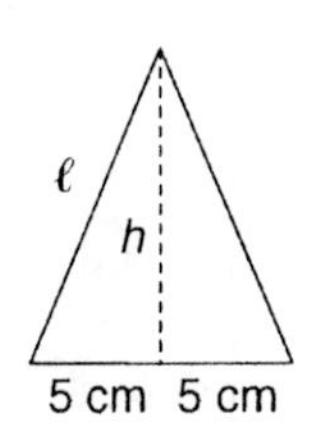

To find the volume of the pyramid you must find the height, h. The cross-section of the pyramid is an isosceles triangle with height h (See figure).

$$5^2 + h^2 = l^2$$
$$25 + h^2 = 231$$
$$h^2 = 206$$
$$h = \sqrt{206} \text{ cm}$$

Then $$V = \frac{1}{3}A \cdot h$$
$$= \frac{1}{3}(100)(\sqrt{206})$$
$$V \approx 478 \text{ cm}^3.$$

21. If the square prism has dimensions x by x by y, then its volume is $V = x^2y$. Doubling each dimension yields a new prism that measures $2x$ by $2x$ by $2y$ and has a volume $V = 2x \cdot 2x \cdot 2y = 8x^2y$. Thus, when the dimensions are doubled, the volume is multiplied by 8.

23. (a) If a square has side length x, then its area is $A = x^2$. A square with side length $3x$ has area $A = (3x)^2 = 9x^2$. When the side length is three times as long. The area is multiplied by 9.

Section 3.5

23. (continued)

(b) If a cube has side length x, then its volume is $V = x^3$. A cube with side length $3x$ has volume, $V = (3x)^3 = 27x^3$. When the side length is three times as long, the volume is multiplied by 27.

(c) If a rectangular box has dimensions x by y by z, then its volume is $V = xyz$. If each side is doubled, the volume is $V = 2x \cdot 2y \cdot 2z = 8xyz$. If all the dimensions of a rectangular box are doubled, the volume is multiplied by 8.

25. Let x = length of longest edge,

$\frac{x}{2}$ = length of medium edge, and

$\frac{x}{9}$ = length of shortest edge.

$$V = x \cdot \frac{x}{2} \cdot \frac{x}{9} = \frac{x^3}{18}$$

$$324 = \frac{x^3}{18}$$

$$5832 = x^3$$

$$x = \sqrt[3]{5832} = 18 \text{ units}$$

$$\frac{x}{2} = 9 \text{ units}$$

$$\frac{x}{9} = 2 \text{ units}$$

The dimensions of the prism are 18 units by 9 units by 2 units.

27. The cylinder with the long side as the height has a circumference of 8.5 in.

$$C = 2\pi r = 8.5$$

$$r = \frac{8.5}{2\pi} \text{ in.}$$

So,

$$V = \pi r^2 h$$

$$= \pi\left(\frac{8.5}{2\pi}\right)^2 (11)$$

$$= \frac{198.6875}{\pi}$$

$$\approx 63.24 \text{ in}^3.$$

The cylinder with the short side as the height has a circumference of 11 in.

$$C = 2\pi r = 11$$

$$r = \frac{11}{2\pi} \text{ in.}$$

Section 3.5

27. (continued)

So,

$$\begin{aligned} V &= \pi r^2 h \\ &= \pi\left(\frac{11}{2\pi}\right)^2 (8.5) \\ &= \frac{257.125}{\pi} \\ &\approx 81.85 \text{ in}^3. \end{aligned}$$

Therefore, the cylinder with the short side of the rectangle as its height has the greater volume.

29. The original volume was

$$V_1 = \frac{1}{3}(230.5^2)(146.5) = 2{,}594{,}527.208 \text{ m}^3.$$

Today's volume is

$$V_2 = \frac{1}{3}(227.5^2)(137) = 2{,}363{,}535.417 \text{ m}^3.$$

Now, $V_1 - V_2 = 2{,}594{,}527.208 - 2{,}363{,}535.417 = 230{,}991.791 \text{ m}^3$.

So, approximately 230,992 m^3 has been lost over time.

31. The slab is in the shape of a cylinder with height 4 in. $= \frac{1}{3}$ ft. and radius 5 ft. The volume of the slab is $V = \pi(5^2)\left(\frac{1}{3}\right) = \frac{25\pi}{3}$ ft^3. Convert ft^3 to yd^3.

$$\frac{\frac{25}{3}\cancel{\text{ft}^3}}{1} \cdot \frac{1 \text{ yd}^3}{3^3 \cancel{\text{ft}^3}} \approx 0.97 \text{ yd}^3$$

Thus you must purchase 1 yd^3 of concrete for \$50.

33. A waffle cone is a right circular cone. For this cone $r = 4.5$ cm and $h = 15$ cm.

$$V = \frac{1}{3}\pi r^2 h = \frac{1}{3}\pi(4.5)^2 \cdot (15) = 101.25\pi \approx 318 \text{ cm}^3$$

35. (a) $\dfrac{43560 \cancel{\text{ft}^2}}{1} \cdot \dfrac{12^2 \text{ in}^2}{1 \cancel{\text{ft}^2}} \cdot 1 \text{ in.} = 6{,}272{,}640 \text{ in}^3$ (Recall: 1 acre = 43,560 ft^2)

$$43{,}560 \text{ ft}^2 \cdot \frac{1}{12} \text{ ft} = 3{,}630 \text{ ft}^3$$

(b) $\dfrac{3630 \cancel{\text{ft}^3}}{1} \cdot \dfrac{62 \cancel{\text{lb}}}{1 \cancel{\text{ft}^3}} \cdot \dfrac{1 \text{ ton}}{2000 \cancel{\text{lb}}} = 112.53$ tons

(c) $\dfrac{3630 \cancel{\text{ft}^3}}{1} \cdot \dfrac{62 \cancel{\text{lb}}}{1 \cancel{\text{ft}^3}} \cdot \dfrac{1 \text{ gal}}{8.3 \cancel{\text{lb}}} \approx 27{,}116$ gal

Section 3.5

37. (a) $V = l \cdot w \cdot h$ where $l = 30$ in., $w = 12$ in., and $h = \frac{1}{4}$ in.

$= 30(12)\left(\frac{1}{4}\right)$

$V = 90 \text{ in}^3$

(b) The volume of space remaining in the aquarium is

$V_R = 30(12)\left(\frac{1}{2}\right) = 180 \text{ in}^3.$

The volume of 200 marbles with radius

$r = \frac{1}{2}(1.5) = 0.75$ cm is $V_m = \frac{4}{3}\pi(0.75^3) \cdot 200 = 112.5\pi \text{ cm}^3.$

Converting to in^3, $\frac{112.5\pi \cancel{\text{cm}^3}}{1} \cdot \frac{1 \text{ in}^3}{2.54^3 \cancel{\text{cm}^3}} \approx 21.57 \text{ in}^3.$

The volume of the marbles is less than the volume remaining in the aquarium. Thus the marbles will not cause the aquarium to overflow.

39. Find the volume of outer sphere less the volume of inner sphere.
$V_1 = \frac{4}{3}\pi(18^3) - \frac{4}{3}\pi(17.8^3) \approx 805.29 \text{ cm}^3$

Find the volume of outer prism less the volume of inner prism.
$V_2 = 40(40)(100) - 39.6(39.6)(99.6) = 3811.264$ cm

Total volume is $V = V_1 + V_2 \approx 805.29 \text{ cm}^3 + 3811.264 \text{ cm}^3$

$V \approx 4616.55 \text{ cm}^3$

Find the mass where $M = V \cdot$ density.

$$M = 4616.55 \text{ cm}^3\left(\frac{7.87 \, g}{\text{cm}^3}\right) \approx 36{,}332.28 \, g$$

Convert to kg. $\frac{36332.28 \, g}{1} \cdot \frac{1 \text{ kg}}{1000 \, g} \approx 36 \text{ kg}$

41. To find the volume of the rubber, find the volume of the entire ball and subtract the volume of the spherical hollow inside the ball. To find the volume you need to find r. Since $C = 22 \text{ cm} = \pi d$, $d = \frac{22}{\pi}$ cm.

Therefore, $r = \frac{d}{2} = \frac{11}{\pi}$ cm.

$$V_{\text{ball}} = \frac{4}{3}\pi r^3$$

$$= \frac{4}{3}\pi\left(\frac{11}{\pi}\right)^3$$

$$= \frac{5324}{3\pi^2}$$

$$\approx 179.811 \text{ cm}^3$$

Section 3.5

41. (continued)

The radius of the hollow part is $\left(\frac{11}{\pi} - 0.6\right)$ cm.

$$V_{\text{hollow}} = \frac{4}{3}\pi\left(\frac{11}{\pi} - 0.6\right)^3 \approx 102.309 \text{ cm}^3$$

$$\approx 180 - 102 = 78 \text{ cm}^3$$

Volume of rubber $\approx 179.881 - 102.309$

$$\approx 78 \text{ cm}^3$$

43. The radius of the sphere is $r = \frac{6}{2} = 3$ ft.

Find volume.

$$V = \frac{4}{3}\pi \cdot 3^3$$

$$= 36\pi \text{ ft}^3$$

Convert to gallons. $\frac{36\pi \text{ ft}^3}{1} \cdot \frac{7.48 \text{ gal}}{1 \text{ ft}^3} = 269.28\pi \approx 846$ gal

There are approximately 846 gal − 200 gal = 646 gallons remaining in the tank.

45. Each step is a triangular prism.

(a) $A = \frac{1}{2} \cdot 15 \cdot 20 = 150 \text{ cm}^2$, and $h = 80$ cm

For 1 step: $V = A \cdot h = 150 \cdot 80 = 12{,}000 \text{ cm}^3$

For 10 steps: $V = 10(12{,}000) = 120{,}000 \text{ cm}^3$

(b) Find carpet needed for 1 step.

$2 \cdot 150 + 20 \cdot 80 + 15 \cdot 80 = 3100 \text{ cm}^2$

Carpet for 10 steps $= 10(3100) = 31{,}000 \text{ cm}^2$

47. Find the area of the cross section.

$$A = 2\left(\frac{5}{8}\right)(8) + \left(\frac{3}{4}\right)(10) = 17.5 \text{ in}^2$$

$h = 25 \text{ ft} = 300$ in.

$V = Ah = 17.5(300) = 5250 \text{ in}^3$

Additional Problems

49. Both the cylindrical and conical portions have circular bases with area $A = \pi(4)^2 = 16\pi$ ft^2.

$$V_{\text{cylinder}} = A \cdot h = 16\pi \cdot 6 = 96\pi \quad \text{ft}^3$$

$$V_{\text{cone}} = \frac{1}{3}Ah = \frac{1}{3} \cdot 16\pi \cdot 4 = \frac{64}{3}\pi \text{ ft}^3$$

$$\begin{aligned} V_{\text{total}} &= 96\pi + \frac{64\pi}{3} \\ &= \frac{352\pi}{3} \\ &\approx 369 \text{ ft}^3 \end{aligned}$$

51. Find the volume of a cylinder of radius 1320 ft and height $\frac{3}{4}$ inch = 0.625 ft.

$$V = \pi(1320^2)(0.0625) = 108{,}900\pi \text{ ft}^3$$

$$\frac{108900\pi \text{ ft}^3}{1} \cdot \frac{7.48 \text{ gal}}{1 \text{ ft}^3} \cdot \frac{1 \text{ min}}{1000 \text{ gal}} \cdot \frac{1 \text{ hr}}{60 \text{ min}} \approx 42.65 \text{ hr}$$

53. (a) First you need to find the volume of the spherical tank. Then you can find the height of the cylinder.

$$r = \frac{d}{2} = \frac{60}{2} = 30 \text{ ft}$$

$$V_{\text{sphere}} = \frac{4}{3}\pi r^3 = \frac{4}{3}\pi (30)^3 \approx 36{,}000\pi \text{ ft}^3$$

Thus

$$\begin{aligned} V_{\text{cylinder}} &= \pi r^2 h \\ 36{,}000\pi &= \pi(30)^2 h \\ 40 \text{ ft} &= h. \end{aligned}$$

(b) $$\frac{36{,}000\pi \text{ ft}^3}{1} \times \frac{7.5 \text{ gal}}{1 \text{ ft}^3} = 270{,}000\pi \text{ gal} = 848{,}000 \text{ gal}$$

(c) $$\begin{aligned} SA_{\text{sphere}} &= 4\pi r^2 = 4\pi (30)^2 = 3600\pi \text{ ft}^2 \\ SA_{\text{cylinder}} &= 2\pi r^2 + 2\pi rh \\ &= 2\pi(30)^2 + 2\pi(30)(40) \\ &= 1800\pi + 2400\pi \\ &= 4200\pi \text{ ft}^2 \end{aligned}$$

The sphere requires less material.

Additional Problems

1. The bottom layer of 144 six-packs is arranged in a 12 by 12 square. To accomplish this, workers adjust the spacing between the six-packs. The next layer would contain 121 six-packs arranged in an 11 by 11 square. Again the workers adjust the spacing between six-packs. Continuing

this same pattern, the greatest number of six-packs that can be stacked on top of the bottom layer to form a square pyramid is $11^2 + 10^2 + 9^2 + 8^2 + 7^2 + 6^2 + 5^2 + 4^2 + 3^2 + 2^2 + 1^2 = 506$. The top 3 layers of the pyramid are shown. The only layer that is not a square is the top layer.

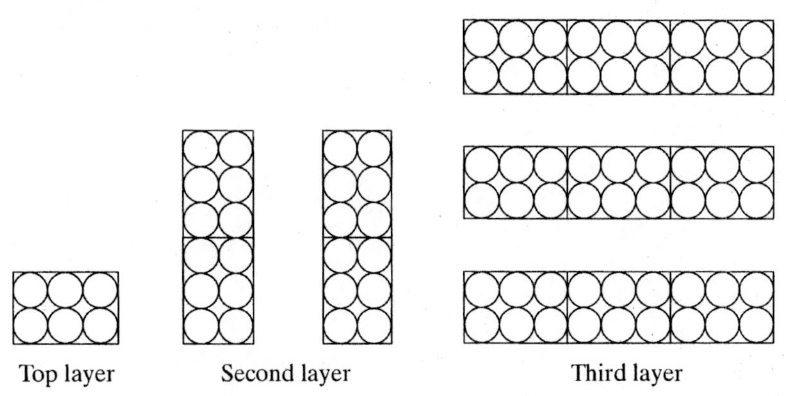

2. A complete solution is given in the answer key at the back of the text.

Solutions to Chapter 3 Review

Section 3.1

1 (a) The perimeter of a rhombus with sides of length s is $4s$.

$P = 4(3.2) = 12.8$ mm

(b) The perimeter of a parallelogram with sides of length a and b is $2a + 2b$.

$P = 2(9.5) + 2(2.1)$

$= 19 + 4.2$

$= 23.2$ cm

(c) The perimeter of a kite with sides of lengths a and b is $2a + 2b$.

$P = 2(12) + 2(20)$

$= 24 + 40$

$= 64$ ft

(d) The perimeter (circumference) of a circle with diameter d is πd.

$C = \pi(6)$

$= 6\pi$

$C \approx 18.8$ cm

(e) $P = 7 + 4.24 + 2.5 + 2.5$

$= 16.24$ ft

Chapter Review

2. (a) $P = 2(360) + 2(160)$

$= 720 + 320$

$= 1040 \text{ ft}$

$= 1040 \cancel{\text{ft}} \times \frac{1 \text{ yd}}{3 \cancel{\text{ft}}} \; (1 \text{ yd} = 3 \text{ ft})$

$P \approx 346.7 \text{ yd}$

(b) $A = 360(160)$

$= 57600 \text{ ft}^2$

$= 57600 \text{ ft}^2 \times \frac{1 \text{ acre}}{43{,}560 \text{ ft}^2} \; (1 \text{ acre} = 43{,}560 \text{ ft}^2)$

$A \approx 1.32 \text{ acres}$

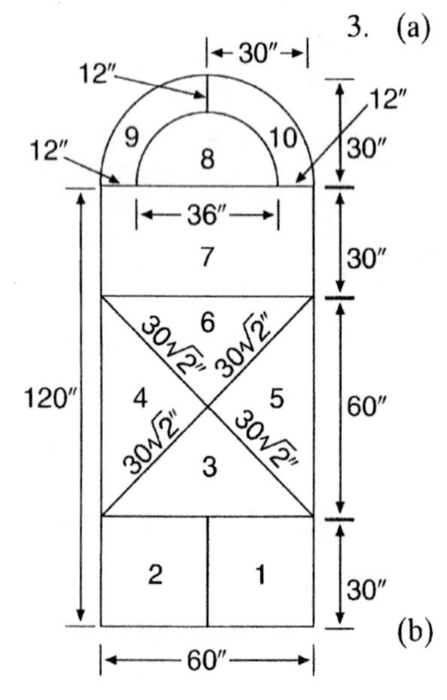

3. (a) Sum the lengths of each straight segment and the two semicircles. The large semicircle has radius $r = 30$ in. and circumference 30π in. The small semicircle has radius $r = 18$ in. and circumference 18π in. The total amount of tape, t, needed is

$t = 522 + 4(30\sqrt{2}) + 30\pi + 18\pi$

$= 522 + 120\sqrt{2} + 48\pi$

$t \approx 843$ in.

(b) The area of triangle 3 is one-fourth of the area of the square containing triangles 3, 4, 5 and 6.

$A_{\Delta 3} = \frac{1}{4}(60)(60)$

$= 900 \text{ in}^2$

$= 900 \cancel{\text{in}^2} \times \frac{1 \cdot 1 \text{ ft}^2}{12 \cdot 12 \cancel{\text{in}^2}}$

$= 6.25 \text{ ft}^2$

4. The perimeter of a rectangular picture frame is $2(10) + 2(8) = 36$ in. The perimeter of a square picture frame with side of x in. is $4x$ in. Set the perimeters equal and solve for x.

$$4x = 36$$

$$x = 9 \text{ in.}$$

So the area inside the square frame is $A = 9(9) = 81 \text{ in}^2$.

Chapter Review

5. $C = 2\pi r = 10$

$$r = \frac{10}{2\pi}$$

$$= \frac{5}{\pi} \text{ in.}$$

6. The circumference of a circle is $C = 2\pi r$. Double the radius.

$$C = 2\pi(2r)$$

$$= 2(2\pi r)$$

The resulting circumference is twice as large.

7 (a) $P = 12 + 5 + 13$

$$= 30 \text{ cm}$$

$A = \frac{1}{2}bh$ where b = base and h = height of right triangle.

$$A = \frac{1}{2}(12)(5)$$

$$= 30 \text{ cm}^2$$

(b) $P = 5 + 6 + \sqrt{109}$

$$P \approx 21.4 \text{ mm}$$

$$A = \frac{1}{2}(6)(3)$$

$$= 9 \text{ mm}^2$$

Section 3.2

1. (a) $A = 8(4.35)$

$$= 34.8 \text{ mm}^2$$

(b) $A = (3.2)(2.9)$

$$= 9.28 \text{ in}^2$$

2. (a) $A = \frac{1}{2}(7.1 + 5.8)5.9$

$$= 38.055 \text{ m}^2$$

(b) $A = \frac{1}{2}(19.1 + 15.25)4.6$

$$= 79.005 \text{ cm}^2$$

Chapter Review

3. A semicircle is one-half a circle.

$$A = \frac{1}{2}(\pi r^2)$$
$$= \frac{1}{2}\pi\left(\frac{5}{2}\right)^2$$
$$= \frac{25}{8}\pi \text{ cm}^2$$
$$A \approx 9.8 \text{ cm}^2$$

4. First, find the radius of the given circle.

$$C = 2\pi r = 19.25$$
$$r = \frac{19.25}{2\pi}$$

Now, use this radius to find the area of the circle.

$$A = \pi r^2$$
$$= \pi\left(\frac{19.25}{2\pi}\right)^2$$
$$= \frac{370.5625}{4\pi}$$
$$A \approx 29.49 \text{ in}^2$$

5. The area of each octagon is found by finding the area of circumscribed square and then subtracting the 4 corners which are right isosceles triangles.

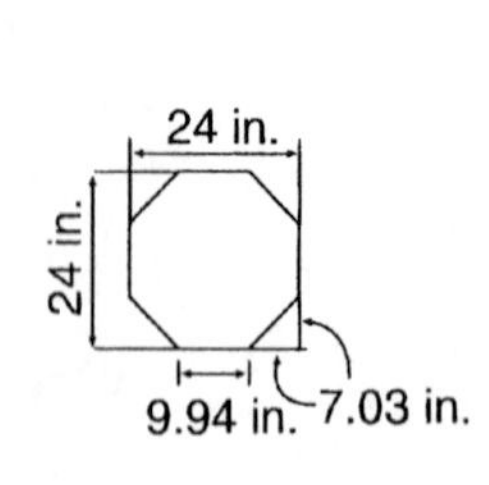

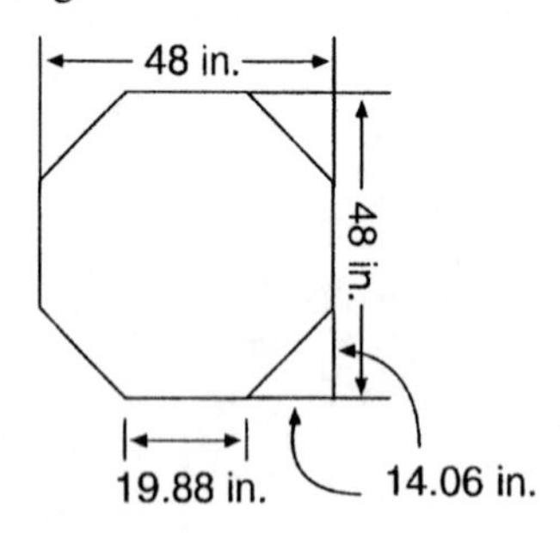

Area of large octagon: $A_1 = 48^2 - 4\left[\frac{1}{2}(14.06)(14.06)\right]$

$= 1908.6328 \text{ in}^2$

Area of small octagon: $A_2 = 24^2 - 4\left[\frac{1}{2}(7.03)(7.03)\right]$

$= 477.1582 \text{ in}^2$

Difference in area: $A_1 - A_2 = 1908.6328 - 477.1582$

$= 1431.4746 \text{ in}^2$

$\approx 1431 \text{ in}^2$

Chapter Review

6. Area of circle:

$$A_{\text{circle}} = \pi(3\sqrt{3})^2$$
$$= 27\pi \text{ mm}^2$$

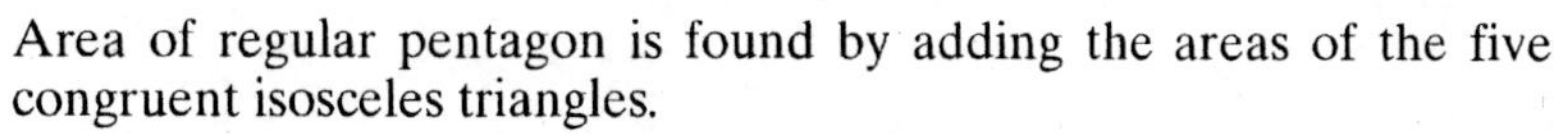

Area of regular pentagon is found by adding the areas of the five congruent isosceles triangles.

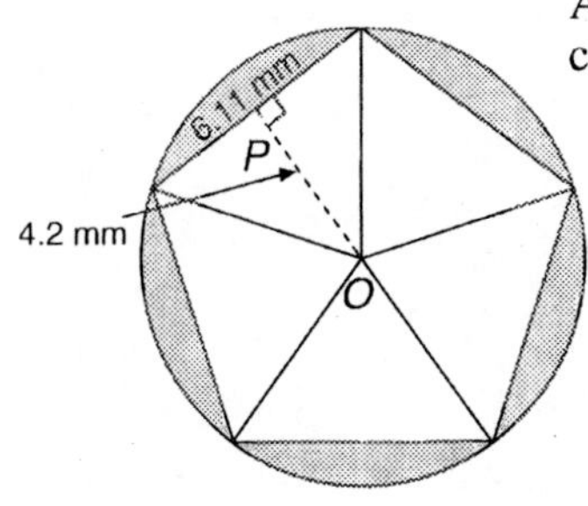

$$A_{\text{pentagon}} = 5\left[\frac{1}{2}(6.11)(4.2)\right]$$
$$= 64.155 \text{ mm}^2$$

The area of the shaded region is

$$A_{\text{pentagon}} - A_{\text{circle}} = 27\pi - 64.155$$
$$\approx 20.668 \text{ mm}^2.$$

Section 3.3

1. A triangle is a right triangle if $a^2 + b^2 = c^2$ where a, b, and c are the lengths of the sides of the triangle.

$$12^2 + 35^2 = 144 + 1225$$
$$= 1369$$
$$= 37^2$$

Therefore, the triangle is a right triangle.

2. $(3.1)^2 + (4.6)^2 = 9.61 + 21.16$

$$= 30.77$$
$$(7.3)^2 = 53.29$$

$(3.1)^2 + (4.6)^2 \neq (7.3)^2$

The triangle is not a right triangle.

Now $(3.1)^2 + (4.6)^2 < (7.3)^2$ so the triangle is obtuse.

3. (a) Let x = missing leg of right triangle in inches.

$$(10.22)^2 + x^2 = (19.5)^2$$
$$x^2 = (19.5)^2 - (10.22)^2$$
$$= 380.25 - 104.4484$$
$$x = \sqrt{275.8016}$$
$$\approx 16.61 \text{ in.}$$

(b) Let x = length of hypotenuse of right triangle in cm.

$$x^2 = (\sqrt{5})^2 + (\sqrt{11})^2$$
$$= 5 + 11$$
$$= 16$$
$$x = 4 \text{ cm}$$

Chapter Review

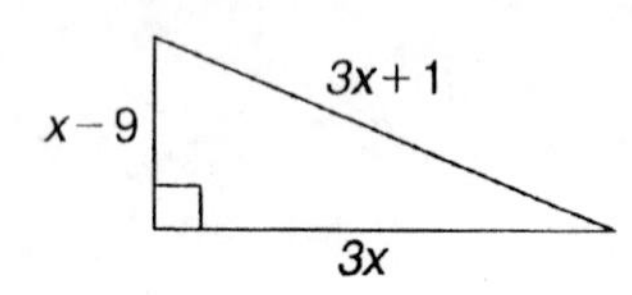

4. Use the Pythagorean Theorem.

$$(x-9)^2 + (3x^2) = (3x+1)^2$$
$$x^2 - 18x + 81 + 9x^2 = 9x^2 + 6x + 1$$
$$x^2 - 24x + 80 = 0$$
$$(x-20)(x-4) = 0$$
$$x - 20 = 0 \qquad x - 4 = 0$$
$$x = 20 \qquad \cancel{x = 4}$$

For $x = 20$ the sides of the triangle are

$3x = 60, x - 9 = 11$, and $3x + 1 = 61$.

The value $x = 4$ must be excluded since it results in one leg of the triangle being negative, which is not possible.

5. (a) A right triangle is a 30°–60° right triangle if one leg of the triangle is one-half of the hypotenuse. Since 2.5 in is one-half of 5 inches the triangle is a 30°–60° right triangle and the length of the other leg is $2.5\sqrt{3}$ in.

   (b) A right triangle is a 45°–45° right triangle if both legs are the same length. Since both legs are the same length we have a 45°–45° right triangle with a hypotenuse of $7\sqrt{2}$ cm.

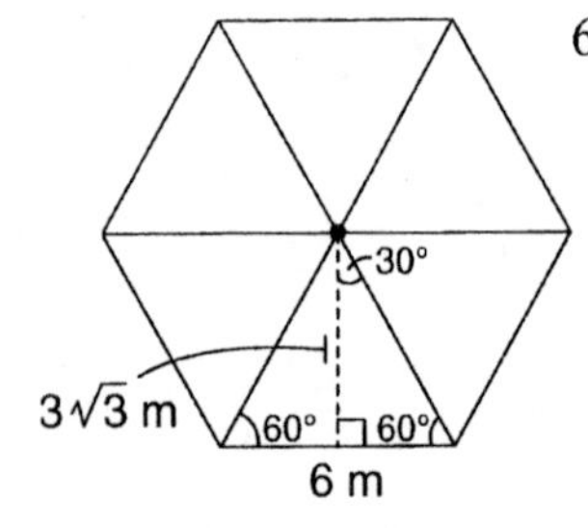

6. Divide the regular hexagon into six equilateral triangles. The altitude of the triangle will divide the triangle into two 30°–60° right triangles. Using Theorem 3.9 the altitude is $3\sqrt{3}$ m. Therefore the area of one equilateral triangle is $A_{\text{triangle}} = \frac{1}{2}(6)(3\sqrt{3})$.

There are six equilateral triangles in the regular hexagon, so the area of the regular hexagon is

$$A_{\text{hexagon}} = 6\left[\frac{1}{2}(6)(3\sqrt{3})\right]$$
$$= 54\sqrt{3}\text{ m}^2$$
$$\approx 93.5\text{ m}^2.$$

7. Insert the altitude to side $\overline{PR}$. Then the triangle consists of two right triangles, one which is a 45°–45° right triangle and the other which is a 30°–60° right triangle. Use Theorems 3.9 and 3.10.

$$PR = 6 + 6\sqrt{3}\text{ in.}$$
$$\approx 16.4\text{ in.}$$
$$\text{and } PQ = 6\sqrt{2}\text{ in.}$$
$$\approx 8.5\text{ in.}$$

Chapter Review **Section 3.4**

1. $SA = 2A + Ph$

$$= (2)(2.8)(3.5) + (2.8 + 3.5 + 2.8 + 3.5)(9.1)$$
$$= 19.6 + 12.6(9.1)$$
$$= 134.26 \text{ m}^2$$

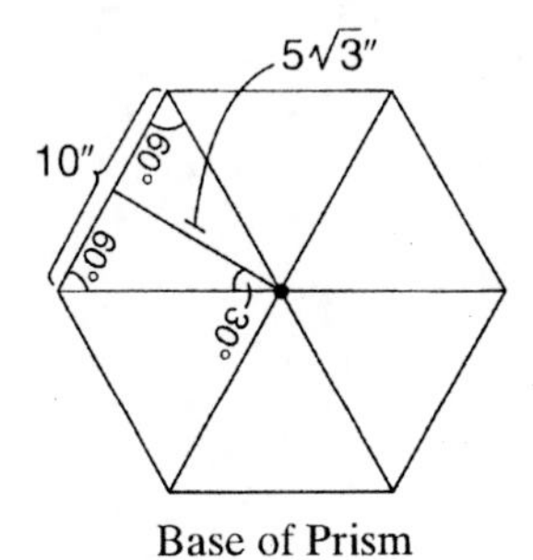

Base of Prism

2. $SA = 2A + Ph$

$$= 2\left[6\left(\frac{1}{2}\right)(10)(5\sqrt{3})\right] + 60(14)$$
$$= 2(150\sqrt{3}) + 840$$
$$= 300\sqrt{3} + 840$$
$$SA \approx 1359.6 \text{ in}^2$$

3. The larger balloon has surface area $SA = 4\pi(5)^2 = 100\pi \text{ cm}^2$.

So, the smaller balloon has surface area $\frac{100\pi}{3}$ cm^2. Now, find the radius of the smaller balloon.

$$4\pi r^2 = \frac{100\pi}{3}$$
$$r^2 = \frac{100\pi}{12\pi}$$
$$= \frac{25}{3}$$
$$r = \frac{5}{\sqrt{3}} \text{ cm}$$

The diameter of the smaller balloon is twice its radius.

$$d = 2\left(\frac{5}{\sqrt{3}}\right)$$
$$= \frac{10}{\sqrt{3}}$$
$$= \frac{10\sqrt{3}}{3} \text{ cm}$$
$$d \approx 5.77 \text{ cm}$$

Chapter Review

4. $SA = A + \frac{1}{2}Pl$

$$= (32.5)^2 + \frac{1}{2}(130)\sqrt{(45)^2 + (16.25)^2}$$

$$= 1056.25 + 65\sqrt{2289.0625}$$

$$SA \approx 4166.12 \text{ m}^2$$

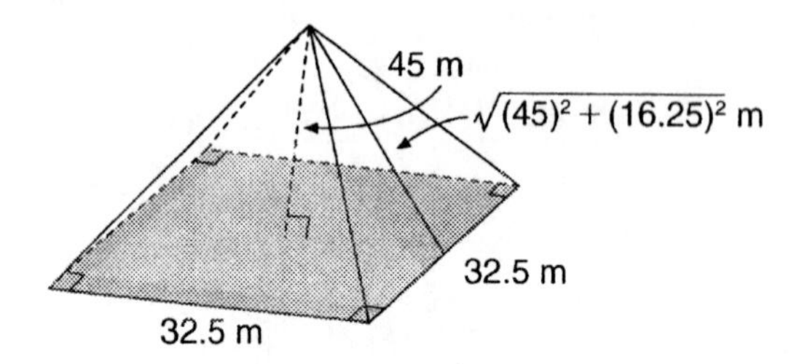

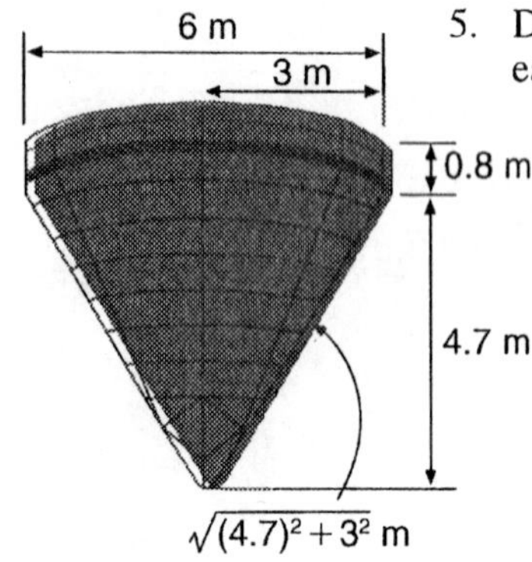

5. Divide the tank into a cylinder and a cone and find the surface area of each of them.

$$SA = SA_{\text{cylinder}} + SA_{\text{cone}}$$

$$= 2\pi rh + \frac{1}{2}Cl$$

$$= 2\pi rh + \pi rl$$

$$= 2\pi(3)(0.8) + \pi \cdot 3\sqrt{(4.7)^2 + 3^2}$$

$$= 4.8\pi + 3\pi\sqrt{(4.7)^2 + 3^2}$$

$$SA \approx 68 \text{ m}^2$$

6. The total surface area is the surface area of the cube less the area of the two circles plus the surface area of the interior cylinder.

$$SA = SA_{\text{cube}} - 2A_{\text{circle}} + SA_{\text{cylinder}}$$

$$= 6(2)(2) - 2\pi\left(\frac{1}{2}\right)^2 + 2\pi\left(\frac{1}{2}\right)(2)$$

$$= 24 - \frac{\pi}{2} + 2\pi$$

$$= 24 + \frac{3\pi}{2}$$

$$SA \approx 28.7 \text{ in}^2$$

7. (a) $\frac{38500\ \cancel{\text{mm}^2}}{1} \times \frac{1 \cdot 1 \text{ m}^2}{1000 \cdot 1000\ \cancel{\text{mm}^2}} = 0.0385 \text{ m}^2$

(b) $\frac{0.00074\ \cancel{\text{mi}^2}}{1} \times \frac{5280 \cdot 5280\ \cancel{\text{ft}^2}}{1 \cdot 1\ \cancel{\text{mi}^2}} \times \frac{12 \cdot 12 \text{ in}^2}{1 \cdot 1\ \cancel{\text{ft}^2}} = 2{,}970{,}722.304 \text{ in}^2$

(c) $\frac{10\ \cancel{\text{m}^2}}{1} \times \frac{100 \cdot 100\ \cancel{\text{cm}^2}}{1 \cdot 1\ \cancel{\text{m}^2}} \times \frac{1 \cdot 1\ \cancel{\text{in}^2}}{2.54 \cdot 2.54\ \cancel{\text{cm}^2}} \times \frac{1 \cdot 1\ \cancel{\text{ft}^2}}{12 \cdot 12\ \cancel{\text{in}^2}} \times \frac{1 \cdot 1 \text{ yd}^2}{3 \cdot 3\ \cancel{\text{ft}^2}}$

$\approx 11.9599 \text{ yd}^2$

Chapter Review **Section 3.5**

1. (a)
$$\begin{aligned} V &= Ah \\ &= \frac{1}{2}(2.1)(6.4)(15) \\ &= 100.8 \text{ cm}^3 \\ &\approx 101 \text{ cm}^3 \end{aligned}$$

(b) Find the area of the regular hexagonal base by dividing the regular hexagon into six congruent equilateral triangles. The area of each equilateral triangle is $\frac{1}{2}(9)\left(\frac{9\sqrt{3}}{2}\right)$ in^2. Now $V = Ah$ so substituting values we have

$$\begin{aligned} V &= 6\left[\left(\frac{1}{2}\right)(9)\left(\frac{9\sqrt{3}}{2}\right)\right](6) \\ &= 36\left(\frac{81\sqrt{3}}{4}\right) \\ &= \frac{729\sqrt{3}}{4} \text{ in}^3 \\ V &\approx 1263 \text{ in}^3. \end{aligned}$$

2. The volume of the sphere is

$$\begin{aligned} V_{\text{sphere}} &= \frac{4}{3}\pi r^3 \\ &= \frac{4}{3}\pi\left(\frac{d}{2}\right)^3 \quad \left(r = \frac{d}{2}, \text{ where } d \text{ is diameter}\right) \\ &= \frac{4}{3}\pi\left(\frac{5}{2}\right)^3 \\ &= \frac{500\pi}{24} \\ &= \frac{125\pi}{6} \text{ cm}^3. \end{aligned}$$

The volume of the cylinder is

$$\begin{aligned} V_{\text{cylinder}} &= \pi r^2 h \\ &= \pi\left(\frac{d}{2}\right)^2 d \quad \text{(diameter of cylinder equals height of cylinder)} \\ &= \frac{\pi d^3}{4}. \end{aligned}$$

Chapter Review

2. (continued)

Set $V_{\text{cylinder}} = V_{\text{sphere}}$ and solve for diameter d.

$$\frac{\pi d^3}{4} = \frac{125\pi}{6}$$

$$d^3 = \frac{500}{6}$$

$$= \frac{250}{3}$$

$$d = \sqrt[3]{\frac{250}{3}} \text{ cm}$$

$$d \approx 4.37 \text{ cm}$$

Since the diameter of the cylinder is the same as its height,

$$h = \sqrt[3]{\frac{250}{3}} \text{ cm}$$

$$\approx 4.37 \text{ cm.}$$

3. (a) $\dfrac{0.015 \text{ km}^3}{1} \times \dfrac{1000 \cdot 1000 \cdot 1000 \text{ m}^3}{1 \cdot 1 \cdot 1 \text{ km}^3} = 15{,}000{,}000 \text{ m}^3$

(b) $\dfrac{650\pi \text{ in}^3}{1} \times \dfrac{1 \cdot 1 \cdot 1 \text{ ft}^3}{12 \cdot 12 \cdot 12 \text{ in}^3} \times \dfrac{1 \cdot 1 \cdot 1 \text{ yd}^3}{3 \cdot 3 \cdot 3 \text{ ft}^3} = 0.044 \text{ yd}^3$

(c) $\dfrac{1800 \text{ mm}^2}{1} \times \dfrac{1 \cdot 1 \cdot 1 \text{ cm}^3}{10 \cdot 10 \cdot 10 \text{ mm}^3} \text{ lts } \dfrac{1 \cdot 1 \cdot 1 \text{ in}^3}{2.54 \cdot 2.54 \cdot 2.54 \text{ cm}^3} = 0.110 \text{ in}^3$

4. The gutter is in the shape of a right trapezoidal prism. Find the volume. All dimensions are in meters.

$$V = Ah$$

$$= \frac{1}{2}(0.115 + 0.115 + 0.015\sqrt{3})(0.065)(20)$$

$$= 0.65(0.23 + 0.015\sqrt{3})$$

$$\approx 0.17 \text{ m}^3$$

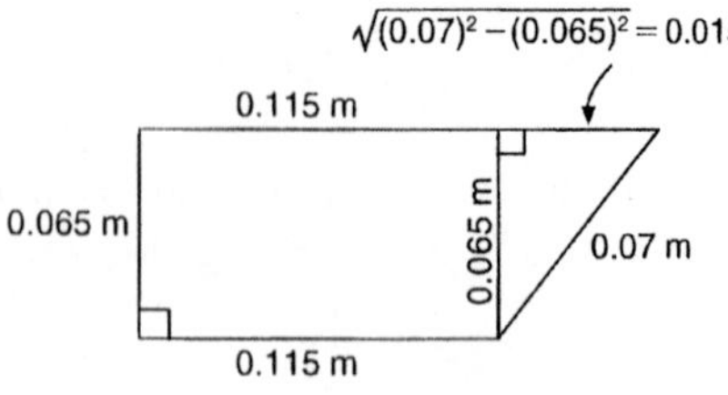

$\sqrt{(0.07)^2 - (0.065)^2} = 0.015\sqrt{3}$ m

0.115 m

0.065 m

0.065 m

0.07 m

0.115 m

Base of prism

5. The total volume is the volume of the cube less the volume of the cylindrical hole

$$V = 2^3 - \pi\left(\frac{1}{2}\right)^2 2$$

$$= 8 - \frac{\pi}{2}$$

$$= \frac{16 - \pi}{2}$$

$$V \approx 6.4 \text{ in}^3$$

Chapter Review

6. (a)
$$V = \frac{1}{3}Ah$$
$$= \frac{1}{3}(544)^2(321)$$
$$= 31{,}665{,}152 \text{ ft}^3$$

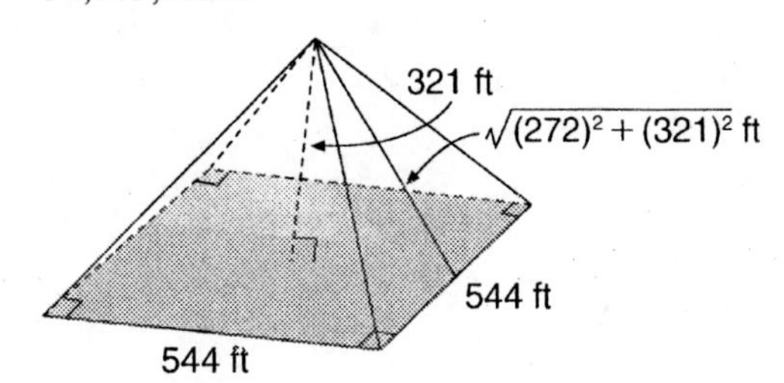

(b) The lateral surface area of a pyramid is $SA = \frac{1}{2}Pl.$

$$SA = \frac{1}{2}Pl.$$
$$= \frac{1}{2}(4)(544)\sqrt{(272)^2 + (321)^2}$$
$$= 1088(5\sqrt{7081})$$
$$= 5440\sqrt{7081}$$
$$SA \approx 457{,}769 \text{ ft}^2$$

7.
$$V = \frac{1}{3}Ah$$
$$250 = \frac{1}{3}\pi\, 5^2\, h$$
$$10 = \frac{1}{3}\pi h$$
$$30 = \pi h$$
$$\frac{30}{\pi} = h$$
$$9.5 \text{ cm} \approx h$$

The height is approximately 9.5 cm.

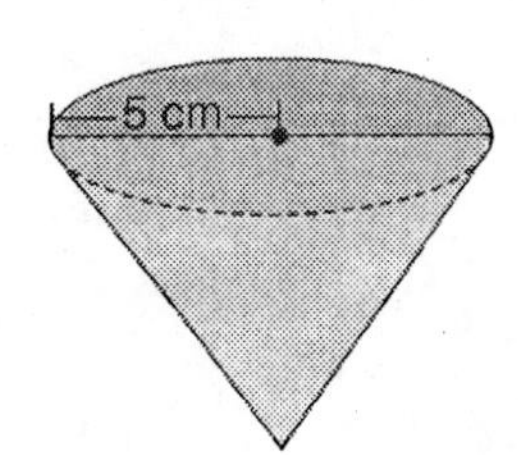

Chapter Test

Solutions to Chapter 3 Test

1. F $A = l \cdot w$ and $P = l + w$.

2. F The Pythagorean Theorem is useful only for right triangles.

3. T $A = \frac{1}{2}(b_1 + b_2)$

4. F Consider a rectangle that is 5 by 10 ($P = 30, A = 50$) and a rectangle that is 7 by 8 ($P = 30, A = 56$).

5. F It could also be a pyramid.

6. F $SA_{\text{sphere}} = 4\pi r^2, SA_{\text{cyl}} = 2\pi r^2 + 2\pi rh$, and when $h = 2r$,

   $SA_{\text{cyl}} = 2\pi r^2 + 2\pi r\,(2r) = 6\pi r^2$.

   The surface area of the cylinder is larger.

7. F Consider any scalene triangle. The altitude changes depending on the side used.

8. T Any circle with the center of the sphere as its center is a great circle.

9. T $6.75^2 + 16.2^2 = 17.55^2$

10. F The circumference is doubled by doubling the radius. Replacing r by $2r$ in $A = \pi r^2$ gives $A = \pi(2r)^2 = 4\pi r^2$, which is four times the original area.

11. (a) $A = \pi r^2$ and $r = \frac{d}{2}$, so $A = \pi\left(\frac{d}{2}\right)^2 = \frac{\pi d^2}{4}$.

    (b) $C = 2\pi r$ so $r = \frac{C}{2\pi}$.

    $$A = \pi r^2 = \pi\left(\frac{C}{2\pi}\right)^2 = \pi \cdot \frac{C^2}{4\pi^2} = \frac{C^2}{4\pi}$$

12.

$$A = \frac{1}{2}s \cdot s = \frac{s^2}{2}$$

$$32 = \frac{s^2}{2}$$

$$64 = s^2$$

$$8 \text{ cm} = s$$

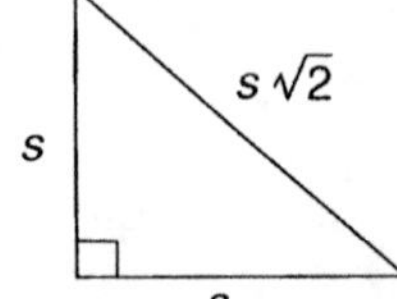

Chapter Test

13. (a) $V = \frac{1}{3}Ah$, where $A = 9 \cdot 9 = 81$ in^2, and $h = 6$ in.

$$V = \frac{1}{3} \cdot 81 \cdot 6 = 162 \text{ in}^3$$

(b) $A = 18 \cdot 18 = 324$ in^2

$$V = \frac{1}{3} \cdot 324 \cdot 6 = 648 = 4 \cdot 162 \text{ in}^3$$

The volume is quadrupled.

(c) $A = 27 \cdot 27 = 729$

$$V = \frac{1}{3} 729 \cdot 6 = 1458 = 9 \cdot 162 \text{ in}^3$$

The volume is multiplied by 9.

(d) $SA = A + \frac{1}{2}Pl,\ P = 4 \cdot 9 = 36$ in.

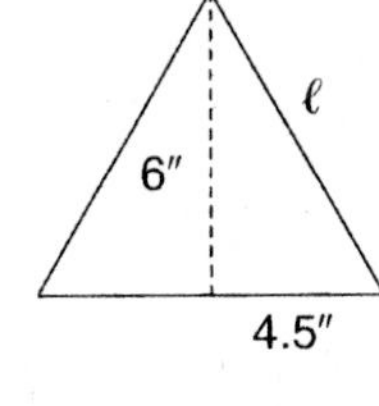

To find l, look at a cross-section of the pyramid (See figure).

$$l^2 = 6^2 + 4.5^2$$

$$l^2 = 56.25$$

$$l = 7.5 \text{ in.}$$

$$SA = 81 + \frac{1}{2}(36)(7.5) = 216 \text{ in}^2$$

14. A hexagon is made up of six equilateral triangles, so the trapezoid that is half of the hexagon has bases of length x and $2x$. To find h, divide one triangle down the middle to get a 30°–60° right triangle. Then $h = \frac{\sqrt{3}}{2}x$.

For a trapezoid,

$$A = \frac{1}{2}h\,(b_1 + b_2) = \frac{1}{2} \cdot \frac{\sqrt{3}}{2}x\,(x + 2x) = \frac{3\sqrt{3}}{4}x^2$$

$$A_{\text{hexagon}} = 2 \cdot \frac{3\sqrt{3}}{4}x^2 = \frac{3\sqrt{3}}{2}x^2$$

15. (a) $C = 56\pi = \pi d$, so $56 = d$.

$$r = \frac{d}{2} = \frac{56}{2} = 28 \text{ units}$$

$$SA = 4\pi r^2 = 4\pi(28)^2 = 3136\pi \text{ square units}$$

$$SA \approx 9852 \text{ square units}$$

$$V = \frac{4}{3}\pi r^3 = \frac{4}{3}\pi(28)^3 = \frac{87808}{3}\pi \text{ cubic units}$$

$$V \approx 91{,}952 \text{ cubic units}$$

Chapter Test

15. (continued)

(b) $SA = 4\pi r^2 = 100\pi$. Dividing by 4π leaves

$r^2 = 25$, so $r = 5$ and $d = 10$.

$$V = \frac{4}{3}\pi r^3 = \frac{4}{3}\pi(5)^3 = \frac{500}{3}\pi \text{ cubic units}$$

$V \approx 524$ cubic units

$C = \pi d = 10\pi$ units ≈ 31.4 units

(c) $V = \frac{4}{3}\pi r^3$ so $\frac{4}{3}\pi r^2 = 972\pi$ and, dividing by $\frac{4}{3}\pi$,

$r^3 = 729$

$r = 9$

Thus, $d = 18$.

$SA = 4\pi r^2 = 4\pi(9)^2 = 324\pi$ square units

$SA \approx 1018$ square units

$C = 2\pi r = 2\pi \cdot 9 = 18\pi \approx 56.5$ units

(d) $d = 6$ so $r = 3$.

$SA = 4\pi r^2 = 4\pi \cdot 3^2 = 36\pi$ square units

$SA \approx 113$ square units

$$V = \frac{4}{3}\pi r^2 = \frac{4}{3}\pi \cdot 3^3 = 36\pi \text{ cubic units}$$

$V \approx 113$ cubic units

$C = \pi d = 6\pi$ units ≈ 18.8 units

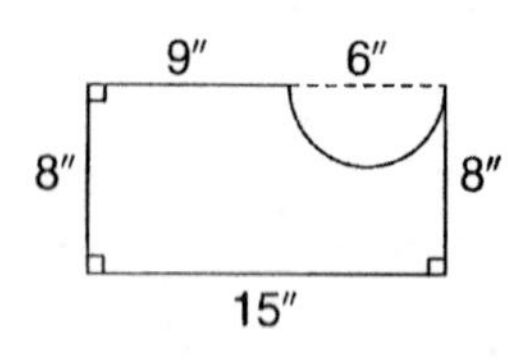

16. $P = 8 + 15 + 8 + 9 + \frac{1}{2} \cdot \pi \cdot 6$

$= 40 + 3\pi$ in. ≈ 49.42 in.

$$A = 8 \cdot 15 - \frac{1}{2} \cdot \pi(3)^2 = 120 - \frac{9}{2}\pi$$

$A \approx 120 - 14.14 \approx 105.86$ in^2

17. $$\frac{3000 \text{ lb}}{1} \times \frac{1 \text{ kg}}{2.2 \text{ lb}} \approx 1364 \text{ kg}$$

$$\frac{4000 \text{ mm}}{1} \times \frac{1 \text{ cm}}{10 \text{ mm}} \times \frac{1 \text{ in.}}{2.54 \text{ cm}} \times \frac{1 \text{ ft}}{12 \text{ in.}} \approx 13.1 \text{ ft}$$

$$\frac{27 \text{ lb}}{1 \text{ in}^2} \times \frac{1 \text{ kg}}{2.2 \text{ lb}} \times \frac{1000 \text{ g}}{1 \text{ kg}} \times \frac{1 \cdot 1 \text{ in}^2}{(2.54)(2.54)\text{cm}^2} \approx 1902 \text{ g/cm}^2$$

18. Divide the figure into shapes for which the base and height are easily found.

Chapter Test

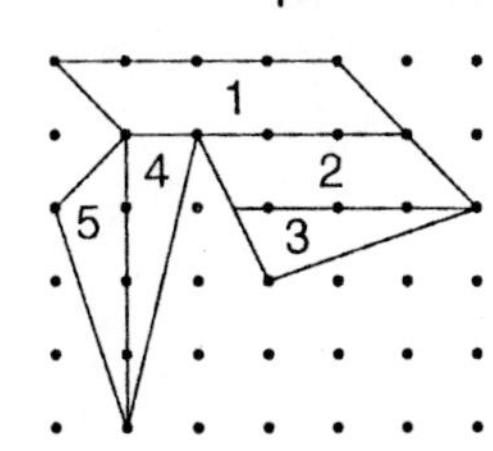

18. (continued)

Area 1 is a parallelogram with $b = 4$ and $h = 1$. Then you have $A = bh = 4 \cdot 1 = 4$.

Area 2 is a trapezoid with $b_1 = 3.5, b_2 = 3$, and $h = 1$.

$$A = \frac{1}{2}h\,(b_1 + b_2) = \frac{1}{2} \cdot 1(3 + 3.5) = 3.25$$

Area 3 is a triangle with $b = 3.5$ and $h = 1$.

$$A = \frac{1}{2}bh = \frac{1}{2}(3.5)(1) = 1.75$$

Area 4 is a triangle with $b = 1$ and $h = 4$.

$$A = \frac{1}{2}bh = \frac{1}{2} \cdot 1 \cdot 4 = 2$$

Area 5 is a triangle with $b = 4$ and $h = 1$

$$A = \frac{1}{2}bh = \frac{1}{2} \cdot 4 \cdot 1 = 2$$

Total area = $4 + 3.25 + 1.75 + 2 + 2 = 13$ square units

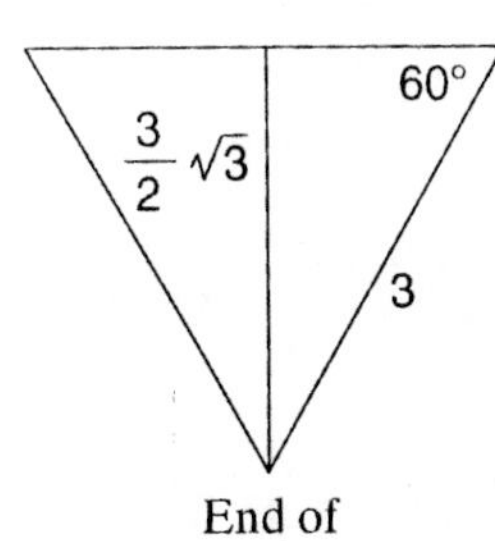

End of cut-out piece

19. This shape is a prism. Use $SA = 2A + Ph$ and $V = Ah$. The base is a rectangle with an equilateral triangle cut out of it. Since the height of an equilateral triangle forms a 30°–60° right triangle with one side and half the base, the height is $\frac{\sqrt{3}}{2}$ times the hypotenuse. Therefore,

$$h = 3 \cdot \frac{\sqrt{3}}{2} \text{ or } \frac{3}{2}\sqrt{3}.$$

$$A = 9 \cdot 8 - \text{area of the triangle} = 72 - \frac{1}{2}(3)\left(\frac{3\sqrt{3}}{2}\right)$$

$$= 72 - \frac{9\sqrt{3}}{4} \approx 68.1 \text{ in}^2$$

$$P = 8 + 8 + 9 + 3 + 3 + 3 + 3 = 37 \text{ in.}$$

$$h = 8 \text{ in.}$$

$$SA = 2\left(72 - \frac{9\sqrt{3}}{4}\right) + 37 \cdot 8$$

$$= 144 - \frac{9\sqrt{3}}{2} + 296$$

$$= 440 - \frac{9\sqrt{3}}{2}$$

$$SA \approx 432 \text{ in}^2$$

$$V = \left(72 - \frac{9\sqrt{3}}{4}\right)8$$

$$= 576 - 18\sqrt{3}$$

$$V \approx 545 \text{ in}^3$$

Chapter Test

20. Surface area is the sum of the areas of the outside, inside, and ends.

Outside area = outside circumference $\cdot h$

$$= 2\pi \cdot \frac{24}{12} \cdot 10 = 40\pi \text{ ft}^2$$

Inside area = inside circumference $\cdot h$

$$= 2\pi \cdot \frac{20}{12} \cdot 10 = \frac{100}{3}\pi \text{ ft}^2$$

End areas = 2 · (area of the outside circle − area of the inside circle)

$$= 2 \cdot \left(\pi \cdot 2^2 - \pi\left(\frac{5}{3}\right)^2\right) = \frac{22}{9}\pi \text{ ft}^2$$

Thus,

$$SA = 40\pi + \frac{100}{3}\pi + \frac{22}{9}\pi$$

$$= \frac{682\pi}{9}$$

$$SA \approx 238 \text{ ft}^2.$$

21. The circumference of Alice's tire is 27π in. Use dimensional analysis to find the number of revolutions her wheel makes in 20 miles.

$$\frac{20 \cancel{\text{mi}}}{1} \times \frac{5280 \cancel{\text{ft}}}{1 \cancel{\text{mi}}} \times \frac{12 \cancel{\text{in.}}}{1 \cancel{\text{ft}}} \times \frac{1 \text{ rev}}{27\pi \cancel{\text{in.}}} = 14{,}939 \text{ revolutions}$$

The circumference of Bob's tire is 26π in.

$$\frac{20 \cancel{\text{mi}}}{1} \times \frac{5280 \cancel{\text{ft}}}{1 \cancel{\text{mi}}} \times \frac{12 \cancel{\text{in.}}}{1 \cancel{\text{ft}}} \times \frac{1 \text{ rev}}{26\pi \cancel{\text{in.}}} = 15{,}514 \text{ revolutions}$$

Bob's wheel makes 15,514 − 14,939 = 575 revolutions more than Alice's wheel.

22. The total volume of the silo is given by

V = volume of hemisphere + volume of cylinder.

You know that the radius of the hemisphere = radius of the cylinder so the total volume of the silo is

$$V = \frac{1}{2} \cdot \frac{4}{3}\pi r^3 + \pi r^2 h$$

$$V = \frac{1}{2} \cdot \frac{4}{3}\pi (10)^3 + \pi(10)^2 \cdot 70$$

$$= \frac{2000}{3}\pi + 7000\pi$$

$$= \frac{23{,}000\pi}{3}$$

$$V \approx 24{,}086 \text{ ft}^3.$$

Chapter Test

15′

25′

25′

23. The wall can be thought of as a very long trapezoidal prism lying on one side. The bases are the same size as any cross-sectional slice of the wall (See figure). The height of the prism is the length of the wall.

Find the area of the cross-section.

$$A = \frac{1}{2}h\,(b_1 + b_2) = \frac{1}{2}(25)(25 + 15) = 500 \text{ ft}^2$$

You need to change the height to feet.

$$h = \frac{1500 \text{ mi}}{1} \times \frac{5280 \text{ ft}}{1 \text{ mi}} = 7{,}920{,}000 \text{ ft}$$

Finally, find the volume and convert it to cubic yards.

$$V = A \cdot h = 500 \cdot 7{,}920{,}000 = 3{,}960{,}000{,}000 \text{ ft}^3$$

Converting to yd^3, $V = \dfrac{3960000000 \text{ ft}^3}{1} \times \dfrac{1 \cdot 1 \cdot 1 \text{ yd}^3}{3 \cdot 3 \cdot 3 \text{ ft}^3}$

$V \approx 146{,}666{,}667$, or $\approx 147{,}000{,}000$ yd^3.

24. Consider the drawings of the kite shown below.

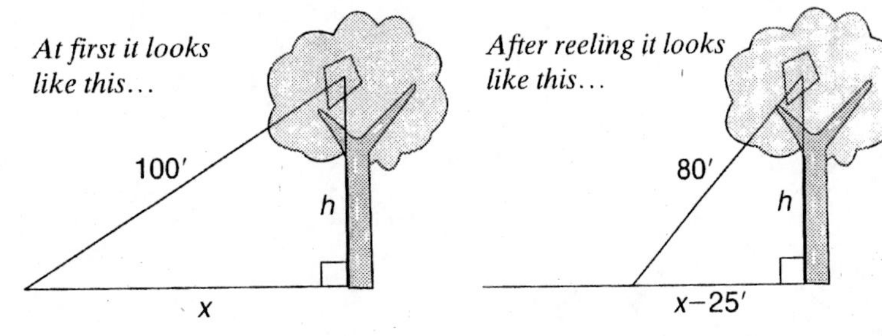

Applying the Pythagorean Theorem to each right triangle you get:

$$x^2 + h^2 = 100^2 \text{ and } (x - 25)^2 + h^2 = 80^2.$$

From the first equation, $h^2 = 100^2 - x^2$.

Now substitute for h^2 in the second equation.

$$(x - 25)^2 + 100^2 - x^2 = 80^2$$
$$x^2 - 50x + 625 + 10{,}000 - x^2 = 6400$$
$$-50x + 10625 = 6400$$
$$-50x = -4225$$
$$x = 84.5 \text{ ft}$$

However, you need to find h.

$$h^2 = 100^2 - x^2 = 10000 - (84.5)^2$$
$$h^2 = 2859.75$$
$$h = \sqrt{2859.75}$$
$$h \approx 53.4766 \approx 53 \text{ ft}$$

The kite is stuck approximately 53 feet up in the tree.

4

Reasoning and Triangle Congruence

Section 4.1

Tips:

- ✔ You may find it easiest to try a paragraph proof first, and then write a statement-reason proof.
- ✔ The two parts of a biconditional statement may be placed in any order.

Solutions to odd-numbered textbook problems

1. Use the Law of Detachment since the second statement is the hypothesis of the first.

   First Statement: If three points are not collinear, then they lie in one and only one plane.

   Second Statement: The three vertices of a triangle are not collinear.

   ∴ The three verticles of a triangle lie in one and only one plane.

3. Use the Law of Syllogism since the conclusion of the first statement is the hypothesis of the second statement.

   First Statement: If a cube has a volume of 27 in^3, then it has an edge of length 3 in.

   Second Statement: If a cube has an edge of length 3 in., then it has a surface area of 54 in^2.

   Therefore (Fourth Statement): If a cube has a volume of 27 in^3, then it has asurface area of 54 in^2.

   Use the Law of Detachment since the third statement is the hypothesis of the first statement and also the hypothesis of the fourth statement.

   Third Statement: ABCDEFGH is a cube with a volume of 27 in^3.

   ∴ ABCDEFGH has an edge of length 3 in.

   ∴ ABCDEFGH has an surface area of 54 in^2.

5. Use the Law of Detachment since the second statement is the hypothesis of the first.

   First Statement: If lines l and m are in the same plane and are not parallel, then l and m intersect.

   Second Statement: Lines l and m are in the same plane and are not parallel.

   ∴ Lines l and m intersect.

Section 4.1

7. Since the second statement is the converse of the first statement, the result is a biconditional.

   Biconditional: The lines are perpendicular if and only if they form right angles.

9. Use the Law of Detachment since the second statement is the hypothesis of the first.

   First statement: If a triangle is scalene, then the triangle has no two sides congruent.

   Second Statement: ΔABC is a scalene triangle.

   $\therefore \Delta ABC$ has no two sides congruent.

11. No deduction is possible, since the second statement is not the hypothesis of the first.

13. If each angle in a triangle is less than 90°, then the triangle is acute.

    ΔQRS is equiangular.

    An equiangular triangle has three 60° angles.

    60° is less than 90°.

    $\therefore \Delta QRS$ is acute.

15. Use the Law of Detachment since the third statement is the hypothesis of the first statement.

    First statement: If ABCD is a square, then it is a quadrilateral.

    Second Statement: If ABCD is a quadrilateral with four congruent sides, then it is a rhombus.

    Third Statement: ABCD is a square.

    Now, since a square has four congruent sides, the third statement is also the hypothesis of the second statement and the Law of Detachment can be used again.

    ∴ ABCD is a quadrilateral with four congruent sides.

    ∴ ABCD is a rhombus.

17. Statement: If a square has an area of 25 cm^2, then the length of its side is 5 cm. True.

    Converse: If the length of the side of a square is 5 cm, then it has an area of 25 cm^2. True.

    Since both the statement and its converse are true, you can write an equivalent biconditional statement.

    Biconditional: A square has an area of 25 cm^2 if and only if the length of its side is 5 cm.

19. Statement: If two angles are adjacent, then they share a common side. True.

    Converse: If two angles share a common side, then they are adjacent. False.

Section 4.1

19. (continued)

The statement is true, but the converse is false, therefore you cannot write a biconditional.

Example to show converse is false:

$\angle ABC$ and $\angle ABD$ have $\overline{AB}$ as a common side, but $\angle ABC$ is not adjacent to $\angle ABD$.

21. Statement: If the side of a square measures 10 in., then the diagonal of the square is $10\sqrt{2}$ in. True.

Converse: If the diagonal of a square is $10\sqrt{2}$ in., then the side of the square measures 10 in. True.

Since both the statement and its converse are true, you can write an equivalent biconditional statement.

Biconditional: The side of a square measures 10 in. if and only if the diagonal of the square is $10\sqrt{2}$ inches.

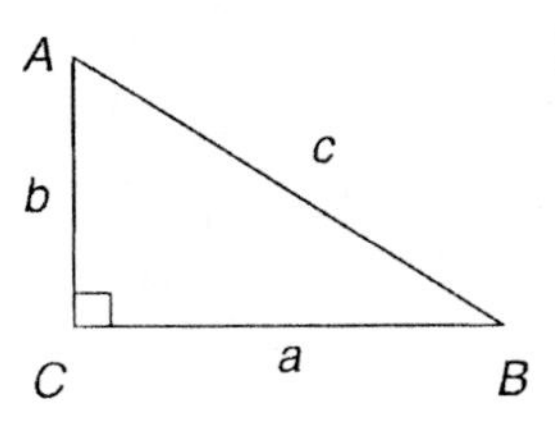

23. Statement: If ΔABC as shown has a right angle at $\angle C$, then $a^2 + b^2 = c^2$. True.

Converse: If $a^2 + b^2 = c^2$ in ΔABC as shown, then ΔABC has a right angle at $\angle C$. True.

Since both the statement and its converse are true you can write an equivalent biconditional statement.

Biconditional: ΔABC has a right angle at $\angle C$ if and only if $a^2 + b^2 = c^2$.

25. Statement: If a triangle has sides of lengths 3, 4, and 5, then it is a right triangle. True.

Converse: If a triangle is a right triangle, then it has sides of lengths 3, 4, 5. False (since sizes and shapes of right triangles vary).

Since the converse is false, you cannot write an equivalent biconditional.

27. Statement: If a triangle is equilateral, then the triangle has rotational symmetry. True.

Converse: If a triangle has rotational symmetry, then the triangle is equilateral. True.

Since both the statement and its converse are true you can write an equivalent biconditional statement.

Biconditional: A triangle is equilateral if and only if it has rotational symmetry.

29. Statement: If a triangle has three equal sides, then the triangle is equiangular. True.

Converse: If a triangle is equiangular, then it has three equal sides. True.

Section 4.1

29. (continued)

Since both the statement and its converse are true, you can write an equivalent biconditional statement.

Biconditional: A triangle has three equal sides if and only if it is equiangular.

31. Statement: If a pyramid has a base with n sides, then the pyramid has $n + 1$ vertices. True.

Converse: If a pyramid has $n + 1$ vertices, then the pyramid has a base with n sides. True.

Since both the statement and its converse are true you can write an equivalent biconditional statement.

Biconditional: A pyramid has a base with n sides if and only if it has $n + 1$ vertices.

33. Use the Law of Detachment.

$r \Rightarrow s$

$\underline{r \quad}$

$\therefore s$

35. Use the Law of Syllogism to conclude any one of $r \Rightarrow t$, $r \Rightarrow p$, $s \Rightarrow t$.

37. Remember, you can change the order of the statements.

$p \Rightarrow \text{not } q$

$\text{not } q \Rightarrow \text{not } r$

$\underline{\text{not } r \Rightarrow t \quad}$

$\therefore p \Rightarrow t$ or $p \Rightarrow \text{not } r$ or $\text{not } q \Rightarrow t$

(All three conclusions are valid.)

39. The first part of the statement (a parallelogram) will become the hypothesis. If $ABCD$ is a parallelogram, then $ABCD$ is a quadrilateral.

41. The first part of the statement (the description of the prism) will become the hypothesis. If a prism has a base with n sides, then the prism has $2n$ vertices.

43. The first part of the statement (a regular polyhedron) will become the hypothesis. If a polyhedron is regular, then its faces are regular polygons.

45.

Statements	Reasons
1. ΔPQR is a right triangle	1. Given
2. $\angle P + \angle Q + \angle R = 180°$	2. Angle sum in a triangle is 180°.
3. $\angle R = 90°$	3. Definition of right angle.
4. $\angle P + \angle Q + 90° = 180°$	4. Substitution [(3) into (2)].
5. $\angle P + \angle Q = 90°$	5. Subtract 90° from both sides.
6. $\angle P$ and $\angle Q$ are complementary.	6. Definition of complementary angles.

47. A complete solution is given in the answer key at the back of the text.

49. A complete solution is given in the answer key at the back of the text.

Section 4.2

Tip:

✔ When you show congruence between triangles, the order of the vertices in one triangle should match the order of the vertices in the second triangle so that congruent sides and angles match up.

✔ Be sure not to make any unwarranted assumptions. For example, do not assume that two angles are congruent just because they *look* congruent.

Solutions to odd-numbered textbook problems

1. Look for matching tick marks and the angles opposite those sides. $\angle P \cong \angle X, \angle Q \cong \angle Y, \angle R \cong \angle Z, \overline{PQ} \cong \overline{XY}, \overline{QR} \cong \overline{YZ}, \overline{PR} \cong \overline{XZ}$

3. $\Delta ABC \cong \Delta VWU$ by matching the tick marks and the angles opposite those sides. (Other possible congruence statements: $\Delta BCA \cong \Delta WUV$, $\Delta CAB \cong \Delta UVW, \Delta ACB \cong \Delta VUW, \Delta CBA \cong \Delta UWV, \Delta BAC \cong \Delta WVU$.)

5. You have $\overline{AB} \cong \overline{DE}, \overline{CA} \cong \overline{FD}, \angle A \cong \angle D$.

   Therefore, $\Delta ABC \cong \Delta DEF$ by SAS.

   (Other possible congruence statements: $\Delta BCA \cong \Delta EFD$, $\Delta CAB \cong \Delta FDE, \Delta ACB \cong \Delta DFE, \Delta CBA \cong \Delta FED, \Delta BAC \cong \Delta EDF$.)

7. The triangles are not necessarily congruent. The right angle is **not** between the sides with the single and double tick marks in ΔMNO, but the right angle is between the sides with the single and double tick marks in ΔRQP. Therefore neither the HL Congruence Theorem nor the LL Congruence Theorem applies.

Section 4.2

9. You have $\overline{QR} \cong \overline{UT}$, $\overline{RS} \cong \overline{TV}$, $\angle R \cong \angle T$.

   Therefore, $\Delta QRS \cong \Delta UTV$ by SAS.

   (Other possible congruence statements: $\Delta RSQ \cong \Delta TVU$, $\Delta SQR \cong \Delta VUT$, $\Delta SRQ \cong \Delta VTU$, $\Delta QSR \cong \Delta UVT$, $\Delta RQS \cong \Delta TUV$, $\Delta QRS \cong \Delta VTU$, $\Delta SRQ \cong \Delta UTV$.)

11. $\angle ABE \cong \angle DBC$ because they are vertical angles. Also, we are given that $\angle EAB \cong \angle CDB$, and $\overline{AB} \cong \overline{BD}$. Therefore, we have $\Delta ABE \cong \Delta DBC$ by ASA.

13. Note that since $\angle WXZ = 90°$ and $\angle YXZ$ is its supplement, $\angle YXZ = 180° - \angle WXZ = 180° - 90° = 90°$.

    So, $\angle WXZ \cong \angle YXZ$.

    Also, $\overline{YX} \cong \overline{XW}$ and $\overline{XZ} \cong \overline{XZ}$. Therefore, $\Delta WXZ \cong \Delta YXZ$ by LL (or SAS).

15. ΔRYV and ΔTWX are right triangles where $\overline{RY} \cong \overline{TW}$ and $\overline{YV} \cong \overline{WX}$. Therefore, $\Delta RYV \cong \Delta TWX$ by LL.

17. You have $\overline{BA} \cong \overline{BC}$, $\angle A \cong \angle C$, $\angle ABE \cong \angle CBD$.

    Therefore, $\Delta CBD \cong \Delta ABE$ by ASA.

    $\angle ABE = \angle ABD + \angle DBE$ and $\angle CBD = \angle CBE + \angle DBE$.

    Since $\angle ABE \cong \angle CBD$ you have $\angle ABD \cong \angle CBE$. Thus, $\Delta CBE \cong \Delta ABD$ by ASA.

19. You have $\overline{QS} \cong \overline{QT}$, $\overline{SR} \cong \overline{TP}$, and $\overline{RQ} \cong \overline{PQ}$.

    Therefore, $\Delta QSR \cong \Delta QTP$ by SSS.

    You have $\overline{ST} \cong \overline{TS}$, $\overline{TP} \cong \overline{SR}$, and $\overline{PS} \cong \overline{RT}$.

    Therefore, $\Delta STP \cong \Delta TSR$ by SSS.

21. $\overline{AB} \cong \overline{DE}$

23. $\angle E \cong \angle B$

25. $\overline{CB} \cong \overline{FE}$ by C.P. (Corresponding parts of congruent triangles are congruent.)

27. The vertex angles of a regular hexagon are congruent since they each have measure 120°.

    Therefore, $\angle B \cong \angle D \cong \angle BCD \cong \angle DEF \cong \angle EFA \cong \angle FAB$.

29. $\overline{AC} \cong \overline{EC}$ by C.P. since $\Delta ACB \cong \Delta ECD$ (by SAS).

31. $\overline{AC} \cong \overline{EC}$ see problem #29, $\overline{FA} \cong \overline{FE}$ since since the sides of a regular hexagon are congruent, and $\overline{CF} \cong \overline{CF}$.

    Thus, $\Delta CFA \cong \Delta CFE$ by SSS.

    Therefore, $\angle FEC \cong \angle FAC$ by C.P.

Section 4.2

33. We are given that $\overline{UW} \cong \overline{YW}$ and $\overline{VW} \cong \overline{XW}$.

    $\angle VWU \cong \angle XWY$ by vertical angles.

    Therefore, $\Delta UVW \cong \Delta YXW$ by SAS.

    By C.P. we have $\angle U \cong \angle Y$.

35. Consider the rhombus $ABCD$ as shown. The four sides are congruent, so $\overline{AB} \cong \overline{CD}$ and $\overline{AD} \cong \overline{CB}$. The diagonal is congruent to itself, so $\overline{BD} \cong \overline{BD}$. Thus $\Delta ABD \approx \Delta CDB$ by SSS.

37. Let $ABCD$ be a quadrilateral and let $\overline{AC}$ and $\overline{BD}$ divide each other in half at E.

    Now $\angle BEC \cong \angle DEA$ by vertical angles and since $\overline{BE} \cong \overline{DE}$ and $\overline{EC} \cong \overline{EA}$ are given, we have $\Delta BEC \cong \Delta DEA$ by SAS.

    Therefore, $\angle DBC \cong \angle BDA$ by C.P.

39. <u>Given</u>: $\overline{AC}$ and $\overline{BD}$ bisect each other at E.

    <u>Subgoal 1</u>: Prove $\Delta BEC \cong \Delta DEA$.

    <u>Subgoal 2</u>: Prove $\Delta DBC \cong \Delta BDA$.

    <u>Proof of Subgoal 1</u>:

    You have both $\overline{BE} \cong \overline{DE}$ and $\overline{AE} \cong \overline{CE}$. Since $\overline{AC}$ and $\overline{BD}$ bisect each other, E is the midpoint of both diagonals. Also $\angle BEC \cong \angle DEA$ since they are vertical angles. Thus, $\Delta BEC \cong \Delta DEA$ by SAS.

    <u>Proof of Subgoal 2</u>:

    Now you know $\overline{BC} \cong \overline{DA}$ and $\angle BDA \cong \angle DBC$ by C.P. Also, $\overline{BD} \cong \overline{DB}$, so $\Delta DBC \cong \Delta BDA$ by SAS.

41. Since $\overline{AB} \cong \overline{AF}$ and $\overline{AE} \cong \overline{AC}$, we have $\Delta ACF \cong \Delta AEB$ by HL.

    Therefore, $\angle ACF \cong \angle AEB$ by C.P.

43. Let $\Delta ABC'$ and $\Delta A'B'C'$ be right triangles with $\overline{AB} \cong \overline{A'B'}$ and $\overline{BC} \cong \overline{B'C'}$.

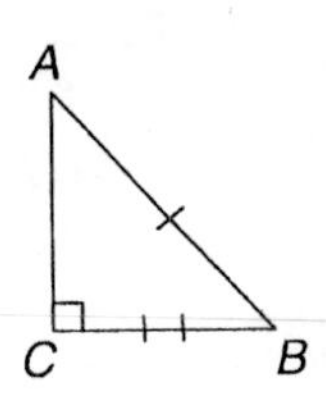

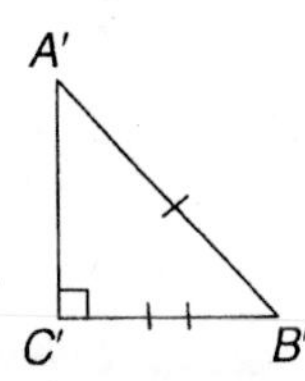

    So $AB = A'B'$ and $BC = B'C'$ as congruent sides have equal measure.

    By the Pythagorean Theorem $AC = A'C'$ and so $\overline{AC} \cong \overline{A'C'}$ as sides with equal measure are congruent.

    Now $\Delta ABC \cong \Delta A'B'C'$ by LL.

45. A complete solution is given in the answer key at the back of the text.

Section 4.3

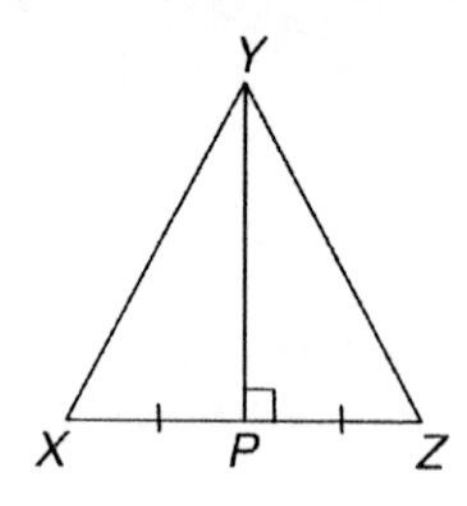

Section 4.3

Solutions to odd-numbered textbook problems

1. You have $\angle XPY = 90°$ and $\angle ZPY = 90°$ because $\overline{YP}$ is perpendicular to $\overline{XZ}$. Then $\overline{XP} \cong \overline{PZ}$ since $\overline{YP}$ bisects $\overline{XZ}$. $\overline{YP} \cong \overline{YP}$ (reflexive property). Therefore, $\Delta XYP \cong \Delta ZYP$ by SAS. That makes $\angle Z \cong \angle X$ and $\angle Z = 70°$.

3. $\angle BEC + \angle CED = 180°$ Supplementary Angles

$$86° + \angle CED = 180°$$

$$\angle CED = 94°$$

ΔCED is isosceles since $\overline{EC} \cong \overline{ED}$. Therefore the base angles of ΔCED are congruent and $\angle EDC = \angle ECD = 43°$.

ΔBCD is isosceles since $\overline{BC} \cong \overline{DC}$. Therefore the base angles of ΔBCD are congruent and $\angle DBC = 43°$.

The angle sum of a triangle is 180°, thus

$$\angle BEC + \angle CBE + \angle BCE = 180°$$

$$86° + 43° + \angle BCE = 180°$$

$$\angle BCE = 51°$$

$$\text{Hence } \angle BCA = 51°.$$

ΔABC is isosceles since $\overline{AB} \cong \overline{AC}$. Therefore the base angles of ΔABC are congruent and $\angle ABC = 51°$.

$$\text{Now } \angle ABD + \angle DBC = \angle ABC$$

$$\angle ABD + 43° = 51°$$

$$\angle ABD = 8°.$$

Thus $\angle ABE = 8°$.

5. Find $\angle GBD$. (part a)

$\overline{BD} \cong \overline{DG}$ implies ΔBDG is isosceles, so $\angle GBD = \angle BGD$.

$$\angle GBD + \angle DGB + \angle BDG = 180°$$

$$2\angle GBD + 28° = 180° \qquad \angle BDG = \angle ADG$$

$$\angle GBD = 76°$$

Find $\angle DAG$.

$\angle AGD + \angle DGE = 180°$ since A, G and E are collinear.

$$\angle AGD + 90° = 180°$$

$$\angle AGD = 90°$$

Now $\angle AGD + \angle GDA + \angle DAG = 180°$ by angle sum of a triangle.

$$90° + 28° + \angle DAG = 180°$$

$$\angle DAG = 62°$$

Section 4.3

5. (continued)

Find $\angle ACG$. (part b)

$\overline{AG} \cong \overline{CG}$ implies ΔGAC is isosceles so $\angle ACG = \angle DAG$. So, $\angle ACG = 62°$.

Find $\angle AGB$. (part c)

$$\angle AGB = \angle AGD - \angle BGD$$
$$\angle AGB = 90° - 76°$$
$$= 14°$$

Before finding part (d), we will find parts (e) and (f)

Find $\angle GCD$.

$$\angle GCD + \angle ACG = 180° \quad \text{supplementary angles}$$
$$\angle GCD + 62° = 180°$$
$$\angle GCD = 118°$$

Find $\angle CGD$. (part e)

$$\angle DCG + \angle GDC + \angle CGD = 180° \quad \text{angle sum of a triangle}$$
$$118° + 28° + \angle CGD = 180°$$
$$\angle CGD = 34°$$

Find $\angle BGC$. (part f)

$$\angle BGC + \angle CGD = \angle BGD$$
$$\angle BGC + 34° = 76° \; (\angle DBG = \angle BGD = 76°)$$
$$\angle BGC = 42°$$

Find $\angle FGE$. (part d)

$\angle FGE = \angle AGC$ vertical angles, since F, G, and C are collinear

$$= \angle AGB + \angle BGC$$
$$= 14° + 42°$$
$$= 56°$$

7. (a) In isosceles ΔABC, $\overline{AB} \cong \overline{BC}$ and $\angle ABD \cong \angle CBD$, so $\overline{BE}$ is the perpendicular bisector of $\overline{AC}$ by Theorem 4.9. Therefore $\angle AEB = \angle CEB = 90°$.

   (b) $\overline{AE} \cong \overline{CE}$ by Theorem 4.9.

   (c) $\overline{AD} \cong \overline{CD}$ by Theorem 4.10.

Section 4.3

9. In the following figure, let $\angle WVY \cong \angle WYV \cong \angle YWX \cong \angle YXW$ and let $\angle VWY \cong \angle XYW$.

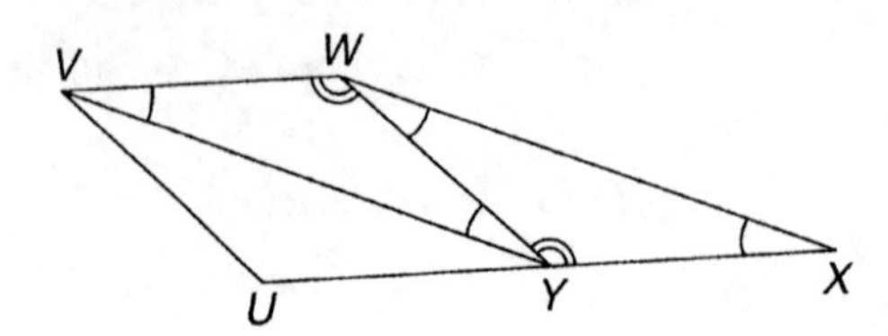

Then $\overline{VW} \cong \overline{YW}$ and $\overline{YW} \cong \overline{YX}$ by Theorem 4.7.

So $\overline{VW} \cong \overline{XY}$ by the transitive property

and $\overline{WY} \cong \overline{WY}$ by the reflexive property.

Hence, $\Delta XYW \cong \Delta VWY$ by SAS.

Therefore, $\overline{WX} \cong \overline{YV}$ by C.P.

11. Given: $\overline{PX} \cong \overline{RY}$, and $\overline{PY}$ and $\overline{RX}$ are altitudes of ΔPQR.

Prove: $\Delta PXR \cong \Delta RYP$

Proof:

Because $\overline{PY}$ and $\overline{RX}$ are altitudes, $\angle PXR = 90° = \angle RYD$. Also, $\overline{PR} \cong \overline{PR}$. So $\Delta PXR \cong \Delta RYP$ by HL.

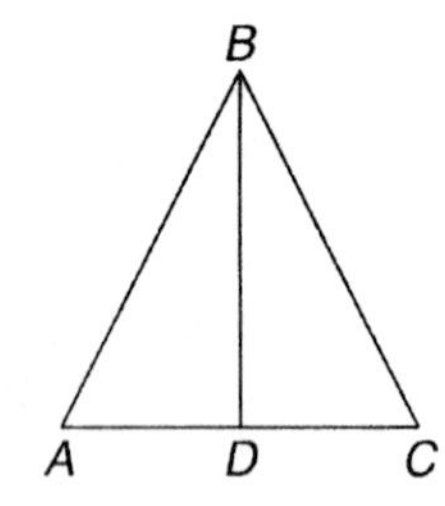

13. Let ΔABC be given where $\overline{BD}$ is an altitude and $\overline{BD}$ bisects $\angle B$.

Since $\overline{BD}$ is an altitude, $\overline{BD} \perp \overline{AC}$.

Now $\angle ABD \cong \angle CBD$ by definition of angle bisector

$\overline{BD} \cong \overline{BD}$ by reflexive property

$\angle ADB = \angle CDB = 90°$ since $\overline{BD} \perp \overline{AC}$.

So $\angle ADB \cong \angle CDB$ angles with equal measure are congruent.

Therefore, $\Delta ABD \cong \Delta CBD$ by ASA.

Then $\overline{AD} \cong \overline{CD}$ by C.P.

and $AD = CB$ congruent segments have equal measure.

So $\overline{BD}$ bisects $\overline{AC}$.

Therefore, $\overline{BD}$ is the perpendicular bisector of $\overline{AC}$.

15. Given: $AB = CB$, and $\overline{BD}$ is a median.

Subgoal 1: Show $\Delta ABD \cong \Delta CBD$.

Subgoal 2: Show $\angle ADB = 90°$.

Section 4.3

15. (continued)

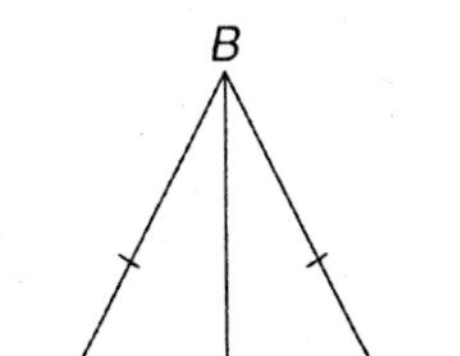

Proof of Subgoal 1:

Consider isosceles ΔABC with median $\overline{BD}$ as shown.

You know that $\overline{AB} \cong \overline{CB}$. $\overline{AD} \cong \overline{CD}$ since D is the midpoint of $\overline{AC}$. Also, $\overline{BD} \cong \overline{BD}$. Therefore, $\Delta ABD \cong \Delta CBD$ by SSS.

Proof of Subgoal 2:

$\angle ADB \cong \angle CDB$ by C.P. Also, $\angle ADB + \angle CDB = 180°$ since their non-adjacent sides form a straight angle.

Substituting $\angle ADB$ for $\angle CDB$ in the equation, you have

$$\angle ADB + \angle CDB = 180°$$

$$2\,\angle ADB = 180°$$

$$\angle ADB = 90°.$$

Therefore, $\angle ADB = \angle CDB = 90°$ and $\overline{BD}$ is perpendicular to $\overline{AC}$.

17. Let ΔABC be equilateral then $\overline{AB} \cong \overline{BC} \cong \overline{AC}$.

Now every equilateral triangle is isosceles, so ΔABC is isosceles.

Hence $\angle A \cong \angle B \cong \angle C$ by Theorem 4.5.

19. Consider equiangular ΔABC. You know that $\angle A \cong \angle C$ and $\overline{CB} \cong \overline{AB}$ since sides opposite congruent angles are congruent (Thm 4.7). Similarly, since $\angle A \cong \angle C$, $\overline{AC} \cong \overline{AB}$. Thus, $\overline{AB} \cong \overline{AC} \cong \overline{CB}$ so ΔABC is equilateral.

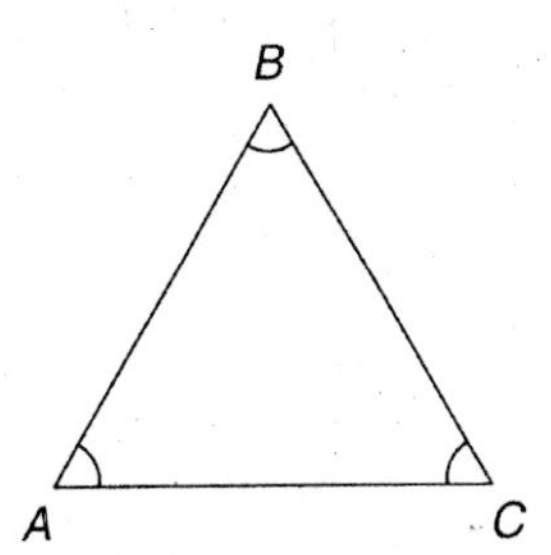

21. A complete solution is given in the answer key at the back of the text.

Section 4.4

Section 4.4

Tips

- ✔ When doing constructions, keep your pencil sharpened.
- ✔ Be very careful on each step of a construction or errors will be compounded.
- ✔ When a problem asks you to *construct* a figure, you need to make the figure with a straightedge and compass. When you are asked to *draw* a figure, you can use any tools to draw a reasonably representative figure; you are not limited to compass and straightedge.

Solutions to odd-numbered textbook problems

1. Step 1: Draw any line segment $\overline{AB}$. Extend the line segment beyond point B to form ray $\overrightarrow{AB}$.

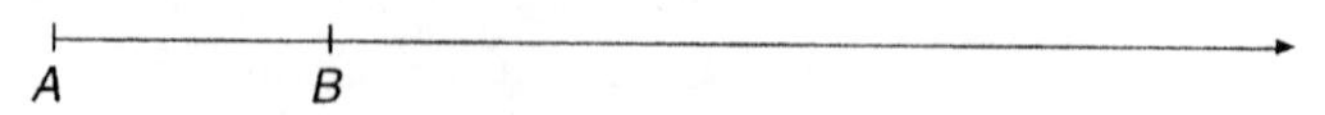

Step 2: Place the point of the compass at A with the compass set to a radius of the length of $\overline{AB}$.

Using this same radius place the point of the compass at B and mark an arc on ray $\overrightarrow{AB}$. Label the point where the arc crosses ray $\overrightarrow{AB}$ point C.

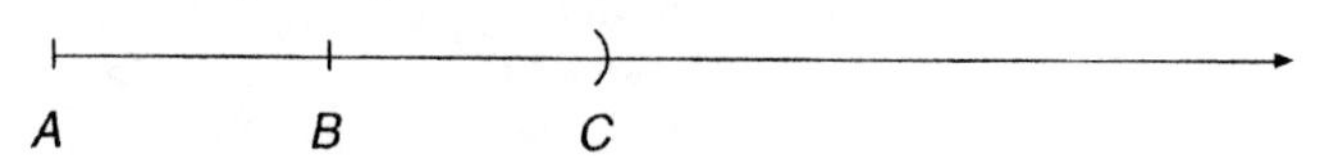

Step 3: Using the same radius place the point of the compass at C and make an arc on ray $\overrightarrow{AB}$.

Label the point where the arc crosses ray $\overrightarrow{AB}$ point D.

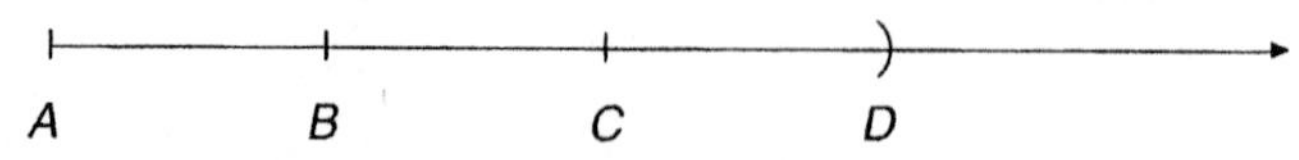

Step 4: Repeat step 3 with point of compass at D.

Label the new point E. $AE = 4(AB)$.

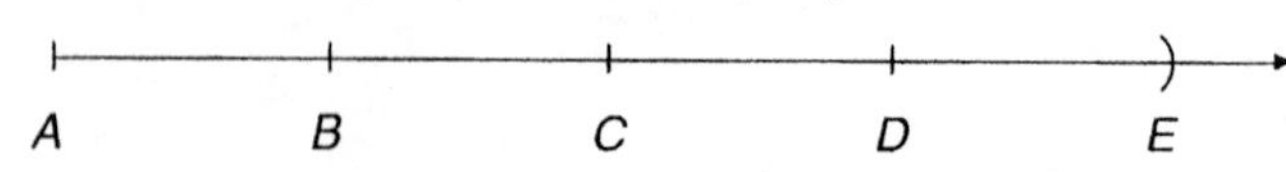

3. Step 1: Draw any acute angle and call it $\angle A$. Place the point of your compass on vertex A and swing an arc which intersects both sides of the angle.

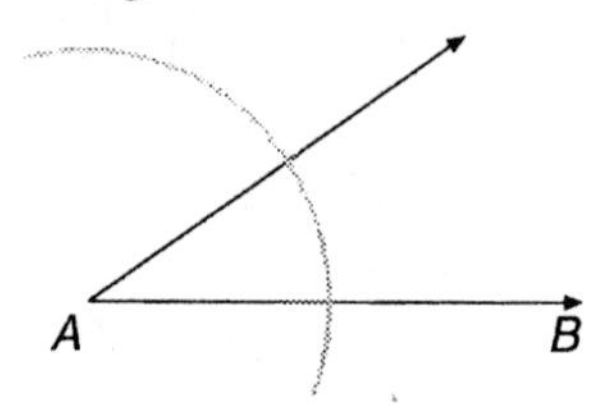

Section 4.4

3. (continued)

Step 2: Set the radius of the compass to length PQ. Keeping that radius, place the point of the compass at Q and mark off the distance PQ on the arc already there, giving point C.

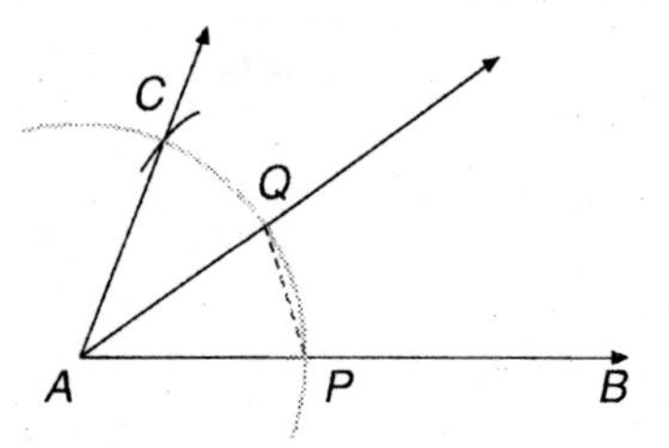

Step 3: Using a straight edge, draw $\overrightarrow{AC}$. Now $\angle BAC$ is twice the original angle.

5. Step 1: First draw any angle and call it $\angle A$. Put the point of your compass on the vertex A and swing an arc which intersects both sides of the angle.

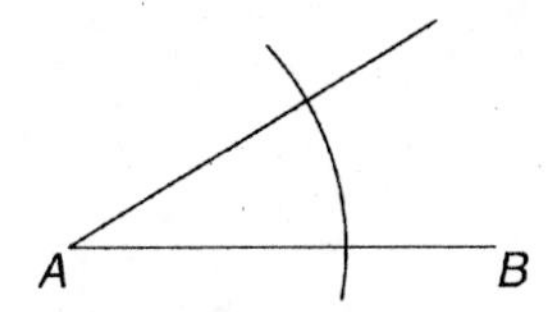

Step 2: Placing the compass point where the arc crosses the side of your angle, swing another arc. You can use the same radius as in Step 1 or you can change the radius if you like. Repeat this procedure from the other side of the angle.

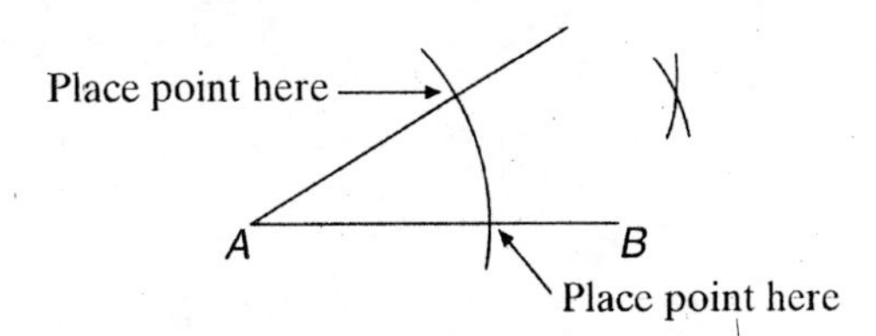

Step 3: Using a straightedge, draw a line joining A and the × you just made by the intersecting arcs.

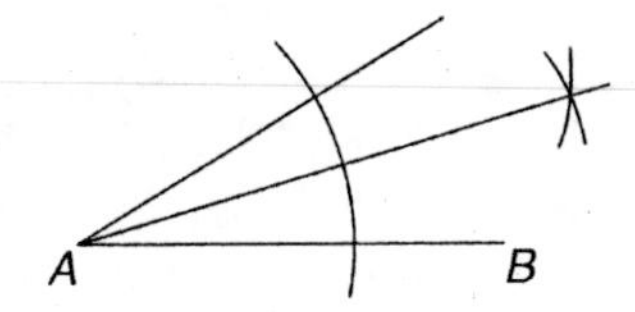

Section 4.4

7. You can do this construction just as you would construct a perpendicular bisector. The point where the bisector crosses $\overline{AB}$ is the midpoint of $\overline{AB}$.

Step 1: First draw any line segment and call it $\overline{AB}$. Place the point of the compass at A with the compass set to a radius of more than half the length of the line segment. Make arcs above and below $\overline{AB}$.

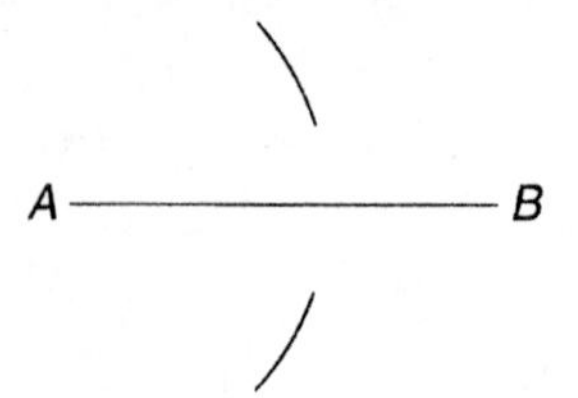

Step 2: Using the same radius, place the point of the compass at B and make arcs above and below, crossing the arcs you just made from point A.

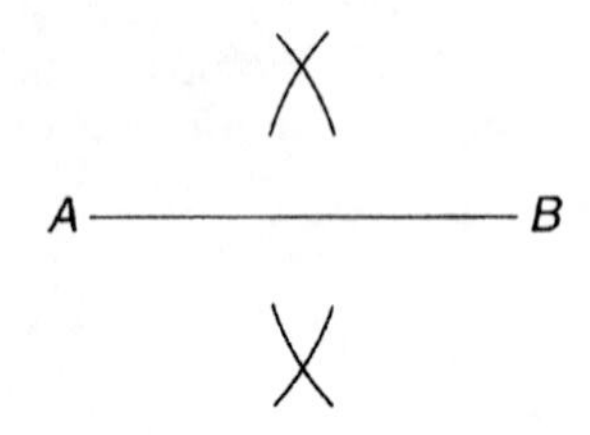

Step 3: Using a straightedge lined up on the X's made by the intersection of the arcs, mark where the straightedge crosses $\overline{AB}$. This is the midpoint.

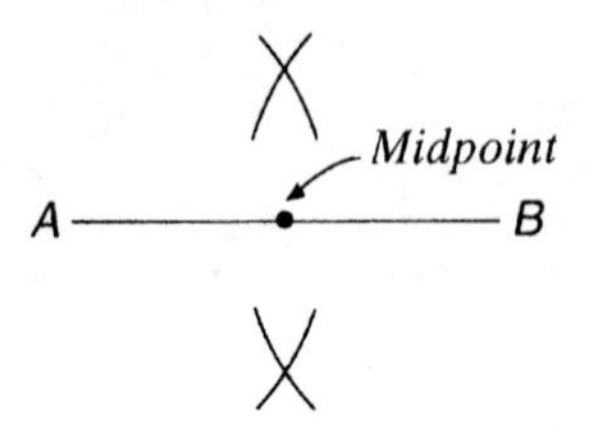

9. Step 1: Draw line $\overleftrightarrow{AB}$ and place any point P that is not on $\overleftrightarrow{AB}$.

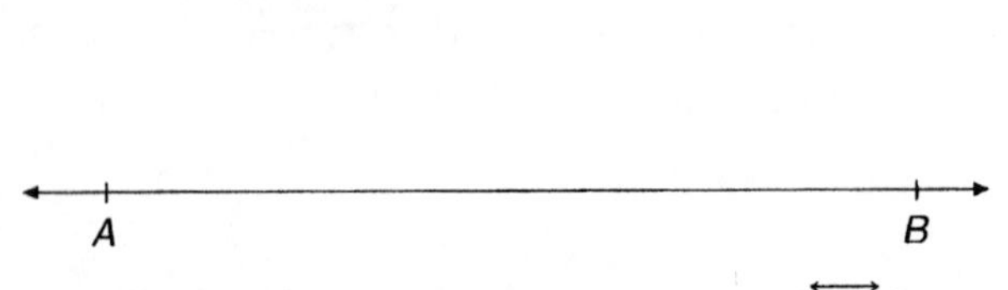

Step 2: Swing an arc with center P that intersects $\overleftrightarrow{AB}$ in two places. Mark the points of intersection C and D.

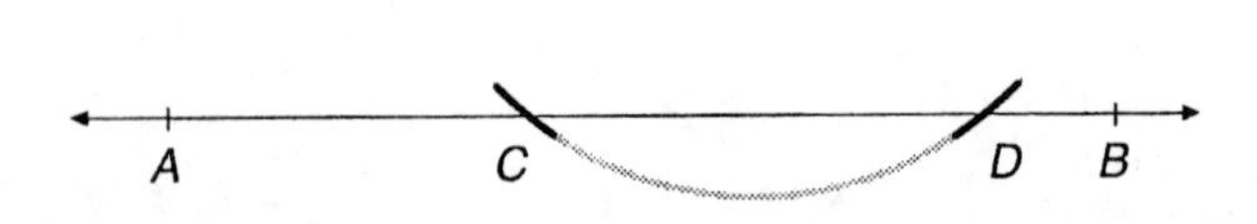

Section 4.4

Step 3: Now swing an arc from C that falls below $\overleftrightarrow{AB}$. Repeat from D, intersecting the two arcs. Connect P and the point of intersection of the arcs, drawing a line. This line is perpendicular to $\overleftrightarrow{AB}$.

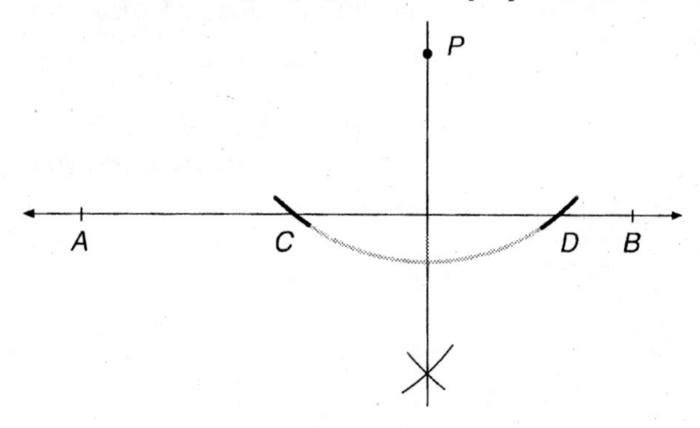

11. Step 1: Draw a line segment, set your compass radius to length a, and mark that length on your line segment.

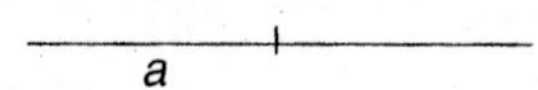

Step 2: Leaving your compass set at radius a, put your compass point on one end of the segment with length a and swing an arc. Then place the compass point at the other end of the segment, and swing an arc crossing the previous one.

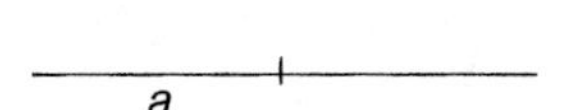

Step 3: Connect the endpoints of the segment with length a to the × made by the arcs.

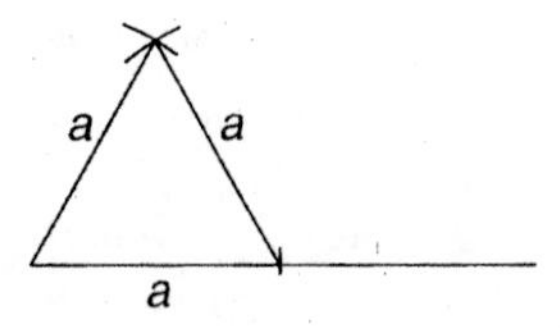

Now you have an equilateral triangle which has sides of length a.

13. Step 1: Copy the line of length a on one side of $\angle C$.

Copy the line of length b on the other side of $\angle C$.

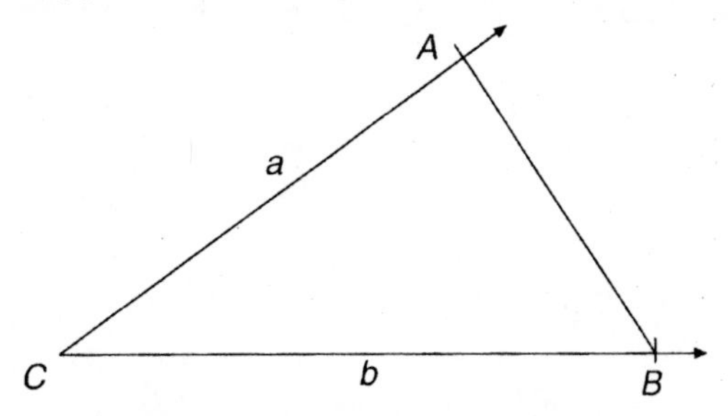

Step 2: Label the two points of intersection found in Step 1 as points A and B. Connect points A and B to form ΔABC.

Section 4.4

15. There are several ways to construct a 60° angle. One is to construct a 30°–60° right triangle, but it is probably easier to proceed as though constructing an equilateral triangle (see problem #11). Since an equilateral triangle is also equiangular, each angle will be 60°.

    Draw a short line segment. Set your compass to a radius equal to the length of the line segment. Put the point on one end of the segment and swing an arc. Repeat from the other end. Connect one end of the segment to the × made by your arcs to make a 60° angle.

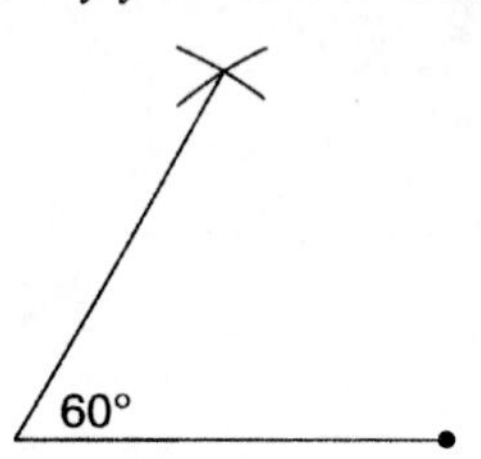

17. Step 1: Draw line $\overleftrightarrow{AB}$ and a point P, that is not on $\overleftrightarrow{AB}$. Construct a line through P perpendicular to $\overleftrightarrow{AB}$ (see problem #9). Label the intersection E.

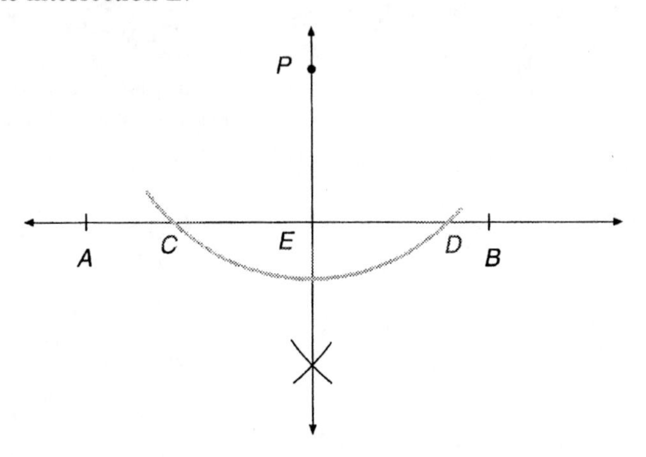

    Step 2: Using Construction 3 (to bisect an angle) in the text, bisect $\angle PEB$ to construct a 45° angle.

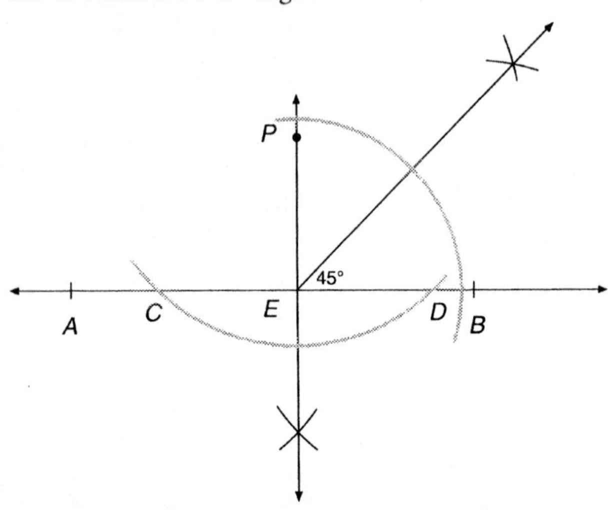

Section 4.4

19. Use the fact that $120° = 2 \cdot 60°$. That is, you can add two $60°$ angles together.

Step 1: Construct a $60°$ angle (see problem #15).

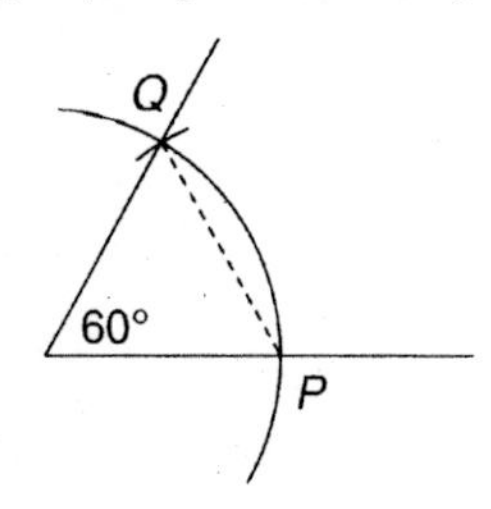

Step 2: Draw a line segment to use as the base of the $120°$ angle. Label the point that will be the vertex A. Now placing your compass point at A, swing an arc. Label as B the point of intersection with your base. Using the same radius and placing the point of your compass at the vertex of the $60°$ angle, swing an arc that intersects the sides of the $60°$ angle at P and Q.

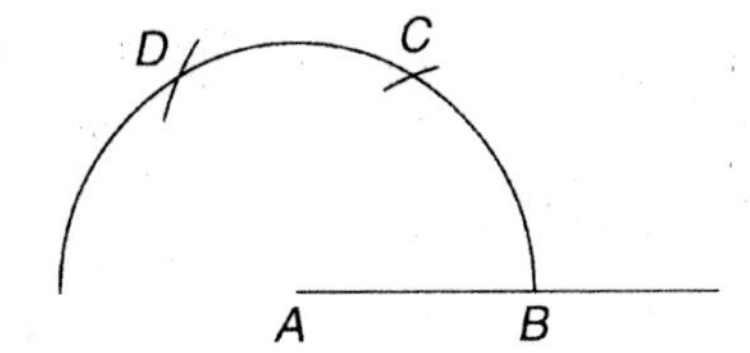

Step 3: Set your compass radius to the length PQ (from the $60°$ angle). Keeping that radius, place the point of the compass at B and mark off the distance PQ on the arc already there, giving point C.

Step 4: Then, still keeping the same radius PQ, place the point of your compass at C and mark the point D further along the arc.

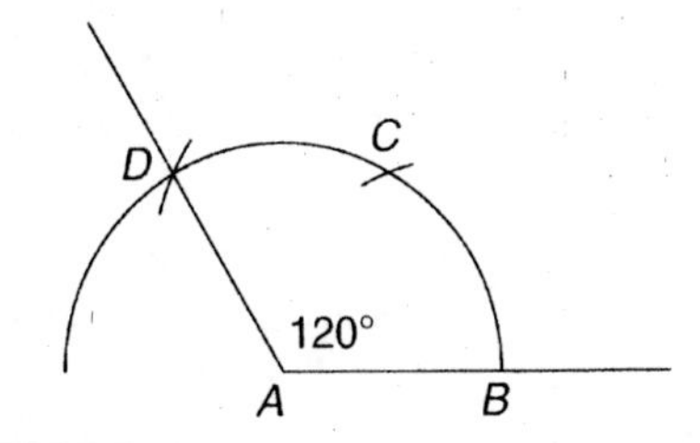

Step 5: Using a straightedge, draw $\overline{AD}$. Now $\angle BAD$ is $120°$.

Section 4.4

21. Step 1: Follow the procedure in Construction 4 in the text to construct a perpendicular to a point on a line.

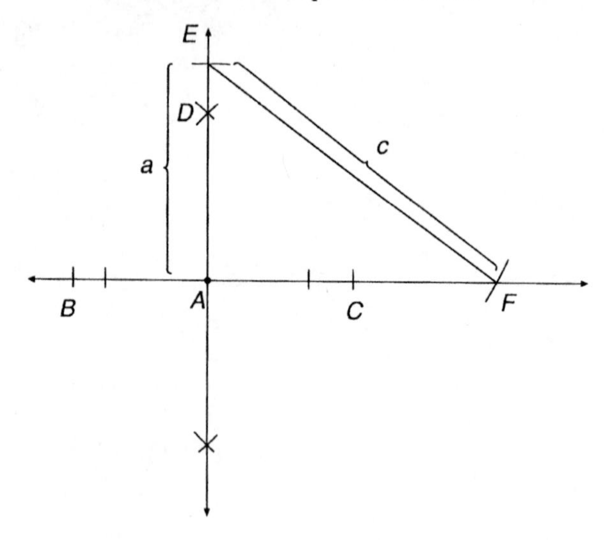

Step 2: Set the radius of the compass to length a and with the point of the compass at A mark off length a on ray $\overrightarrow{AD}$. Label the point E.

Set the radius of the compass to length c and with the point of the compass at E swing an arc to intersect $\overrightarrow{AC}$. Label the point F. ΔAEF is the desired triangle.

23. You can trace the points A, B, and C and then join them using a straightedge to make a good copy of ΔABC. To construct the altitude from B, construct a perpendicular from B to $\overline{AC}$.

Step 1: Swing an arc with center B that intersects $\overline{AC}$ in two places. Mark the points of intersection D and E.

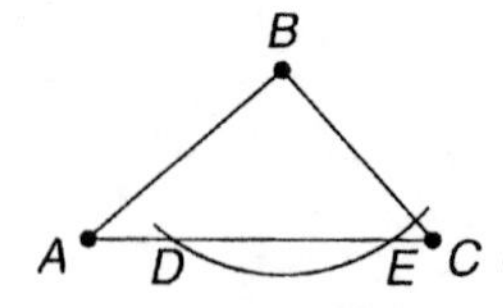

Step 2: Now swing an arc from D that falls below $\overline{AC}$. Repeat from E, intersecting the two arcs. Connect B and the point of intersection of the arcs, drawing the segment from B to $\overline{AC}$ to form the altitude.

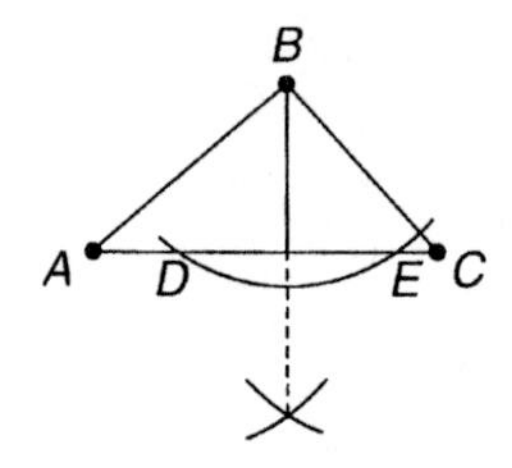

Section 4.4

25. Step 1: Follow the procedure of Construction 6 in the text to construct the perpendicular bisector of $\overline{AC}$.

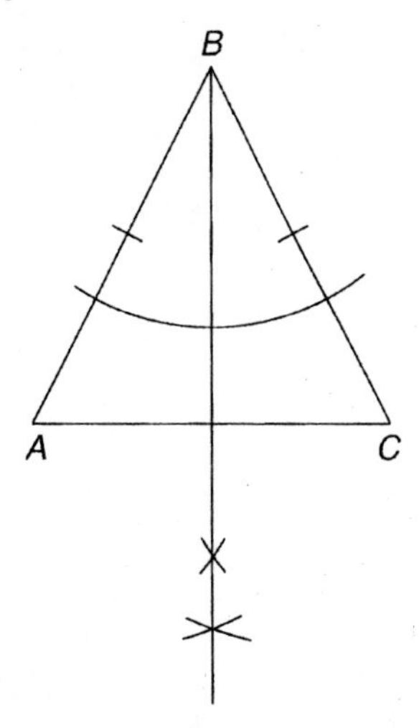

Step 2: Follow the procedure of Construction 3 in the text to bisect $\angle B$.

Observation: The perpendicular bisector of $\overline{AC}$ also bisects $\angle B$ in ΔABC.

27. Step 1: To construct the perpendicular bisector of $\overline{AC}$, set the compass radius to an opening that is greater than half the length of $\overline{AC}$. Placing the point of the compass at A, make arcs above and below $\overline{AC}$. Using the same radius, repeat from C to form two $\times$'s with the intersecting arcs. Draw a line segment through the $\times$'s.

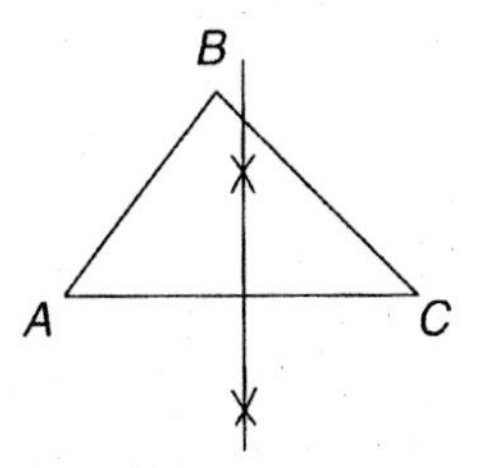

Step 2: Now draw the perpendicular bisectors of $\overline{AB}$ and $\overline{BC}$ in the same way. For clarity, you may need to erase construction marks when finished with them or use different-colored pencils to keep track of your arcs.

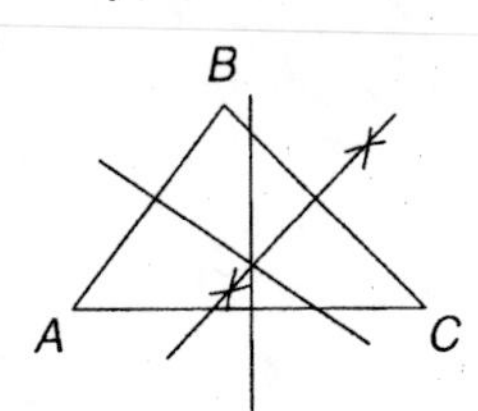

Section 4.4 27. (continued)

Repeat this construction for two more triangles such as those shown below. Extend the bisectors if necessary so that they intersect.

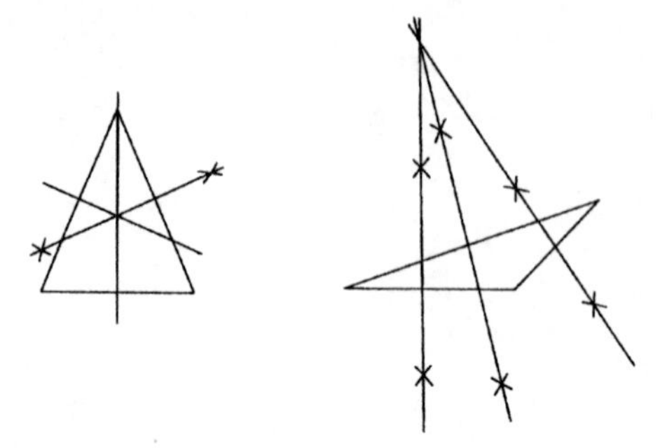

You should find that the three perpendicular bisectors of the sides of a triangle intersect in a single point. That point, called the circumcenter, will sometimes lie inside the triangle and sometimes lie outside the triangle.

29. Follow the procedure given in problem #9 to construct the altitude to each side of ΔABC.

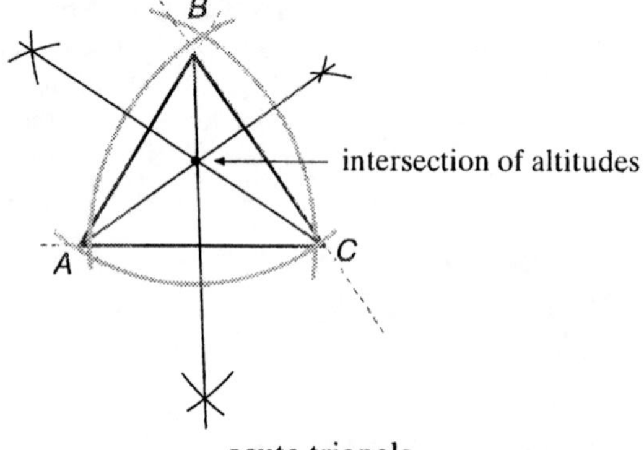

acute triangle

Repeat this construction for two more triangles. We will construct the altitude of an obtuse triangle and a right triangle.

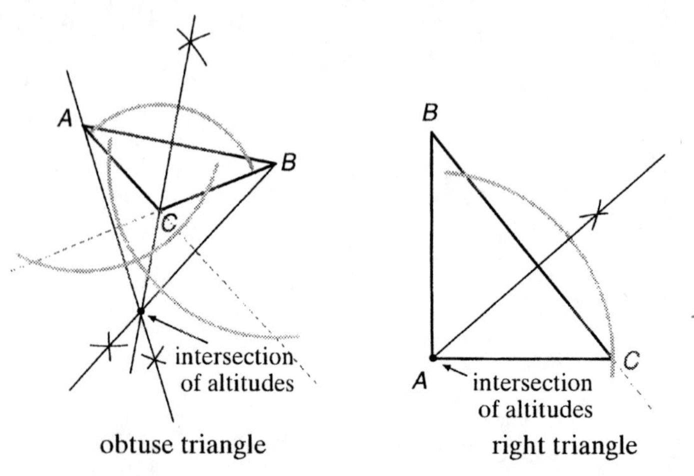

obtuse triangle

right triangle

Additional Problems

29. (continued)

The three altitudes are concurrent (they intersect in a single point). That point, called the orthocenter, will sometimes lie inside the triangle, sometimes lie outside the triangle, and sometimes lie on a vertex of the triangle.

31. Since you used the length a as the radius for all of your arcs, all three sides of the triangle are the same length. Thus, it is an equilateral triangle.

33. A complete solution is given in the answer key at the back of the text.

35. Since an altitude of a triangle is perpendicular to the base, constructing a perpendicular from B to $\overline{AC}$ forms an altitude. See the justification of Construction 5 in the text.

37. A complete solution is given in the answer key at the back of the text.

39. (a) Any point on the perpendicular bisector of a line segment is equidistant from its ends (see Theorem 4.10 in Section 4.3). Thus, to be equidistant from A and B, the well must be on the perpendicular bisector of $\overline{AB}$.

(b) To be equidistant from A and C, the well must be on the perpendicular bisector of $\overline{AC}$.

(c) Locate the well at the intersection of the perpendicular bisectors of $\overline{AB}$ and $\overline{AC}$. Then it will be equidistant from A, B and C.

Additional Problems

1. Subgoal 1: Find CD.

Subgoal 2: Find AC.

Subgoal 3: Find EC.

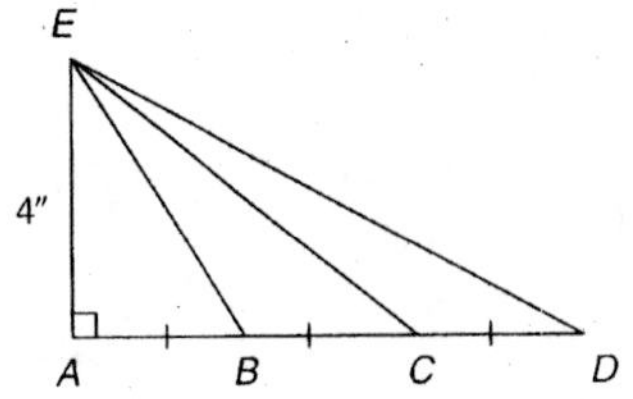

E
4"
A B C D

The area of $\Delta CDE = \frac{1}{2}(CD)(4) = 6$

$$2CD = 6$$

$$CD = 3 \text{ in.}$$

Since $AB = BC = CD = 3$, $AC = 6$ in.

ΔAEC is a right triangle. Use the Pythagorean Theorem.

$$\begin{aligned}(EC)^2 &= (AC)^2 + (AE)^2 \\ &= 6^2 + 4^2 \\ &= 36 + 16 \\ &= 52 \\ EC &= \sqrt{52} \\ &= 2\sqrt{13} \text{ in.}\end{aligned}$$

Chapter Review

2. Subgoal 1: The outer sheet of a roll of felt has a larger circumference than the inner sheet of a roll of felt. Find the "average" circumference of the sheets of felt.

   Subgoal 2: Find the number of square inches in the average sheet of felt.

   Subgoal 3: Find the approximate thickness of a sheet of felt.

   To find the average circumference of the sheets of felt, first find the average radius.

   outside radius: 4 inches

   inside radius: $4 - 1\frac{5}{8} = \frac{19}{8}$ in.

   average radius: $\dfrac{4 + \frac{19}{8}}{2} = 3.1875$ in.

   The average circumference is $2\pi(3.1875) = 6.375\pi$ in.

   The number of square inches in the average sheet of felt is $(6.375\pi)(36) = 229.5\pi$ in^2.

   The approximate number of sheets in the roll is;

$$\frac{324 \text{ ft}^2 \cdot \dfrac{12 \cdot 12 \text{ in}^2}{1 \cdot 1 \text{ ft}^2}}{229.5\pi \text{ in}^2} \approx 64.71 \text{ sheets.}$$

   The approximate thickness of a sheet of felt is the thickness of the entire roll divided by the approximate number of sheets.

$$\frac{1\frac{5}{8}}{64.71} \approx 0.025 \text{ in.}$$

Solutions to Chapter 4 Review

Section 4.1

1. (a) A triangle is equilateral.

   (b) The triangle is isosceles.

   (c) The triangle is isosceles. This can be deduced by the Law of Detachment.

   (d) No conclusion is possible.

2. First Statement: If an equilateral triangle has perimeter 18 cm, then it has a base of 6 cm and a height of $3\sqrt{3}$ cm.

   Second Statement: If a triangle has a base of 6 cm and a height of $3\sqrt{3}$ cm, then it has an area of $9\sqrt{3}$ cm^2.

Chapter Review

2. (continued)

   Using the Law of Syllogism we can conclude the following (Fourth Statement): If an equilateral triangle has perimeter 18 cm, then it has an area of $9\sqrt{3}$ cm^2.

   Third Statement: ABC is an equilateral triangle with perimeter 18 cm.

   Now, use the Law of Detachment since the third statement is the hypothesis of the fourth statement to conclude the following: The triangle has an area of $9\sqrt{3}$ cm^2.

3. (a) Use the Law of Syllogism to conclude $t \Rightarrow w$. Use the Law of Detachment to conclude w.

   (b) Use the Law of Syllogism to conclude $s \Rightarrow t$. Use the Law of Detachment to conclude t.

   (c) Use the Law of Syllogism to conclude $s \Rightarrow t$, and $t \Rightarrow s$. Since a statement and its converse are true we conclude the biconditional $s \Leftrightarrow t$.

4. A complete solution is given in the answer key at the back of the text.

5. A complete solution is given in the answer key at the back of the text.

6. A complete solution is given in the answer key at the back of the text.

Section 4.2

1. (a) $\angle R \cong \angle U$ by C.P.

   (b) $\overline{PR} \cong \overline{SU}$ by C.P.

   (c) $\overline{UT} \cong \overline{RQ}$ by C.P.

2. A complete solution is given in the answer key at the back of the text.

3. A complete solution is given in the answer key at the back of the text.

4. A complete solution is given in the answer key at the back of the text.

5. A complete solution is given in the answer key at the back of the text.

Section 4.3

1. ΔABC is isosceles since $\overline{AB} \cong \overline{CB}$. Therefore the base angles are congruent so $\angle A = 50°$.

   Then $\angle B = 180° - 2(50°) = 80°$.

2. ΔPQR is isosceles and $\overline{RS}$ is the perpendicular bisector. Therefore, $SQ = 7.5$ cm and $\angle RSQ = 90°$.

Chapter Review

3. A complete solution is given in the answer key at the back of the text.

4. Find $\angle ABC$. (part d)

$$\angle BAD \cong \angle ABC \text{ so } \angle ABC = 24°.$$

Find $\angle BAC$.

ΔABC is isoscles with vertex $\angle ABC = 24°$.

So $\angle BAC = 78°$.

Find $\angle DAC$. (part a)

$$\angle DAC = \angle BAC - \angle BAD$$
$$= 78° - 24°$$
$$= 54°.$$

Find $\angle ADC$. (part b)

ΔCAD is isosceles with vertex $\angle DAC = 54°$.

So $\angle ADC = \dfrac{180° - 54°}{2} = 63°$.

Find $\angle ACD$. (part c)

$\angle ACD \cong \angle ADC$ since ΔCAD is isosceles.

So $\angle ACD = 63°$.

Find $\angle DCB$. (part e)

$$\angle BAC \cong \angle ACB \text{ so } \angle ACB = 78°$$
$$\angle DCB = \angle ACB - \angle ACD$$
$$= 78° - 63°$$
$$= 15°.$$

5. A complete solution is given in the answer key at the back of the text.

Section 4.4

1. Step 1: Draw any line segment $\overline{AB}$. Extend the line segment beyond point B to form ray $\overrightarrow{AB}$.

Step 2: Place the point of the compass at A with the compass set to a radius of the length of $\overline{AB}$.

Using this same radius place the point of the compass at B and make an arc on ray $\overrightarrow{AB}$. Label the point where the arc crosses ray $\overrightarrow{AB}$, point C.

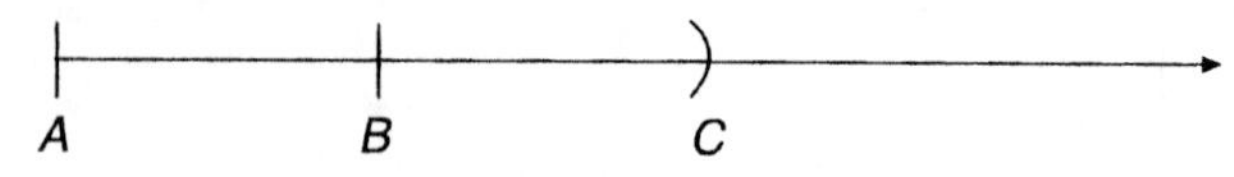

Chapter Review

1. (continued)

Step 3: Using the same radius, place the point of the compass at C and make an arc on ray $\overrightarrow{AB}$. Label the point where the arc crosses ray $\overrightarrow{AB}$ point D. $AE = 3(AB)$.

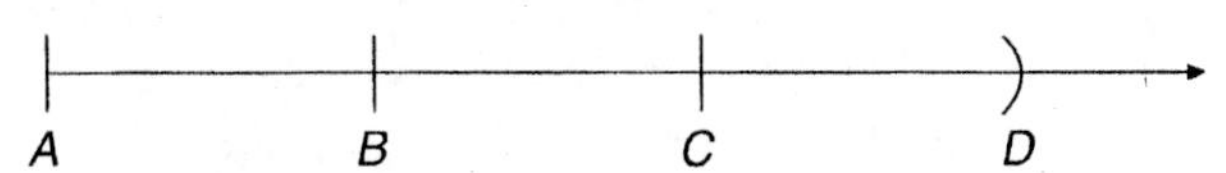

2. Step 1: Follow the procedure in Construction 2 of the text to copy $\angle A$.

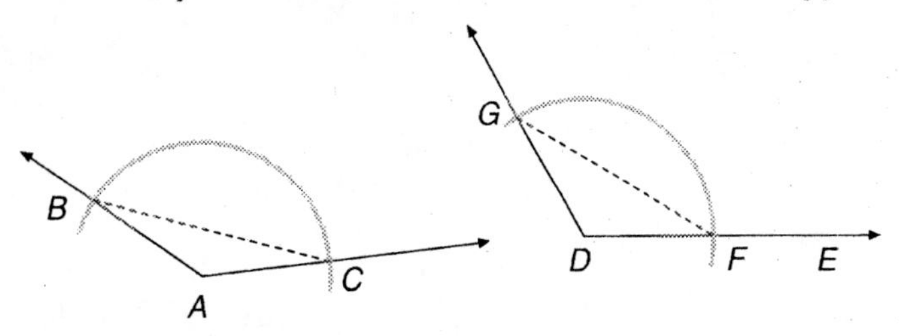

Step 2: Follow the procedure in Construction 3 of the text to bisect $\angle D$, the copied angle.

$\overrightarrow{DH}$ bisects $\angle D$.

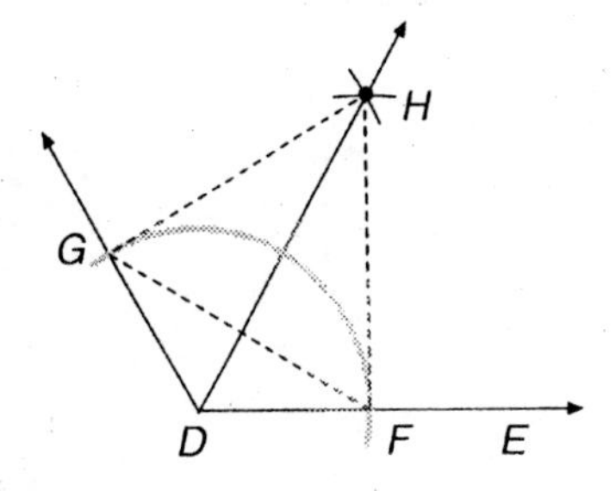

3. Step 1: Construct a 60° angle (see problem #15 of Section 4.4) and construct a 45° angle (see problem #17 of Section 4.4).

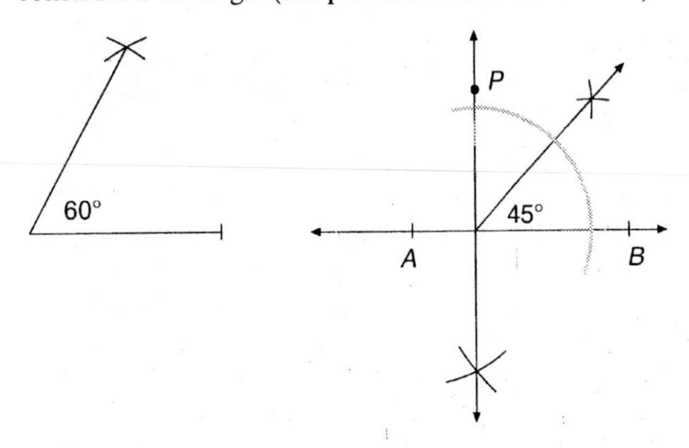

3. (continued)

Step 2: Follow the procedure of Construction 2 in the text to copy the 45° angle onto the 60° so that the two angles are adjacent. $\angle ABC = 105°$.

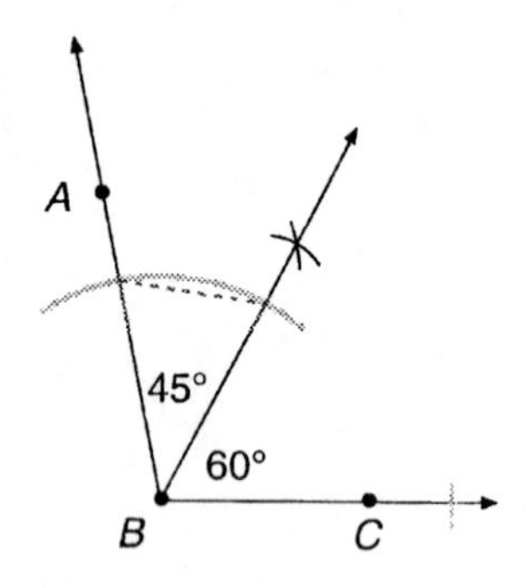

4. A right isosceles triangle will have angle measures of 45°, 45° and 90°.

Step 1: Construct a 45° angle (see problem #17 of Section 4.4). Label the vertex of the 45° angle, point C.

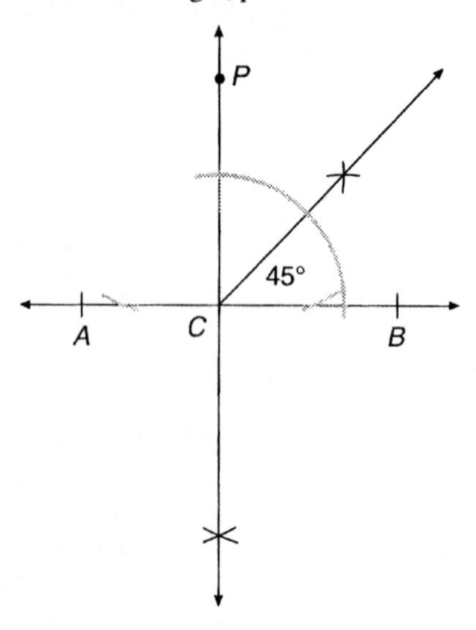

Step 2: Set the radius of the compass to length a and place the point of the compass at C and mark off length a on ray $\overrightarrow{CB}$. Label the point D. Now $\overline{CD}$ has length a.

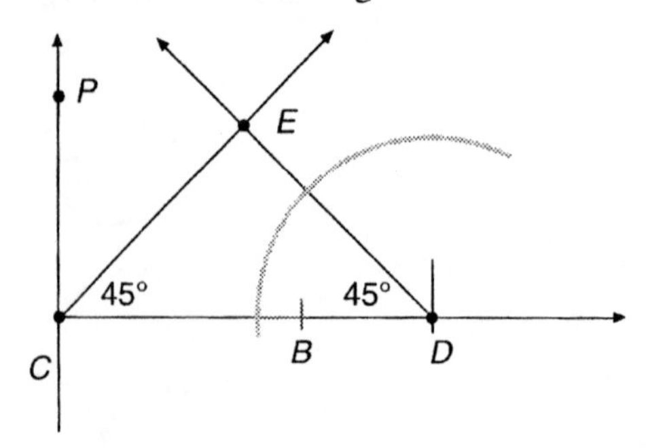

Chapter Review

4. (continued)

Step 3: Follow the procedure of Construction 2 in the text to copy the 45° angle with the vertex at point D. Label the intersection point of the two intersecting rays of the 45° angles E.

Now ΔCED is an isosceles right triangle.

5. Step 1: Construct an equilateral triangle (see problem #11 of Section 4.4).

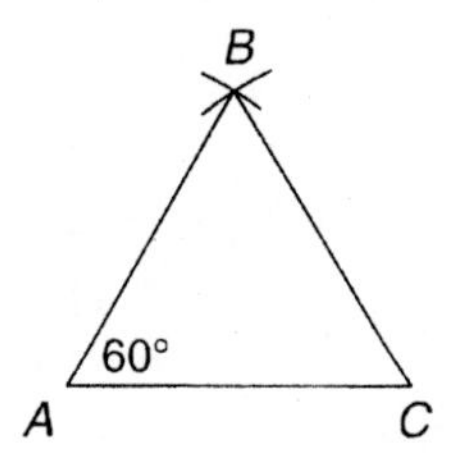

Step 2: Follow the procedure of Construction 3 in the text to bisect $\angle B$. Label the intersection of the angle bisector and $\overline{AC}$, point D.

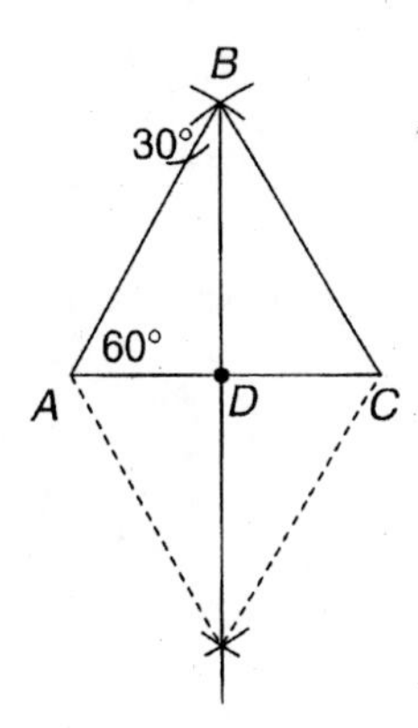

ΔABD is a 30°–60° right triangle.

Note: Instead of constructing the angle bisector in Step 2, we could have constructed the perpendicular to $\overline{AC}$ from point B or constructed the altitude to $\overline{AC}$.

6. Step 1: Copy ΔABC.

(a) Copy $\overline{BA}$. See Construction 1 of text.

(b) Copy $\angle A$ and $\angle C$. See Construction 2 of text.

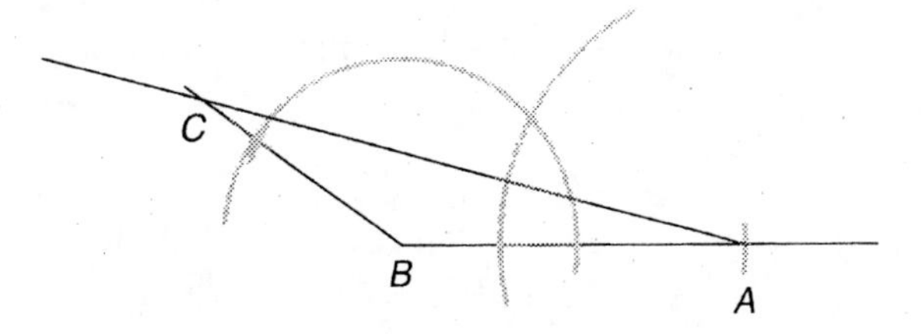

Chapter Test

6. (continued)

Step 2: To construct the altitude from B, follow the procedure of Construction 5 in text to construct the line perpendicular to $\overline{AC}$ from point B.

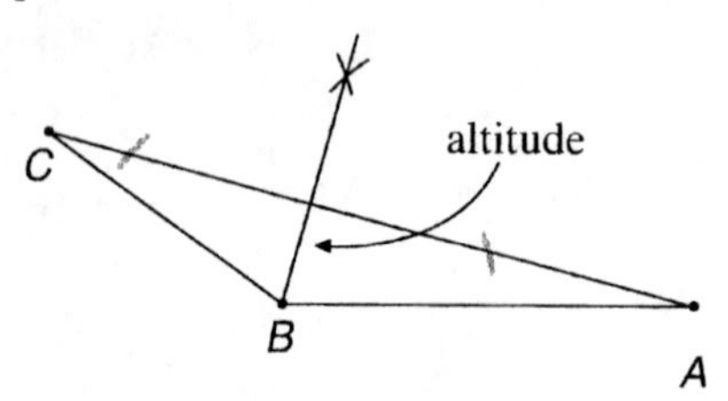

Step 3: To construct the median from A, find the midpoint of $\overline{BC}$ (see problem #7 of Section 4.4). Connect the midpoint to vertex A to construct the median.

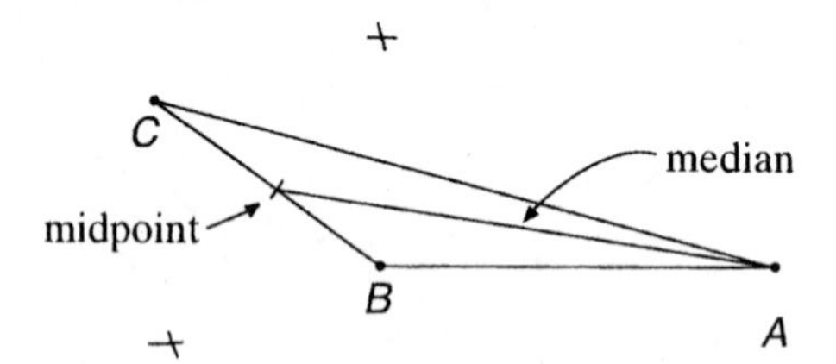

Solutions to Chapter 4 Test

1. T Since you have an equilateral triangle, the median forms two congruent triangles (SSS). Therefore, it forms two congruent 90° angles with the base. Since it is perpendicular to the base, the median is also the altitude.

2. T See Theorem 4.7.

3. F A true biconditional statement can be written only when both a statement and its *converse* are true.

4. T If the four congruent parts contain
   - Three pairs of congruent sides, then the two triangles are congruent by SSS.
   - Two pairs of congruent sides and two pairs of congruent angles, then the third pair of angles are also congruent. Hence we have congruent triangles by ASA.
   - One pair of congruent sides and three apirs of congruent angles, then we have congruent triangles by ASA.

5. T $16^2 + 30^2 = 34^2$, so the first triangle must have a leg that is 30 inches long, and the second triangle must have a hypotenuse that is 34 inches long. The triangles are congruent by SSS.

6. F The "if" part of the conditional is the hypothesis. The "then" part is the conclusion.

Chapter Test

7. F There is no AAA congruence. The triangles will have the same shape but different sizes unless they also have a pair of corresponding congruent sides.

8. T A point on the perpendicular bisector of a segment is equidistant from its endpoints. (Theorem 4.10)

9. T Intersecting lines always form congruent vertical angles. (Theorem 4.1)

10. F The congruent angles must be *between* the congruent sides.

11. (a) If a number is an integer, then it is a real number.

    (b) If a number is a real number, then it is an integer.

    (c) No, the converse is not true.

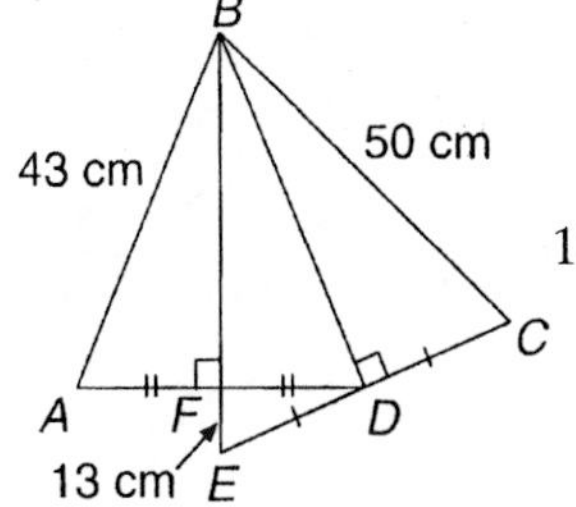

12. You know $BC = 50$ cm, $EF = 13$ cm, and $AB = 43$ cm.

Since $\overline{BE}$ bisects $\overline{AD}$, $\overline{AF} \cong \overline{FD}$.

Since $\overline{BD}$ bisects $\overline{EC}$, $\overline{ED}, \cong \overline{DC}$.

Also, $\overline{BF}, \cong \overline{BF}$ and $\angle AFB \cong \angle BFD = 90°$.

By LL, $\Delta AFB \cong \Delta DFB$. Then $BD = 43$ cm by C.P.

Similarly, since $\overline{BD} \cong \overline{BD}$ and $\angle EDB = \angle CDB = 90°$, $\Delta EBD \cong \Delta CBD$ by LL. Thus, $BE = 50$ cm by C.P.

$BF = BE - FE = 50 - 13 = 37$ cm.

Since ΔFBD is a right triangle.

$$(FD)^2 + (BF)^2 = (BD)^2$$

$$(FD)^2 + 37^2 = 43^2$$

$$(FD)^2 = 43^2 - 37^2$$

$$(FD)^2 = 480$$

$$FD = \sqrt{480}$$

$$= 4\sqrt{30}$$

$$FD \approx 21.9 \text{ cm}$$

13. Valid conclusions include: She gains weight. She carries her work shoes. She doesn't carry a lunch. She eats out for lunch. She overeats.

14. Because $\angle VWZ \cong \angle YWX$ by vertical angles, $\Delta VWZ \cong \Delta YWX$ by SAS.

15. Since $\overline{AC} \cong \overline{AC}$, $\Delta ABC \cong \Delta ADC$ by HL.

16. $\overline{PQ} \cong \overline{PT}$ because the sides opposite congruent angles are congruent (Theorem 4.7). Thus, $\Delta PQR \cong \Delta PTS$ by SAS. Since you know that $TR = QR$, $TS = TR - SR$, and $QR = QS - SR$, then $TR = QS$ and $\overline{TR} \cong \overline{QS}$. Therefore, $\Delta PTR \cong \Delta PQS$ by SAS.

Chapter Test

17. Use the fact that $22.5° = \frac{1}{2}(45°)$ and $45° = \frac{1}{2}(90°)$. You can bisect a 90° angle twice to perform this construction.

Step 1: Construct a perpendicular to a line segment to create a 90° angle.

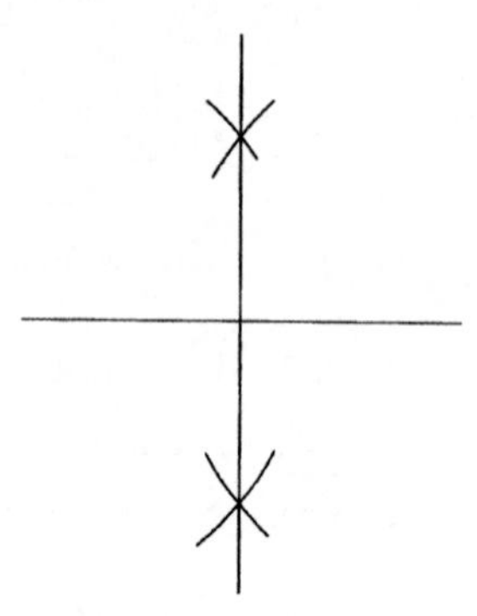

Step 2: Bisect the 90° angle to make a 45° angle. Then bisect the 45° angle to make a 22.5° angle.

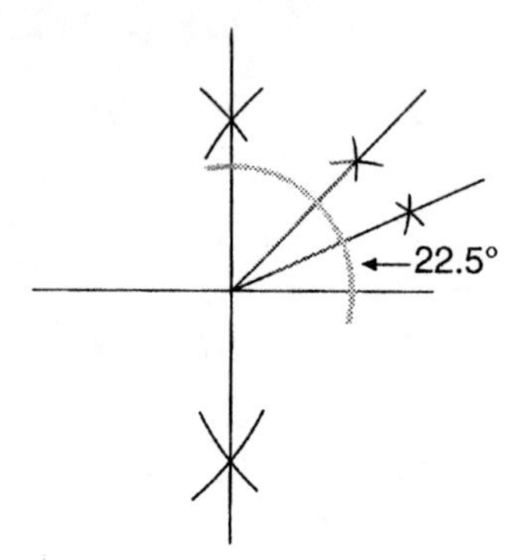

18. Step 1: To construct a median you must first find the midpoint of the opposite side by constructing the perpendicular bisector. You do not have to draw the perpendicular bisector; just mark the point at which it intersects the side.

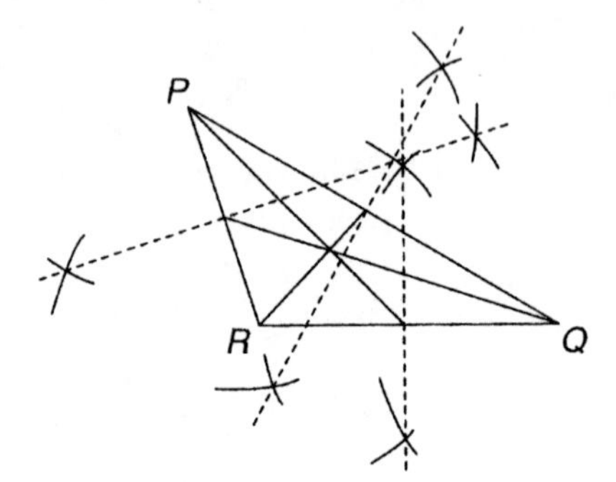

Step 2: Connect each vertex to the midpoint of the opposite side.

Notice that the three medians of the triangle intersect in a single point.

Chapter Test

19. Step 1: Draw a line segment at least $4a$ long. Set your compass radius at a and mark off four segments of length a on the line segment drawn.

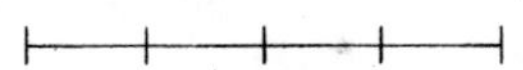

Step 2: Then set your compass radius for $2a$ and make an arc from one end of the segment of length $4a$. Reset it for $3a$ and make an arc from the other end of the segment so that it intersects the first arc. Join the ends of the $4a$ base to the × made by the intersection of the arcs.

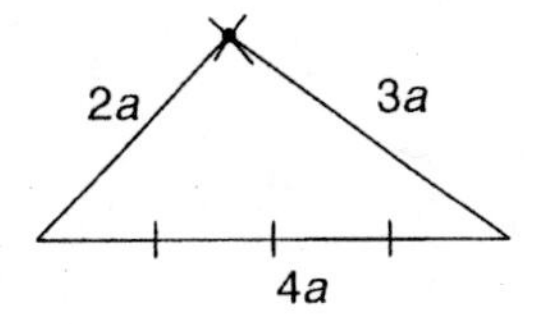

(The triangle you construct may be a reflection of the one shown here.)

20. One possible construction is described here.

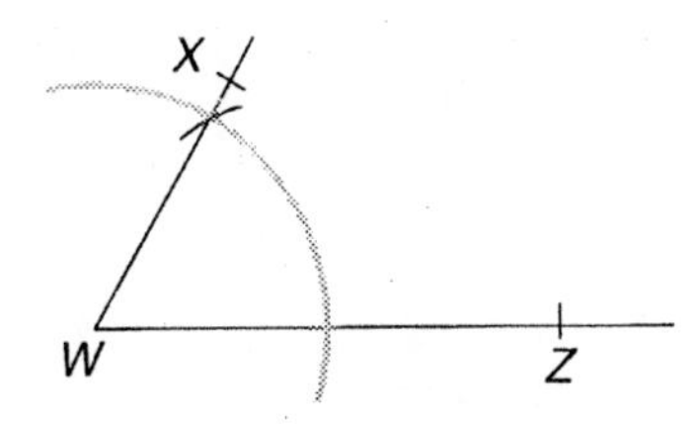

Step 1: Copy $\angle XWZ$ and extend the sides of the angle.

Step 2: Measure off $\overline{WZ}$ on the bottom ray of $\angle XWZ$ and measure off $\overline{WX}$ on the other ray of $\angle XWZ$, using your compass radius to measure.

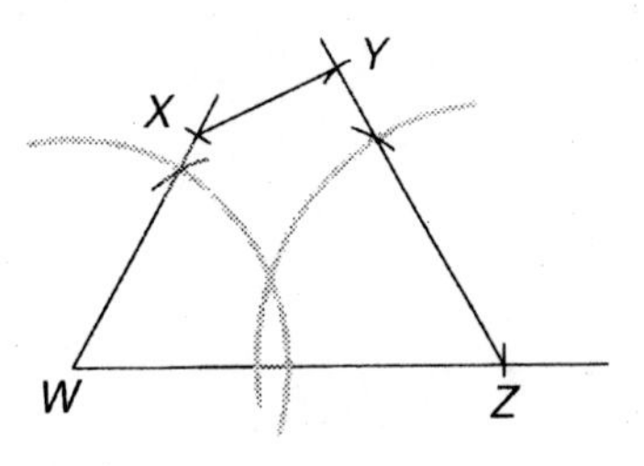

Step 3: Copy $\angle WZY$ using $\overline{WZ}$ as a base with Z as the vertex.

Step 4: Measure off $\overline{YZ}$ on the right-hand ray of $\angle WZY$.

Step 5: Join points X and Y.

21. A complete solution is given in the answer key at the back of the text.

22. A complete solution is given in the answer key at the back of the text.

Chapter Test

23. Find the areas of the equilateral triangles with sides of 54 feet and 17 feet.

An altitude of an equilateral triangle is also the perpendicular bisector of a side. Using the Pythagorean Theorem, you can determine the height of the triangle.

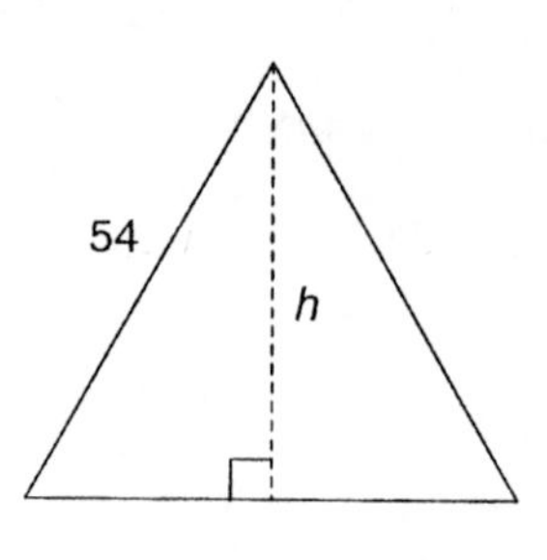

$$\left(\frac{54}{2}\right)^2 + h^2 = 54^2$$

$$h^2 = 2187$$

$$h = \sqrt{2187}$$

$$= 27\sqrt{3}$$

$$h \approx 46.8 \text{ ft}$$

At the base,

$$A = \frac{1}{2}bh$$

$$= \frac{1}{2}(54)(27\sqrt{3})$$

$$= 729\sqrt{3}$$

$$A \approx 1263 \text{ ft}^2.$$

Similarly, for the triangle at the top of the arch,

$$\left(\frac{17}{2}\right)^2 + h^2 = 17^2$$

$$h^2 = 17^2 - \left(\frac{17}{2}\right)^2$$

$$= \frac{867}{4}$$

$$h = \frac{1}{2}\sqrt{867}$$

$$h \approx 14.7 \text{ ft.}$$

At the top,

$$A = \frac{1}{2}bh$$

$$= \frac{1}{2}(17)\left(\frac{1}{2}\sqrt{867}\right)$$

$$= \frac{17\sqrt{867}}{4}$$

$$\approx 125 \text{ ft}^2.$$

The difference in the cross-sectional areas is approximately $1263 - 125 = 1138 \text{ ft}^2$.

24. Remember that the points on a perpendicular bisector of a line segment are equidistant from the ends of the line segment. The runner should travel along the path that is the perpendicular bisector of the line segment joining his home and his friend's house.

5

Parallel Lines and Quadrilaterals

Section 5.1

Tips:

✔ Don't assume that lines are parallel based solely on their appearance. They must be given as parallel or shown to be parallel.

✔ You *can* use the fact that sides of parallelograms, rectangles, trapezoids, etc., are parallel.

Solutions to odd-numbered textbook problems

1. Alternate interior angles are between the parallel lines but on opposite sides of a transversal. Pairs of alternate interior angles are: $\angle 5$ and $\angle 3$, $\angle 11$ and $\angle 9$.

3. Corresponding angles are angles in the same position relative to the parallel lines and the transversal. Pairs of corresponding angles are:

   $\angle 3$ and $\angle 8$

   $\angle 11$ and $\angle 6$

   $\angle 1$ and $\angle 5$

   $\angle 9$ and $\angle 13$.

5. You know that $\angle 6 = \angle 11$ by corresponding angles.

   Now $\angle 11 + \angle 12 = 180°$, since they form a straight line.

   Substituting $\angle 6$ and $\angle 11$, you get $\angle 6 + \angle 12 = 180°$.

7. $\angle 3 = \angle 5$ by alternate interior angles. So $\angle 3 = 65°$.

   $\angle 5 = \angle 8$ by vertical angles. So $\angle 8 = 65°$.

9. First, determine some sets of congruent angles.

$\angle 1 = \angle 3 = \angle 5 = \angle 8$	vertical and corresponding angles
$\angle 2 = \angle 4$	vertical angles
$\angle 6 = \angle 9 = \angle 11 = \angle 13$	alternate interior and vertical angles
$\angle 12 = \angle 14$	vertical angles

   Since $\angle 5 = 60°$, you also have $\angle 1 = \angle 3 = \angle 8 = 60°$.

   Since $\angle 6 = 45°$, you have $\angle 9 = \angle 11 = \angle 13 = 45°$.

   Since $\angle 3 = 60°$, $\angle 2 = 180° - \angle 3 = 180° - 60° = 120°$.

Section 5.1

9. (continued)

Thus, $\angle 4 = 120°$ since $\angle 2$ and $\angle 4$ are vertical angles.

You have $\angle 5 + \angle 6 + \angle 7 = 180°$. Substituting for $\angle 5$ and $\angle 6$, you have $60° + 45° + \angle 7 = 180°$, so $\angle 7 = 75°$.

$\angle 10 = \angle 7$ by vertical angles, so $\angle 10 = 75°$.

$\angle 11 + \angle 12 = 180°$, so $45° + \angle 12 = 180°$ and $\angle 12 = 135°$.

Since $\angle 12 = \angle 14$ by vertical angles, you have $\angle 14 = 135°$.

Summarizing,

$\angle 1 = 60°$	$\angle 5 = 60°$	$\angle 9 = 45°$	$\angle 13 = 45°$
$\angle 2 = 120°$	$\angle 6 = 45°$	$\angle 10 = 75°$	$\angle 14 = 135°$.
$\angle 3 = 60°$	$\angle 7 = 75°$	$\angle 11 = 45°$	
$\angle 4 = 120°$	$\angle 8 = 60°$	$\angle 12 = 135°$	

11. $\angle 16$ and $\angle 11$ are a pair of alternate interior angles because they are both between the parallel lines but on opposite sides of the transversal.

13. $\angle 2$ and $\angle 17$ are a pair of corresponding angles because they are both to the right of the parallel lines and above the same transversal.

15. | | |
|---|---|
| $\angle 4 = \angle 6$ | vertical angles |
| $\angle 4 = \angle 13$ | corresponding angles |
| Thus $\angle 13 = \angle 6 = 140°$. | transitive property |
| $\angle 6 + \angle 5 = 180°$ | supplementary angles |
| $140° + \angle 5 = 180°$ | |
| Thus $\angle 5 = 40°$. | |

17. It helps here to work with one transversal at a time.

$\angle 1 = 80°$, so $\angle 2 = 180° - 80° = 100°$.	supplementary angles
$\angle 11 = \angle 1$	vertical angles
$\angle 16 = \angle 1$	corresponding angles
$\angle 18 = \angle 1$	alternate exterior angles

Thus, $\angle 1 = \angle 11 = \angle 16 = \angle 18 = 80°$.

Similarly,

$\angle 12 = \angle 2$	vertical angles
$\angle 17 = \angle 2$	corresponding angles
$\angle 15 = \angle 2$	alternate interior angles.

Thus, $\angle 2 = \angle 12 = \angle 17 = \angle 15 = 100°$.

Section 5.1

17. (continued)

Using $\angle 4 = 125°$, you find that $\angle 7 = 180° - 125° = 55°$ by supplementary angles.

$\angle 6 = \angle 4$	vertical angles,
$\angle 13 = \angle 4$	corresponding angles
$\angle 19 = \angle 4$	alternate interior angles

Thus, $\angle 4 = \angle 6 = \angle 13 = \angle 19 = 125°$.

$\angle 5 = \angle 7$	vertical angles,
$\angle 14 = \angle 7$	alternate interior angles
$\angle 20 = \angle 7$	corresponding angles

Thus, $\angle 7 = \angle 5 = \angle 14 = \angle 20 = 55°$.

To find $\angle 3$, $\angle 8$, $\angle 9$ and $\angle 10$, notice that $\angle 11$, $\angle 14$ and $\angle 10$ are the three angles of a triangle.

Since $\angle 11 = 80°$ and $\angle 14 = 55°$, $\angle 10 = 180° - 80° - 55° = 45°$.

$\angle 8 = \angle 10 = 45°$	vertical angles.
$\angle 3 = 180° - \angle 10 = 180° - 45° = 135°$	supplementary angles
$\angle 9 = \angle 3 = 135°$	vertical angles

Summarizing,

$\angle 1 = 80°$	$\angle 8 = 45°$	$\angle 15 = 100°$
$\angle 2 = 100°$	$\angle 9 = 135°$	$\angle 16 = 80°$
$\angle 3 = 135°$	$\angle 10 = 45°$	$\angle 17 = 100°$
$\angle 4 = 125°$	$\angle 11 = 80°$	$\angle 18 = 80°$
$\angle 5 = 55°$	$\angle 12 = 100°$	$\angle 19 = 125°$
$\angle 6 = 125°$	$\angle 13 = 125°$	$\angle 20 = 55°$.
$\angle 7 = 55°$	$\angle 14 = 55°$	

19. F Since no congruent corresponding angles nor alternate interior angles are marked, no conclusion can be made.

21. $m \parallel l$ is true because there is a pair of congruent corresponding angles.

23. $\overline{AB}$ and $\overline{FE}$ are parallel. Using transversal $\overline{BC}$, $\angle B$ and $\angle FEC$ are corresponding angles. Thus $\angle B = 80°$.

Section 5.1

25. To find $\angle ADF$, first note that $\angle DBE = \angle FEC = 80°$ (corresponding angles with $\overline{DB}$ and $\overline{EF}$ as the parallel line segments and $\overline{BE}$ the transversal). Looking at $\overline{DF}$ and $\overline{BE}$ as the parallel line segments and $\overline{AB}$ as the transversal, you can see that $\angle ADF$ and $\angle DBE$ are corresponding angles. Thus, $\angle ADF = 80°$ also.

27. $\overline{DF}$ and $\overline{BC}$ are parallel. Using transversal $\overline{EF}$, $\angle DFE$ and $\angle FEC$ are alternate interior angles. Thus $\angle DFE = 80°$.

29. You know that $\angle 2 = 67°$ by vertical angles and $\angle 4 = 42°$ by alternate interior angles. Since $\angle 2$, $\angle 1$ and $42°$ are the angles of a triangle,

    $\angle 1 = 180° - \angle 2 - 42° = 180° - 67° - 42° = 71°.$

    Then you have that $\angle 3 = \angle 1 = 71°$ by alternate interior angles.

    Summarizing, $\angle 1 = 71°, \angle 2 = 67°, \angle 3 = 71°, \angle 4 = 42°$.

31. In the diagram given here more angles are numbered to help in finding $\angle 1, \angle 2, \angle 3$, and $\angle 4$.

    $\angle 5 = \angle 3$ by alternate interior angles. Thus $\angle 3 = 55°$.

    $\angle 6 = \angle 7$ by corresponding angles.

$\angle 7 + \angle 4 = 180°$	supplementary angles
$\angle 6 + \angle 4 = 180°$	
$80° + \angle 4 = 180°$	
Thus $\angle 4 = 100°$.	
$\angle 7 + \angle 3 + \angle 8 = 180°$	angle sum of a triangle
$80° + 55° + \angle 8 = 180°$	
$135° + \angle 8 = 180°$	
Thus $\angle 8 = 45°$.	
$\angle 8 + \angle 2 = 180°$	supplementary angles
$45° + \angle 2 = 180°$	
Thus $\angle 2 = 135°$.	
$\angle 1 = \angle 8$	corresponding angles
Thus, $\angle 1 = 45°$.	

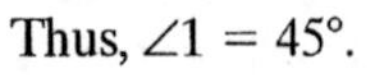

33. A complete solution is given in the answer key at the back of the text.

35. A complete solution is given in the answer key at the back of the text.

37. A complete solution is given in the answer key at the back of the text.

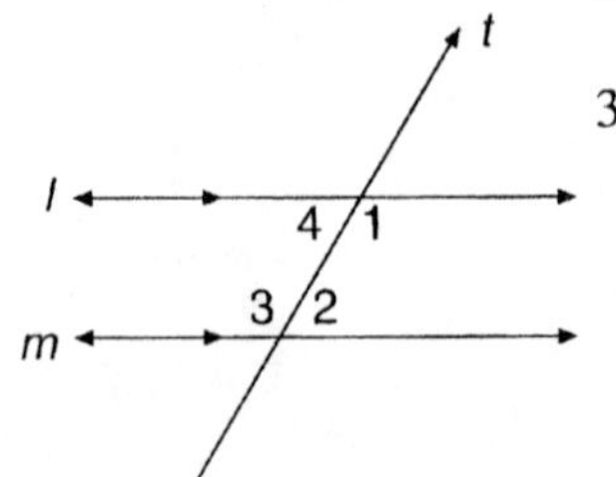

39. Let l and m be parallel lines cut by transversal t as shown.

    Prove: $\angle 1 + \angle 2 = 180°$ and $\angle 3 + \angle 4 = 180°$.

$\angle 1 = \angle 3$ and $\angle 2 = \angle 4$	Theorem 5.5 (alternate interior angles)
$\angle 1 + \angle 4 = 180°$	supplementary angles
$\angle 2 + \angle 3 = 180°$	supplementary angles

Section 5.1

39. (continued)

Thus $\angle 1 + \angle 2 = 180°$ by substitution. Hence $\angle 1$ and $\angle 2$ are supplementary.

$\angle 4 + \angle 3 = 180°$ by substitution. Hence $\angle 3$ and $\angle 4$ are supplementary.

Therefore, both pairs of interior angles on the same side of the transversal are supplementary.

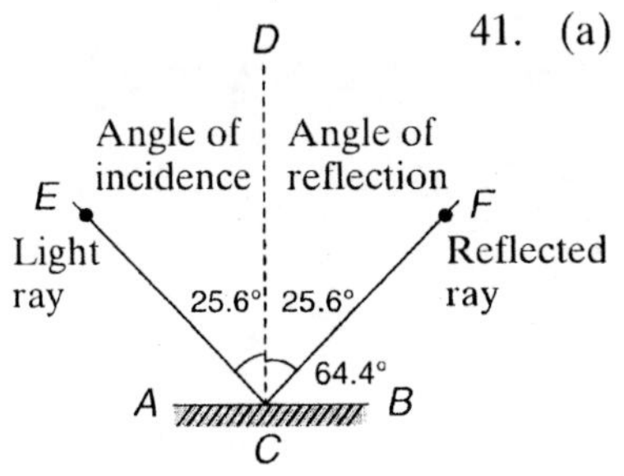

41. (a) $\angle ECD = \angle DCF = 25.6°$ as the angle of incidence equals the angle of reflection.

$\angle DCF + \angle FCB = \angle DCB$ by adjacent angles.

$\angle DCB = 90°$ since $\overline{CD}$ is perpendicular to $\overline{AB}$.

$\angle DCF + \angle FCB = 90°$by substitution.

$25.6° + \angle FCB = 90°$by substitution.

Thus $\angle FCB = 64.4.°$

(b) Sketch in perpendicular lines $\overline{HB}$ and $\overline{EK}$.

$\angle ABF + \angle FBH = 90°$ since $\overline{AB}$ is perpendicular to $\overline{HB}$.

$70° + \angle FBH = 90°$ substitution

Thus $\angle FBH = 20°$.

$\angle FBH = \angle HBD$ as the angle of incidence is equal to the angle of reflection.

So $\angle HBD + \angle DBG = 90°$ since $\overline{HB}$ is perpendicular to $\overline{AB}$.

$20° + \angle DBG = 90°$ substitution

Thus $\angle DBG = 70°$.

Hence the ray of light now strikes mirror $\overline{DE}$ with an angle of 70°.

Similarly, the ray of light alternately strikes mirrors $\overline{AC}$ and $\overline{DE}$ with an angle of 70°.

That is, $\angle BDH = \angle EDC = \angle ECG = 70°$.

Since $\angle ECG = 70°$, the angle of incidence, $\angle CEK$, measures 20°.

Also the angle of reflection, $\angle KEG$, measures 20°.

Thus$\angle CEG = \angle CEK + \angle KEG$

$= 20° + 20°$

$\angle CEG = 40°$.

Section 5.2

Tip:

✔ In complex pictures, look for triangles and parallel lines with transversals.

Section 5.2 Solutions to odd-numbered textbook problems

1. T The exterior angle of a triangle is equal to the sum of the two nonadjacent interior angles. (Corollary 5.10)

3. T $\angle 2 + \angle 3 + \angle 6 = 180°$ and $\angle 8 + \angle 10 + \angle 11 = 180°$, so $\angle 2 + \angle 3 + \angle 6 = \angle 8 + \angle 10 + \angle 11$. Since $\angle 6 = \angle 8$ (vertical angles), $\angle 2 + \angle 3 = \angle 10 + \angle 11$. An alternate justification is given in the answer key at the back of the text.

5. T $\angle 2 + \angle 3 + \angle 4 + \angle 12$ is the sum of the angles in ΔABC. Thus $\angle 2 + \angle 3 + \angle 4 + \angle 12 = 180°$.

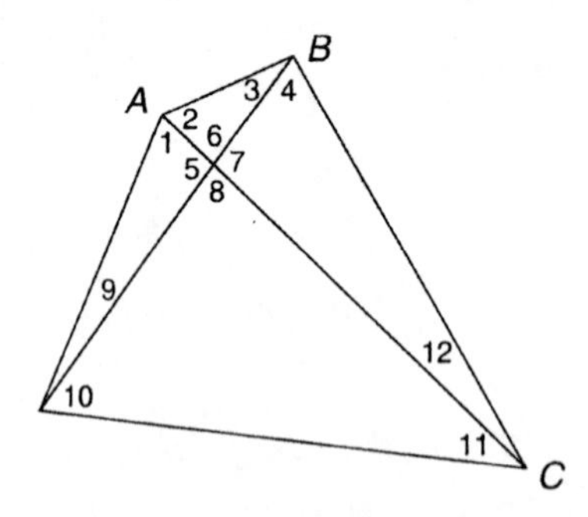

7. T The exterior angle of a triangle is equal to the sum of the two nonadjacent interior angles. (Corollary 5.10)

9. T $\angle 1 \cong \angle 5$ by alternate interior angles.

   $\angle 5 + \angle 2 + \angle 6 = 180°$ by angle sum of triangle.

   $\angle 1 + \angle 2 + \angle 6 = 180°$ by substitution.

11. F All three angles of a triangle have a sum of 180°. A supplement of $\angle 6$ with respect to the lines l and m is the angle composed of $\angle 1 + \angle 2$.

13. F These are not congruent alternate interior angles with respect to *parallel* lines.

15. T $\angle 15 = \angle 8 + \angle 9 + \angle 11$ by the exterior angle theorem for ΔABC.

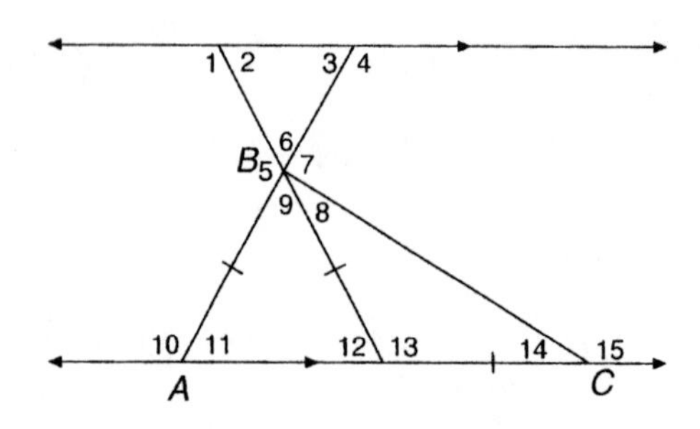

Section 5.2

17. T $\angle 8 \cong \angle 14$ since they are opposite congruent sides of an isosceles triangle (Theorem 4.7). Then you have $\angle 12 = \angle 8 + \angle 14$ because as an exterior angle, $\angle 12$ is equal to the sum of the nonadjacent interior angles. Since $\angle 14 = \angle 8$, by substitution you find that $\angle 12 = \angle 8 + \angle 8 = 2(\angle 8)$.

19. A complete solution is given in the answer key at the back of the text.

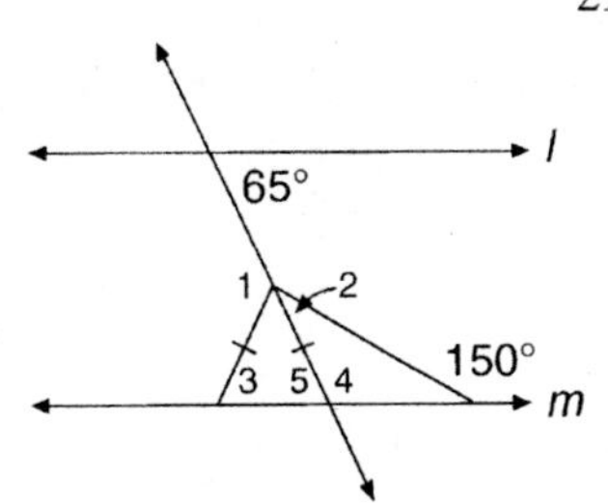

21. Name the angle adjacent and to the left of $\angle 4$ as $\angle 5$. Then $\angle 4 + 65° = 180°$ since interior angles on the same side of a transversal are supplementary. Subtracting, you find that $\angle 4 = 115°$. Then $\angle 2 + \angle 4 = 150°$ since 150° is an exterior angle of the triangle. Thus, $\angle 2 = 150° - \angle 4 = 150° - 115° = 35°$.

$\angle 4 + \angle 5 = 180°$, so $\angle 5 = 180° - \angle 4 = 180° - 115° = 65°$.

$\angle 3 = \angle 5 = 65°$ (base angles of an isosceles triangle)

$\angle 1 = \angle 3 + \angle 5$ (exterior angle), so $\angle 1 = 65° + 65° = 130°$.

Summarizing, $\angle 1 = 130°, \angle 2 = 35°, \angle 3 = 65°, \angle 4 = 115°$.

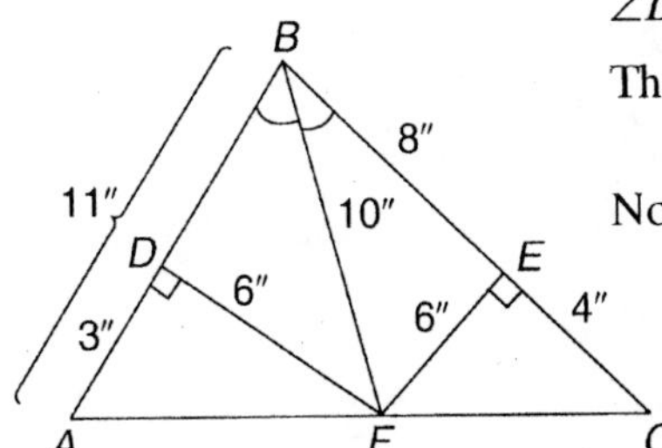

23. ΔDBF and ΔEBF are right triangles, each with hypotenuse $\overline{BF}$. Also $\angle DBF \cong \angle EBF$.

Thus $\Delta DBF \cong \Delta EBF$ by HA.

Now $\overline{DB} \cong \overline{EB}$ by C.P. So $DB = EB$.

$$\begin{aligned} DB &= AB - AD \\ &= 11 - 3 \\ &= 8 \text{ in.} \end{aligned}$$

Thus $DB = EB = 8$ in.

Use the Pythagorean Theorem.

$$\begin{aligned} EF^2 + EB^2 &= FB^2 \\ EF^2 + 8^2 &= 10^2 \\ EF^2 &= 36 \\ EF &= 6 \end{aligned}$$

Since $\overline{EF} \cong \overline{DF}$ by C.P., $EF = DF = 6$ in.

Find the area.

$$\begin{aligned} A_{\Delta ABC} &= A_{\Delta_{ABF}} + A_{\Delta_{FBC}} \\ &= \frac{1}{2}(11)(6)\ 1\ \frac{1}{2}(12)(6) \\ &= 33 + 36 \\ &= 69 \text{ in}^2 \end{aligned}$$

Note: The diagram given may be misleading. Do not make assumptions based on the diagram. Only use the information that is given.

Section 5.2

25. A complete solution is given in the answer key at the back of the text.

27. A complete solution is given in the answer key at the back of the text.

29. *Note:* The distance from a point to a line is always measured along a perpendicular path.

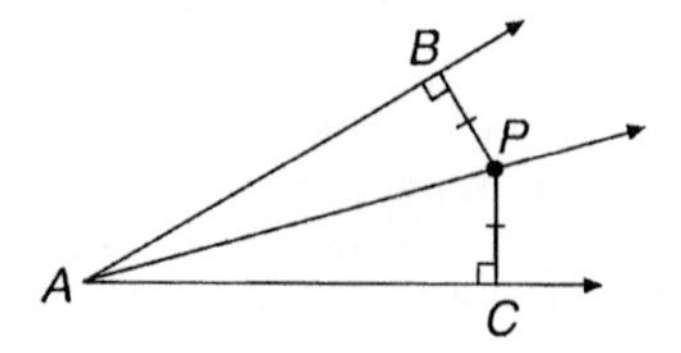

Given: $BP = CP$.

Prove: $\angle BAP \cong \angle PAC$

Proof: $\overline{AP} \cong \overline{AP}$ by the reflexive property. Because they are both right angles, $\angle ABP \cong \angle ACP$, and $\overline{BP} \cong \overline{CP}$ as given. Therefore, $\Delta ABP \cong \Delta ACP$ by HL or SAS. Then $\angle BAP \cong \angle PAC$ by C.P.

31. Find the angle bisector of $\angle ABC$ and $\angle BCD$. The marker is to be placed at the intersection of the two angle bisectors. This is a result of Theorem 5.13 Angle Bisector Theorem.

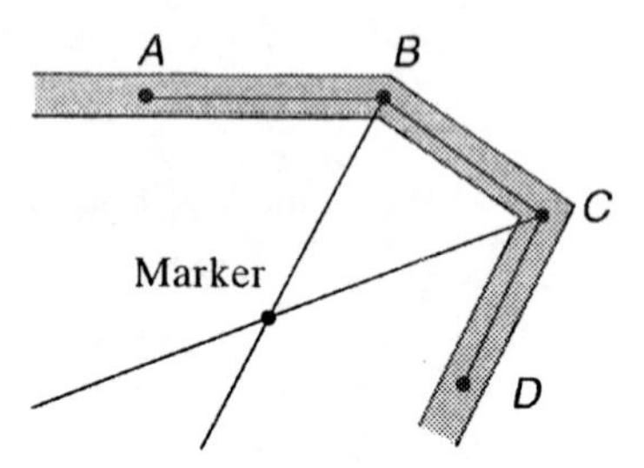

Section 5.3

Tips:

✔ You will not always be able to solve for numerically-labeled angles in numerical order. Start with any convenient angle.

✔ It is helpful to write angle measures on your diagram as you find them. In that way, you can keep track of what angles you still need and can use the angle measures you have already found.

Solutions to odd-numbered textbook problems

1. F Diagonals of a parallelogram are not necessarily congruent.

3. T The opposite sides of a parallelogram are congruent. (Corollary 5.15)

5. F Although the diagonals bisect each other, they may not be the same length.

7. F Consecutive angles of a parallelogram are not necessarily congruent.

Section 5.3

9. F However, $\Delta AED \cong \Delta CEB$ by SAS. The order in which the vertices are listed makes the difference.

11. T The diagonal $\overline{AC}$ of the parallelogram forms two congruent triangles.

13. $\angle X = \angle Z = n + 50°$ and $\angle W = \angle Y = n$.

$\angle X + \angle Y + \angle Z + \angle W = 360°$ since $XYZW$ is a quadrilateral, and the sum of the measures of the angles in a quadrilateral is 360°.

$$\begin{aligned} 2(n + 50°) + 2n &= 360° \\ 2n + 100° + 2n &= 360° \\ 4n + 100° &= 360° \\ 4n &= 260° \\ n &= 65° = \angle W = \angle Y \\ n + 50° &= 115° = \angle X = \angle Z \end{aligned}$$

15. Refer to the diagram given when solving this problem.

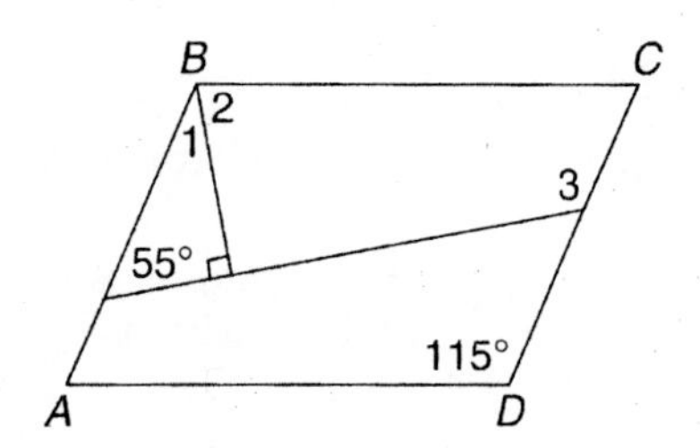

$\angle 1 + \angle 2 = \angle B = D = 115°$. Using the small triangle, you have $\angle 1 + 55° + 90° = 180°$, so $\angle 1 = 35°$. Substituting for $\angle 1$ in $\angle 1 + \angle 2 = 115°$, you have that $35° + \angle 2 = 115°$, so $\angle 2 = 80°$.

$\angle C = 180° - 115° = 65°$ since interior angles on the same side of a transversal of parallel lines are supplementary. Now using the quadrilateral containing $\angle 2$ and $\angle 3$, you have $90° + \angle 2 + \angle 3 + \angle C = 360°$ since the interior angles of a quadrilateral add to 360°. Substituting, you have $90° + 80° + \angle 3 + 65° = 360°$, so $\angle 3 = 125°$.

Summarizing, $\angle 1 = 35°$, $\angle 2 = 80°$, and $\angle 3 = 125°$.

17. The parallelogram given has consecutive angles in the ratio 2:3.

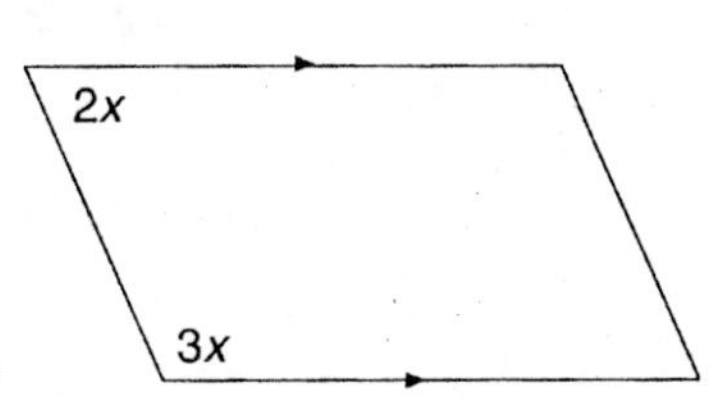

Section 5.3

17. (continued)

The angles with measure $2x$ and $3x$ are a pair of interior angles on the same side of the transversal. Thus the two angles are supplementary.

$$2x + 3x = 180°$$
$$5x = 180°$$
$$x = 36°$$
$$2x = 72°$$
$$3x = 108°$$

The measures of the angles of the parallelogram are 72°, 72°, 108° and 108°.

19. T The diagonals of any rhombus bisect each other since a rhombus is also a parallelogram.

21. T The sides of a rhombus are congruent.

Thus $\overline{AB} \cong \overline{AD}, \overline{BC} \cong \overline{DC}, \overline{CA} \cong \overline{CA}$.

So $\Delta ADC \cong ABC$ by SSS.

23. T ΔBEC is a right triangle since $\overline{BD} \perp \overline{AC}$, so $\angle BEC = 90°$. Thus, the other two angles of ΔBEC must add up to the remaining 90°.

25. T Every rhombus is a parallelogram and the opposite angles of a parallelogram are congruent.

Thus $\angle ABC \cong \angle CDA$.

27. The diagonals of a rhombus are perpendicular, so the rhombus is divided into four right triangles. The diagonals bisect the opposite angles, so $\angle 1 = 30°$ and $\angle 2 = \angle 3$.

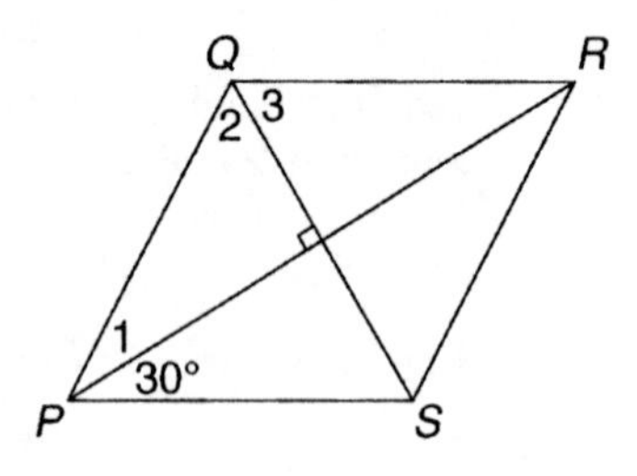

$$\text{Thus } \angle 1 + \angle 2 + 90° = 180°$$
$$30° + \angle 2 + 90° = 180°$$
$$\angle 2 = 60°$$
$$\text{and } \angle 3 = 60°.$$

Section 5.3

29. AB = $2x$ and $\angle AEB = 90°$ since the sides of a rhombus are congruent and the diagonals of a rhombus are perpendicular.

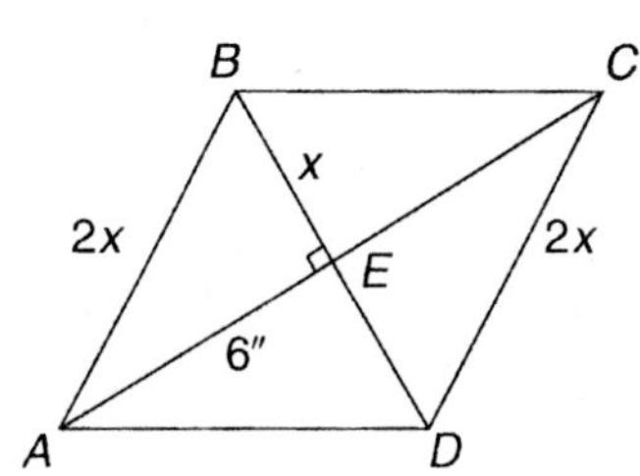

Use the Pythagorean Theorem.

$$(2x)^2 = x^2 + 6^2$$
$$4x^2 = x^2 + 36$$
$$3x^2 = 36$$
$$x^2 = 12$$
$$x = 2\sqrt{3}\text{ in.}$$

Thus the length of a side of rhombus $ABCD$ is $2x = 4\sqrt{3}$ in.

31. $\angle 4 = 40°$ by alternate interior angles for $\overline{CB}$ and $\overline{DE}$.

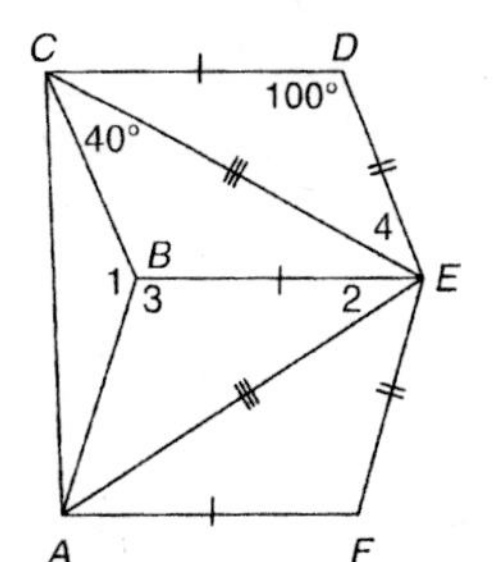

$\Delta CDE \cong \Delta EBC \cong \Delta AFE \cong \Delta EBA$ by SSS since $BCDE$ and $ABEF$ are parallelograms.

Thus, $\angle 3 = \angle D = 110°$ and $\angle EBC = \angle D = 110°$.

You know that $\angle 1 + \angle 3 + \angle EBC = 360°$. Substituting for $\angle 3$ and $\angle EBC$, you have $\angle 1 + 110° + 110° = 360°$, so $\angle 1 = 140°$.

To find $\angle 2$, first find $\angle ECD$. You know that $\angle ECD + 110° + \angle 4 = 180°$. Since $\angle 4 = 40°$, $\angle ECD + 110° + 40° = 180°$, so you have $\angle ECD = 30°$. Now $\angle 2 \cong \angle ECD$ by C.P. Therefore, $\angle 2 = 30°$.

Summarizing,

$\angle 1 = 140°$, $\angle 2 = 30°$, $\angle 3 = 110°$, and $\angle 4 = 40°$.

33. First, find the height of the parallelogram.

$\angle P = 45°$ as opposite angles in a parallelogram are congruent.

ΔPQT is a 45°–45° right triangle which results in $h = 8$ cm.

Thus the area of the parallelogram is

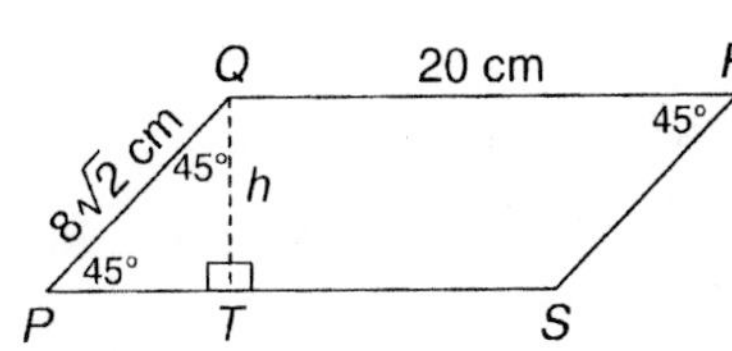

$$A = bh$$
$$= 20(8)$$
$$= 160\text{ cm}^2.$$

35. Since a diagonal measures 12 cm, half of that diagonal measures 6 cm (See diagram).

The diagonals of a rhombus are perpendicular and bisect each other. Therefore, the half-diagonals are the legs of a right triangle with a side of the rhombus as the hypotenuse. Let x = half of the unknown diagonal. Using the Pythagorean Theorem, you have

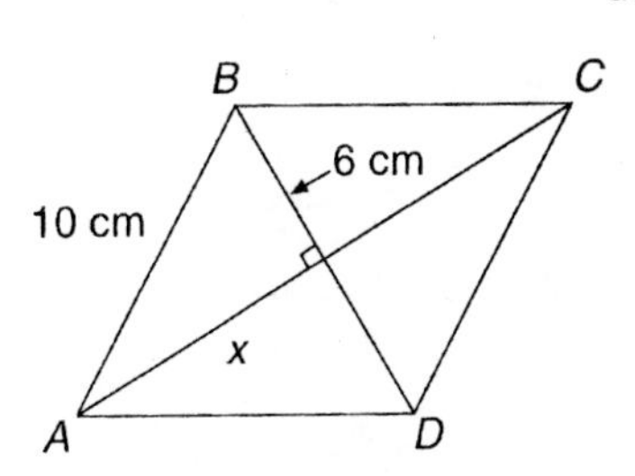

$$6^2 + x^2 = 10^2$$
$$36 + x^2 = 100$$
$$x^2 = 64$$
$$x = 8\text{ cm} = \text{half of the diagonal}$$
$$2x = 16\text{ cm} = \text{length of the diagonal}$$

Section 5.3

37. A complete solution is given in the answer key at the back of the text.

39. A complete solution is given in the answer key at the back of the text.

41. Consider the figure given.

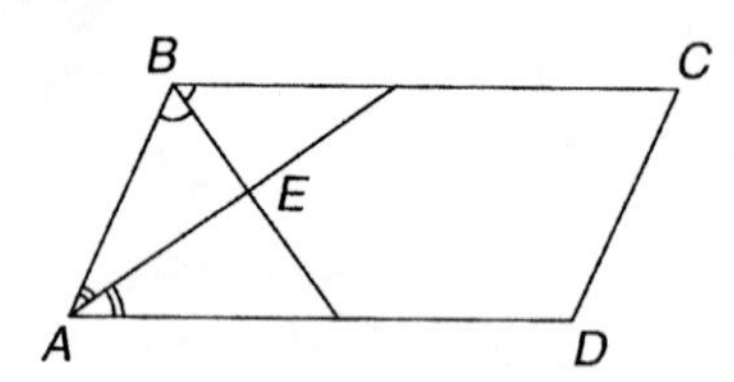

Given: $ABCD$ is a parallelogram, $\overline{BE}$ bisects $\angle ABC$, and $\overline{AE}$ bisects $\angle BAD$.

Prove: $\overline{AE} \perp \overline{BE}$

Proof:

Statements	Reasons
1. $\angle BAE = \frac{1}{2}\angle BAD$	1. $\overline{AE}$ bisects $\angle BAD$.
2. $\angle ABE = \frac{1}{2}\angle ABC$	2. $\overline{BE}$ bisects $\angle ABC$.
3. $\angle ABC + \angle BAD = 180°$	3. Any two consecutive angles of a parallelogram are supplementary.
4. $\frac{1}{2}\angle ABC + \frac{1}{2}\angle BAD = 90°$	4. Multiply both sides of the equation in step 3 by $\frac{1}{2}$.
5. $\angle BAE + \angle ABE = 90°$	5. Substitution
6. $\angle BAE + \angle ABE + \angle AEB = 180°$	6. Angle sum of a triangle is 180°.
7. $90° + \angle AEB = 180°$	7. Substitution
8. $\angle AEB = 90°$	8. Subtract 90° from both sides of the equation in step 7.
9. $\overline{AE} \perp \overline{BE}$	9. Definition

43. Proof:

Let $ABCD$ be a parallelogram as shown where $\overline{AE}$ bisects $\angle A$ and $\overline{CF}$ bisects $\angle C$.

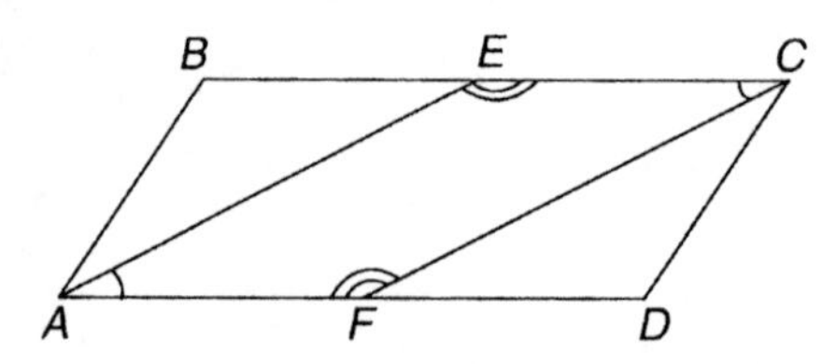

Section 5.3

43. (continued)

Show $AECF$ is a parallelogram.

1. $\angle BAE + \angle EAD = \angle BCD$	1. Opposite angles in parallelogram are congruent.
2. $2\angle EAD = \angle BCD$	2. $\overline{AE}$ bisects $\angle A$.
3. $\angle BCF + \angle FCD = \angle BAD$	3. Opposite angles in parallelogram are congruent.
4. $2\angle BCF = \angle BAD$	4. $\overline{CF}$ bisects $\angle C$.
5. $2\angle EAD = 2\angle BCF$	5. Substitution
6. $\angle EAD = \angle BCF$	6. Divide both sides by 2.
7. $\angle EAD + \angle CEA = 180°$	7. $\overline{EC} \parallel \overline{AF}$ so both pairs of interior angles on the same side of the transversal are supplementary (Corollary 5.8).
8. $\angle EAD + \angle CEA = \angle BCF + \angle CFA$	8. Substitution
9. $\angle BCF + \angle CEA = \angle BCF + \angle CFA$	9. Substitution
10. $\angle CEA = \angle CFA$	10. Subtract $\angle BCF$ from both sides of the equation.
11. $AECF$ is a parallelogram	11. Both pairs of opposite angles are congruent.

45. Imagine that the legs are the diagonals of a quadrilateral. Since the diagonals bisect each other, the quadrilateral must be a parallelogram (Theorem 5.21). Thus, the ironing board and the floor are a pair of parallel sides of the parallelogram.

47. The angle of incidence always equals the angle of reflection. The scale drawing below shows that the ball hits a side 11 times before finally landing in the corner pocket.

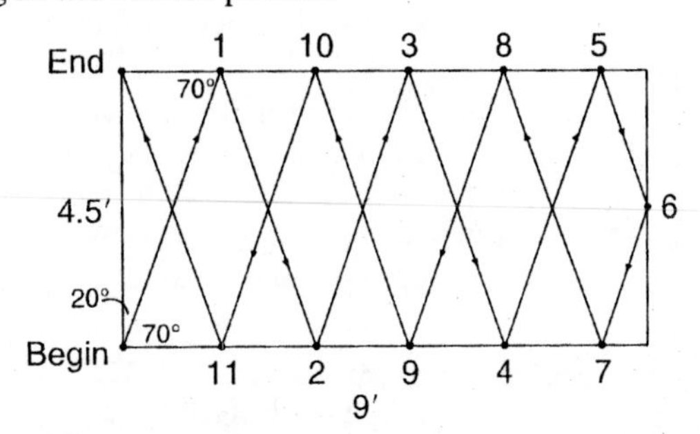

(Note: If the ball is thought of as a point, then the point will not hit the upper left corner of the rectangle. It will, however, be very close to the corner. The size of the opening of the corner pocket is large enough so that the ball will land in the pocket after it hits a side 11 times.)

Section 5.3

Section 5.4

Solutions to odd-numbered textbook problems

1. F This is true only if the rectangle is also a square.

3. T Every rectangle is a parallelogram (Theorem 5.26) and the diagonals of a parallelogram bisect each other (Theorem 5.20).

5. T The diagonals of a rectangle are congruent (Theorem 5.28).

7. F This is true only if the rectangle is also a square.

9. T Since $\overline{AC} \cong \overline{BD}$, $\overline{BC} \cong \overline{AD}$ and $\overline{AB} \cong \overline{BA}$, you have $\Delta ABD \cong \Delta BAC$ by SSS.

11. The diagonal and two adjacent sides of the rectangle form a right triangle. Let x be the length of the rectangle and then $x - 3$ is the width. Use the Pythagorean Theorem to solve for x.

$$x^2 + (x - 3)^2 = 15^2$$
$$x^2 + x^2 - 6x + 9 = 225$$
$$2x^2 - 6x - 216 = 0 \text{ and dividing by 2,}$$
$$x^2 - 3x - 108 = 0$$
$$(x - 12)(x + 9) = 0$$
$$\text{So } x = 12 \text{ or } x = -9$$

(*Note:* You could also use the quadratic formula to solve the equation above.)

Therefore, $x = 12$ cm = length and $x - 3 = 9$ cm = width. (Ignore the solution $x = -9$ since it is not physically possible.)

The area of the rectangle is $9 \cdot 12 = 108$ cm^2.

13. T Every square is a rhombus since all sides are congruent and the diagonals of a rhombus are perpendicular (Theorem 5.23).

15. T The diagonals of any parallelogram bisect each other, and a square is a parallelogram.

17. T $\overline{BD}$ is the transversal of parallel lines $\overline{AB}$ and $\overline{CD}$. Thus alternate interior angles are congruent.

19. T Since a square is also a rectangle, the diagonals are congruent and bisect each other. Therefore, the half-diagonals are also congruent; so $\overline{AE} \cong \overline{BE} \cong \overline{CE}$. The sides of the square are congruent, so $\overline{AB} \cong \overline{BC}$.

Thus, $\Delta ABE \cong \Delta BCE$ by SSS.

Section 5.4

21. T Every square is a rhombus and every square has four right angles. Thus, $ABCD$ is a rhombus with four right angles. The diagonals of a rhombus bisect the opposite angles, so $\angle BDA = 45°$.

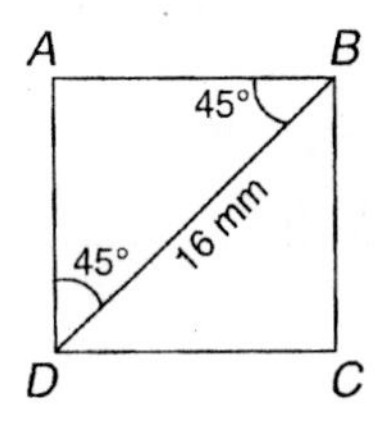

23. The diagonal bisects the opposite angles into 45° angles. ΔABD is a 45°–45° right triangle, thus

$$\begin{aligned} AB &= \frac{16}{\sqrt{2}} \cdot \frac{\sqrt{2}}{\sqrt{2}} \\ &= \frac{16\sqrt{2}}{2} \\ &= 8\sqrt{2} \text{ mm} \\ &\approx 11.31 \text{ mm.} \end{aligned}$$

25. T The base angles of an isosceles trapezoid are congruent.

27. T $\overline{AB} \cong \overline{AD} \cong \overline{AD}$. $\angle BAD \cong \angle CDA$ because the base angles of an isosceles trapezoid are congruent. (Theorem 5.30). Thus $\Delta BAD \cong \Delta CDA$ by SAS.

29. T $\Delta ABD \cong \Delta DCA$ by SAS since $\overline{AB} \cong \overline{DC}$, $\overline{AD} \cong \overline{DA}$, and $\angle BAD \cong \angle CDA$. Thus, $\overline{BD} \cong \overline{CA}$ by C.P.

31. $\Delta BAD \cong \Delta CDA$ by problem #27, thus $\angle ABD \cong \angle DCA$ by C.P. Also $\angle ABC \cong \angle DCA$ as the base angles of an isosceles trapezoid are congruent (Theorem 5.30). So $\angle DBC \cong \angle ACB$, which implies ΔBCE is isosceles. Therefore $BE = CE$.

33. The right triangle formed by one leg and the height has a base equal to half the difference in the bases of the trapezoid.

The base of the triangle $= \frac{1}{2}(10 - 6) = 2$ in.

Using the Pythagorean Theorem,

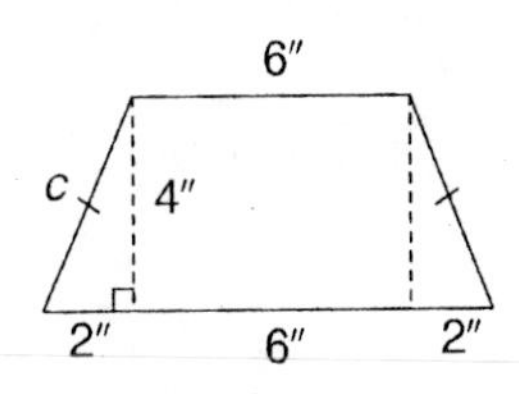

$$\begin{aligned} 2^2 + 4^2 &= c^2, \text{ where } c \text{ is the length of a leg} \\ 4 + 16 &= c^2 \\ 20 &= c^2 \\ \sqrt{20} &= c \\ 2\sqrt{5} &= c \\ 4.47 \text{ in.} &\approx c \end{aligned}$$

Each leg of the trapezoid measures approximately 4.47 in.

35. Each right triangle formed by a leg and the height is a 30°–60° right triangle. So the length of side opposite the 30° angle is 10 in. (one-half of the hypotenuse).

Thus $$\begin{aligned} EH &= 10 + 25 + 10 \\ &= 45 \text{ in.} \end{aligned}$$

Section 5.4

37. $\angle WXY = 90° + \angle PXY$ and $\angle ZYX = 90° + \angle PYX$. Since $\angle WXY \cong \angle ZYX$, $90° + \angle PXY = 90° + \angle PYX$. Thus, $\angle PXY \cong \angle PYX$. Then ΔXYP is an isosceles triangle, and $XP = YP = 5$ in.

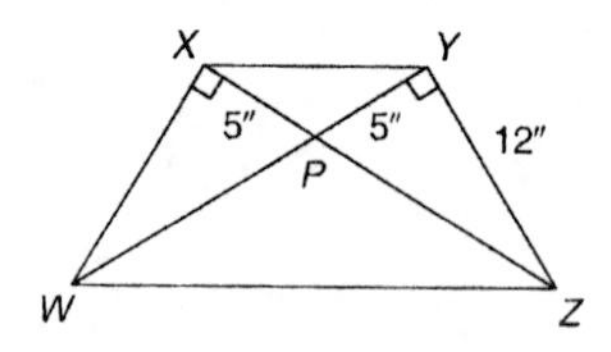

ΔYPZ is a right triangle with $\overline{PZ}$ as the hypotenuse. By the Pythagorean Theorem,

$$5^2 + 12^2 = (PZ)^2$$
$$25 + 144 = (PZ)^2$$
$$169 = (PZ)^2$$
$$13 \text{ in.} = PZ$$

Since $PW = PZ = 13$ in., you also have $WY = 13 + 5 = 18$ in. Then for ΔWZY,

$$18^2 + 12^2 = (WZ)^2$$
$$468 = (WZ)^2$$
$$\sqrt{468} = WZ$$
$$6\sqrt{13} = WZ$$
$$21.63 \text{ in.} \approx WZ$$

39. Let $ABCD$ be a rhombus with one right angle. Since every rhombus is a parallelogram, $ABCD$ is a parallelogram with one right angle. By Theorem 5.27, $ABCD$ is a rectangle. However, since all of its sides are congruent, $ABCD$ is also a square.

41. Let $ABCD$ be a rhombus with congruent diagonals. Since $ABCD$ is a rhombus it is a parallelogram (Theorem 5.22). Since $ABCD$ is a parallelogram and since it has congruent diagonals, $ABCD$ is a rectangle (Theorem 5.29).

Thus $ABCD$ has all sides congruent and four right angles which implies that $ABCD$ is a square.

43. A complete solution is given in the answer key at the back of the text.

45. A complete solution is given in the answer key at the back of the text.

47. A complete solution is given in the answer key at the back of the text.

Section 5.5

Tips:

✔ Always use a straightedge and be very careful in each step of your construction. Small errors are magnified.

✔ There is often more than one method for a construction. (For example, there are several ways to construct a rectangle.) The method that involves the smallest number of basic constructions is usually the most accurate.

✔ Many times you can use a property of quadrilaterals to devise a method of construction.

Section 5.5 Solutions to odd-numbered textbook problems

1. Trace line l and point P. Then use Construction 7 as described in the text.

3. Step 1: Draw a line and mark off length a.

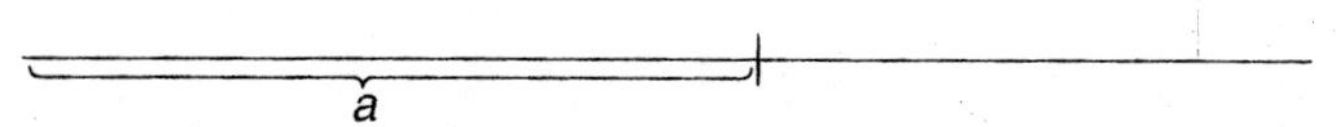

Step 2: Label the endpoints of segment A and D and construct a line perpendicular to $\overline{AD}$ passing through point A.

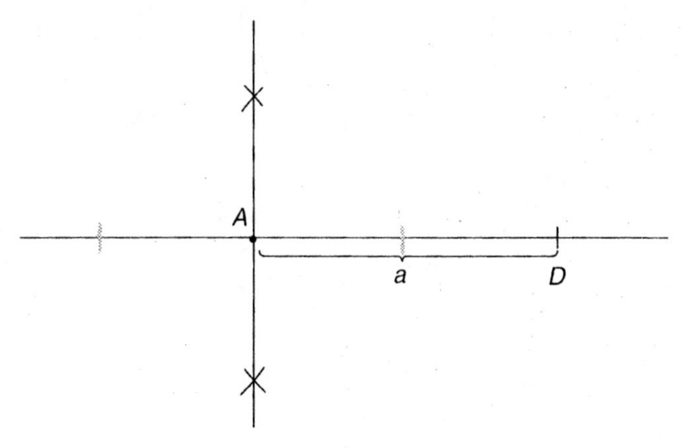

Step 3: Set the compass to radius a and with point at A mark off length a on the perpendicular line. Label the point B.

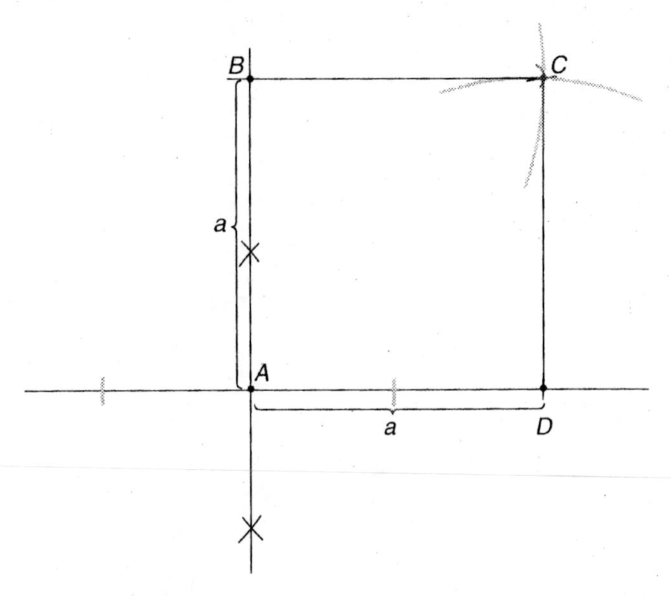

Step 4: Keeping compass set to radius a and with point at B swing an arc. Again, keeping the compass set to radius a and with point at C, swing an arc. Label the point of intersection of the arcs C.

Quadrilateral $ABCD$ is a square with sides of length a.

Section 5.5

5. More than one method of construction is possible. One is shown here.

   Step 1: Draw a line and mark off length a on one end.

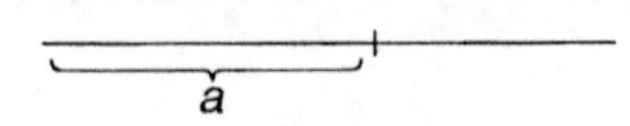

   Step 2: Copy angle A at each end of line segment with length a. Both angles should open in the same direction.

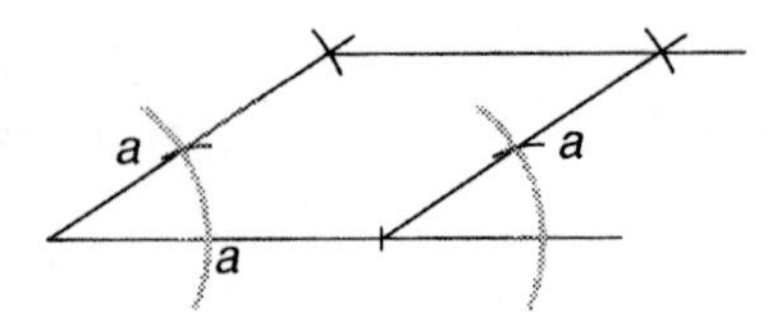

   Step 3: Mark off the length a on the upper ray of each angle and join the two endpoints with your straightedge.

7. Step 1: Draw a line. Construct a perpendicular line through point A.

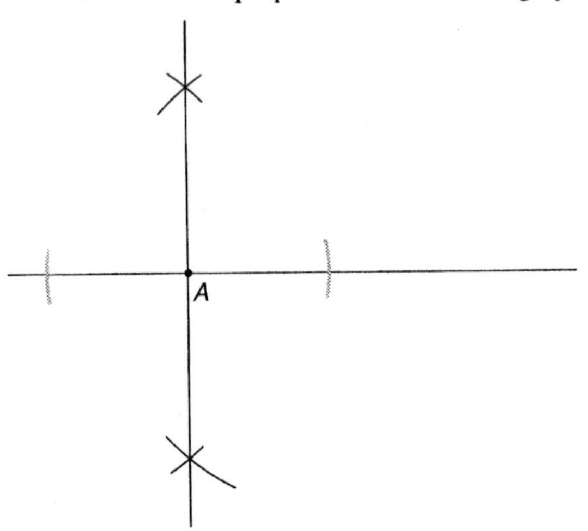

   Step 2: Bisect $\angle A$. Setting the compass to radius a, mark off length a on the angle bisector. Label point C.

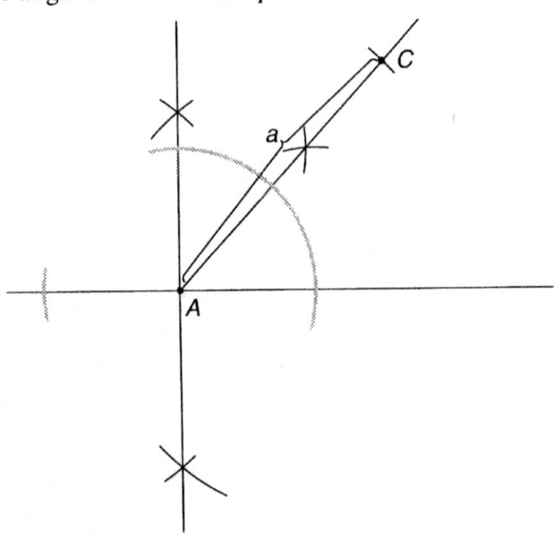

Section 5.5

7. (continued)

Step 3: Construct the two lines through point C that are perpendicular to the two sides of the square. Label points B and D.

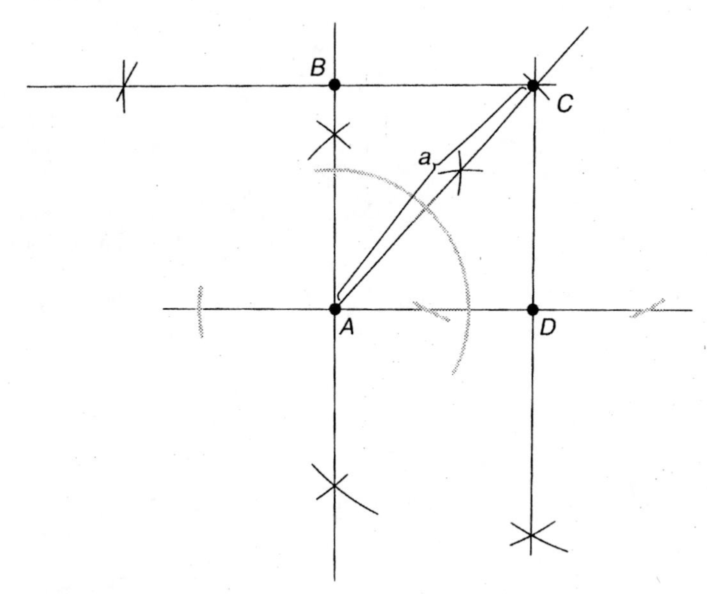

Quadrilateral $ABCD$ is a square with diagonal of length a.

9. Step 1: There are many possibilities here since the length of the legs is not specified. One method of construction is to first copy a and use it as the lower base. Then copy $\angle A$ at each end of the segment. Construct angles opening to the right on the left end of the segment and opening to the left on the right end of the segment.

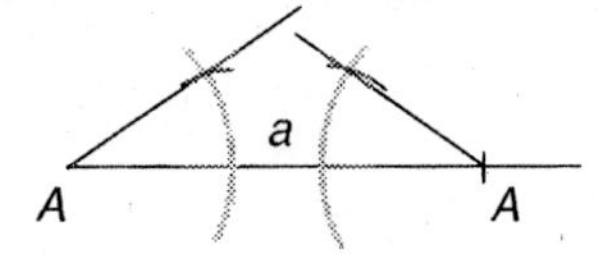

Step 2: Use any convenient radius on your compass and mark off an equal length on the upper ray of each angle. Draw the upper base through those marks.

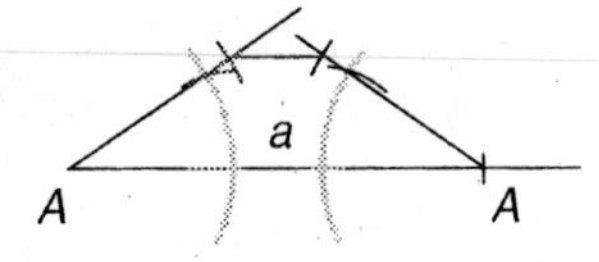

Different trapezoids can be constructed by varying the lengths of the legs.

Section 5.5

11. Step 1: Construct a 60° angle.

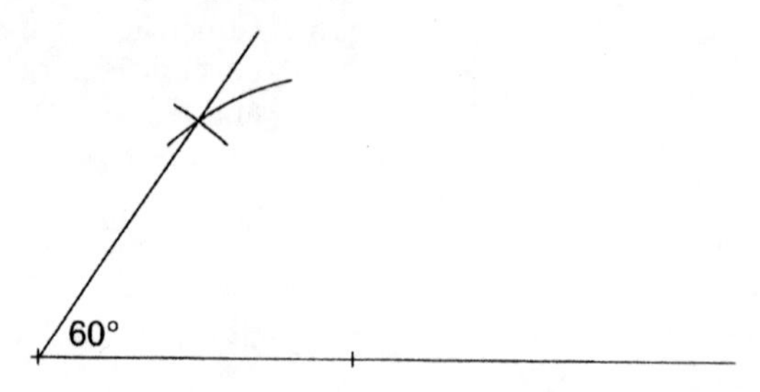

Step 2: Mark off length a on one side of angle and length b on the other side of angle. Label points B and D.

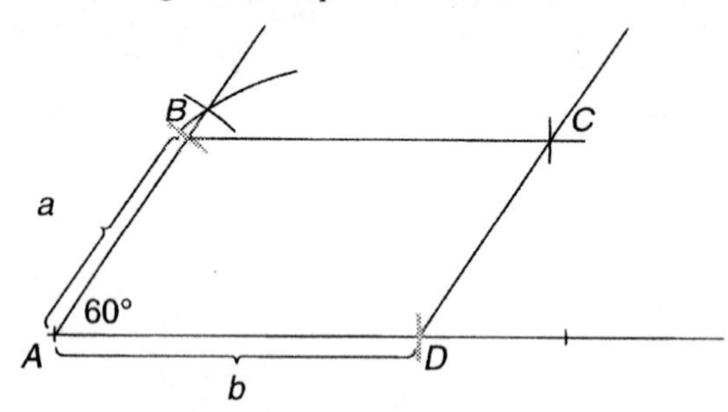

Step 3: Follow the procedure in Construction 7 to construct a line parallel to $\overline{AD}$ through point B.

Step 4: Mark off length b on parallel line that was constructed in step 3. Label the point C.

Step 5: Connect points C and D to complete construction of parallelogram $ABCD$.

13. To perform this construction, use the fact that the diagonals of a rhombus are perpendicular bisectors of each other.

Step 1: Copy angle P and mark off the length of the diagonal on its upper ray, making $\overline{PA}$.

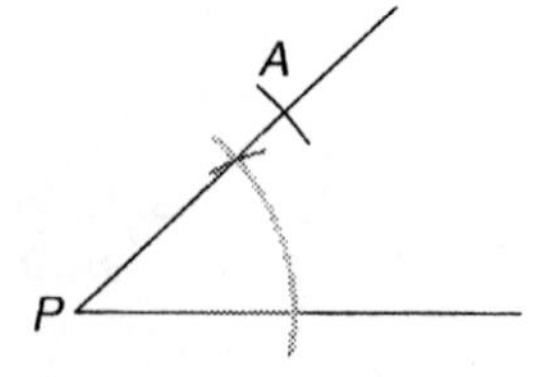

Step 2: Construct the perpendicular bisector of $\overline{PA}$. The other diagonal of the rhombus will lie on this segment. Mark the point where the perpendicular bisector intersects the lower ray of $\angle P$ as B and the point where it intersects $\overline{PA}$ as C.

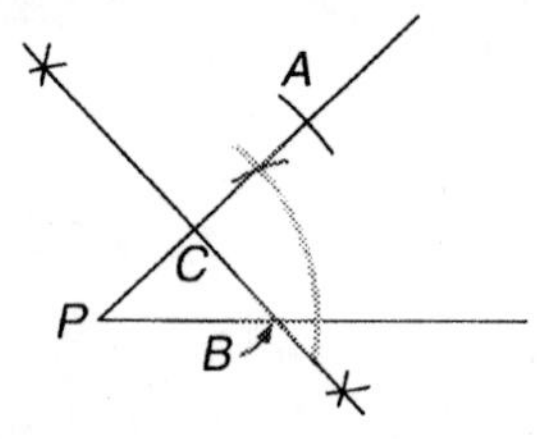

Section 5.5

Step 3: Measure $\overline{CB}$ with your compass, and, on the other side of C, mark off length CB on the perpendicular bisector. This gives you the other end of the second diagonal. Then join the endpoints of the diagonals to make the rhombus.

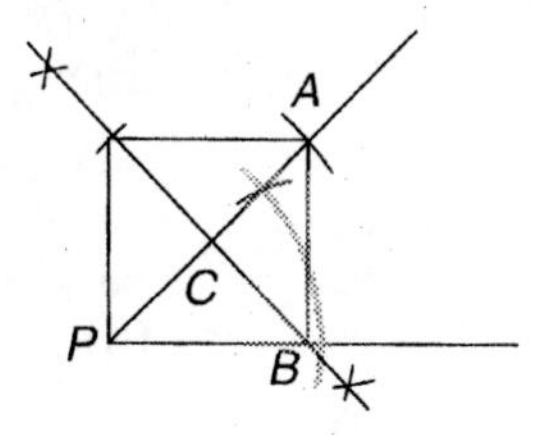

15. Step 1: Copy diagonal c and construct midpoint. Label endpoints A and C.

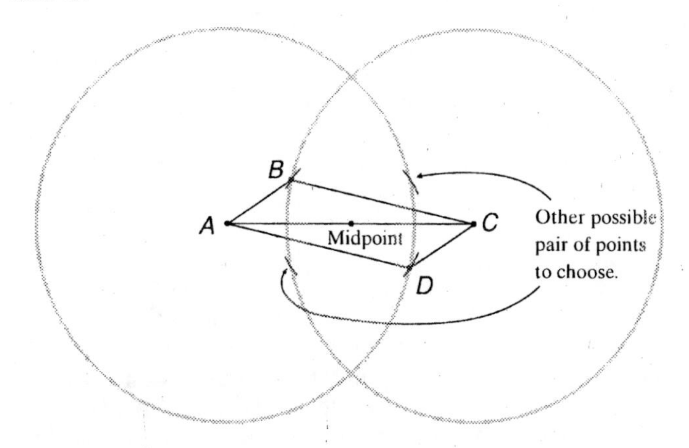

Step 2: Set compass to radius a and swing an arc from point A and from point C.

Step 3: Set compass to radius $\frac{1}{2}b$ and swing an arc from the midpoint of $\overline{AC}$.

Step 4: Where the arcs from steps 2 and 3 intersect will give the location of the other two vertices of the parallelogram. Label the points B and D. Note: There are two possible pairs of points to choose.

Step 5: Connect the points to form parallelogram $ABCD$.

17. Copy line segment $\overline{AB}$. Then use Construction 8 as described in the text.

19. Step 1: Copy $\overline{MN}$ using Construction 1.

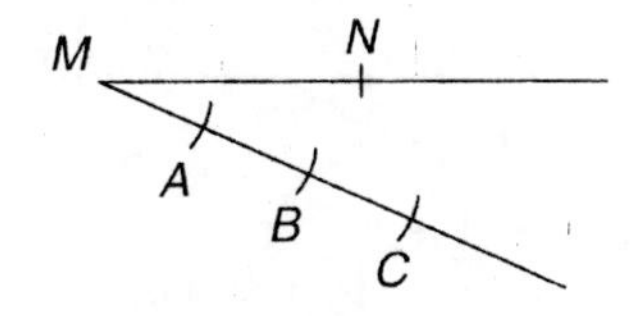

Section 5.5

19. (continued)

Step 2: Draw a ray from endpoint M forming some acute angle with $\overline{MN}$.

Step 3: Setting the compass at any convenient radius, mark off three congruent segments on the ray you drew in Step 2.

Step 4: Join point C to N. Construct an angle at B that is congruent to $\angle BCN$. Extend the side of the angle if necessary so that it intersects $\overline{MN}$ at P. Point P divides $\overline{MN}$ into two segments, one twice as long as the other. That is, $MP = 2PN$.

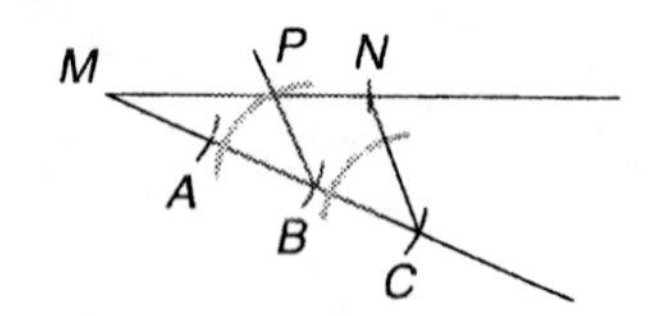

21. If T is the midpoint of $\overline{PS}$, then U must be the midpoint of $\overline{RP}$ since the parallel lines divide the two transversals proportionally. Thus $RP = 2RU = 2(5.8) = 11.6$ cm. Using RP as the base of the triangle and SQ as its height, find the area of ΔPRS.

$$A = \frac{1}{2}(11.6)(7.4) = 42.92 \text{ cm}^2$$

23. Trace ΔABC. Find the midpoint of $\overline{AC}$ using straight edge and compass. Note that the midpoint of $\overline{AC}$ will lie on the perpendicular bisector of $\overline{AC}$. Use construction 6 to find the perpendicular bisector. Label the midpoint D. Now ΔABD and ΔDBC have the same altitude, h, and the same base measure. Thus, ΔABD and ΔDBC have the same area.

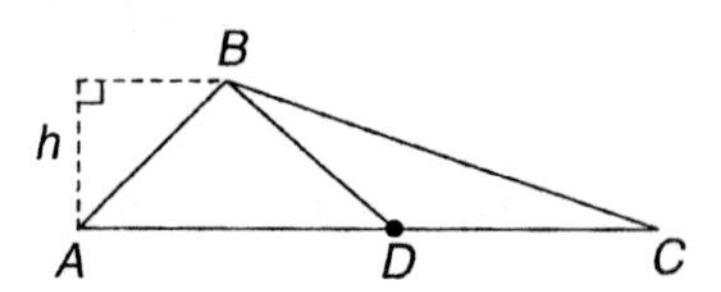

25. In problem #6 you were asked to construct a parallelogram given two adjacent sides and the included angle. One method of construction is described below. It uses congruent corresponding angles to construct the parallel sides. Other constructions, such as one using congruent alternate interior angles, are also possible.

Step 1: Copy the given angle A and mark off sides a and b on the rays forming it.

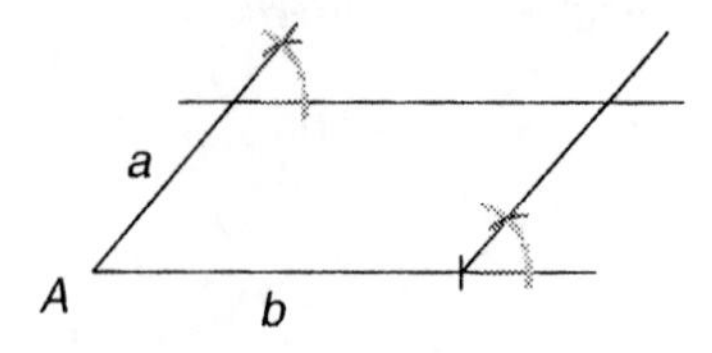

Step 2: At the upper end of side a, construct a line parallel to side b. At the right end of side b, construct a line parallel to side a.

Now you have a quadrilateral with two pairs of parallel sides. By definition the quadrilateral is a parallelogram and it contains the two given sides and the given angle.

Section 5.5

27. The justification provided in problem #25 with $\angle A = 90°$ applies. The resulting parallelogram will have four right angles, so it will be a rectangle.

29. In problem #12 you were asked to construct a rhombus with vertex angle 60° and a diagonal of the given length d. One method of construction is described below.

Step 1: Construct a 60° angle by constructing one angle of an equilateral triangle. (See Section 4.4, problem #11.)

Step 2: Bisect the angle you constructed.

Step 3: Locate point D at distance d from A on the angle bisector.

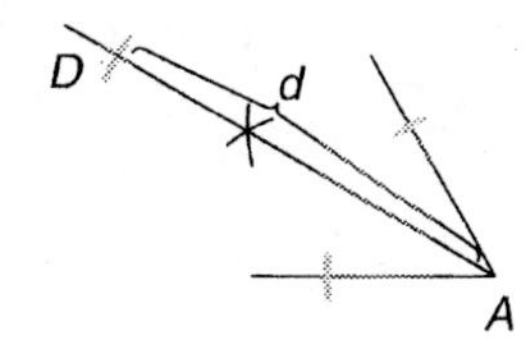

Step 4: Construct the perpendicular bisector of the diagonal and extend it to intersect the sides of the 60° angle. The points of intersection, B and C, are the other two vertices of the rhombus. Draw the other sides of the rhombus from the points C and B to D.

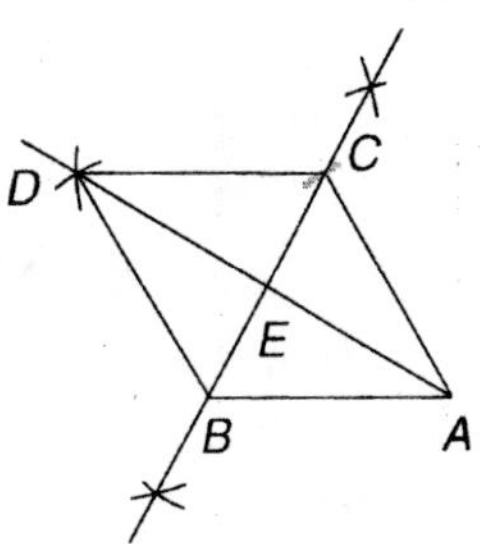

$\overline{BC}$ is the perpendicular bisector of $\overline{AD}$ by construction. Therefore, $\angle DEC = \angle AEC = 90°$ and $DE = AE$. Since $CE = CE$, $\Delta DEC \cong \Delta AEC$ by SAS. Similarly, $\Delta DEB \cong \Delta AEB$.

Since $\overline{AD}$ is the bisector of $\angle BAC$ by construction, you have $\angle BAE = \angle CAE$. You also have $EA = EA$. Thus, $\Delta AEB \cong \Delta AEC$ by ASA. Therefore, $\Delta AEC \cong \Delta AEB \cong \Delta DEB \cong \Delta DEC$. Then $CD = BD = CA = BA$ by C.P., and $ABDC$ is a rhombus.

31. In the construction of problem #15 the diagonals of the quadrilateral $ABCD$ intersect at their midpoints. Thus the diagonals bisect each other. If the diagonals of a quadrilateral bisect each other, then the quadrilateral is a parallelogram (Theorem 5.21).

33. $CU = UQ$ because $\overline{CQ}$ and $\overline{SQ}$ are transversals of the parallel line segments $\overline{SC}$, $\overline{RU}$, and $\overline{QD}$. The transversals are divided proportionally by the parallel lines.

35. A complete solution is given in the answer key at the back of the text.

Section 5.5

Additional Problems

1. A complete solution is given in the answer key at the back of the text.

2. A complete solution is given in the answer key at the back of the text.

Solutions to Chapter 5 Review

Section 5.1

1. A complete solution is given in the answer key at the back of the text.

2. A complete solution is given in the answer key at the back of the text.

3. $\angle 13 = 130°$

   $\angle 13 = \angle 8$ by vertical angles, $\angle 8 = \angle 15$ by alternate interior angles, $\angle 15 = \angle 17$ by vertical angles.

   Thus $\angle 13 = \angle 8 = \angle 15 = \angle 17 = 130°$.

   $\angle 13 + \angle 14 = 180°$ supplementary angles

   $130° + \angle 14 = 180°$ substitution

   $\angle 14 = 50°$

   $\angle 14 = \angle 7$ by vertical angles, $\angle 7 = \angle 2$ by alternate interior angles, $\angle 2 = \angle 16$ by vertical angles.

   Thus $\angle 14 = \angle 7 = \angle 2 = \angle 16 = 50°$.

   $\angle 1 = 90°$

   $\angle 1 = \angle 19$ vertical angles

   $\angle 1 + \angle 20 = 180°$ supplementary angles

   $90° + \angle 20 = 180°$ substitution

   $\angle 20 = 90°$

   $\angle 20 = \angle 18$ vertical angles

   Thus $\angle 1 = 19 = 90°, \angle 20 = \angle 18 = 90°$.

   $\angle 11 = 75°$

   $\angle 11 = \angle 9$ by vertical angles, so $\angle 11 = \angle 9 = 75°$.

   $\angle 11 + \angle 12 = 180°$ supplementary angles

   $75° + \angle 12 = 180°$ substitution

   $\angle 12 = 105°$

   $\angle 12 = \angle 10$ by vertical angles, so $\angle 12 = \angle 10 = 105°$.

   $\angle 1 + \angle 2 + \angle 3 = 180°$ angle sum of a triangle

   $90 + 50 + \angle 3 = 180°$ substitution

   $\angle 3 = 40°$

Chapter Review

3. (continued)

$$\angle 7 + \angle 9 + \angle 5 = 180^\circ \quad \text{angle sum of a triangle}$$
$$50^\circ + 75^\circ + \angle 5 = 180^\circ \quad \text{substitution}$$
$$\angle 5 = 55^\circ$$

$$\angle 3 + \angle 6 = 180^\circ \quad \text{supplementary angles}$$
$$40^\circ + \angle 6 = 180^\circ \quad \text{substitution}$$
$$\angle 6 = 140^\circ$$

$$\angle 4 + \angle 5 = 180^\circ \quad \text{supplementary angles}$$
$$\angle 4 + 55^\circ = 180^\circ \quad \text{substitution}$$
$$\angle 4 = 125^\circ$$

To summarize

$\angle 1 = 90^\circ$	$\angle 6 = 140^\circ$	$\angle 11 = 75^\circ$	$\angle 16 = 50^\circ$
$\angle 2 = 50^\circ$	$\angle 7 = 50^\circ$	$\angle 12 = 105^\circ$	$\angle 17 = 130^\circ$
$\angle 3 = 40^\circ$	$\angle 8 = 130^\circ$	$\angle 13 = 130^\circ$	$\angle 18 = 90^\circ$
$\angle 4 = 125^\circ$	$\angle 9 = 75^\circ$	$\angle 14 = 50^\circ$	$\angle 19 = 90^\circ$
$\angle 5 = 55^\circ$	$\angle 10 = 105^\circ$	$\angle 15 = 130^\circ$	$\angle 20 = 90^\circ$.

4. Let l and m be parallel lines cut by transversal t.

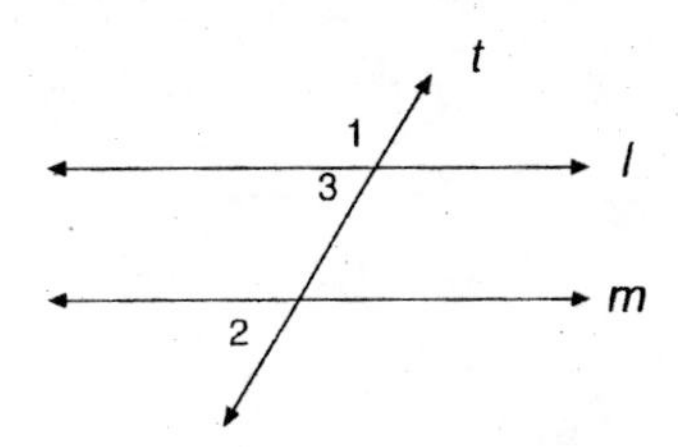

Let $\angle 1$ and $\angle 2$ be a pair of exterior angles on the same side of the transversal. Let $\angle 3$ be as shown in diagram.

$\angle 2 \cong \angle 3$ by corresponding angles.

$\angle 1 + \angle 3 = 180^\circ$ by supplementary angles.

Therefore, $\angle 1 + \angle 2 = 180^\circ$ by substitution, and the pair of exterior angles on the same side of the transversal are supplementary.

Section 5.2

1. An exterior angle is formed by one side of a triangle and the extension of an adjacent side. The exterior angles are $\angle CBD$, $\angle BDG$, and $\angle ADF$.

2. The exterior angle of a triangle is equal to the sum of the nonadjacent interior angles.

Chapter Review

2. (continued)

$$2x - 15° + 3x + 30° = x + 69°$$
$$5x + 15° = x + 69°$$
$$4x = 54°$$
$$x = \frac{54°}{4} = 13.5°$$

3. (a) T $\angle AEB \cong \angle DEB$ since $\overline{BE}$ bisects $\angle AED$, $\overline{BE} \cong \overline{BE}$ by the reflexive property.

   Thus $\Delta BAE \cong \Delta BDE$ by HA.

   (b) F

   (c) T $\angle ABD = \angle CDB + \angle BCD$ Exterior Angle Theorem (Corollary 5.10)

   Also $\angle CDB = 90° = \angle BAE$.

   So $\angle ABE + \angle EBD = \angle BAE + \angle BCD$.

4. Let points A, B, C, D, and E be collinear and let $\overline{BC} \cong \overline{DC}$.

   So ΔBCD is isosceles, by definition of isosceles triangle.

   Then $\angle CBD \cong \angle CDB$ as the angles opposite the congruent sides of an isosceles triangle are congruent (Theorem 4.5).

   Now $\angle CBA + \angle CBD = 180°$ by supplementary angles

   $\angle CDB + \angle CDE = 180°$ by supplementary angles.

   Hence $\angle CBA + \angle CBD = \angle CDB + \angle CDE$ by substitution.

   Thus $\angle CBA = \angle CDE$ since $\angle CBD = \angle CDB$.

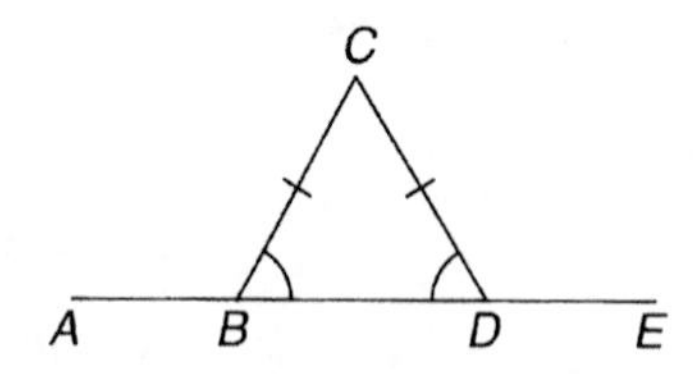

Section 5.3

1. The opposite sides of a parallelogram are congruent, and the diagonals of a parallelogram bisect each other. Therefore, $\overline{AB} \cong \overline{DC}$, $\overline{AD} \cong \overline{BC}$, $\overline{BE} \cong \overline{DE}$, and $\overline{AE} \cong \overline{CE}$.

2. Consecutive angles in a parallelogram are supplementary since the consecutive angles are a pair of interior angles on the same side of a transversal.

   Thus, $4n - 2° + n - 3° = 180°$

$$5n - 5° = 180°$$
$$5n = 185°$$
$$n = 37°$$

Chapter Review

2. (continued)

The measures of the consecutive angles are

$$4n - 2° = 4(37°) - 2°$$
$$= 146°$$
$$\text{and } n - 3° = 37° - 3° = 34°.$$

Thus the angles are 34°, 146°, 34° and 146°.

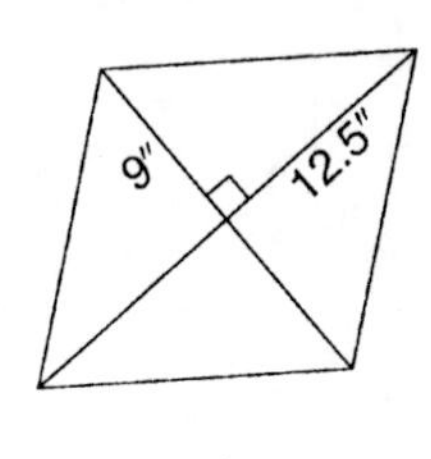

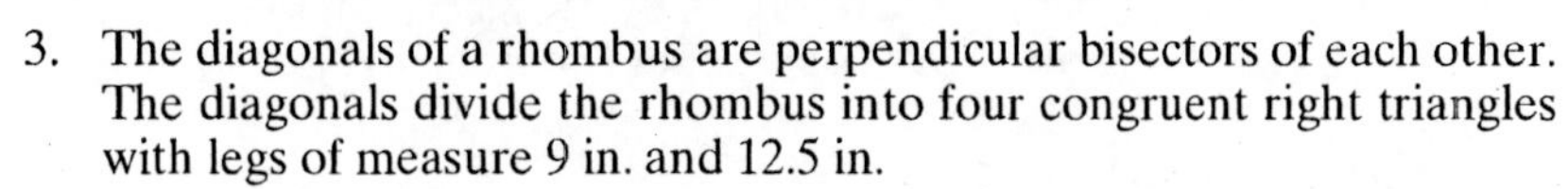

3. The diagonals of a rhombus are perpendicular bisectors of each other. The diagonals divide the rhombus into four congruent right triangles with legs of measure 9 in. and 12.5 in.

   The area of the rhombus is four times the area of one of the congruent triangles.

$$A = 4\left(\frac{1}{2}\right)(9)(12.5)$$
$$= 18(12.5)$$
$$= 225 \text{ in}^2$$

4. The diagonals of a rhombus bisect the opposite angles and the diagonals of a rhombus are perpendicular bisectors of each other. Now, the sum of the angles of one of the four congruent right triangles can be expressed as

$$n - 4° + 2n + 1° + 90° = 180°.$$

So $\quad 3n + 87° = 180°$

$$3n = 93°$$

and $\quad n = 31°.$

Thus, $\angle WZX = 2(31°) + 1° = 63°$

and $\angle XWY = 31° - 4° = 27°.$

Therefore $\angle XWZ = 54°$ and $\angle WZY = 126°.$

Since opposite angles of a rhombus are congruent we conclude $\angle XYZ = 54°$ and $\angle WXY = 126°.$

5. Let rhombus $PQRS$ be given with midpoints A, B, C, and D as shown in the diagram.

   $\angle Q \cong \angle S$ as opposite angles are congruent.

   $\overline{AQ} \cong \overline{QB} \cong \overline{SD} \cong \overline{SC}$ since

   A, B, C, and D are midpoints of the sides of the rhombus.

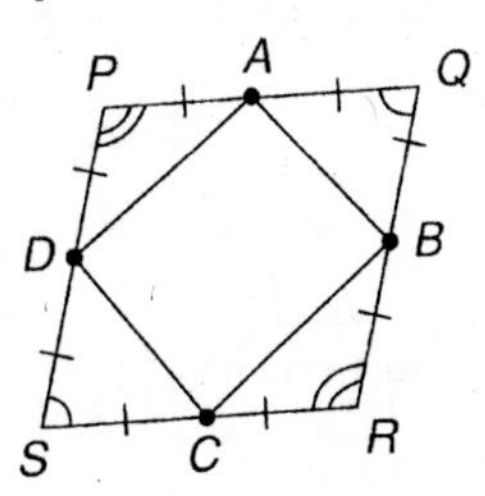

Chapter Review

5. (continued)

Thus $\Delta AQB \cong \Delta CSD$ by SAS and $\overline{AB} \cong \overline{DC}$ by C.P. Similarly, $\angle P \cong \angle R$ as opposite angles are congruent.

$\overline{AP} \cong \overline{PD} \cong \overline{BR} \cong \overline{RC}$ since A, B, C, and D are midpoints of the sides of a rhombus.

Thus $\Delta DPA \cong \Delta CRB$ by SAS and $\overline{AD} \cong \overline{BC}$ by C.P.

Therefore, $ABCD$ is a parallelogram as both pairs of opposite sides are congruent (Theorem 5.17).

Section 5.4

1. $ABCD$ can be a rectangle, a square, or an isosceles trapezoid. See the summary of attributes of quadrilaterals on page 275 of the text.

2. Rectangles and squares. See the summary of attributes of quadrilaterals on page 275 of the text.

3. $AE = BD = 12$ cm — as opposite sides of a parallelogram are congruent.

$BD = CE = 12$ cm — as the diagonals of a rectangle are congruent.

$CF = BF = 6$ cm — as the diagonals of a rectangle bisect each other.

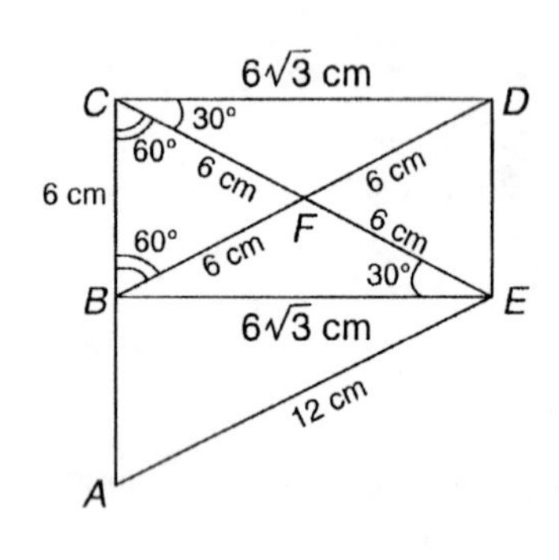

$\angle CEB = \angle ECD = 30°$ — by alternate interior angles.

$\angle ECB = 60°$ — as $\angle ECB$ and $\angle ECD$ are complementary.

So $\angle CBF = 60°$ — as the angles opposite the congruent sides have the same measure.

Therefore ΔCBF is an equilateral triangle, so $CF = BF = BC = 6$ cm.

Now ΔCBE is a 30°–60° right triangle with hypotenuse 12 cm so leg $BE = 6\sqrt{3}$ cm.

Therefore $CD = BE = 6\sqrt{3}$ cm.

To summarize

$$CF = 6 \text{ cm}$$
$$CB = 6 \text{ cm}$$
$$CD = 6\sqrt{3} \text{ cm}.$$

4. $AE = BC = 9$ cm since $EABC$ is an isosceles trapezoid. $\angle AEC = \angle BCE = 45°$ as the base angles of an isosceles trapezoid are congruent and $\overline{ED}$ bisects right angle $\angle AEF$.

$\angle EAB = \angle CBA = 135°$ as the base angles of an isosceles trapezoid are congruent and $\angle ABC$ and $\angle BCE$ are supplementary.

Chapter Review

4. (continued)

$\angle BAD = 45°$ since $\angle EAD + \angle DAB = 135°$ and $\angle DAB = 90°$.

Therefore, $ABCD$ is a parallelogram since the opposite angles are congruent.

Since $ABCD$ is a parallelogram with $AB = 9$ cm we have $CD = 9$ cm.

To find the area of the trapezoid we need base EC and the altitude.

ΔEAD is a $45° - 45°$ right triangle with legs 9 cm,

so $ED = 9\sqrt{2}$ cm. Also, $AF = ED = 9\sqrt{2}$ cm.

Since $\overline{AF} \perp \overline{ED}$, $\overline{AG}$ is the altitude of $EABC$.

$$AG = \frac{1}{2}AF = \frac{1}{2}9\sqrt{2}.$$

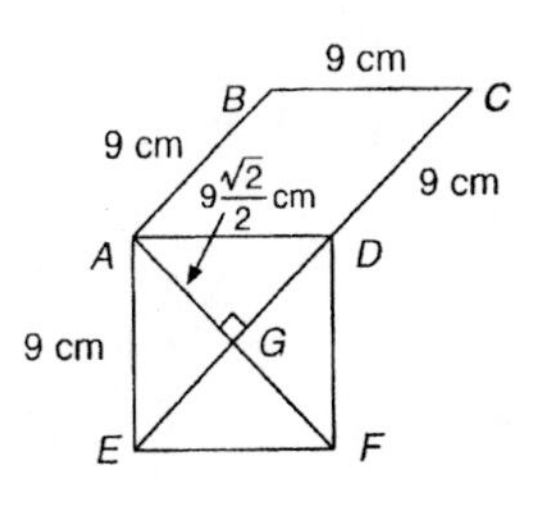

Now we know $AB = 9$ cm,

$$EC = ED + DC$$
$$= (9\sqrt{2} + 9) \text{ cm,}$$

$$\text{and } AG = \frac{9\sqrt{2}}{2} \text{ cm}$$

Therefore, the area of isosceles trapezoid $EABC$ is

$$A_{EABC} = \frac{1}{2}(AB + EC)AG$$
$$= \frac{1}{2}[9 + (9\sqrt{2} + 9)]\frac{9\sqrt{2}}{2}$$
$$= \frac{9\sqrt{2}}{4}(18 + 9\sqrt{2})$$
$$= \frac{162\sqrt{2} + 162}{4}$$
$$= \frac{81\sqrt{2} + 81}{2}$$
$$\approx 97.78 \text{ cm}^2.$$

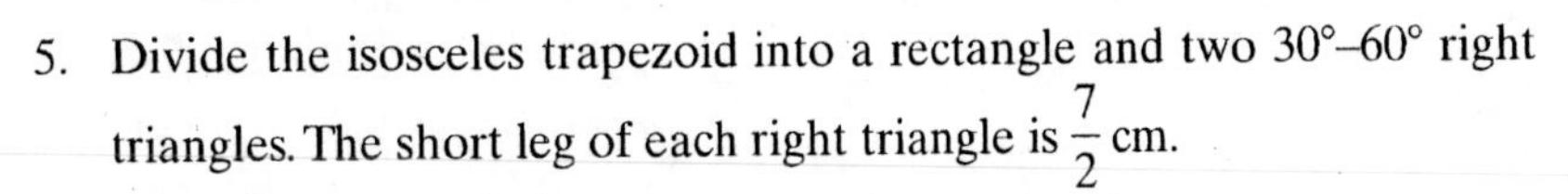

5. Divide the isosceles trapezoid into a rectangle and two 30°–60° right triangles. The short leg of each right triangle is $\frac{7}{2}$ cm.

Therefore, the longer base of the trapezoid is 22 cm.

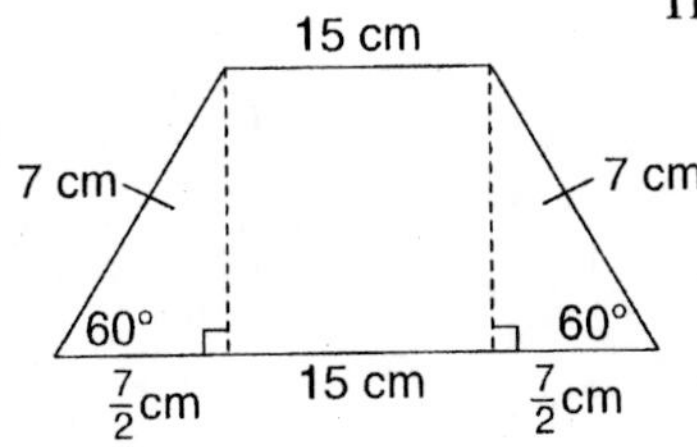

Chapter Review

Section 5.5

1. Trace line m and point Q. Then use Construction 7 as described in the text.

2. Use the construction as described in problem #7 of Section 5.4 in the Student Solutions Manual.

3. Step 1: Construct a right angle.

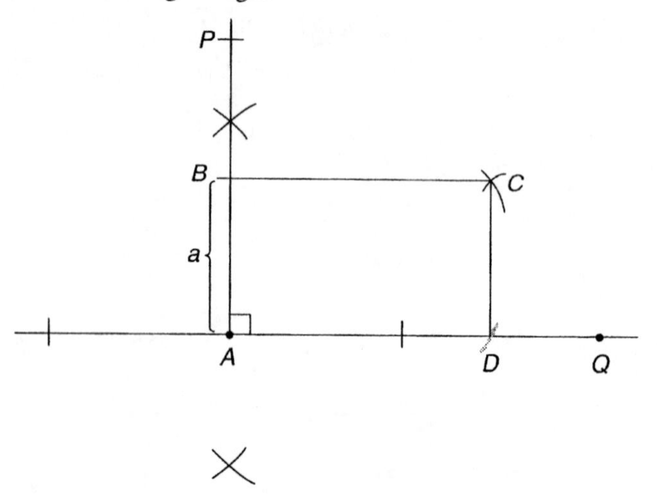

Step 2: Set the compass to radius a. With compass point at A mark off length a on ray $\overrightarrow{AP}$. Label the intersection B.

Step 3: Set compass to radius c. With compass point at B swing an arc to intersect ray $\overrightarrow{AQ}$. Label the intersection D.

Step 4: Keeping the compass with radius c, place point of compass on A and swing an arc.

Step 5: Set compass to radius a. With compass point at D, swing an arc. Label the point of intersection of step 4 and step 5 point C. Join points B and C and points C and D with a straight edge. $ABCD$ is the desired rectangle.

4. Step 1: Construct a 60° angle (See procedure in Section 4.4 problem #15 of Student Solutions Manual.)

   Bisect 60° angle. Use Construction 3 in text.

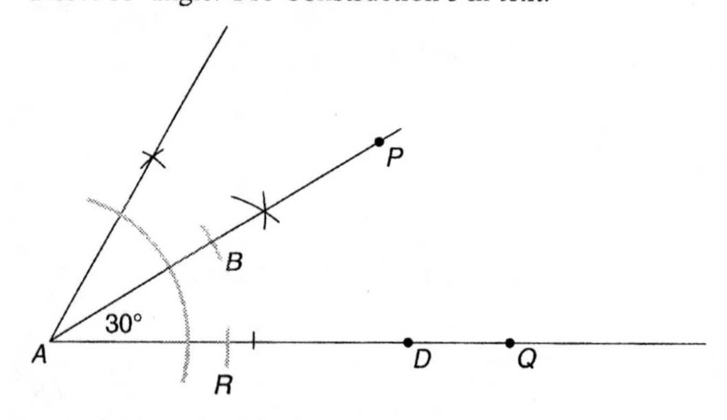

Step 2: Mark off any desired length on ray $\overrightarrow{AP}$. Label intersection B.

Step 3: Set compass to radius AB. With point on A mark off length AB on $\overrightarrow{AQ}$. Label intersection R.

Step 4: Keeping compass to radius AB, set point on R and mark off length AB on $\overrightarrow{RQ}$. Label intersection D.

Step 5: Copy $\angle BAD$ at D. Use Construction 2 in text.

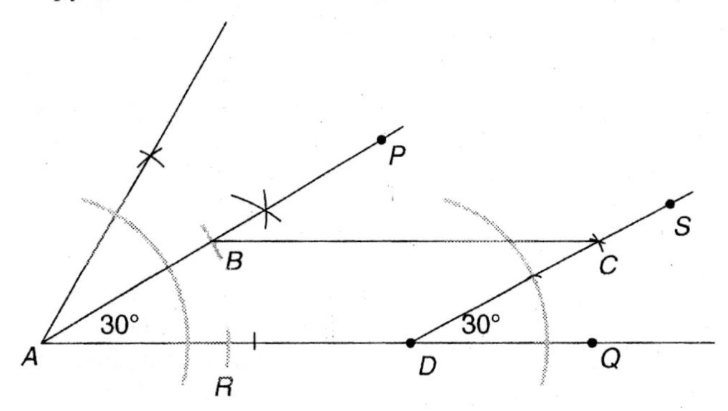

Step 6: Set compass to radius AB. With point at D mark off length AB on $\overrightarrow{DS}$. Label intersection C.

Step 7: Connect points B and C with straight edge. Now $ABCD$ is the desired parallelogram.

5. Step 1: Copy $\angle A$. Use Construction 2 in text.

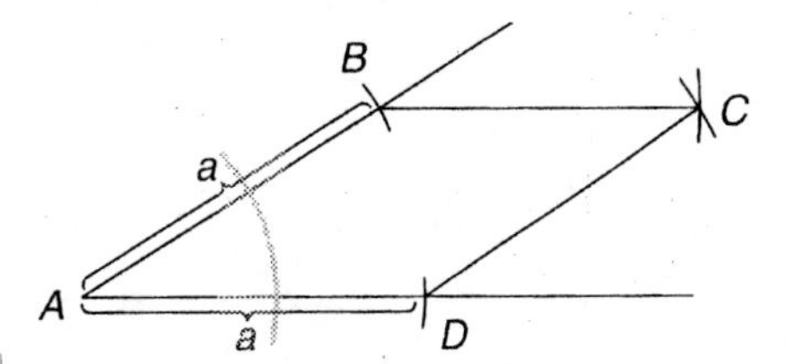

Step 2: Set compass to radius of length a and with point at vertex A mark off length a on both sides of $\angle A$. Label the intersection points B and D.

Step 3: Keeping compass set to radius a, set point on B and swing an arc. Similarly, set point of compass on D and swing an arc. Label the intersection of the two arcs C.

Step 4: Connect points B and C and points C and D with straight edge. $ABCD$ is the desired rhombus.

Justification:

Statement	Reason
1. $\angle BAD \cong \angle A$	1. Construction 2
2. $AD = AB = BC = CD = a$	2. Construction 1
3. $ABCD$ is a rhombus with angle congruent to $\angle A$	3. Definition of a rhombus

Chapter Test

6. Follow steps like those of Construction 8 in the text as if you were going to divide $\overline{AB}$ into five congruent pieces, but do not fill in all of the parallel lines.

Step 1: Copy segment $\overline{AB}$ and then draw any ray with endpoint at A, but not containing $\overline{AB}$.

Step 2: Mark off 5 congruent segments on the ray. Connect the last point D to B.

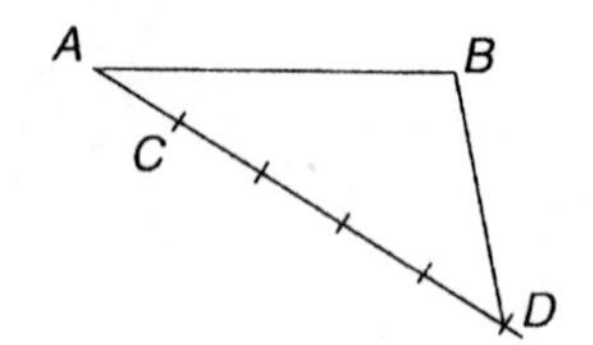

Step 3: Label the first mark to the right of A on the ray as C. Construct an angle at C congruent to $\angle BDA$. Let E be the point of intersection of the side of the angle with $\overline{AB}$. Then you have $EB = 4AE$.

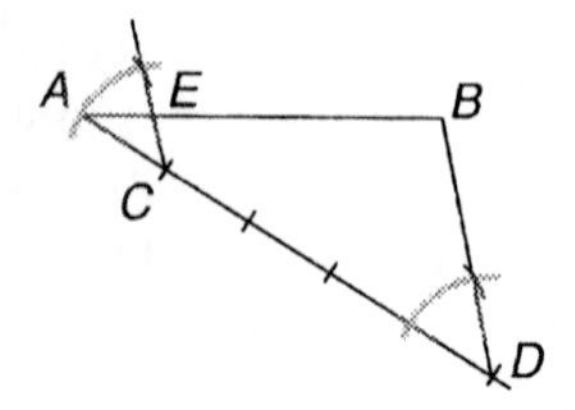

Solutions to Chapter 5 Test

1. T   See problem #40 in Section 5.1.

2. F   Unless the two angles are each 90°, the lines will not be parallel. Corresponding angles must be *congruent* for the lines to be parallel.

3. F   The angle sum in a quadrilateral is always 360°.

4. F   Since the bases are not congruent, the two triangles will not be congruent.

5. F   The opposite angles of any parallelogram are congruent.

6. T   Corollary 5.15 and Theorem 5.19

7. T   Theorem 5.21

8. F   The diagonals of an isosceles trapezoid are also congruent. If the diagonals of a *parallelogram* are congruent, then it is a rectangle.

Chapter Test

9. T Half-diagonals and one side of a rhombus form a right triangle with the side of the rhombus as hypotenuse.

$$a = \frac{1}{2}(16) = 8 \text{ and } c = 10.$$

Using the Pythagorean Theorem, you have

$$8^2 + b^2 = 10^2$$
$$64 + b^2 = 100$$
$$b^2 = 36$$
$$b = 6 = \text{half of the second diagonal.}$$

Since b is only half of the second diagonal, the full length is $2b = 12$ cm.

10. F The points on an angle bisector are equidistant from the *sides* of the angle, not necessarily from the other vertices of the triangle.

11. If the parallelogram is $ABCD$ with $\angle A = 20°$ then $\angle C = 20°$, and $\angle B = \angle D = 180° - 20° = 160°$.

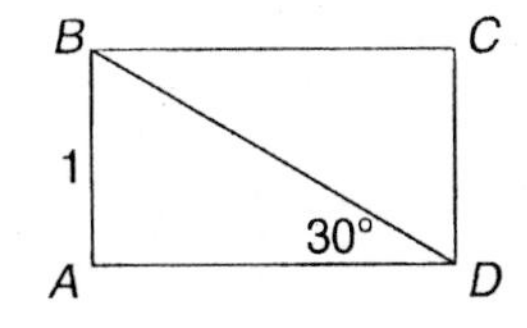

12. Since $\overline{BC} \parallel \overline{AD}$, you know $\angle CBD = 30°$ since alternate interior angles are congruent. ΔABD is a 30°–60° right triangle, so $BD = 2$ and $AD = \sqrt{3}$. Since the diagonals of a rectangle are congruent, $AC = 2$.

13. Label the point of intersection of the diagonals as E. Since the alternate interior angles are congruent, $\overline{AD} \parallel \overline{BC}$. The angles between the diagonals are 90° since (from ΔBCE) $62° + 28° + 90° = 180°$. Then $\angle ACD = 28°$ (from ΔCDE). $\Delta AED \cong \Delta CED$ by ASA. Also, $\Delta CED \cong \Delta CEB$ by ASA. $\overline{AE} \cong \overline{CE}$ and $\overline{EB} \cong \overline{ED}$ by C.P.

Since the diagonals bisect each other, $ABCD$ is a parallelogram. The diagonals are also perpendicular, so $ABCD$ is a rhombus. However, the vertex angles are not 90°, so this rhombus is not a square.

2w + 3

45 Yards

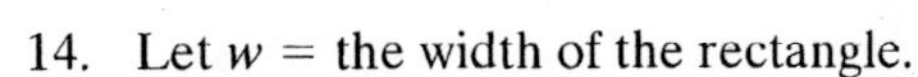

14. Let w = the width of the rectangle.

Use the Pythagorean Theorem to find w.

$$45^2 + w^2 = (2w + 3)^2$$
$$2025 + w^2 = 4w^2 + 12w + 9$$
$$0 = 3w^2 + 12w - 2016$$

The quadratic formula gives $w = 24$ or $w = -28$.

Ignore the solution $w = -28$ since the width of the rectangle must be positive. Thus, $w = 24$ yards.

15. There are two transversals and each one gives two sets of alternate interior angles: $\angle 6$ and $\angle 11$, $\angle 7$ and $\angle 10$, $\angle 8$ and $\angle 13$, $\angle 9$ and $\angle 12$.

Chapter Test

16. There are only six pairs of corresponding angles since the left transversal does not extend above the upper parallel line: $\angle 6$ and $\angle 14$, $\angle 7$ and $\angle 15$, $\angle 4$ and $\angle 12$, $\angle 5$ and $\angle 13$, $\angle 8$ and $\angle 16$, $\angle 9$ and $\angle 17$.

17. You can use some of the pairs of congruent angles you found in problems 15 and 16.

$$\angle 3 = 180° - \angle 2 = 180° - 170° = 10°$$

$$\angle 4 = \angle 12 = 140°$$

Then using the triangle at the top of the figure, you have

$\angle 1 = 180° - \angle 3 - \angle 4 = 180° - 10° - 140° = 30°$.

Now $\angle 5 = 180° - \angle 4 = 180° - 140° = 40°$, and

$\angle 8 = \angle 5 = 40°$.

Using the large triangle containing $\angle 7$, $\angle 8$, and $\angle 18$,

$\angle 7 = 180° - \angle 8 - \angle 18 = 180° - 40° - 65° = 75°$.

Then $\angle 10 = \angle 7 = 75°$

Summarizing, $\angle 1 = 30°$, $\angle 5 = 40°$, and $\angle 10 = 75°$.

18. There is more than one method to complete the construction.

Method one: Follow the procedure given in problem #6 of the Chapter 5 Review, Section 5.5 with the following modification. Divide $\overline{AB}$ into four congruent pieces instead of five congruent pieces.

or

Method 2:

Step 1: Copy $\overline{AB}$. Find the perpendicular bisector of $\overline{AB}$. Label as C the midpoint of $\overline{AB}$.

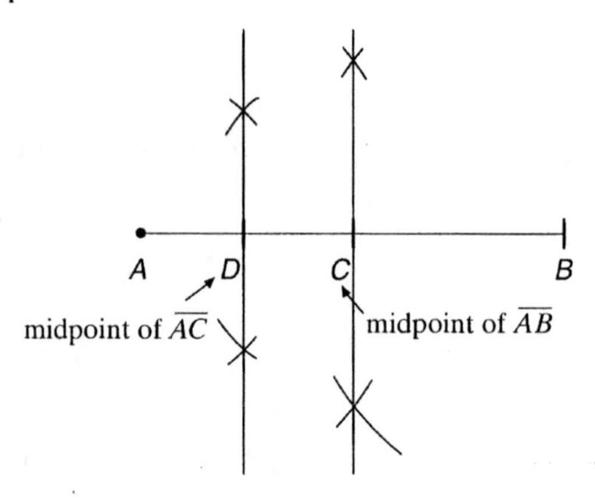

Step 2: Find the perpendicular bisector of $\overline{AC}$. Label as D the midpoint of $\overline{AC}$.

Now $DB = 3(AD)$.

Chapter Test

19. Step 1: Construct an equilateral triangle and bisect one of its 60° angles to get a 30° angle to copy. (See Section 4.4, problem #11.) You might instead make a 30°–60° right triangle. (See Example 4.14 in Chapter 4.)

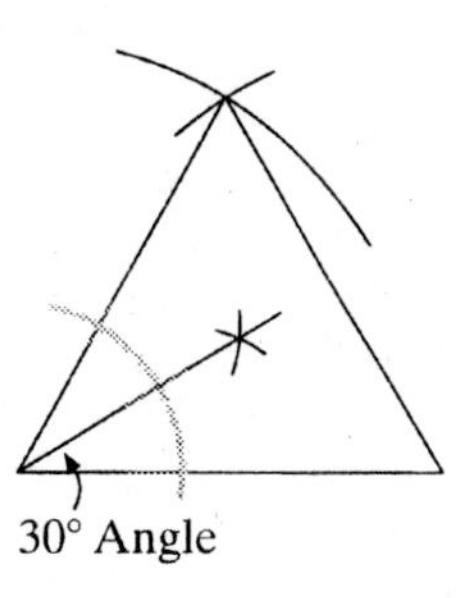

Step 2: Copy the 30° angle.

Step 3: Mark length a on one ray of the angle and length d on the other ray. Label points B and C.

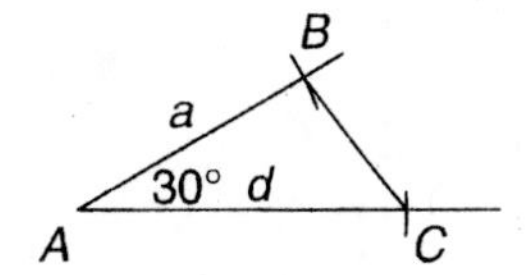

Step 4: Connect points C and B.

Step 5: Copy another 30° angle at C but below $\overline{AC}$ as shown.

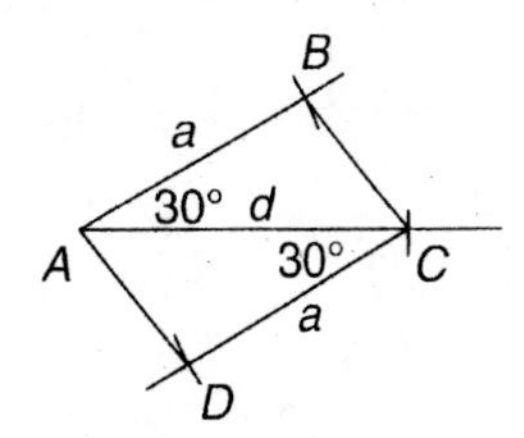

Step 6: Mark off length a on the side other than $\overline{AC}$ of this new 30° angle. Label the end of this segment as point D.

Step 7: Join D to A. Quadrilateral $ABCD$ is now a parallelogram with side length a, diagonal length d, and a 30° angle between side and diagonal.

Chapter Test

20. For this construction, take advantage of the fact that the diagonals of a rhombus are perpendicular bisectors of each other.

Step 1: Copy line segment b and construct its perpendicular bisector.

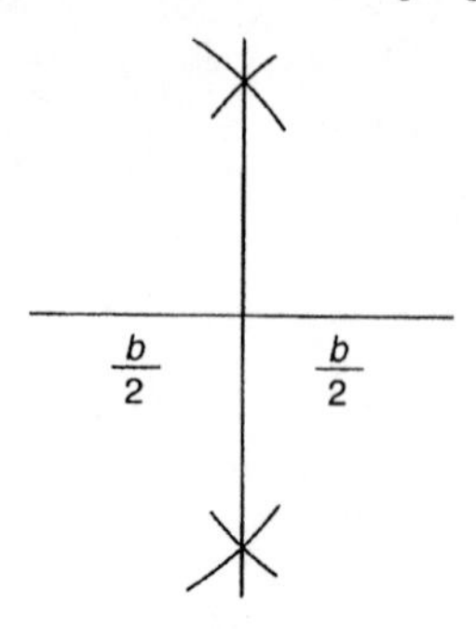

Step 2: Find the midpoint of line segment a and mark it.

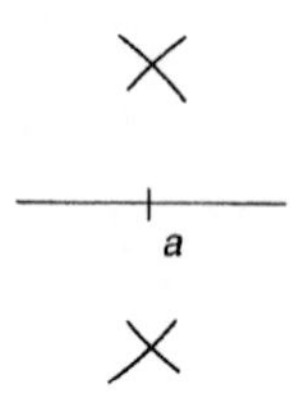

Step 3: Set your compass at a radius equal to half of a. On the perpendicular bisector of the segment in Step 1, mark off segments of length $\frac{a}{2}$ both above and below the segment of length b.

Step 4: Join the ends of the segments marked off on the perpendicular bisector with the endpoints of the segment of length b to finish the rhombus.

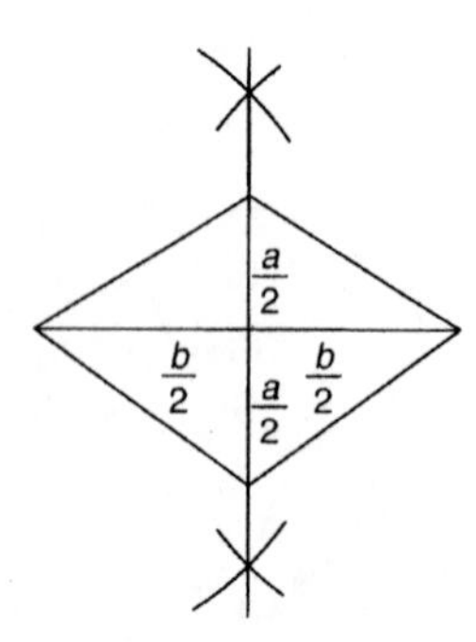

21. A complete solution is given in the answer key at the back of the text.

22. A complete solution is given in the answer key at the back of the text.

Chapter Test

23. Draw the height from each end of the 10-foot base to the 50-foot base as shown.

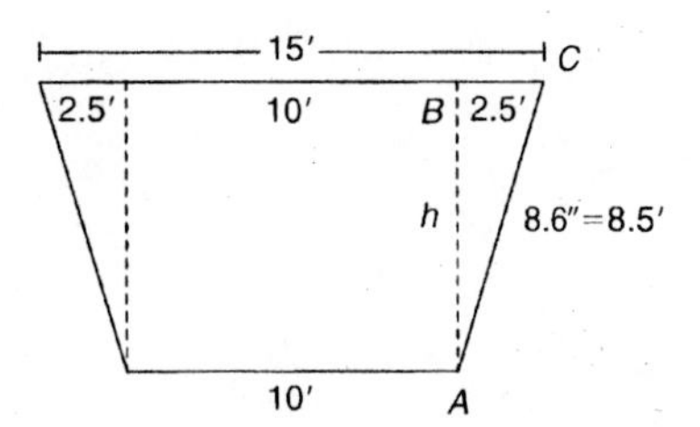

To find BC, divide the difference between the lengths of the bases. Now you have a right triangle ABC with a leg and the hypotenuse given. Use the Pythagorean Theorem to find h.

$$2.5^2 + h^2 = 8.5^2$$
$$6.25 + h^2 = 72.25$$
$$h^2 = 66$$
$$h = \sqrt{66} \approx 8.124 \text{ ft}$$

The concrete layer is in the form of a trapezoidal prism where the height is the thickness of the concrete. The volume is the area of the trapezoid times the thickness of the concrete (in feet).

$$A_{\text{trapezoid}} = \frac{1}{2}h\,(b_1 + b_2)$$
$$= \frac{1}{2}\sqrt{66}(10 + 15)$$
$$= \frac{25}{2}\sqrt{66}$$
$$\approx 101.55 \text{ ft}^2$$

$$\text{Thickness} = 4 \text{ in.} = \frac{1}{3} \text{ ft}$$

$$\text{Volume} = \left(\frac{25}{2}\sqrt{66}\right) \cdot \frac{1}{3}$$
$$= \frac{25}{6}\sqrt{66}$$
$$\approx 33.85$$
$$\approx 34 \text{ ft}^3 \text{ of concrete}$$

24. A complete explanation is given in the answer key at the back of the text.

6

Similarity

Section 6.1

Tips:

✔ Make sure that units in corresponding parts of proportions match.

✔ There are usually several correct ways to set up a proportion. Be consistent both horizontally across the proportion and vertically in each ratio when you set up the proportion. For example, you might write

$$\frac{\text{length}_1}{\text{length}_2} = \frac{\text{width}_1}{\text{width}_2} \text{ or } \frac{\text{length}_1}{\text{width}_1} = \frac{\text{length}_2}{\text{length}_2},$$

$$\textit{but not } \frac{\text{length}_1}{\text{width}_1} = \frac{\text{width}_2}{\text{length}_2}.$$

Solutions to odd-numbered textbook problems

1. 16 to 64 is $\dfrac{16 \div 16}{64 \div 16} = \dfrac{1}{4}$

3. 82.5 to 16.5 is $\dfrac{82.5 \div 16.5}{16.5 \div 16.5} = \dfrac{5}{1}$

5. A complete solution is given in the answer key at the back of the text.

7. A complete solution is given in the answer key at the back of the text.

9. $\dfrac{57}{95} = \dfrac{18}{n}$ if and only if $57n = 95(18)$, or $n = \dfrac{95(18)}{57} = 30$.

11. $\dfrac{n}{70} = \dfrac{6}{21}$ if and only if

$$\begin{aligned} 21n &= 70(6) \\ \text{or } n &= \frac{70(6)}{21} \\ &= 20. \end{aligned}$$

13. $\dfrac{7}{5} = \dfrac{42}{s}$ if and only if $7s = 5(42)$, or $s = \dfrac{5(42)}{7} = 30$.

Section 6.1

15. $\frac{\frac{3}{5}}{6} = \frac{d}{25}$ if and only if

$$6d = \frac{3}{5}(25)$$

$$d = \frac{\frac{3}{5}(25)}{6}$$

$$= \frac{15}{6}$$

$$= \frac{5}{2}.$$

17. $\frac{x}{100} = \frac{4.8}{1.5}$ if and only if $1.5x = 100(4.8)$, or $x = \frac{100(4.8)}{1.5} = 320.$

19. $\frac{5}{8} = \frac{x}{x+9}$ if and only if

$$8x = 5(x + 9)$$

$$8x = 5x + 45$$

$$3x = 45$$

$$x = 15.$$

21. The geometric mean, b, of a and c is $\sqrt{ac}$ for positive values of a and c.

$$b = \sqrt{2 \cdot 18} = \sqrt{36} = 6$$

23. The geometric mean, b, of a and c is $\sqrt{ac}$ for positive values of a and c.

$$b = \sqrt{10\pi} \approx 5.6$$

25. For a cylinder with a height, h, and a base with radius r, its volume is $\pi r^2 h$.

The new cylinder has volume $\pi(2r)^2(3h) = 12\pi r^2 h$.

Thus the ratio of the volume of the original cylinder to the volume of the new cylinder is

$$\pi r^2 h : 12\pi r^2 h \text{ or } 1:12.$$

27. Prove: $\frac{a}{b} = \frac{c}{d}$ if and only if $\frac{d}{b} = \frac{c}{a}$.

Proof: If $\frac{a}{b} = \frac{c}{d}$, then $ad = bc$ (by cross multiplication).

Dividing both sides of the equation by the same nonzero number, ab, you get $\frac{d}{b} = \frac{c}{a}$.

Similarly, if you start with $\frac{d}{b} = \frac{c}{a}$, then $ad = bc$ by cross multiplication.

Dividing by bd yields $\frac{a}{b} = \frac{c}{d}$.

Section 6.1

29. Prove: $\frac{a}{b} = \frac{c}{d}$ if and only if $\frac{b}{a} = \frac{d}{c}$.

Proof: If $\frac{a}{b} = \frac{c}{d}$, then $ad = bc$, and dividing by ac gives $\frac{d}{c} = \frac{b}{a}$.

Starting with $\frac{d}{c} = \frac{b}{a}$, cross multiplying yields $ad = bc$. Dividing by bd leaves $\frac{a}{b} = \frac{c}{d}$.

31. A complete solution is given in the answer key at the back of the text.

33. If $\frac{a}{b} = \frac{c}{b}$, then $ab = bc$. Dividing both sides by b yields $a = c$.

35. Given: $\frac{a}{b} = \frac{c}{d}$

Prove: $\frac{a}{b} = \frac{a+c}{b+d}$

Proof:

Statement	Reason
1. $\frac{a}{b} = \frac{c}{d}$	1. Given
2. $ad = bc$	2. Cross Multiplication
3. $ab + ad = ab + bc$	3. Addition
4. $a(b + d) = b(a + c)$	4. Distributive property
5. $\frac{a}{b} = \frac{a+c}{b+d}$	5. Cross Multiplication

37. Generalizing the result in problem #36 to n ratios, you are given that

$$\frac{a_1}{b_1} = \frac{a_2}{b_2} = \ldots = \frac{a_{n-1}}{b_{n-1}} = \frac{a_n}{b_n}, \text{ for } n > 3$$

and you must show that

$$\frac{a_1 + a_2 + \ldots + a_{n-1} + a_n}{b_1 + b_2 + \ldots + b_{n-1} + b_n} = \frac{a_n}{b_n} = \frac{a_1}{b_1}.$$

Proof (for $n = 4$ and $n = 5$):

From the result in problem #36, if $\frac{a_1}{b_1} = \frac{a_2}{b_2} = \frac{a_3}{b_3}$, then

$$\frac{a_1 + a_2 + a_3}{b_1 + b_2 + b_3} = \frac{a_1}{b_1}.$$

By hypothesis $\frac{a_1}{b_1} = \frac{a_4}{b_4}$, so you have $\frac{a_1 + a_2 + a_3}{b_1 + b_2 + b_3} = \frac{a_4}{b_4}$.

Section 6.1

37. (continued)

Applying the result from problem #35, you have

$$\frac{(a_1 + a_2 + a_3) + a_4}{(b_1 + b_2 + b_3) + b_4} = \frac{a_4}{b_4} = \frac{a_1}{b_1},$$

which completes the proof for n = 4.

Now since you also know that $\frac{a_1}{b_1} = \frac{a_5}{b_5}$, by substitution

$$\frac{(a_1 + a_2 + a_3) + a_4}{(b_1 + b_2 + b_3) + b_4} = \frac{a_5}{b_5}.$$

Using $\frac{a_1 + a_2 + a_3 + a_4}{b_1 + b_2 + b_3 + b_4} = \frac{a_5}{b_5}$ and problem #35, you have

$$\frac{(a_1 + a_2 + a_3 + a_4) + a_5}{(b_1 + b_2 + b_3 + b_4) + b_5} = \frac{a_5}{b_5} = \frac{a_1}{b_1}.$$

which completes the proof for $n = 5$.

Continuing in this manner, it can be shown that for any n,

$$\frac{a_1 + a_2 + a_3 + \dots + a_{n-1}}{b_1 + b_2 + b_3 + \dots + b_{n-1}} = \frac{a_n}{b_n},$$

and thus, $\frac{a_1 + a_2 + \dots + a_{n-1} + a_n}{b_1 + b_2 + \dots + b_{n-1} + b_n} = \frac{a_n}{b_n} = \frac{a_1}{b_1}.$

39. Use the proportion $\frac{\text{inches}}{\text{miles}} = \frac{\text{inches}}{\text{miles}}$.

(a) $\frac{1}{35} = \frac{3}{x}$ if and only if $x = 3(35)$

$x = 105$ miles.

$\frac{1}{35} = \frac{10}{x}$ if and only if $x = 10(35)$

$x = 350$ miles.

$\frac{1}{35} = \frac{n}{x}$ if and only if $x = 35n$ miles

(b) $\frac{1}{35} = \frac{x}{1000}$ if and only if $35x = 1000$

$x = 28.6$ inches.

41. Let x = distance in miles your car has gone in 2.75 years. Then one possible proportion you could use is $\frac{\text{miles}}{\text{years}} = \frac{\text{miles}}{\text{years}}$.

$\frac{4460}{0.5} = \frac{x}{2.75}$ if and only if $0.5x = 2.75(4460)$, or

$x = \frac{2.75(4460)}{0.5} = 24{,}530$ miles.

Section 6.1

43. Use the proportion $\frac{\text{audited returns}}{\text{individual returns}} = \frac{\text{audited returns}}{\text{individual returns}}$.

(a) Let x = the number of audited returns in 2004.

$$\frac{65}{10{,}000} = \frac{x}{34{,}000} \text{ if and only if } 10{,}000x = 65(34{,}000)$$

$$x = \frac{65(34{,}000)}{10{,}000}$$

$$x = 221 \text{ audited returns.}$$

(b) Let y = the number of audited returns in 2003.

$$\frac{57}{10{,}000} = \frac{y}{34{,}000} \text{ if and only if } 10{,}000y = 57(34{,}000)$$

$$y = \frac{57(34{,}000)}{10000}$$

$$y \approx 194 \text{ audited returns.}$$

Therefore, we can expect $221 - 194 = 27$ more returns to be audited in 2004 than 2003.

45. First you need to convert both of the depths to the same unit. 18 inches = 1.5 feet. Now set up your proportion using p = pressure at 8 feet. One proportion that can be used is $\frac{\text{pressure}}{\text{depth}} = \frac{\text{pressure}}{\text{depth}}$.

$$\frac{93.6}{1.5} = \frac{p}{8} \text{ if and only if } 1.5p = 8(93.6), \text{ or } p = \frac{8(93.6)}{1.5} = 499.2 \text{ lb/ft}^2.$$

47. Convert both the length and wing span of the "Spruce Goose" to inches.

319 feet 11 inches = 3839 inches (wing span)

218 feet 8 inches = 2624 inches (length)

Now set up your proportion using x = wing span in inches.

One proportion that can be used is $\frac{\text{wing span}}{\text{length}} = \frac{\text{wing span}}{\text{length}}$.

$$\frac{3839}{2624} = \frac{x}{20} \text{ if and only if } 2624x = 20(3839)$$

$$x = \frac{20(3839)}{2624}$$

$$x \approx 29 \text{ inches.}$$

The wing span of the scale model is approximately 29 inches.

49. First convert all measurements to the same units. 5′6″ = 66″. Let x = the waist measurement in inches for a life size model with the same proportions as the doll. Using $\frac{\text{height}}{\text{waist}} = \frac{\text{height}}{\text{waist}}$, you can solve for x.

$\frac{3}{11.5} = \frac{x}{66}$ if and only if $11.5x = 3(66)$, or $x = \frac{3(66)}{11.5} \approx 17$ in.

A 17-inch waist is extremely unlikely for a woman 5′6″ tall.

51. Let x = speed of the lead car in mph. Using the proportion

$\frac{\text{feet}}{\text{mph}} = \frac{\text{feet}}{\text{mph}}$, you can solve for x.

$\frac{1.738}{x} = \frac{1.67}{198}$ if and only if $1.67x = 1.738(198)$

$$x = \frac{1.738(198)}{1.67}$$

$$x \approx 206 \text{ mph.}$$

The lead car is traveling at approximately 206 mph.

Section 6.2

Tips:

✔ As with congruence, the order in which you list the vertices of similar triangles is important.

✔ Look for pairs of congruent angles or proportional sides to show that triangles are similar.

Solutions to odd-numbered textbook problems

1. $\angle A \cong \angle D$ and $\angle B \cong \angle E$ since they are sets of alternate interior angles. Therefore $\Delta ABC \sim \Delta DEC$ by AA Similarity.

3. ΔLMN and ΔOPQ are right triangles with $\frac{MN}{PQ} = \frac{NL}{QO}$. Thus $\Delta LMN \sim \Delta OPQ$ by LL Similarity.

5. By corresponding parts of similar triangles you have $\frac{DE}{5} = \frac{6}{3}$, so $3(DE) = 30$ or $DE = 10$.

You also have $\frac{DF}{7} = \frac{6}{3}$, so $3(DF) = 42$, or $DF = 14$.

Section 6.2

7. $\Delta UVW \sim \Delta XYZ$ by AA Similarity. By corresponding parts of similar triangles $\frac{UV}{UW} = \frac{XY}{XZ}$.

$$\text{Thus } \frac{6}{16} = \frac{XY}{9}$$
$$16(XY) = 9(6)$$
$$XY = \frac{54}{16}$$
$$= \frac{27}{8} \text{ cm.}$$

Also, by corresponding parts of similar triangles $\frac{VW}{UW} = \frac{YZ}{XZ}$.

$$\text{Thus } \frac{VW}{16} = \frac{8}{9}$$
$$9(VW) = 16(8)$$
$$VW = \frac{128}{9} \text{ cm.}$$

9. $\Delta GHI \sim \Delta EFI$ by AA Similarity.

By corresponding parts of similar triangles $\frac{EG + 6}{6} = \frac{5}{3}$ so $3(EG + 6) = 30$ or $EG + 6 = 10$. Then $EG = 4$.

Use the Pythagorean Theorem to find HI.

$$HI^2 + 3^2 = 6^2$$
$$HI^2 + 9 = 36$$
$$HI^2 = 27$$
$$HI = \sqrt{27}$$
$$= 3\sqrt{3}$$

Then using similarity again to find FH, you have

$\frac{FH + \sqrt{27}}{\sqrt{27}} = \frac{5}{3}$, so $3(FH + \sqrt{27}) = 5\sqrt{27}$ and

$FH + \sqrt{27} = \frac{5}{3}\sqrt{27}$ or $FH = \frac{5}{3}\sqrt{27} - \sqrt{27}$

$$= 5\sqrt{3} - 3\sqrt{3}$$
$$= 2\sqrt{3}.$$

Section 6.2

11. Show $\Delta BAD \sim \Delta ACD$.

$$\angle ACD = \angle B + \angle BAC \qquad \text{Exterior Angle Theorem (Corollary 5.10)}$$
$$\angle ACD = \angle CAD + \angle BAC \; (\angle B \cong \angle CAD)$$
$$= \angle BAD$$

Therefore, $\Delta BAD \sim \Delta ACD$ by AA Similarity.

Now find the missing measures.

(i) $\dfrac{BA}{AC} = \dfrac{AD}{CD}$ by C.P. of similar triangles.

Thus, $\dfrac{14}{AC} = \dfrac{9}{4}$

$$9(AC) = 14(4)$$
$$AC = \frac{56}{9} \text{ in.}$$

(ii) $\dfrac{AD}{CD} = \dfrac{BD}{AD}$ by C.P. of similar triangles.

Thus, $\dfrac{9}{4} = \dfrac{BC + 4}{9}$

$$4(BC + 4) = 9(9)$$
$$4(BC) + 16 = 81$$
$$4(BC) = 65$$
$$BC = \frac{65}{4} \text{ in.}$$

13. (a) The ratio of base to base is 4:8 or 1:2.

The ratio of height to height is 3:6 or 1:2.

$$\text{Area}_{\text{small}} = \frac{1}{2}(4)(3) = 6 \text{ ft}^2.$$

$$\text{Area}_{\text{large}} = \frac{1}{2}(8)(6) = 24 \text{ ft}^2.$$

Thus, the ratio of $\text{area}_{\text{small}}$ to $\text{area}_{\text{large}}$ is 6:24 or 1:4.

(b) The ratio of base to base is 6:9 or 2:3.

The ratio of height to height is 3:4.5 or 2:3.

$$\text{Area}_{\text{small}} = \frac{1}{2}(6)(3) = 9 \text{ cm}^2.$$

$$\text{Area}_{\text{large}} = \frac{1}{2}(9)(4.5) = 20.25 \text{ cm}^2.$$

Thus, the ratio of $\text{area}_{\text{small}}$ to $\text{area}_{\text{large}}$ is 9:20.25 or 4:9 (dividing by 2.25 to simplify).

(c) The ratio of the areas is equal to the square of the ratios of the linear dimensions.

Section 6.2

15. There is more than one way to prove that all equilateral triangles are similar to each other. Here is one proof:

    Let ΔABC and $\Delta A'B'C'$ be equilateral triangles.

    Then ΔABC and $\Delta A'B'C'$ are equiangular triangles.

    Hence $\angle A = \angle B = \angle C = 60°$ and $\angle A' = \angle B' = \angle C' = 60°$.

    Thus $\angle A \cong \angle A'$ and $\angle B \cong \angle B'$.

    Therefore, $\Delta ABC \sim \Delta A'B'C'$ by AA Similarity.

17. In any isosceles triangle the two base angles are congruent.

    Thus in ΔABC, suppose that $\angle A \cong \angle B$ and in ΔDEF, that $\angle D \cong \angle E$. If $\angle A \cong \angle D$ also, then you have that $\angle A \cong \angle B \cong \angle D \cong \angle E$. Since $\angle A \cong \angle D$ and $\angle B \cong \angle E$, $\Delta ABC \sim \Delta DEF$ by AA Similarity.

19. Let $ABCD$ be an isosceles trapezoid with diagonals intersecting in point E.

    Then $\overline{BD}$ and $\overline{AC}$ are transversals of parallel lines $\overline{AD}$ and $\overline{BC}$.

    Thus $\angle EAD \cong \angle ECB$ and $\angle CBE \cong \angle ADE$ by alternate interior angles.

    Therefore $\Delta AED \sim \Delta CEB$ by AA Similarity.

21. A complete solution is given in the answer key at the back of the text.

23. Let ΔABC and $\Delta A'B'C'$ be similar triangles.

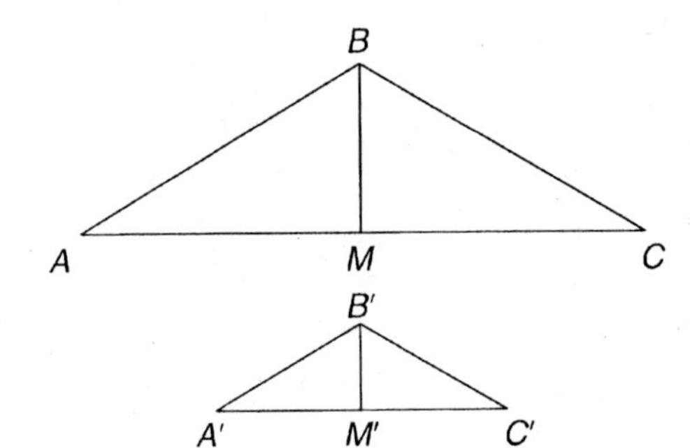

Then $\dfrac{AB}{A'B'} = \dfrac{BC}{B'C'} = \dfrac{AC}{A'C'} = k$, for some positive constant k.

Now let $\overline{BM}$ and $\overline{B'M'}$ be the altitudes of ΔABC and $\Delta A'B'C'$ respectively.

Then $\dfrac{BM}{B'M'} = k$ (problem #22).

Now, show the ratio of the areas of the two triangles is k^2.

$$\frac{\text{Area}_{\Delta ABC}}{\text{Area}_{\Delta A'B'C'}} = \frac{\frac{1}{2}(AC)(BM)}{\frac{1}{2}(A'C')(B'M')}$$

$$= \frac{AC}{A'C'} \cdot \frac{BM}{B'M'}$$

$$= k \cdot k$$

$$= k^2.$$

Section 6.2

25. Given: $\Delta ABC \sim \Delta A'B'C'$, V = the volume of the cone generated by rotating ΔABC about $\overline{BC}$ and V' = the volume of the cone generated by rotating $\Delta A'B'C'$ about $\overline{B'C'}$.

Prove: $\frac{V}{V'} = \left(\frac{AC}{A'C'}\right)^3 = \left(\frac{BC}{B'C'}\right)^3 = \left(\frac{AB}{A'B'}\right)^3$

Proof:

$\frac{AC}{A'C'} = \frac{BC}{B'C'}$ since the triangles are similar.

Use AC as the base of ΔABC. Then its height is BC. Therefore, the volume of the right circular cone formed by rotating ΔABC is $\frac{1}{3}\pi(AC)^2(BC)$. Similarly, the base of $\Delta A'B'C'$ is $A'C'$ and its height is $B'C'$, so the volume of the cone formed by rotating the triangle is given by $\frac{1}{3}\pi(A'C')^2(B'C')$.

The ratio of the two volumes is $\frac{V}{V'} = \frac{\frac{1}{3}\pi(AC)^2(BC)}{\frac{1}{3}\pi(A'C')^2(B'C')}$.

Divide both numerator and denominator by $\frac{1}{3}\pi$.

$$\frac{V}{V'} = \frac{(AC)^2(BC)}{(A'C')^2(B'C')} = \left(\frac{AC}{A'C'}\right)^2 \cdot \frac{BC}{B'C'}$$

Since $\frac{BC}{B'C'} = \frac{AC}{A'C'}$, you can make a substitution.

$$\frac{V}{V'} = \left(\frac{AC}{A'C'}\right)^2 \cdot \frac{AC}{A'C'} = \left(\frac{AC}{A'C'}\right)^3$$

Since the triangles are similar, you know that

$$\frac{AC}{A'C'} = \frac{BC}{B'C'} = \frac{AB}{A'B'}.$$

By substitution,

$$\frac{V}{V'} = \left(\frac{AC}{A'C'}\right)^3 = \left(\frac{BC}{B'C'}\right)^3 = \left(\frac{AB}{A'B'}\right)^3.$$

Therefore, the ratio of the volumes is the cube of the ratio of the lengths of any two corresponding sides.

Section 6.2

27. In the figure shown, let $\overline{AB} \perp \overline{A'B'}$, $\overline{BC} \perp \overline{B'C'}$ and $\overline{AC} \perp \overline{A'C'}$.

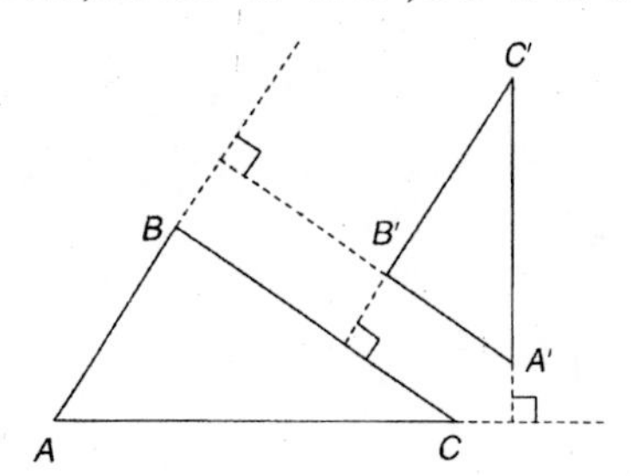

Prove: $\Delta ABC \sim \Delta A'B'C'$

Goal 1: $\angle C \cong \angle C'$

Extend $\overline{B'C'}$ and $\overline{C'A'}$ as shown in diagram 1.

$\Delta PC'R \sim \Delta PCQ$ by AA Similarity. Both triangles have a right angle and both triangles contain $\angle QPC$.

Thus $\angle PCQ \cong \angle PC'R$ as corresponding angles of similar triangles are congruent.

Therefore, $\angle C \cong \angle C'$.

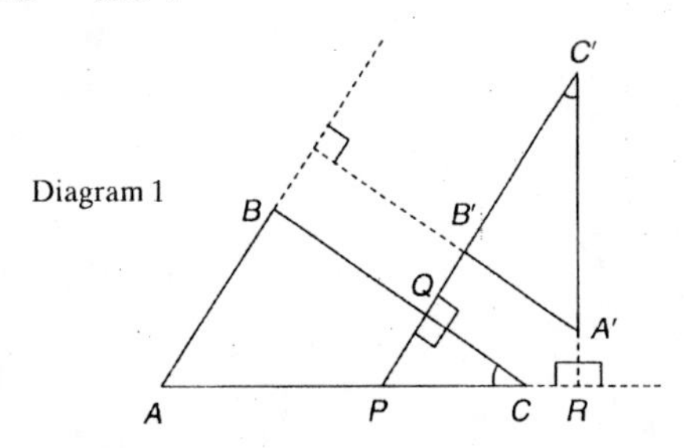

Diagram 1

Goal 2: $\angle A \cong \angle A'$

Extend $\overline{B'A'}$ as shown in diagram 2.

$\Delta ATS \sim \Delta A'RS$ by AA Similarity. Both triangles have a right angle and both triangles contain $\angle S$.

Thus, $\angle BAC \cong \angle RA'S$ as corresponding angles of similar triangles and congruent.

Now $\angle RA'S \cong \angle B'A'C'$ by vertical angles, so $\angle BAC \cong \angle B'A'C'$ by the transitive property.

Therefore $\angle A \cong \angle A'$.

Since $\angle C \cong \angle C'$ and $\angle A \cong \angle A'$ we have $\Delta ABC \sim \Delta A'B'C'$ by AA Similarity.

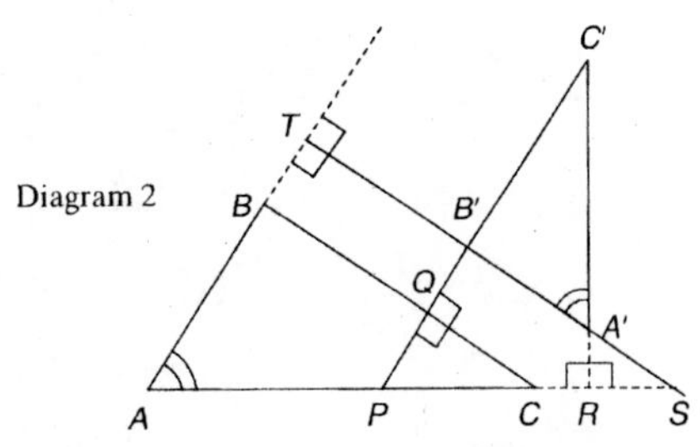

Diagram 2

Section 6.2

29. A complete solution is given in the answer key at the back of the text.

31. Use similarity to find the height of the tree.

Let x = height of tree in meters.

Then,

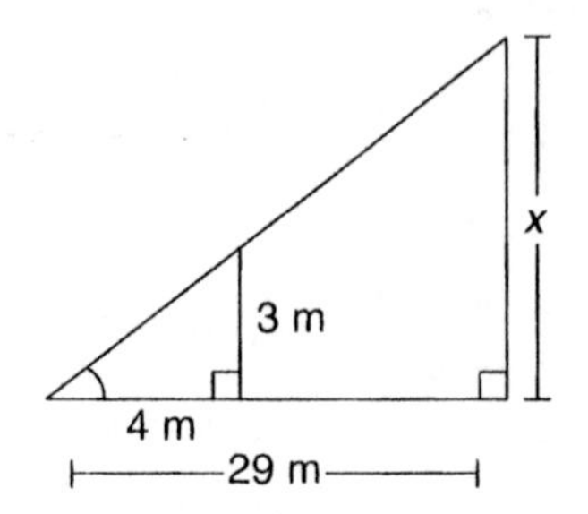

$$\frac{3}{4} = \frac{x}{29}$$
$$4x = 29(3)$$
$$x = \frac{29(3)}{4}$$
$$\approx 22 \text{ m.}$$

33. A complete solution is given in the answer key at the back of the text.

35. Use similarity to find the distance Kathy is from building.

Let x = distance Kathy is from building in feet.

Convert 5 ft 6 inches to 5.5 feet.

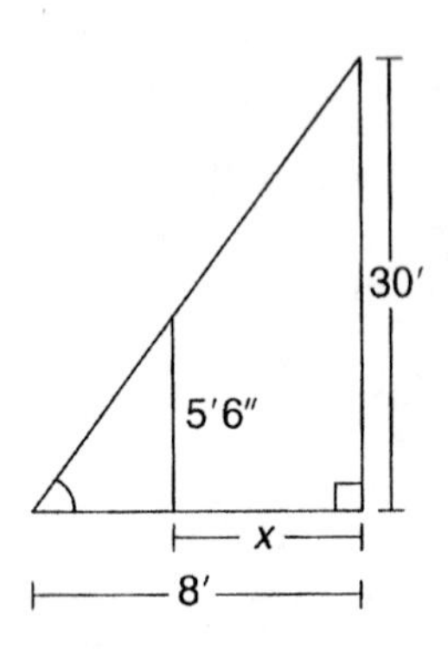

Then,
$$\frac{30}{8} = \frac{5 \cdot 5}{8 - x}$$
$$8(5.5) = 30(8 - x)$$
$$44 = 240 - 30x$$
$$30x = 196$$
$$x = \frac{196}{30}$$
$$x \approx 6.53 \text{ feet}$$

37. (a) Using similarity to find r, you have $\frac{4}{12} = \frac{r}{8}$, so

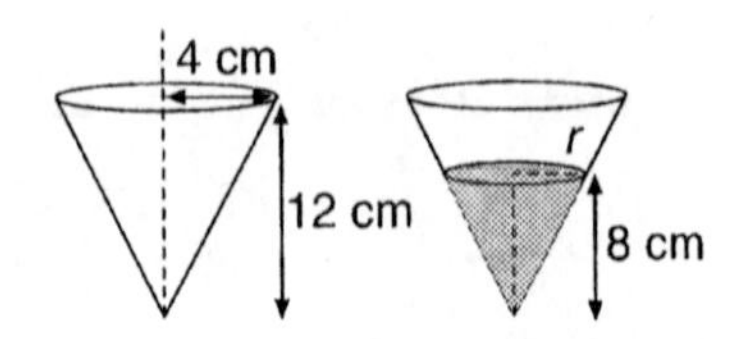

$$12r = 32, \text{ and } r = \frac{32}{12} = \frac{8}{3} \text{ cm.}$$

The volume of water in the cup is given by

$$V = \frac{1}{3}\pi r^2 h = \frac{1}{3}\pi\left(\frac{8}{3}\right)^2 (8) = \frac{512\pi}{27} \approx 60 \text{ cm}^3.$$

(b) When the cup is full, the volume is

$$V = \frac{1}{3}\pi \cdot 4^2 \cdot 12 = 64\pi \approx 201 \text{ cm}^3.$$

When the cup is half full, the volume of water in it is

$$\frac{1}{2}V = \frac{64\pi}{2} \approx 100.5 \text{ cm}^3.$$

Section 6.2

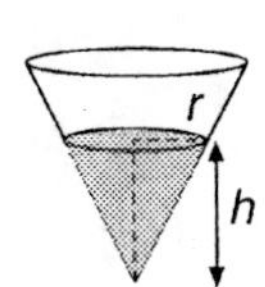

37. (continued)

Now find the height when the volume is 100.5 cm^3. Let h and r be the height and radius of the cone with half the volume. Then h and r form a triangle similar to the triangle that is half the cross-section of the cone. Using the result from problem #25, you can find h. $\frac{V_{\text{half}}}{V_{\text{total}}} = \frac{1}{2} = \left(\frac{h}{12}\right)^3$, so $\frac{1}{2} = \frac{h^3}{1728}$.

Then $2h^3 = 1728$ and $h^3 = 864$. Thus, $h = \sqrt[3]{864}$

$$= 6\sqrt[3]{4}$$

$$h \approx 9.5 \text{ cm.}$$

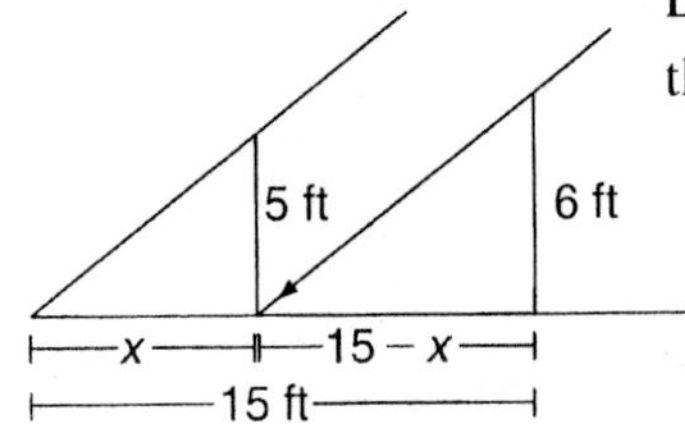

39. Use similarity to find the length of each shadow.

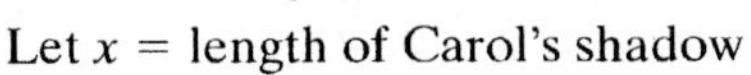

Let x = length of Carol's shadow

then $15 - x$ = length of Tom's shadow.

$$\frac{6}{15 - x} = \frac{5}{x}$$

$$6x = 5(15 - x)$$

$$6x = 75 - 5x$$

$$11x = 75$$

Carol's Shadow: $x = \frac{75}{11}$

$$= 6\frac{9}{11} \text{ ft}$$

$$\approx 6 \text{ ft } 10 \text{ in.}$$

Tom's Shadow: $15 - x = 15 - 6\frac{9}{11}$ ft

$$= 8\frac{2}{11} \text{ ft}$$

$$\approx 8 \text{ ft } 2 \text{ in.}$$

41. (a) Use similarity. Let x = distance from statue to camera in feet.

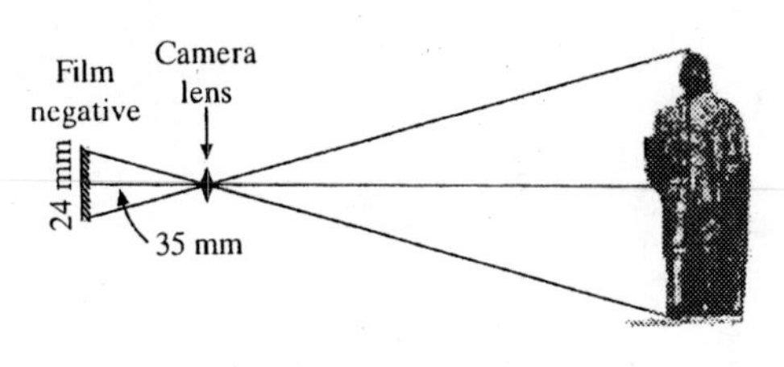

$$\frac{24}{35} = \frac{6}{x}$$

$$24x = 6(35)$$

$$x = \frac{6(35)}{24}$$

$$= 8.75 \text{ feet}$$

41. (continued)

(b) Set up the proportion

$$\frac{\text{Top of negative not printed}}{\text{height of negative}} = \frac{\text{Top of statue not in photo}}{\text{height of statue}}.$$

Let x = Top of statue not in photo.

$$\text{Then } \frac{0.5}{24} = \frac{x}{6}$$

$$24x = 3$$

$$x = \frac{3}{24}$$

$$= 0.125 \text{ feet}$$

$$= 1.5 \text{ in.}$$

Section 6.3

Solutions to odd-numbered textbook problems

1. ΔABC is a right triangle where $\overline{CD}$ is the altitude to $\overline{AB}$. By Theorem 6.9 CD is the mean proportional between AD and DB. Thus, $\frac{AD}{CD} = \frac{CD}{BD}$.

So
$$\frac{24}{CD} = \frac{CD}{9}$$
$$(CD)^2 = 24(9)$$
$$(CD)^2 = 216$$
$$CD = \sqrt{216}$$
$$= 6\sqrt{6}.$$

3. By Theorem 6.9, CD is the mean proportional between AD and DB. Thus, $\frac{AD}{CD} = \frac{CD}{BD}$, so $\frac{6}{4} = \frac{4}{BD}$. Then you have $6BD = 16$, or $BD = \frac{16}{6} = \frac{8}{3}$.

5. ΔABC is a right triangle where $\overline{CD}$ is the altitude to $\overline{AB}$.

Thus, $\frac{AD}{CD} = \frac{CD}{BD}$.

So
$$\frac{AB - BD}{6} = \frac{6}{BD}$$
$$\frac{20 - BD}{6} = \frac{6}{BD}$$
$$BD(20 - BD) = 6(6)$$
$$20(BD) - (BD)^2 = 36$$
$$BD^2 - 20(BD) + 36 = 0$$
$$(BD - 18)(BD - 2) = 0$$
$$BD = 18 \text{ or } BD = 2.$$

Section 6.3

5. (continued)

Reject $BD = 18$ as it does not make sense here because from the diagram we see that $BD < AD$. (If $BD = 18$, then $AD = 2$.)

Therefore $BD = 2$.

7. By Theorem 6.10, $\Delta ABC \sim \Delta ADE$. Since corresponding sides are proportional, $\frac{AB}{AD} = \frac{AC}{AE}$. Substituting given values, $\frac{5}{4} = \frac{AC}{3}$, or $4AC = 15$, so $AC = \frac{15}{4} = 3.75$.

9. By Theorem 6.10, $\Delta ABC \sim \Delta ADE$. Since corresponding sides are proportional, $\frac{AD}{AB} = \frac{AE}{AC}$.

$$\text{Now } AB = AD + DB$$
$$= AD + 3$$
$$\text{and } AE = AC - EC$$
$$= 9 - 4$$
$$= 5.$$

Substituting values, $\frac{AD}{AD + 3} = \frac{5}{9}$

$$9(AD) = 5(AD + 3)$$
$$9(AD) = 5(AD) + 15$$
$$4(AD) = 15$$
$$AD = \frac{15}{4}.$$

11. Since $\Delta ABC \sim \Delta ADE$, $\frac{AB}{AD} = \frac{BC}{DE}$. Substituting given values, $\frac{AB}{4} = \frac{7}{5}$, or $5AB = 28$, and $AB = \frac{28}{5} = 5.6$.

13. By the Side Splitting Theorem, since $\Delta ABG \sim \Delta ACF$, $\frac{AB}{BC} = \frac{AG}{GF}$. Thus, $\frac{4}{BC} = \frac{3}{2}$, $3BC = 8$ and $BC = \frac{8}{3} \approx 2.67$.

Similarly, because $\Delta ADE \sim \Delta ACF$, you have $\frac{AF}{EF} = \frac{AC}{CD}$.

$AF = AG + GF = 3 + 2 = 5$, and $AC = AB + BC = 4 + \frac{8}{3} = \frac{20}{3}$.

Substituting into the proportion gives $\frac{5}{EF} = \frac{20/3}{3}$ or $\frac{20}{3}EF = 15$, so

$EF = \frac{15}{1} \cdot \frac{3}{20} = \frac{9}{4}$.

Section 6.3

15. $\Delta ABG \sim \Delta ADE$ and $\dfrac{AG}{GE} = \dfrac{AB}{BD}$ by the Side Splitting Theorem.

Thus, $\dfrac{x-2}{7} = \dfrac{x}{10.5}$

$$10.5(x-2) = 7x$$
$$10.5x - 21 = 7x$$
$$3.5x = 21$$
$$x = 6$$
$$x - 2 = 4.$$

Therefore $AB = 6$ and $AG = 4$.

17. Solution #1: By the Midsegment Theorem, $MN = \frac{1}{2}PR = \frac{1}{2}(12) = 6$ cm.

Solution #2: Using the Pythagorean Theorem,

$$(PR)^2 + (PQ)^2 = (QR)^2$$
$$12^2 + 20^2 = (QR)^2$$
$$544 = (QR)^2$$
$$\sqrt{544} = QR.$$

Since $QN = \frac{1}{2}QR$, $QN = \frac{1}{2}\sqrt{544}$. Similarly, you have $QM = \frac{1}{2}QP = 10$. Applying the Pythagorean Theorem to ΔMQN, $(MN)^2 + (QM)^2 = (QN)^2$

$$(MN)^2 + 10^2 = \left(\frac{1}{2}\sqrt{544}\right)^2$$
$$(MN)^2 + 100 = \frac{1}{4}(544)$$
$$(MN)^2 = 36$$
$$MN = 6 \text{ cm}.$$

19. Since M and N are midpoints of $\overline{AB}$ and $\overline{CD}$ respectively, P is the midpoint of $\overline{BD}$ by Theorem 5.32.

Thus $\Delta BMP \sim \Delta BAD$ and $\Delta DNP \sim \Delta DCB$ by AA Similarity.

Set up the proportion $\dfrac{\text{altitude}}{\text{base}} = \dfrac{\text{altitude}}{\text{base}}$ in each pair of similar triangles.

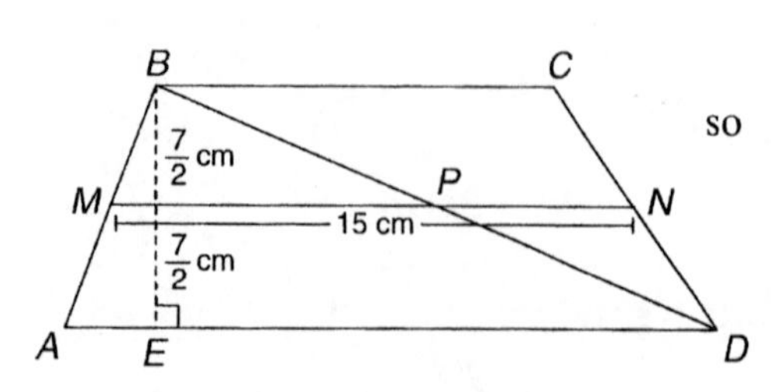

so

$$\frac{\frac{7}{2}}{MP} = \frac{7}{AD}$$
$$\frac{7}{2}(AD) = 7(MP)$$
$$AD = 2(MP)$$

Section 6.3 19. (continued)

and
$$\frac{\frac{7}{2}}{PN} = \frac{7}{BC}$$
$$\frac{7}{2}(BC) = 7(PN)$$
$$BC = 2(PN).$$

Now find the area of the trapezoid.

$$\text{Area} = \frac{1}{2}(AD + BC)(BE) = \frac{1}{2}[2(MP) + 2(PN)] \cdot 7$$
$$= \frac{1}{2}[2(MN)] \cdot 7$$
$$= \frac{1}{2}(2 \cdot 15) \cdot 7$$
$$= 105 \text{ cm}^2$$

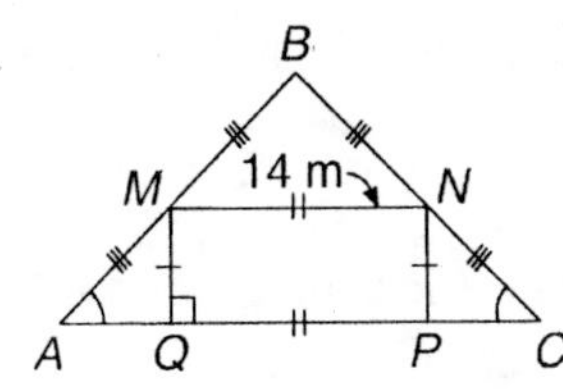

21. $\angle A \cong \angle C$, $\angle MQA = 90° = \angle NPC$. Since two angles of ΔMQA are congruent to two angles of ΔNPC, $\Delta MQA \sim \Delta NPC$ and the third pair of angles is also congruent. Since $MQ = NP$, $\Delta MQA \cong \Delta NPC$ by ASA.

MN is the midsegment of ΔABC. Therefore, $MN = \frac{1}{2}AC$. Since $MN = 14$ m, $AC = 28$ m. $QP = 14$ m also.

$AQ + PC + QP = 28$ m, so $AQ + PC = 14$ m. Since $AQ \cong PC$ by C.P., $AQ + PC = 2PC = 14$ cm and $PC = 7$ m.

Since N is the midpoint of $\overline{BC}$, $NC = \frac{1}{2}BC = 11$ m.

ΔNPC is a right triangle so you can use the Pythagorean Theorem to find NP.

$$(PC)^2 + (NP)^2 = (NC)^2$$
$$7^2 + (NP)^2 = 11^2$$
$$(NP)^2 = 11^2 - 7^2$$
$$(NP)^2 = 72$$
$$NP = \sqrt{72} = 6\sqrt{2}$$

The area of $MNPQ$ is $(QP)(NP) = 14(6\sqrt{2}) = 84\sqrt{2}\text{ m}^2 \approx 118.79\text{ m}^2$.

23. A complete solution is given in the answer key at the back of the text.

25. A complete solution is given in the answer key at the back of the text.

27. A complete solution is given in the answer key at the back of the text.

Section 6.3

29. A complete solution is given in the answer key at the back of the text.

31. A complete solution is given in the answer key at the back of the text.

33. A complete solution is given in the answer key at the back of the text.

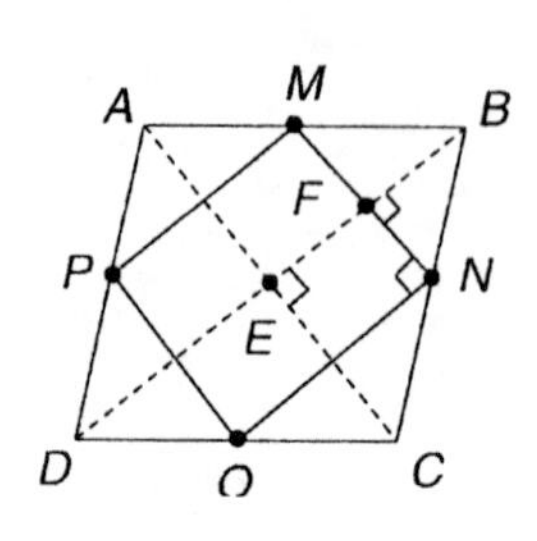

35. Let M, N, O and P be the midpoints of the sides of rhombus $ABCD$ as shown in the given figure.

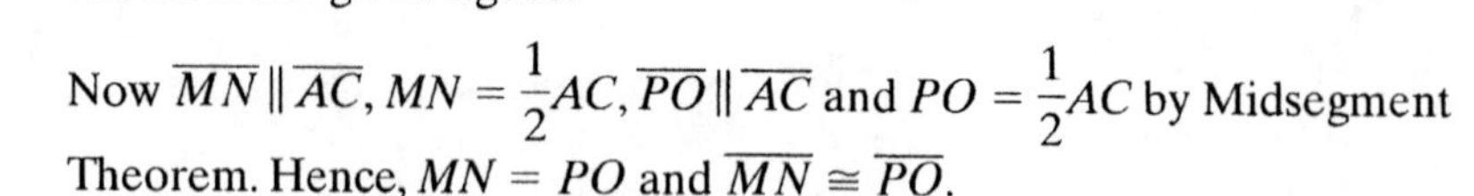

Now $\overline{MN} \parallel \overline{AC}$, $MN = \frac{1}{2}AC$, $\overline{PO} \parallel \overline{AC}$ and $PO = \frac{1}{2}AC$ by Midsegment Theorem. Hence, $MN = PO$ and $\overline{MN} \cong \overline{PO}$.

Also $PM = \frac{1}{2}BD$, $\overline{PM} \parallel \overline{BD}$, $\overline{ON} \parallel \overline{BD}$, and $ON = \frac{1}{2}BD$ by Midsegment Theorem.

Hence, $PM = ON$ and $\overline{PM} \cong \overline{ON}$.

Thus we have shown $MNOP$ is a parallelogram since opposite sides are parallel.

Now $\overline{AC} \perp \overline{BD}$, so $\angle BEC = 90°$ as the diagonals of a rhombus are perpendicular.

Since $\overline{MN} \parallel \overline{AC}$, $\angle BFN \cong \angle BEC$ by corresponding angles.

Since $\overline{ON} \parallel \overline{BD}$, $\angle BFN \cong \angle FNO$ by alternate interior angles.

Therefore $\angle FNO = 90°$. Hence $MNOP$ is a parallelogram with one right angle which implies $MNOP$ is a rectangle.

37. Let M, N, O and P be the midpoints of the sides of isosceles trapezoid $ABCD$ as shown in the given figure.

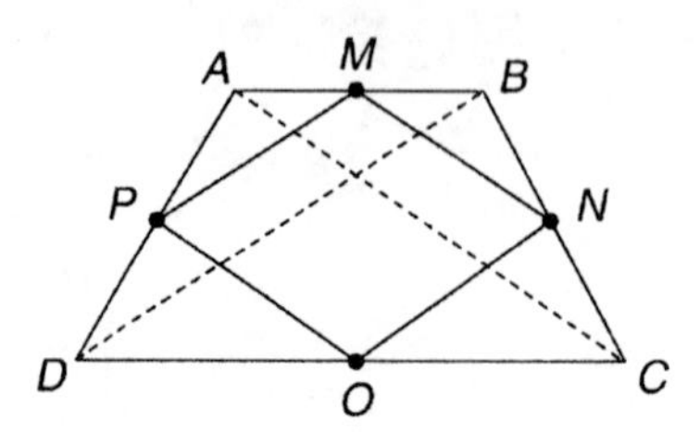

Now $\overline{MN} \parallel \overline{AC}$, $MN = \frac{1}{2}AC$, and $PO = \frac{1}{2}AC$ by Midsegment Theorem.

Hence, $MN = PO$ and $\overline{MN} \cong \overline{PO}$.

Also, $PM = \frac{1}{2}BD$, $\overline{PM} \parallel \overline{BD}$, $\overline{ON} \parallel \overline{BD}$, and $ON = \frac{1}{2}BD$ by Midsegment Theorem.

Hence, $PM = ON$ and $\overline{PM} \cong \overline{ON}$.

Thus, the opposite sides of the midquad are congruent.

Section 6.3

37. (continued)

$\angle A \cong \angle B$ as they are base angles of an isosceles trapezoid. Also $\overline{AM} \cong \overline{MB}$ and $\overline{PA} \cong \overline{NB}$ since $ABCD$ is an isosceles trapezoid with midpoints M, N, O and P.

Thus $\Delta MAP \cong \Delta MBN$ by SAS and $\overline{PM} \cong \overline{NM}$ by C.P.

Therefore $\overline{MN} \cong \overline{NO} \cong \overline{OP} \cong \overline{PM}$ which implies that $MNOP$ is a rhombus.

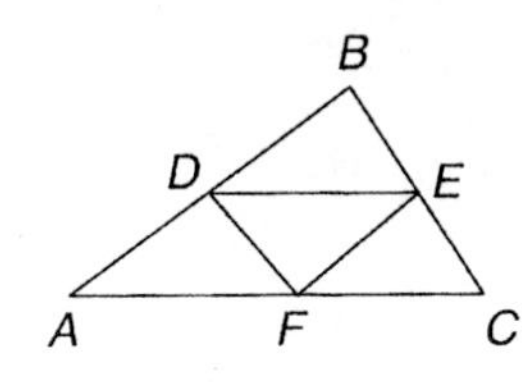

39. Given: D, E and F are midpoints.

Prove: $\Delta ADF \cong \Delta FEC \cong \Delta DBE \cong \Delta EFD$

Proof:

Statement	Reason
1. $AD = \frac{1}{2}AB,\ DB = \frac{1}{2}AB,$ $BE = \frac{1}{2}BC,\ EC = \frac{1}{2}BC,$ $FA = \frac{1}{2}AC,\ CF = \frac{1}{2}AC$	1. Definition of midpoint
2. $AD = DB,\ BE = EC,$ $AF = FC$	2. Definition of midpoint
3. $DE = \frac{1}{2}AC$	3. Midsegment Thm
4. $DE = FA = CF$	4. Substitution
5. $FE = \frac{1}{2}AB$	5. Midsegment Thm
6. $FE = AD = DB$	6. Substitution
7. $DF = \frac{1}{2}BC$	7. Midsegment Thm
8. $DF = BE = EC$	8. Substitution
9. $\Delta ADF \cong \Delta FEC \cong \Delta DBE \cong \Delta EFD$	9. SSS

41. In right triangle ABC, $\overline{CD}$ is the altitude to hypotenuse $\overline{AB}$, hence $\frac{AD}{CD} = \frac{CD}{BD}$ by Theorem 6.9.

Now, $(AC)^2 + (CB)^2 = (AB)^2$ by the Pythagorean Theorem

$$7^2 + 5^2 = (AB)^2$$

$$\sqrt{74} = AB.$$

Section 6.3

41. (continued)

Also $(CD)^2 + (BD)^2 = (CB)^2$ by the Pythagorean Theorem.

$$(CD)^2 + (BD)^2 = 25$$

$$CD = \sqrt{25 - (BD)^2}.$$

Now, $AD = AB - BD$

$= \sqrt{74} - BD.$

Using $\dfrac{AD}{CD} = \dfrac{CD}{BD}$ we have $\dfrac{\sqrt{74} - BD}{\sqrt{25 - (BD)^2}} = \dfrac{\sqrt{25 - (BD)^2}}{BD}$

$$BD(\sqrt{74} - BD) = 25 - (BD)^2$$

$$\sqrt{74}\,BD = 25$$

$$BD = \frac{25}{\sqrt{74}} \approx 2.906 \text{ inches.}$$

Since $AD = AB - BD$, $AD = \sqrt{74} - BD$

$$= \sqrt{74} - \frac{25}{\sqrt{74}}$$

$$\approx 5.696 \text{ inches.}$$

Also, $CD = \sqrt{25 - (BD)^2}$

$$= \sqrt{25 - \left(\frac{25}{\sqrt{74}}\right)^2}$$

$$\approx 4.096 \text{ in.}$$

Congruent segments are marked.

Therefore the dimensions of the original cardboard are

length $= AD + DB \approx 5.696 + 2.906 \approx 8.60$ inches

width $= CD + CD \approx 2(4.069) \approx 8.14$ inches.

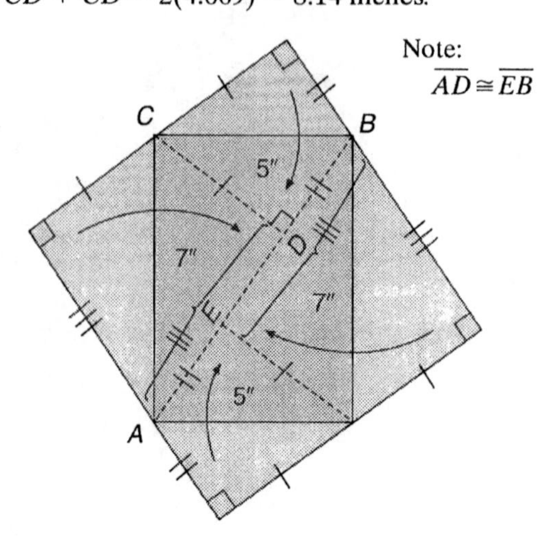

Section 6.4

Section 6.4

Tips:

✔ A mnemonic to help you remember the sine, cosine and tangent ratios is SOHCAHTOA:

SOH stands for $\sin A = \dfrac{\text{opposite}}{\text{hypotenuse}}$,

CAH stands for $\cos A = \dfrac{\text{adjacent}}{\text{hypotenuse}}$,

TOA stands for $\tan A = \dfrac{\text{opposte}}{\text{adjacent}}$.

✔ Remember that the adjacent side is the leg of the right triangle that forms a side of the angle you are using. The opposite side is the leg opposite the angle (i.e., the leg that does *not* form a side of the angle).

✔ Since opposite and adjacent are relative terms, these sides will change when you refer to different angles. The hypotenuse, however, is always the side opposite the right angle.

✔ The -1 in $\sin^{-1}$, $\cos^{-1}$, and $\tan^{-1}$ is *not* an exponent. It indicates the inverse of a function. For example, $\sin^{-1}$ accepts as input a trigonometric ratio and gives you the corresponding angle.

✔ The square of a trigonometric ratio such as $(\sin A)^2$ is usually written as $\sin^2 A$.

✔ Make sure your calculator is set in degree mode. If your answers are consistently wrong, check the mode.

Solutions to odd-numbered textbook problems

1. $\sin A = \dfrac{\text{opposite}}{\text{hypotenuse}} = \dfrac{5}{13}$

3. $(AC)^2 + (CB)^2 = (AB)^2$ by the Pythagorean Theorem.

$$(AC)^2 + 5^2 = 13^2$$
$$(AC)^2 = 169 - 26$$
$$(AC)^2 = 144$$
$$AC = 12$$

Now, $\tan A = \dfrac{\text{opposite}}{\text{adjacent}} = \dfrac{5}{12}$.

5. $\cos B = \dfrac{\text{adjacent}}{\text{hypotenuse}} = \dfrac{5}{13}$

Notice that $\cos B = \sin A$ from problem #1.

Section 6.4

7. $\tan A = \frac{\text{opposite}}{\text{adjacent}} = \frac{7}{5} = 1.4$

9. $\cos A = \frac{\text{adjacent}}{\text{hypotenuse}} = \frac{b}{c} = \frac{7}{11} \approx 0.64$

11. If $a = b$ then
$$\begin{aligned} c^2 &= a^2 + b^2 \\ &= a^2 + a^2 \\ &= 2a^2 \\ c &= a\sqrt{2}. \end{aligned}$$
$$\begin{aligned} \cos B &= \frac{a}{c} \\ &= \frac{a}{a\sqrt{2}} \\ &= \frac{1}{\sqrt{2}} \\ &= \frac{\sqrt{2}}{2} \\ &\approx 0.71 \end{aligned}$$

13. $\sin A = \frac{a}{c}$, so $0.503 = \frac{a}{10}$ or $a = 5.03$.

15.
$$\begin{aligned} 0.6 = \tan A &= \frac{a}{b} = \frac{5}{b} \\ 0.6b &= 5 \\ b &= \frac{5}{0.6} = \frac{25}{3} \end{aligned}$$
So
$$\begin{aligned} c^2 &= a^2 + b^2 \\ c^2 &= 5^2 + \left(\frac{25}{3}\right)^2 \quad \text{substituting values} \\ c &= \sqrt{5^2 + \left(\frac{25}{3}\right)^2} \\ &\approx 9.72. \end{aligned}$$

17. You are given a, which is the length of the side opposite $\angle A$ and b, which is the length of the side adjacent to $\angle A$. Therefore, use the tangent ratio.
$$\tan A = \frac{a}{b} = \frac{5}{7}$$
Thus, $\angle A = \tan^{-1}\left(\frac{5}{7}\right) \approx 35.5°$.

Section 6.4

19. $\cos A = \dfrac{\text{adjacent}}{\text{hypotenuse}}$

$= \dfrac{\pi}{\sqrt{21}}$

Thus, $\angle A = \cos^{-1}\left(\dfrac{\pi}{\sqrt{21}}\right)$

$\approx 46.7°.$

21. $\cos B = \dfrac{\text{adjacent}}{\text{hypotenuse}} = \dfrac{a}{c}$, but you are not given the value of c. Using the Pythagorean Theorem and $b = a\sqrt{3}$ (as given), you can determine c.

$$a^2 + b^2 = c^2$$
$$a^2 + (a\sqrt{3})^2 = c^2$$
$$a^2 + 3a^2 = c^2$$
$$4a^2 = c^2$$
$$2a = c$$

Thus, $\cos B = \dfrac{a}{c} = \dfrac{a}{2a} = \dfrac{1}{2} = 0.5.$

23.
$$\sin B = \frac{\text{opposite}}{\text{hypotenuse}}$$
$$\sin 70° = \frac{0.8}{c}$$
$$c \sin 70° = 0.8$$
$$c = \frac{0.8}{\sin 70°}$$
$$\approx 0.85$$

25. Using the Pythagorean Theorem, find CA.

$$4^2 + (CA)^2 = 12^2$$
$$16 + (CA)^2 = 144$$
$$(CA)^2 = 128$$
$$CA = \sqrt{128} = 8\sqrt{2} \approx 11.3 \text{ in.}$$

To find angle A you can use $\sin A = \dfrac{4}{12} = \dfrac{1}{3}$.

$$\angle A = \sin^{-1}\left(\frac{1}{3}\right) \approx 19.5°.$$

Then you have $\angle B \approx 180° - 90° - 19.5° \approx 70.5°$.

Section 6.4

27. $\angle A = 35°$ by angle sum of triangle.

To find AB, you can use $\sin 55° = \dfrac{11.2}{AB}$.

$$AB \sin 55° = 11.2$$
$$AB = \frac{11.2}{\sin 55°}$$
$$\approx 13.7 \text{ m}$$

To find BC, it is best to use values that have not been rounded so as not to compound round off error.

$$\tan 35° = \frac{BC}{11.2}$$
$$BC = 11.2 \tan 35°$$
$$\approx 7.8 \text{ m}$$

To summarize: $\angle A = 35°, AB \approx 13.7$ m, $BC \approx 7.8$ m.

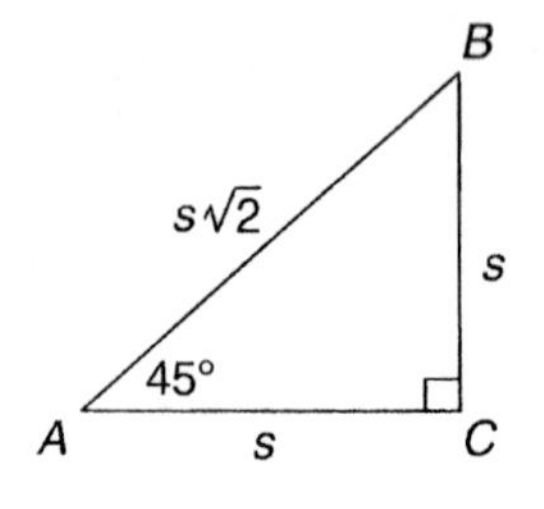

29. Remember the general 45°–45° right triangle from Section 3.3 (See diagram). Use this general triangle to find the *exact* values for sin 45° and cos 45°.

If $\angle A = 45°$, then

$$\sin 45° = \frac{s}{s\sqrt{2}} = \frac{1}{\sqrt{2}}, \text{ and } \cos 45° = \frac{s}{s\sqrt{2}} = \frac{1}{\sqrt{2}}.$$

Thus, $\dfrac{\sin 45°}{\cos 45°} = \dfrac{\frac{1}{\sqrt{2}}}{\frac{1}{\sqrt{2}}} = 1$. From the triangle shown you also know that

$\tan 45° = \dfrac{s}{s} = 1$, so $\tan 45° = \dfrac{\sin 45°}{\cos 45°}$, which verifies Theorem 6.13.

Also, $\sin^2 45° + \cos^2 45° = \left(\dfrac{1}{\sqrt{2}}\right)^2 + \left(\dfrac{1}{\sqrt{2}}\right)^2 = \dfrac{1}{2} + \dfrac{1}{2} = 1$, which verifies Theorem 6.14.

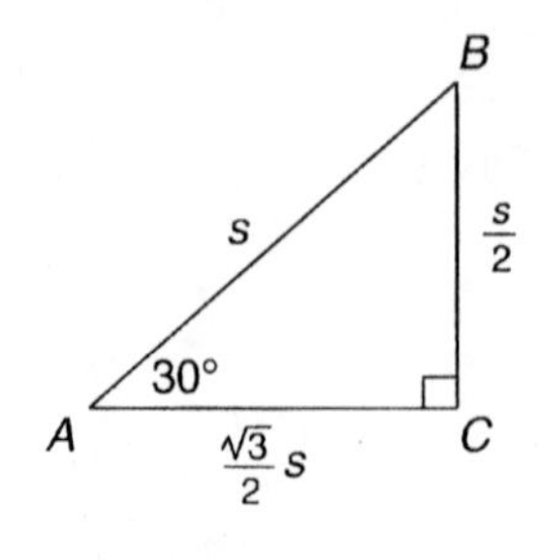

31. Remember the general 30°–60° right triangle from Section 3.3 (See diagram). Use this general triangle to find the exact values for sin 30° and cos 30°.

If $\angle A = 30°$, then

$$\sin 30° = \frac{\frac{s}{2}}{s} = \frac{1}{2}, \text{ and } \cos 30° = \frac{\frac{\sqrt{3}}{2}s}{s} = \frac{\sqrt{3}}{2}.$$

$$\text{Thus, } \frac{\sin 30°}{\cos 30°} = \frac{\frac{1}{2}}{\frac{\sqrt{3}}{2}} = \frac{1}{\sqrt{3}} = \frac{\sqrt{3}}{3}.$$

Section 6.4

31. (continued)

From the triangle shown we know that $\tan 30° = \dfrac{\frac{s}{2}}{\frac{\sqrt{3}}{2}s} = \dfrac{1}{\sqrt{3}} = \dfrac{\sqrt{3}}{3}$.

So $\tan 30° = \dfrac{\sin 30°}{\cos 30°}$ which verifies Theorem 6.13.

Also, $\sin^2 30° + \cos^2 30° = \left(\dfrac{1}{2}\right)^2 + \left(\dfrac{\sqrt{3}}{2}\right)^2$

$= \dfrac{1}{4} + \dfrac{3}{4}$

$= 1.$

This verifies Theorem 6.14.

33. You are given the hypotenuse and one angle of the triangle. To find side a, use $\sin A$ which involves the hypotenuse and the side opposite $\angle A$.

$$\sin 30° = \frac{a}{10}, \text{ so } a = 10 \sin 30° = 10\left(\frac{1}{2}\right) = 5.$$

To find b, either use the Pythagorean Theorem or use the cosine ratio.

$$\cos 30° = \frac{b}{10}, \text{ so } b = 10 \cos 30° = 10\left(\frac{\sqrt{3}}{2}\right) = 5\sqrt{3} \approx 8.7$$

Finally, $\angle B = 180° - 90° - 30° = 60°$.

35. Two legs of a right triangle are given. Use the Pythagorean Theorem to find hypotenuse c.

$$\begin{aligned} c^2 &= a^2 + b^2 \\ &= 10^2 + 24^2 \\ &= 676 \\ c &= 26 \end{aligned}$$

One way to find $\angle A$ is $\sin A = \dfrac{24}{26}$

Thus, $\angle A = \sin^{-1}\dfrac{24}{26} \approx 67.4°$

and

$$\begin{aligned} \angle B &\approx 180° - 90° - 67.4° \\ &\approx 22.6°. \end{aligned}$$

37. You are given $\angle A$ and its adjacent leg. To find the length of the opposite side, use the tangent ratio.

$$\tan 23° = \frac{a}{4}, \text{ so } a = 4 \tan 23° \approx 1.7.$$

Section 6.4

37. (continued)

Find c using the cosine ratio.

$$\cos 23° = \frac{4}{c}, \text{ so } c\,(\cos 23°) = 4, \text{ and } c = \frac{4}{\cos 23°} \approx 4.3.$$

Then $\angle B = 180° - 90° - 23° = 67°$.

39. Given is $\angle B$ and the hypotenuse. To find the length of the adjacent side use the cosine ratio.

$$\cos 12° = \frac{a}{10.3}, \text{ so } a = 10.3 \cos 12° \approx 10.1$$

To find the length of the opposite side, use the sine ratio.

$$\sin 12° = \frac{b}{10.3}, \text{ so } b = 10.3 \sin 12° \approx 2.1$$

Then $\angle A = 180° - 90° - 12° = 78°$.

41. Use QR as the height of ΔPQR and RP as the base. You must find RP, which is opposite the given angle, and you were given the adjacent side. Use the tangent ratio.

$$\tan 51.3° = \frac{RP}{25}, \text{ so } RP = 25 \tan 51.3° \approx 31.21 \text{ ft.}$$

The area of ΔPQR is $\frac{1}{2}(31.21)(25) = 390.1 \text{ ft}^2$.

43. $\overline{CD}$ is the altitude of ΔABC to base $\overline{AB}$. CD can be found using the cosine ratio with $\angle ACD$ in right triangle ACD.

B, 8″, D, 78°, A, 17″, C

$$\angle ACD = 180° - 90° - 78° = 12°.$$

$$\cos 12° = \frac{CD}{17}$$

$$CD = 17 \cos 12°$$

The area of $\Delta ABC = \frac{1}{2}(AB)(CD)$

$$= \frac{1}{2}(8)(17 \cos 12°).$$

$$\approx 66.5 \text{ in}^2.$$

45. The height and side of this parallelogram form a right triangle with a 64° angle. You can use the sine ratio to determine the height.

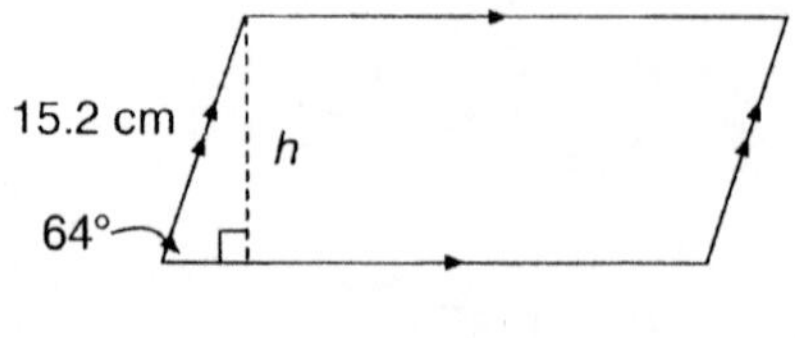

15.2 cm, h, 64°

$$\sin 64° = \frac{h}{15.2}, \text{ so } h = 15.2 \sin 64° \approx 13.66 \text{ cm.}$$

Then the area is $A = bh = (24.7)(13.66) \approx 337.4 \text{ cm}^2$.

Section 6.4

47. A complete solution is given in the answer key at the back of the text.

49. A complete solution is given in the answer key at the back of the text.

51. A complete solution is given in the answer key at the back of the text.

53. A complete solution is given in the answer key at the back of the text.

55. A complete solution is given in the answer key at the back of the text.

57. A complete solution is given in the answer key at the back of the text.

top of hill →
150 ft
x
38°

59. The height of the hill can be found using the sine ratio with the 38° angle.

Let x = height of hill in feet.

$\sin 38° = \dfrac{x}{150}$, so $x = 150 \sin 38° \approx 92$ ft.

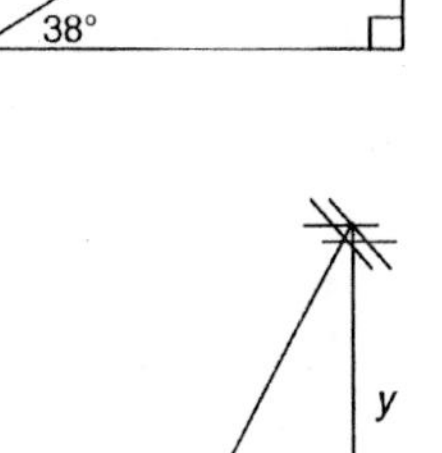

61. The guy wire, radio antenna, and level ground form a right triangle. You are given the length of the side adjacent to the 65° angle and want the length of the opposite side, so use the tangent ratio.

$\tan 65° = \dfrac{y}{20}$, so $y = 20 \tan 65° \approx 42.89$ ft.

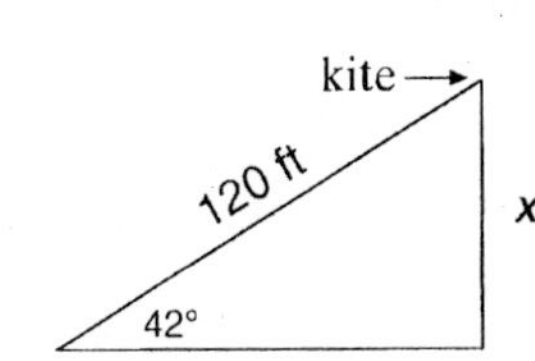

63. The height of the kite can be found using the sine ratio with the 42° angle.

Let x = height of kite in feet.

$\sin 42° = \dfrac{x}{120}$, so $x = 120 \sin 42° \approx 80.30$ ft.

65. (a) The conveyor is the hypotenuse of a right triangle and the height is the side opposite the given 30° angle, so use the sine ratio.

$$\sin 30° = \frac{\text{height}}{100}, \text{ or } h = 100 \sin 30° = 50 \text{ ft.}$$

(b) $\sin 28° = \dfrac{h}{100}$, or $h = 100 \sin 28° = 46.9$ ft.

67. There are two right triangles formed. Find the length of the opposite side to the given angle in each right triangle.

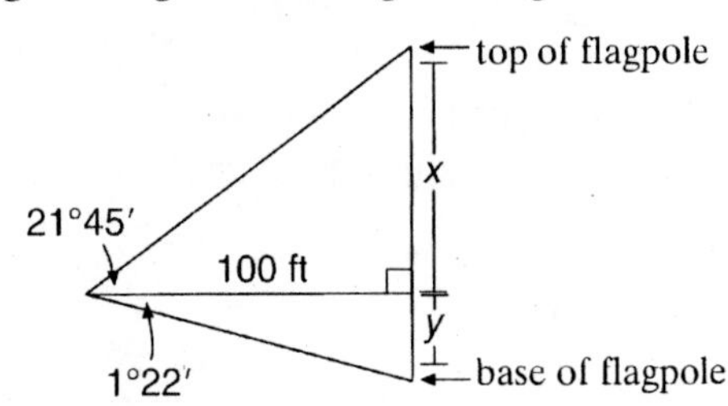

Section 6.4

67. (continued)

Let x = length of opposite side to the $21° 45'$ angle and y = length of opposite side to the $1° 22'$ angle.

Convert angle measures to degrees.

$$21° 45' = 21\frac{45°}{60} = 21\frac{3°}{4} = \frac{87°}{4}$$

$$1° 22' = 1\frac{22°}{60} = 1\frac{11°}{30} = \frac{41°}{30}$$

The adjacent side is given and the opposite side is the unknown, so use the tangent ratio

$$\tan\left(\frac{87°}{4}\right) = \frac{x}{100}, \text{ so } x = 100 \tan\frac{87°}{4}.$$

$$\tan\left(\frac{41°}{30}\right) = \frac{y}{100}, \text{ so } y = 100 \tan\frac{41°}{30}.$$

$$\text{Height of flagpole} = x + y = 100 \tan\frac{87°}{4} + 100 \tan\frac{41°}{30}$$

$$= 100\left(\tan\frac{87°}{4} + \tan\frac{41°}{30}\right)$$

$$\approx 42.3 \text{ ft}$$

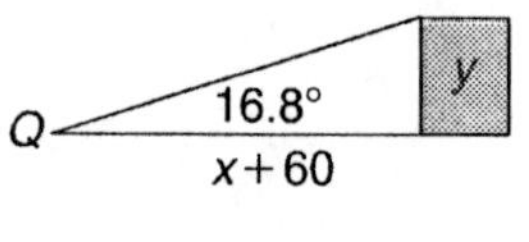

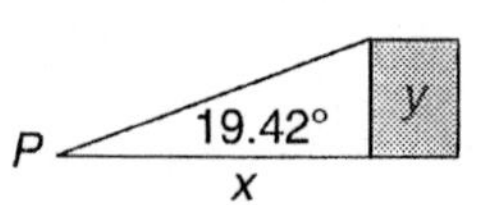

69. There are two right triangles formed here. It may be easier to solve the problem if you look at the two triangles separately.

Let y be the height of the building in meters and let x be the distance from P to the building in meters. Then $x + 60$ is the distance from Q to the building in meters.

$$\tan 16.8° = \frac{y}{x + 60}, \text{ or } y = (\tan 16.8°)(x + 60).$$

Also, $\tan 19.42° = \frac{y}{x}$, or $y = (\tan 19.42°)x$.

Substituting for y, you have

$$(\tan 16.8°)(x + 60) = (\tan 19.42°)x$$

$$0.3019(x + 60) = 0.3525x.$$

$$0.3019x + 18.114 = 0.3535x$$

$$18.114 = 0.0506x$$

$$358 \text{ m} \approx x.$$

To find y, substitute for x in one of the equations involving x and y. You have $y = (\tan 19.42°)(358) \approx 126$ m.

Section 6.5

Section 6.5

Tips

✔ Use the Law of Sines to solve an oblique triangle when

(1) the measure of at least one side and the measures of any two angles are known.

or

(2) the measures of any two sides and the measure of an angle opposite one of them are known.

✔ Use the Law of Cosines to solve an oblique triangle when

(1) the measures of two sides and the included angle of the triangle are known.

or

(2) the measures of the three sides are known.

✔ If possible, do not use a rounded value to find another missing part of the triangle, as it may compound round off error.

Solutions to odd-numbered textbook problems

1. $$\frac{10}{\sin 27°} = \frac{c}{\sin 70°}$$

$$c = \frac{10 \sin 70°}{\sin 27°} \approx 20.7$$

3. $$\frac{11}{\sin 73°} = \frac{b}{\sin 19°}$$

$$b = \frac{11 \sin 19°}{\sin 73°} \approx 3.7$$

5. $$\frac{9.5}{\sin 12°} = \frac{a}{\sin 85°}$$

$$a = \frac{9.5 \sin 85°}{\sin 12°} \approx 45.5$$

7. $$\frac{\sin B}{8.5} = \frac{\sin 33°}{14}$$

$$\sin B = \frac{8.5 \sin 33°}{14}$$

Thus, $\angle B = \sin^{-1}\left[\frac{8.5 \sin 33°}{14}\right] \approx 19.3°$.

Section 6.5

9. $\dfrac{\sin A}{43} = \dfrac{\sin 61.3°}{44.2}$

$\sin A = \dfrac{43 \sin 61.3°}{44.2}$

Thus, $\angle A = \sin^{-1}\left[\dfrac{43 \sin 61.3°}{44.2}\right] \approx 58.6°$.

11. Find side c.

$\dfrac{c}{\sin 37°} = \dfrac{7\sqrt{3}}{\sin 89.5°}$

$c = \dfrac{7.3 \sin 37°}{\sin 89.5°} \approx 7.3$

Now find $\angle B$.

$$\angle A + \angle B + \angle C = 180°$$
$$89.5° + \angle B + 37° = 180°$$
$$\angle B = 53.5°$$

Finally, find side b.

$\dfrac{b}{\sin 53.5°} = \dfrac{7\sqrt{3}}{\sin 89.5°}$

$b = \dfrac{7\sqrt{3}\sin 53.5°}{\sin 89.5°} \approx 9.7$

13. Find $\angle C$.

$\dfrac{\sin C}{67} = \dfrac{\sin 68°}{91.1}$

$\sin C = \dfrac{67 \sin 68°}{91.1}$

Thus $\angle C = \sin^{-1}\left[\dfrac{67 \sin 68°}{91.1}\right] \approx 43.0°$.

Next, find $\angle A$.

$$\angle A \approx 180° - 68° - 43°$$
$$\approx 69.0°$$

Finally, find side a.

$\dfrac{a}{\sin 69°} = \dfrac{91.1}{\sin 68°}$

$a = \dfrac{91.1 \sin 69°}{\sin 68°} \approx 91.7$

Section 6.5

15. Find $\angle B$.

$$\frac{\sin 73^\circ}{7.2} = \frac{\sin B}{\sqrt{29}}$$

$$\sin B = \frac{\sqrt{29}\sin 73^\circ}{7.2}$$

Thus, $\angle B = \sin^{-1}\left[\frac{\sqrt{29}\sin 73^\circ}{7.2}\right] \approx 45.7^\circ$.

Next, find $\angle A$.

$$\angle A \approx 180^\circ - 73^\circ - 45.7^\circ$$
$$\approx 61.3^\circ$$

Finally, find side a.

$$\frac{a}{\sin 61.3^\circ} = \frac{7.2}{\sin 73^\circ}$$

$$a = \frac{7.2 \sin 61.3^\circ}{\sin 73^\circ} \approx 6.6$$

17.
$$\begin{aligned} a^2 &= b^2 + c^2 - 2bc\cos A \\ &= (36.2)^2 + (54.6)^2 - 2(36.2)(54.6)\cos 85^\circ \\ &= 4291.6 - 3953.04\cos 85^\circ \\ a &= \sqrt{4291.6 - 3953.04\cos 85^\circ} \approx 62.8 \end{aligned}$$

19.
$$\begin{aligned} b^2 &= a^2 + c^2 - 2bc\cos B \\ &= 47^2 + 47^2 - 2(47)(47)\cos 51^\circ \\ &= 4418 - 4418\cos 51^\circ \\ b &= \sqrt{4418 - 4418\cos 51^\circ} \approx 40.5 \end{aligned}$$

21.
$$\begin{aligned} c^2 &= a^2 + b^2 - 2ab\cos C \\ &= (18.75)^2 + (21.30)^2 - 2(18.75)(21.30)\cos 60.26^\circ \\ &= 805.2525 - 798.75\cos 60.26^\circ \\ c &= \sqrt{805.2525 - 798.75\, cos\, 60.26^\circ} \approx 20.2 \end{aligned}$$

23.
$$\begin{aligned} b^2 &= a^2 + c^2 - 2ac\cos B \\ 8^2 &= 14^2 + 11^2 - 2(14)(11)\cos B \\ 64 &= 196 + 121 - 308\cos B \\ -253 &= -308\cos B \\ \cos B &= \frac{253}{308} \end{aligned}$$

Thus, $\angle B = \cos^{-1}\left(\frac{253}{308}\right) \approx 34.8^\circ$.

Section 6.5

25.

$$
\begin{aligned}
c^2 &= a^2 + b^2 - 2ab \cos C \\
(2.4)^2 &= (2.3)^2 + (1.9)^2 - 2(2.3)(1.9) \cos C \\
5.76 &= 5.29 + 3.61 - 8.74 \cos C \\
-3.14 &= -8.74 \cos C \\
\cos C &= \frac{3.14}{8.74}
\end{aligned}
$$

Thus, $\angle C = \cos^{-1}\left(\frac{3.14}{8.74}\right) \approx 68.9°$.

27. Find side a.

$$
\begin{aligned}
a^2 &= b^2 + c^2 - 2bc \cos A \\
&= (36.2)^2 + (54.6)^2 - 2(36.2)(54.6) \cos 85° \\
&= 4291.6 - 3953.04 \cos 85° \\
a &= \sqrt{4291.6 - 3953.04 \cos 85°} \\
&\approx 62.8
\end{aligned}
$$

Next find $\angle B$.

$$
\begin{aligned}
\frac{\sin A}{a} &= \frac{\sin B}{b} \\
\frac{\sin 85°}{62.8} &= \frac{\sin B}{36.2} \\
\sin B &= \frac{36.2 \sin 85°}{62.8}
\end{aligned}
$$

Thus, $\angle B = \sin^{-1}\left[\frac{36.2 \sin 85°}{62.8}\right] \approx 35.0°$.

Finally find $\angle C$.

$$
\begin{aligned}
\frac{\sin A}{a} &= \frac{\sin C}{c} \\
\frac{\sin 85°}{62.8} &= \frac{\sin C}{54.6} \\
\sin C &= \frac{54.6 \sin 85°}{62.8}
\end{aligned}
$$

Thus, $\angle C = \sin^{-1}\left[\frac{54.6 \sin 85°}{62.8}\right] \approx 60.0°$.

Note: $\angle C$ could be found by $\angle C \approx 180° - 85° - 35.0° \approx 60.0°$.

29. Find $\angle A$.

$$
\begin{aligned}
a^2 &= b^2 + c^2 - 2bc \cos A \\
55^2 &= 67^2 + 43^2 - 2(67)(43) \cos A \\
3025 &= 6338 - 5762 \cos A \\
\cos A &= \frac{3313}{5762}
\end{aligned}
$$

Section 6.5 29. (continued)

Thus, $\angle A = \cos^{-1}\left(\frac{3313}{5762}\right) \approx 54.9°$.

Next, find $\angle B$.

$$\frac{\sin B}{b} = \frac{\sin A}{A}$$

$$\frac{\sin B}{67} = \frac{\sin 54.9°}{55}$$

$$\sin B = \frac{67 \sin 54.9°}{55}$$

Thus, $\angle B = \sin^{-1}\left[\frac{67 \sin 54.9°}{55}\right] \approx 85.3°$.

Finally find $\angle C$.

$$\frac{\sin A}{a} = \frac{\sin C}{c}$$

$$\frac{\sin 54.9°}{55} = \frac{\sin C}{43}$$

$$\sin C = \frac{43 \sin 54.9°}{55}$$

Thus, $\angle C = \sin^{-1}\left[\frac{43 \sin 54.9°}{55}\right] \approx 39.8°$.

Note: $\angle C$ could be found by $\angle C \approx 180° - 54.9° - 85.3° \approx 39.8°$.

31. Find $\angle A$.

$$a^2 = b^2 + c^2 - 2bc \cos A$$

$$(75.6)^2 = (92.3)^2 + (69.9)^2 - 2(92.3)(69.9) \cos A$$

$$5715.36 = 13405.3 - 12903.54 \cos A$$

$$\cos A = \frac{7689.94}{12903.54}$$

Thus, $\angle A = \cos^{-1}\left[\frac{7689.94}{12903.54}\right. \approx 53.4°$.

Next, find $\angle B$.

$$\frac{\sin B}{b} = \frac{\sin A}{a}$$

$$\frac{\sin B}{92.3} = \frac{\sin 53.4°}{75.6}$$

$$\sin B = \frac{92.3 \sin 53.4°}{75.6}$$

Thus, $\angle B = \sin^{-1}\left[\frac{92.3 \sin 53.4°}{75.6}\right] \approx 78.6°$.

Section 6.5

31. (continued)

Finally, find $\angle C$.

$$\frac{\sin A}{a} = \frac{\sin C}{c}$$

$$\frac{\sin 53.4°}{75.6} = \frac{\sin C}{69.9}$$

$$\sin C = \left[\frac{69.9 \sin 53.4°}{75.6}\right]$$

Thus, $\angle C = \sin^{-1}\left[\frac{69.9 \sin 53.4°}{75.6}\right] \approx 47.9°$

Note: If $\angle C$ was found by using the angle sum of a triangle then $\angle C \approx 180° - 53.4° - 78.6° \approx 4.8°$. This measure is not as accurate. This illustrates the concept of round off error.

When finding $\angle C$ using the Law of Sines only one approximate value ($\angle A \approx 53.4°$) was used in the computation.

When finding $\angle C$ using the angle sum of a triangle two approximate measures ($\angle A \approx 53.4$ and $\angle B \approx 78.6$) were used. In this case, there was enough rounding in each of the angle measures to make a difference in the angle measure for $\angle C$.

Also note that $\angle A + \angle B + \angle C = 53.4° + 78.6° + 47.9° \neq 180°$.

This is because each angle measure was rounded down when rounded to the nearest tenth.

33. Use the Law of Sines since the measure of one side and any two angles are known.

$$\frac{b}{\sin 81°} = \frac{15}{\sin 26°}$$

$$b = \frac{15 \sin 81°}{\sin 26°} \approx 33.8 \text{ cm}$$

35. Use the Law of Cosines since the measures of three sides are known.

$$a^2 = b^2 + c^2 - 2bc \cos A$$

$$(2.1)^2 = (3.5)^2 + (2.6)^2 - 2(3.5)(2.6) \cos A$$

$$4.41 = 19.01 - 18.2 \cos A$$

$$\cos A = \frac{14.6}{18.2}$$

Thus, $\angle A = \cos^{-1}\left(\frac{14.6}{18.2}\right) \approx 36.7°$.

37. The regular heptagon has a vertex angle measure of

$$\frac{180(7-2)}{7} = \frac{900°}{7} \text{ (See Section 2.3).}$$

Divide the regular heptagon into seven congruent isosceles triangles.

Each base angle of the isosceles triangle has measure $\frac{1}{2}\left(\frac{900°}{7}\right) = \frac{450°}{7}$.

Section 6.5

37. (continued)

Next, draw the altitude, x, to the vertex angle of each isosceles triangle. This divides the regular heptagon into 14 congruent right triangles with angles of measure $90°, \frac{450°}{7}$ and $180° - 90° - \frac{450°}{7} = \frac{180°}{7}$. The base of each right triangle has measure 5 inches.

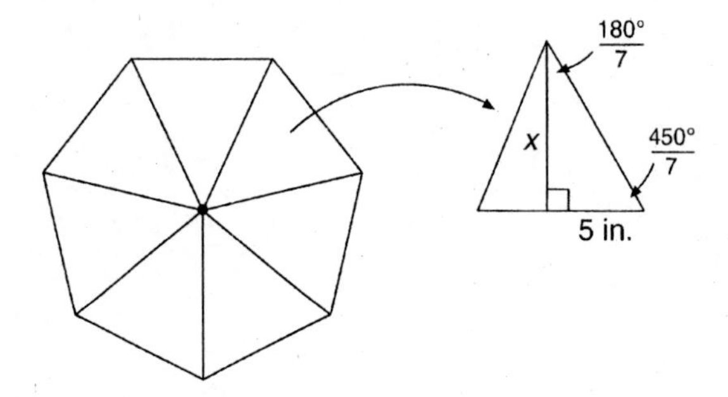

Use the Law of Sines to find altitude x.

$$\frac{x}{\sin\left(\frac{450°}{7}\right)} = \frac{5}{\sin\left(\frac{180°}{7}\right)}$$

$$x = \frac{5\sin\left(\frac{450°}{7}\right)}{\sin\left(\frac{180°}{7}\right)}$$

Now find area of the regular heptagon by finding the area of the 14 right triangles.

$$A = 14\left(\frac{1}{2}\right)(5)\left[\frac{5\sin\left(\frac{450°}{7}\right)}{\sin\left(\frac{180°}{7}\right)}\right]$$

$$= \frac{175\sin\frac{450°}{7}}{\sin\left(\frac{180°}{7}\right)}$$

$$\approx 363.39 \text{ in}^2$$

39. A complete solution is given in the answer key at the back of the text.

41. A complete solution is given in the answer key at the back of the text.

Section 6.5

43. Use the Law of Sines since the measures of two sides and an angle opposite one of them are known. First find $\angle A$ using the Law of Sines.

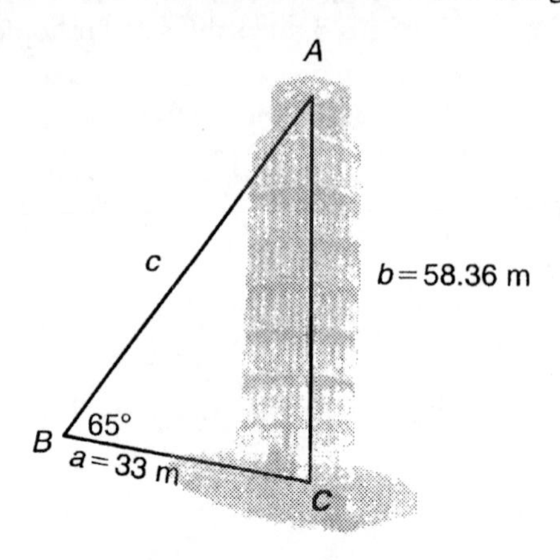

$$\frac{\sin A}{33} = \frac{\sin 65°}{58.36}$$

$$\sin A = \frac{33 \sin 65°}{58.36}$$

Thus, $\angle A = \sin^{-1}\left[\frac{33 \sin 65°}{58.36}\right] \approx 31°$.

Now find $\angle C$, the desired angle.

$$\angle C = 180° - 65° - 31° \approx 84°$$

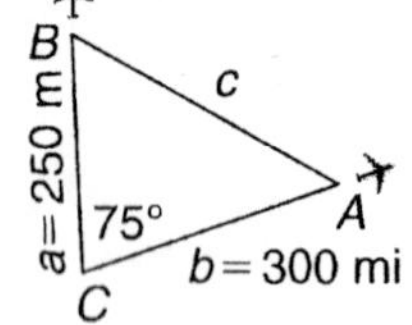

45. Use the Law of Cosines since the measures of two sides are known.

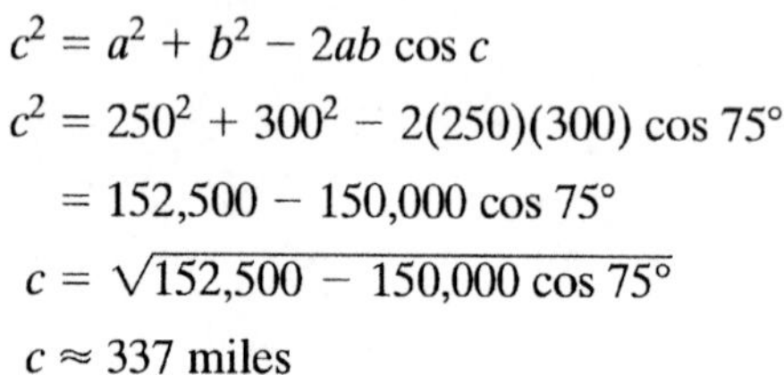

$$c^2 = a^2 + b^2 - 2ab \cos c$$

$$c^2 = 250^2 + 300^2 - 2(250)(300) \cos 75°$$

$$= 152{,}500 - 150{,}000 \cos 75°$$

$$c = \sqrt{152{,}500 - 150{,}000 \cos 75°}$$

$$c \approx 337 \text{ miles}$$

47. Use of the Law of Sines to find x.

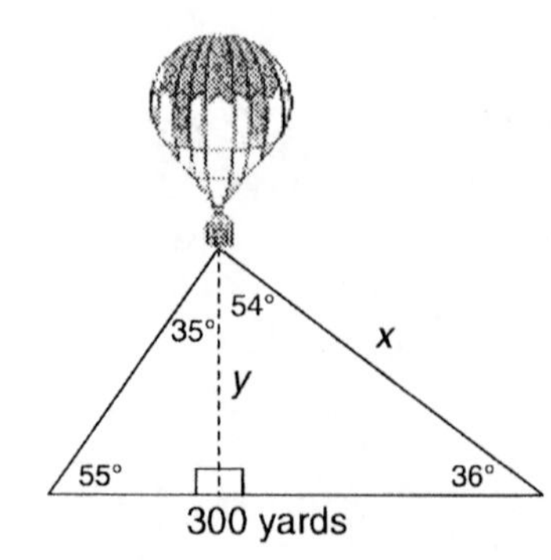

$$\frac{100}{\sin 89°} = \frac{x}{\sin 55°}$$

$$x = \frac{100 \sin 55°}{\sin 89°} \approx 81.9 \text{ yards}$$

Use the sine ratio of a right triangle to find height y.

$$\sin 36° = \frac{y}{81.9}$$

$$y = 81.9 \sin 36° \approx 48 \text{ yards}$$

Additional Problems

49. Use the Law of Cosines to find the resultant speed.

$$x = \text{resultant speed}$$
$$x^2 = 450^2 + 80^2 - 2(450)(80)\cos 85°$$
$$= 208{,}900 - 72{,}000 \cos 85°$$
$$x = \sqrt{208{,}900 - 72{,}000 \cos 85°}$$
$$\approx 450.14 \text{ mph}$$

Use the Law of Sines to find the angle of direction.

Let A = resultant angle of direction.

$$\frac{\sin A}{80} = \frac{\sin 85°}{450.14}$$

Thus, $\angle A = \sin^{-1}\left[\frac{80 \sin 85°}{450.14}\right] \approx 10.20°$.

Hence the angle of direction is N 10.20° E.

Therefore the resultant velocity is 450.14 mph in the direction N 10.20° E.

Additional Problems

1. Use the Pythagorean Theorem with

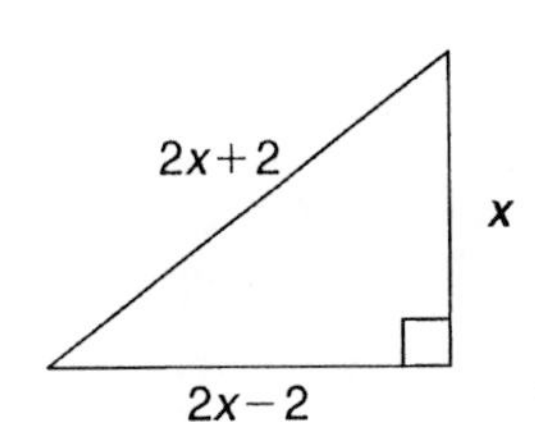

x = length of shorter leg.

$2x - 2$ = length of longer leg

$2x + 2$ = length of hypotenuse

$$x^2 + (2x - 2)^2 = (2x + 2)^2$$
$$x^2 + \cancel{4x^2} - 8x + \cancel{4} = \cancel{4x^2} + 8x + \cancel{4}$$
$$x^2 - 16x = 0$$
$$x(x - 16) = 0$$
$$x = 0 \; x = 16$$

$x = 0$ is not a possible solution as it makes one leg of the triangle have negative measure.

Therefore, the sides of the triangle are $x = 16$ cm, $2x - 2 = 30$ cm and $2x + 2 = 34$ cm.

2. The sides of a rhombus are congruent so $AB = 5$ in. and $BC = 5$ in.

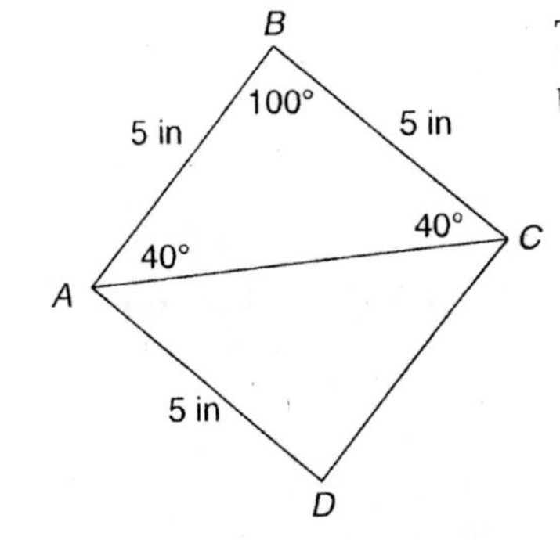

Thus ΔABC is an isosceles triangle with base angles of 40°.

Use the Law of Sines to find AC.

$$\frac{AC}{\sin 100°} = \frac{5}{\sin 40°}, \text{ so } AC = \frac{5 \sin 100°}{\sin 40°}$$
$$\approx 7.66 \text{ in.}$$

Chapter Review

Solutions to Chapter 6 Review

Section 6.1

1. 4 to 5, $\frac{4}{5}$, and 4/5.

2. $\frac{10}{75} = \frac{2(5)}{15(5)} = \frac{2}{15}$

3. $\frac{18}{32} = \frac{27}{48}$ if and only if $18(48) = 32(27)$.

$$864 = 864$$

Yes, it is a proportion.

4. Set up the proportion.

$$\frac{\text{Actual height of Eiffel Tower}}{\text{Scale height of Eiffel Tower}} = \frac{\text{Actual height of Empire State Building}}{\text{Scale height of Empire State Building}}$$

Let x = scale height of Eiffel Tower in cm.

$$\frac{986}{x} = \frac{1250}{5.6}$$

$$1250x = 986(5.6)$$

$$x = \frac{986(5.6)}{1250}$$

$$\approx 4.4 \text{ cm}$$

5. The geometric mean, c, of two positive numbers a and b is $c = \sqrt{ab}$. Hence the geometric mean is $\sqrt{6.2(24.8)} = 12.4$

6. A complete solution is given in the answer key at the back of the text.

Section 6.2

1. $\angle A \cong \angle P, \angle B \cong \angle Q, \angle C \cong \angle R$

$$\frac{BC}{QR} = \frac{AC}{PR}$$

See Definition of Similar Triangles in text.

2. (a) MNO

(b) MNO

See Reflexive, Symmetric, and Transitive relationships in Section 6.2 of the text.

Chapter Review

3. (a) If $\frac{DF}{PQ} = \frac{DE}{PR} = \frac{FE}{QR}$, then $\Delta DFE \sim \Delta PQR$ by SSS Similarity.

$$\frac{DF}{PQ} = \frac{5}{12.5} = 0.4$$
$$\frac{DE}{PR} = \frac{6.8}{17} = 0.4$$
$$\frac{FE}{QR} = \frac{3.1}{7.75} = 0.4$$

Hence $\Delta DFE \sim \Delta PQR$.

(b) $\Delta MNO \sim \Delta TUS$ by SSS Similarity.

Since the sides of ΔMNO are congruent and the sides of ΔTUS are congruent, corresponding sides are proportional.

Note: Other similarity statements are possible.

(e.g. $\Delta NOM \sim \Delta TUS$ using the same SSS Similarity argument.)

(c) Not similar as no similarity property applies.

(d) $\angle ACB \cong \angle CED$ by vertical angles. $\frac{AC}{CE} = \frac{8.3}{9.96} = 0.8\overline{3}$ and

$\frac{BC}{CD} = \frac{9.7}{11.64} = 0.8\overline{3}$ thus $\frac{AC}{CE} = \frac{BC}{CD}$.

Therefore $\Delta ABC \sim \Delta EDC$ by SAS Similarity.

4. $\overline{BD} \parallel \overline{AE}$ results in $\angle A \cong \angle B$ and $\angle E \cong \angle D$ as they are corresponding angles. Also $\angle C \cong \angle C$ by reflexive property. Thus $\Delta ACE \sim \Delta BCD$ by AAA Similarity.

So $\frac{BD}{DC} = \frac{AE}{EC}$ or $\frac{BD}{DC} = \frac{AE}{ED + DC}$

That is, $\frac{9.5}{15} = \frac{14.25}{ED + 15}$

and
$$9.5(ED + 15) = 15(14.25)$$
$$9.5(ED) + 142.5 = 213.75$$
$$9.5(ED) = 71.25$$
$$ED = \frac{71.25}{9.5}$$

Thus, $ED = 7.5$ cm.

5. Since $\overline{AB} \parallel \overline{DE}$, we have $\angle A \cong \angle E$ and $\angle B \cong \angle D$ (alternate interior angles).

$\Delta ABC \sim \Delta EDC$ by AA Similarity. Thus, $\frac{AB}{DE} = \frac{BC}{DC}$.

Substituting values, $\frac{AB}{42.5} = \frac{65}{20}$

$$AB = 42.5\left(\frac{65}{20}\right)$$
$$= 138.125 \text{ yd.}$$

Chapter Review **Section 6.3**

1. In right triangle ABC, $\overline{BD}$ is the altitude to $\overline{AC}$, thus by Theorem 6.9 $\frac{DC}{BD} = \frac{BD}{AD}$. Substitute values to solve part *a* and part *b*.

   (a) $\frac{11}{BD} = \frac{BD}{9}$

   $$\begin{aligned}(BD)^2 &= 99\\ BD &= \sqrt{99}\\ &= 3\sqrt{11}\\ &\approx 9.95\end{aligned}$$

   Use the Pythagorean Theorem to find BC.

   $$\begin{aligned}(BC)^2 &= (BD)^2 + (DC)^2\\ (BC)^2 &= (\sqrt{99})^2 + 11^2\\ &= 99 + 121\\ &= \sqrt{220}\\ &= 2\sqrt{55}\\ &\approx 14.83\end{aligned}$$

   (b) $\frac{DC}{4} = \frac{4}{2.5}$, $2.5(DC) = 16$

   $$DC = 6.4$$

   Use the Pythagorean Theorem to find AB.

   $$\begin{aligned}(AB)^2 &= (BD)^2 + (AD)^2\\ &= 4^2 + (2.5)^2\\ &= 16 + 6.25\\ AB &= \sqrt{22.25}\\ &\approx 4.72\end{aligned}$$

2. Since $\overline{AB} \parallel \overline{CE}$, $\Delta ABD \sim \Delta ECD$ by AAA Similarity.

   (a) $\frac{CP}{BC} = \frac{DE}{AE}$ by Side Splitting Theorem 6.10.

   Substituting values $\frac{7}{2.5} = \frac{DE}{4}$

   $$\begin{aligned}2.5(DE) &= 28\\ DE &= \frac{28}{2.5}\\ &= 11.2.\end{aligned}$$

Chapter Review

2. (continued)

(b) $\dfrac{ED}{AE} = \dfrac{CD}{BC}$ by Side Splitting Theorem 6.10.

Substituting values $\dfrac{18}{14} = \dfrac{16 - BC}{BC}$

$$18(BC) = 14(16 - BC)$$
$$18(BC) = 224 - 14(BC)$$
$$32(BC) = 224$$
$$BC = 7.$$

3. Since $\overline{CE} \parallel \overline{BF} \parallel \overline{AG}$ we have $\Delta ADG \sim \Delta BDF \sim \Delta CDE$,

and $\dfrac{CD}{CE} = \dfrac{BD}{BF}$ by Side Splitting Theorem 6.10.

So, $$\frac{x+4}{9.5} = \frac{x+4+7}{19}$$
$$19(x+4) = 9.5(x+11)$$
$$19x + 76 = 9.5x + 104.5$$
$$9.5x = 28.5$$
$$x = 3.$$

Also $\dfrac{CD}{CE} = \dfrac{AD}{AG}$ by Side Splitting Theorem 6.10.

So, $$\frac{x+4}{9.5} = \frac{x+4+7+9x+1}{AG}$$
$$\frac{3+4}{9.5} = \frac{3+4+7+9(3)+1}{AG}$$
$$\frac{7}{9.5} = \frac{42}{AG}$$
$$7(AG) = 42(9.5)$$
$$AG = \frac{42(9.5)}{7}$$
$$= 57.$$

4. Use the Pythagorean Theorem to find AC and BD.
$AC = BD = \sqrt{10^2 + 5^2} = \sqrt{125} = 5\sqrt{5}$

By the Midsegment Theorem $MN = PO = ON = PM = \dfrac{5\sqrt{5}}{2}$.

The perimeter of $MNOP = 4\left(\dfrac{5\sqrt{5}}{2}\right) = 10\sqrt{5} \approx 22.36$ cm.

Chapter Review

4. (continued)

The area of $MNOP$ is the sum of the areas of the two congruent triangles ΔMNO and ΔMPO.

$$\text{Area} = \frac{1}{2}5(5) + \frac{1}{2}5(5) = 25 \text{ cm}^2.$$

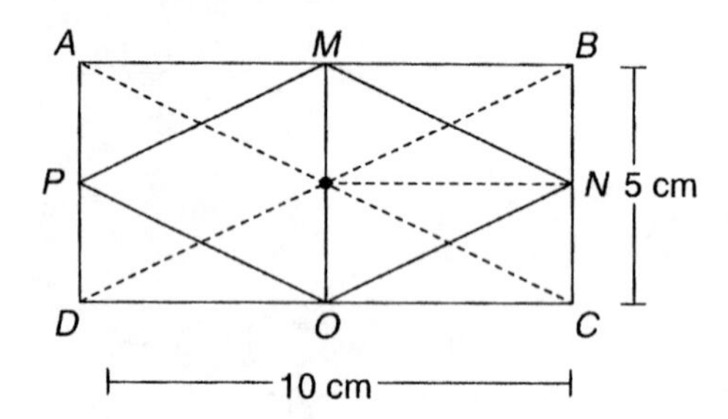

5. A complete solution is given in the answer key at the back of the text.

6. A complete solution is given in the answer key at the back of the text.

Section 6.4

1. Use the Pythagorean Theorem to find side c.

$$\begin{aligned} c^2 &= 17^2 + 21^2 \\ &= 289 + 441 \\ &= 730 \\ c &= \sqrt{730} \\ &\approx 27 \text{ in.} \end{aligned}$$

Since $c \approx 27$ is an approximate value, it is best to find $\angle A$ and $\angle B$ using the given exact values.

$$\tan A = \frac{21}{17}$$

Thus, $\angle A = \tan^{-1}\left(\frac{21}{17}\right) \approx 51°$.

$$\tan B = \frac{17}{21}$$

Thus, $\angle B = \tan^{-1}\left(\frac{17}{21}\right) \approx 39°$.

2. (a) First find a using the Pythagorean Theorem.

$$\begin{aligned} a^2 + 2^2 &= 5^2 \\ a^2 &= 25 - 4 \\ a^2 &= 21 \\ a &= \sqrt{21} \end{aligned}$$

Since the tangent ratio is $\frac{\text{opposite}}{\text{adjacent}}$, $\tan A = \frac{\sqrt{21}}{2}$.

Chapter Review

2. (continued)

(b) The sine ratio is $\frac{\text{opposite}}{\text{hypotenuse}}$. Thus, $\sin 22° = \frac{16.3}{c}$.

$$c \sin 22° = 16.3$$

$$c = \frac{16.3}{\sin 22°}$$

$$\approx 43.51$$

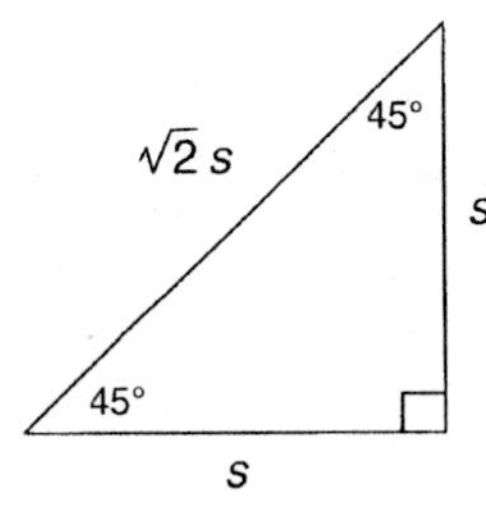

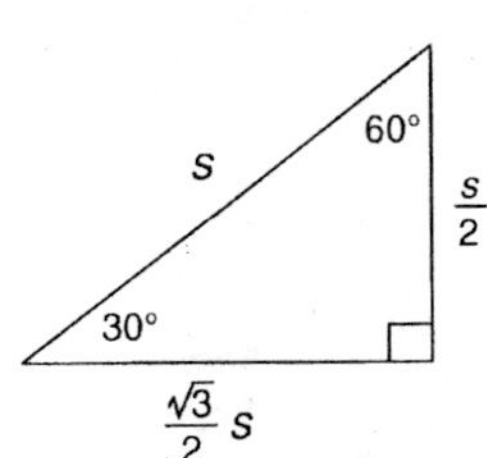

3. The 45°–45° right triangle and the 30°–60° right triangle are given here.

(a) $\sin 45° = \frac{\text{opposite}}{\text{hypotenuse}} = \frac{s}{\sqrt{2}s} = \frac{1}{\sqrt{2}} = \frac{\sqrt{2}}{2}$

(b) $\cos 30° = \frac{\text{adjacent}}{\text{hypotenuse}} = \frac{\frac{\sqrt{3}}{2}s}{s} = \frac{\sqrt{3}}{2}$

(c) $\tan 60° = \frac{\text{opposite}}{\text{adjacent}} = \frac{\frac{\sqrt{3}}{2}s}{\frac{s}{2}} = \sqrt{3}$

4. First find the height of ΔABC. Let $\overline{BD}$ be the altitude to base $\overline{AD}$ in ΔABD. Then $\overline{BD}$ is also the height of ΔABC. Since ΔABD is a right triangle, use the sine ratio to find BD.

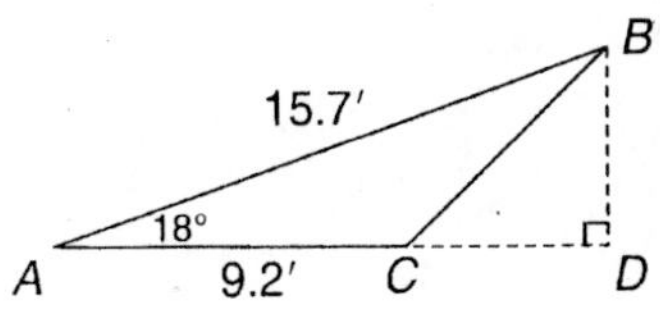

$$\sin 18° = \frac{BD}{15.7}$$

$$BD = 15.7 \sin 18°$$

Thus, $\text{Area}_{\Delta ABC} = \frac{1}{2}bh$

$$= \frac{1}{2}9.2(15.7 \sin 18°)$$

$$\approx 22.32 \text{ ft}^2.$$

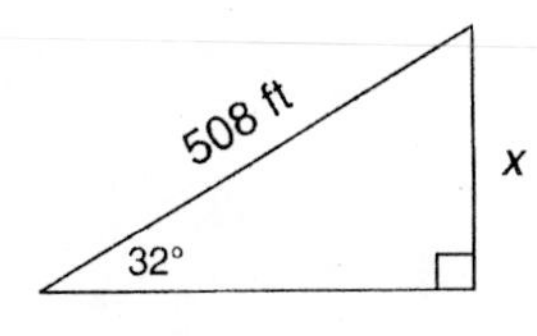

5. Let x = vertical distance from bottom to top of escalator in feet.

Use the sine ratio.

$$\sin 32° = \frac{x}{508}$$

$$x = 508 \sin 32°$$

$$\approx 269.2 \text{ ft}$$

Chapter Review **Section 6.5**

1. Law of Sines: $\frac{\sin A}{a} = \frac{\sin B}{b} = \frac{\sin C}{c}$

First find side c.

$$\frac{\sin 38°}{18} = \frac{\sin 54°}{c}$$

$$c = \frac{18 \sin 54°}{\sin 38°}$$

$$\approx 23.65 \text{ cm.}$$

Now find $\angle B$.

$$\angle B = 180° - 58° - 34°$$

$$= 88°$$

Finally find side b.

$$\frac{\sin 38°}{18} = \frac{\sin 88°}{b}$$

$$b = \frac{18 \sin 88°}{\sin 38°}$$

$$\approx 29.22 \text{ cm}$$

2. Law of Cosines: $a^2 = b^2 + c^2 - 2bc \cos A$

$$a^2 = (13.54)^2 + (15.32)^2 - 2(13.54)(15.32) \cos 73.2°$$

$$= 183.3316 + 234.7024 - 414.8656 \cos 73.2°$$

$$a = \sqrt{183.3316 + 234.7024 - 414.8656 \cos 73.2°}$$

$$= \sqrt{418.034 - 414.8656 \cos 73.2°}$$

$$\approx 17.27 \text{ mm}$$

Use the Law of Sines to find $\angle B$.

$$\frac{\sin \text{B}}{13.54} = \frac{\sin 73.2}{a}$$

To obtain the most accurate results use the exact value of a instead of the approximate value of $a = 17.27$ mm

$$\sin B = 13.54 \frac{\sin 73.2}{a}$$

Thus, $\angle B = \sin^{-1}\left[\frac{13.54 \sin 73.2}{\sqrt{418.034 - 414.8656 \cos 73.2°}}\right] \approx 48.65°$

Now $\angle C \approx 180° - 73.2° - 48.65°$

$\approx 58.15°.$

Chapter Review

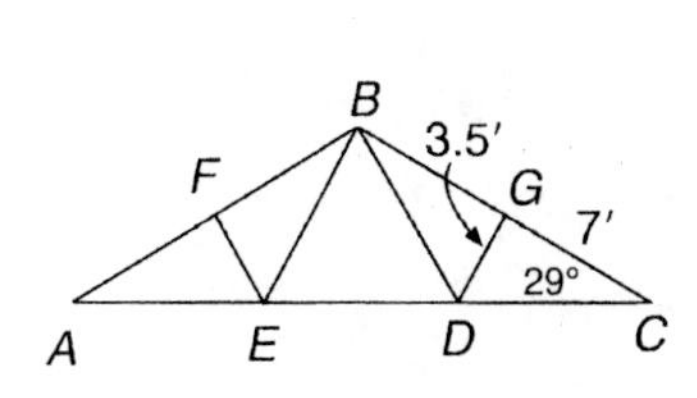

3. Use the Law of Sines to find $\angle GDC$.

$$\frac{\sin 29^\circ}{3.5} = \frac{\sin \angle GDC}{7}$$

$$\sin \angle GDC = \frac{7 \sin 29^\circ}{3.5}$$

$$\angle GDC = \sin^{-1}[2 \sin 29^\circ]$$

$$\approx 75.8^\circ$$

Thus $\angle DGC \approx 180^\circ - 29^\circ - 75.8^\circ$

$$\approx 75.2^\circ.$$

Again use the Law of Sines; this time to find DC.

$$\frac{3.5}{\sin 29^\circ} = \frac{DC}{\sin 75.2^\circ}$$

$$DC = \frac{3.5 \sin 75.2}{\sin 29^\circ}$$

$$\approx 7.0 \text{ feet}$$

4. Use the Law of Cosines to find $\angle A$, $\angle B$, and $\angle C$.

If at all possible use exact values, so as not to compound round off error.

Find $\angle A$.

$$a^2 = b^2 + c^2 - 2bc \cos A$$

$$19^2 = 25^2 + 22^2 - 2(25)(22) \cos A$$

B
a = 19
c = 22
C
b = 25
A

$$1100 \cos A = 748$$

$$\cos A = \frac{748}{1100}$$

Thus, $\angle A = \cos^{-1}\left[\frac{748}{1100}\right] \approx 47.2^\circ$.

Find $\angle B$.

$$b^2 = a^2 + c^2 - 2ac \cos B$$

$$25^2 = 19^2 + 22^2 - 2(19)(22) \cos B$$

$$836 \cos B = 220$$

$$\cos B = \frac{220}{836}$$

Thus, $\angle B = \cos^{-1}\left[\frac{220}{836}\right] \approx 74.7^\circ$.

Chapter Test

4. (continued)

Find $\angle C$.

$$c^2 = a^2 + b^2 - 2ab \cos C$$
$$22^2 = 19^2 + 25^2 - 2(19)(25) \cos C$$
$$950 \cos C = 19^2 + 25^2 - 22^2$$
$$\cos C = \frac{502}{950}$$

Thus, $\angle C = \cos^{-1}\left[\frac{502}{950}\right] \approx 58.1°$.

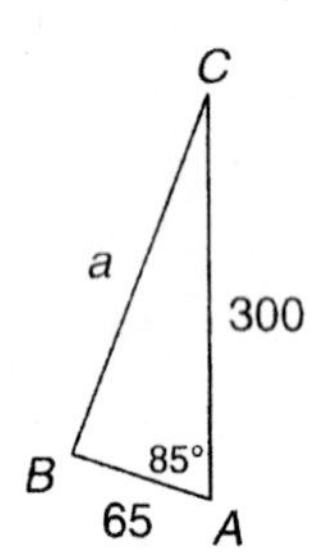

5. Use the Law of Cosines to find the resultant speed.

$$a^2 = 300^2 + 65^2 - 2(65)(300) \cos 85°$$
$$= 90{,}000 + 4225 - 39{,}000 \cos 85°$$
$$= 94{,}225 - 39{,}000 \cos 85°$$
$$a = \sqrt{94{,}225 - 39{,}000 \cos 85°}$$
$$\approx 301.4 \text{ mph}$$

Next use the Law of Sines with the exact value of a to find the direction of the plane.

$$\frac{\sin c}{65} = \frac{\sin 85°}{a}$$

$$\sin C = \frac{65 \sin 85°}{a}$$

Thus $\angle C = \sin^{-1}\left[\frac{65 \sin 85°}{\sqrt{94{,}225 - 39{,}000 \cos 85°}}\right] \approx 12.4°$.

The resultant speed is 301.4 mph and the direction of the plane is S 12.4° W.

Solutions to Chapter 6 Test

1. F $\quad \frac{a}{b} = \frac{c}{d}$ yields $ad = bc$, but $\frac{a}{c} = \frac{d}{b}$ gives $ab = cd$ instead. In order for the ratios to be equal, the cross products must be equal.

2. T $\quad$ Cross-Multiplication Theorem (Theorem 6.1).

3. F $\quad$ If the angles between the two sides are different, the third sides will not be proportional and the triangles will not be similar.

Chapter Test

4. F The midsegment is *parallel* to one side of the triangle and will form with the other sides angles congruent to the base angles, which may or may not be 90°.

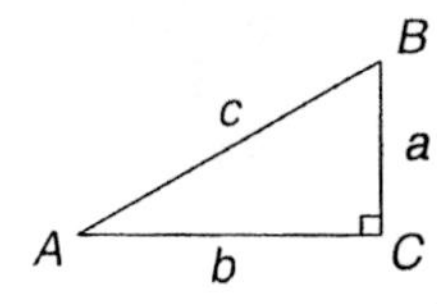

5. T If $\angle A < 45°$, then side a will be shorter than side b and $\frac{a}{c} < \frac{b}{c}$.

6. T They are 45°–45° right triangles, so they are similar by AA Similarity.

7. F Rectangles and squares both have four 90° angles, but not all rectangles are similar to a square.

8. T Each of the smaller triangles contains a 90° angle and one of the angles of the original triangle, so it is similar to the original triangle by AA Similarity.

9. T See Section 6.2, problem #13.

10. F $\cos A = \frac{\text{adjacent}}{\text{hypotenuse}}$ and $\sin A = \frac{\text{opposite}}{\text{hypotenuse}}$

Now, $\csc A = \frac{\text{hypotenuse}}{\text{opposite}}$, so the cosecant ratio is the reciprocal of the sine ratio. (See problem #51 of 6.4.)

11. F The Law of Cosines must be used in that situation.

12. T Use the Law of Cosines to find the exact measures of the vertex angles.

13. $\frac{x}{24} = \frac{5}{16}$, so $16x = 120$ and $x = 7.5$.

14. $\frac{7}{x} = \frac{x}{13}$ so $x^2 = 91$ and $x = \sqrt{91} \approx 9.54$.

15. $\Delta ABC \sim \Delta DEF$, $AB = 5$, $BC = 9$, and $DE = 35$. Then $\frac{AB}{DE} = \frac{BC}{EF}$, so $\frac{5}{35} = \frac{9}{EF}$, or $5EF = 315$ and $EF = 63$.

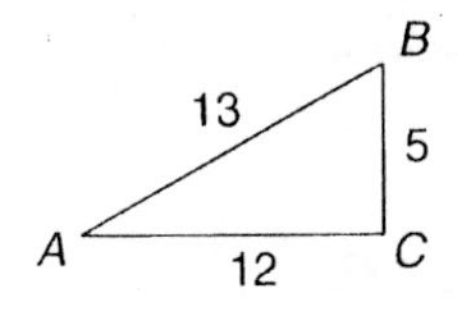

16. (a) Since 13 is the length of the longest side, it is the hypotenuse, and 5 and 12 are the lengths of the legs. Since $5^2 + 12^2 = 169 = 13^2$ satisfies the Pythagorean Theorem, the triangle is a right triangle.

To construct the triangle, make a base 12 units long. With a radius of 13 units, make an arc from one side of the base. With a radius of 5 units make an arc from the other side of the base. Join the ends of the base to the × formed by the two arcs.

Chapter Test

16. (continued)

(b) Using the triangle shown, $\sin A = \frac{5}{13}$, $\cos A = \frac{12}{13}$, $\sin B = \frac{12}{13}$, and $\cos B = \frac{5}{13}$.

(c) $$\sin^2 A + \cos^2 A = \left(\frac{5}{13}\right)^2 + \left(\frac{12}{13}\right)^2 = \frac{25}{169} + \frac{144}{169} = \frac{169}{169} = 1$$

$$\sin^2 B + \cos^2 B = \left(\frac{12}{13}\right)^2 + \left(\frac{5}{13}\right)^2 = \frac{144}{169} + \frac{25}{169} = \frac{169}{169} = 1$$

17. Since AC is a diagonal of a square, it forms a 45°–45° right triangle with two sides of square $ABCD$. If one side of square $ABCD$ is s, then $s^2 + s^2 = 20^2$, so $2s^2 = 400$ or $s^2 = 200$, and $s = \sqrt{200} = 10\sqrt{2}$.

Now, $GD = HD = \frac{1}{2}s = 5\sqrt{2}$.

ΔGDH is an isosceles right triangle, so its hypotenuse is $\sqrt{2}$ times the length of one of its legs. Thus, the hypotenuse $GH = (5\sqrt{2})(\sqrt{2}) = 5 \cdot 2 = 10$.

Similarly, you can show that $FG = 10$, $EF = 10$ and $EH = 10$. Thus, the perimeter of $EFGH$ is $10 + 10 + 10 + 10 = 40$ cm.

18. Let h = hypotenuse of right triangle. Then $\sin 48° = \frac{10}{h}$, so $h(\sin 48°) = 10$, or $h = \frac{10}{\sin 48°} \approx 13.5$.

19. By the Side Splitting Theorem (Theorem 6.10), you have $\Delta ABC \sim \Delta ADE$. Thus, $\frac{EA}{CA} = \frac{DE}{BC}$ or $\frac{9}{x+9} = \frac{15}{25}$, and $15(x + 9) = 225$. Then

$15x + 135 = 225$, or $15x = 90$, so $x = 6$.

20. A complete solution is given in the answer key at the back of the text.

21. Form a right triangle by extending $\overline{AD}$ and drawing a perpendicular from B to $\overline{AD}$. Then BE is the height of the parallelogram.

$\angle EAB = 180° - 109° = 71°$

$\sin 71° = \frac{BE}{22.4}$, so $BE = 22.4 \sin 71° = 21.18$ m.

Then the area of the parallelogram is given by

$A = bh = (48.3)(21.18) \approx 1023 \text{ m}^2$.

Chapter Test

22. Use the Law of Cosines.

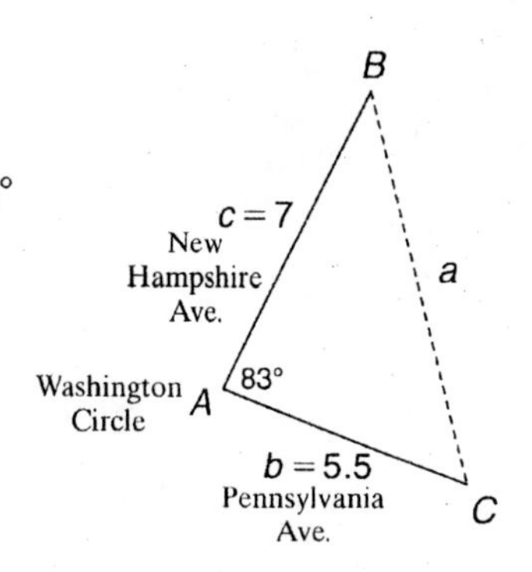

$$a^2 = b^2 + c^2 - 2bc \cos 83°$$
$$a^2 = (5.5)^2 + 7^2 - 2(5.5)\,7 \cos 83°$$
$$= 30.25 + 49 - 77 \cos 83°$$
$$= 79.25 - 77 \cos 83°$$
$$a = \sqrt{79.25 - 77 \cos 83°}$$
$$\approx 8.4 \text{ blocks}$$

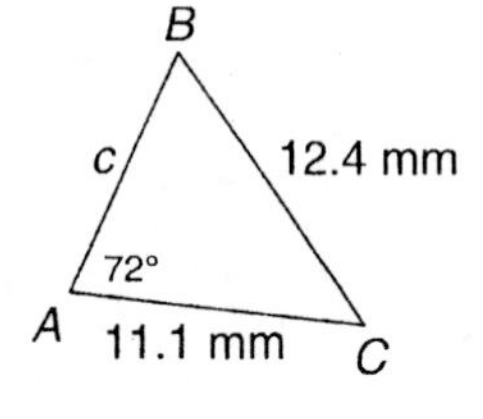

23. Use the Law of Sines to find $\angle A$.

$$\frac{\sin A}{a} = \frac{\sin B}{b}$$
$$\frac{\sin 72°}{12.4} = \frac{\sin B}{11.1}$$
$$\sin B = \frac{11.1 \sin 72°}{12.4}$$

Thus, $\angle B = \sin^{-1}\left[\frac{11.1 \sin 72°}{12.4}\right] \approx 58.4°$.

Next find $\angle C$.

$$\angle C \approx 180° - 72° - 58.4°$$
$$\approx 49.6°$$

Finally find $\angle C$ using the Law of Sines.

$$\frac{c}{\sin C} = \frac{a}{\sin A}$$
$$\frac{c}{\sin 49.6°} = \frac{12.4}{\sin 72°}$$
$$c = \frac{12.4 \sin 49.6°}{\sin 72°}$$
$$\approx 9.9 \text{ mm}$$

24. Use the Pythagorean Theorem to find a.

$a^2 + 55^2 = 73^2$, so $a^2 + 3025 = 5329$ and $a^2 = 2304$, or $a = 48$.

Then $\cos A = \frac{b}{c} = \frac{55}{73}$, so $\angle A = \cos^{-1}\left(\frac{55}{73}\right) \approx 41.1°$ and

$\angle B \approx 180° - 90° - 41.1° = 48.9°$.

25. A complete solution is given in the answer key at the back of the text.

26. A complete solution is given in the answer key at the back of the text.

Chapter Test

27. Let d be the horizontal distance of the observer from the whale. Then $\tan 16.2° = \frac{70}{d}$, or $d\,(\tan 16.2°) = 70$, and thus, $d = \frac{70}{\tan 16.2°} \approx 241$ feet.

28. It may be helpful to draw a perpendicular from point B to $\overline{ED}$ (See diagram).

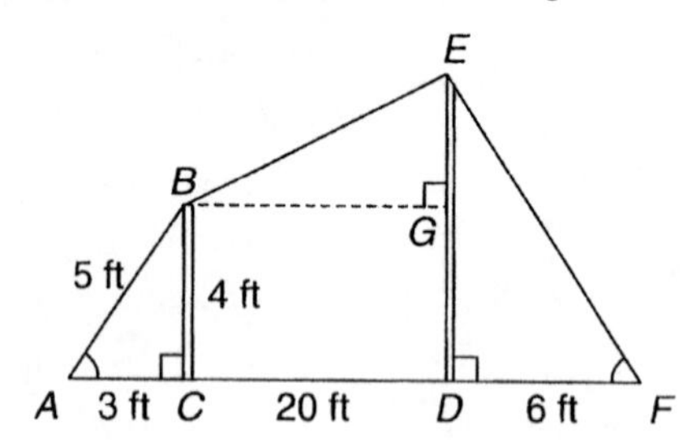

By AA Similarity, $\Delta ABC \sim \Delta FED$, so $\frac{3}{6} = \frac{4}{ED}$, and $3ED = 24$, so $ED = 8$ feet. Similarly, $\frac{3}{6} = \frac{5}{EF}$, so $3EF = 30$ and $EF = 10$ feet.

To find BE, note that $GE = ED - 4 = 8 - 4 = 4$ feet.

Applying the Pythagorean Theorem to ΔBEG, you have

$$(BE)^2 = (BG)^2 + (GE)^2 = 20^2 + 4^2 = 400 + 16 = 416,$$

so $BE = \sqrt{416}$ feet, and the total length of the wire is
$5 + \sqrt{416} + 10 \approx 35.4$ feet.

29. Use the Law of Sines to find $\angle P$.

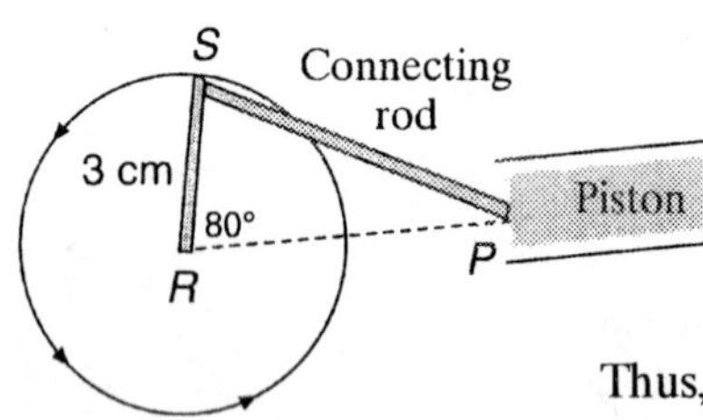

$$\frac{\sin 80°}{9} = \frac{\sin P}{3}$$

$$\sin P = \frac{3 \sin 80°}{9}$$

Thus, $\angle P = \sin^{-1}\left[\frac{3 \sin 80°}{9}\right] \approx 19.16°$.

Next, find $\angle S$.

$$\angle S \approx 180° - 80° - 19.16°$$
$$\approx 80.84°$$

Finally find RP using the Law of Sines.

$$\frac{9}{\sin 80°} = \frac{RP}{\sin 80.84°}$$

$$RP = \frac{9 \sin 80.84°}{\sin 80°}$$
$$\approx 9.02 \text{ cm}$$

7

Circles

Section 7.1

Tips:

✔ A central angle has its vertex at the center of a circle and an inscribed angle has its vertex on the circle.

✔ Be careful not to *assume* an angle is a central angle. For example, see $\angle BED$ in the diagram for problems 19–30.

✔ The measure of an inscribed angle is only *half* the measure of its intercepted arc.

Solutions to odd-numbered textbook problems

1. An arc of a circle that is shorter than a semicircle is a minor arc. Five minor arcs are $\overparen{AB}$, $\overparen{AD}$, $\overparen{BC}$, $\overparen{BD}$ and $\overparen{CD}$.

3. $\overparen{AB} + \overparen{BC} = \overparen{ABC}$

5. $\angle BOC$ is a central angle since O is the center of the circle and $\overline{OB}$ and $\overline{OC}$ are radii.

7. $\angle COD$ intercepts $\overparen{CD}$ since the sides $\overline{OC}$ and $\overline{OD}$ of $\angle COD$ intersect circle O at points C and D.

9. $\angle AOD = \angle AOC - \angle COD$

   $= 180° - 43°$ $\overparen{CD} = \angle COD$ by definition of Measure of an Arc

   $= 137°$

11. $\overparen{BC} = \angle BOC$ Definition of Measure of an Arc

    $= 65°$

13. $\overparen{ADB} = \overparen{ADC} + \overparen{CB}$

    $= \angle AOC + \angle COB$ Definition of Measure of an Arc

    $= 180° + 65°$

    $= 245°$

Section 7.1

15. $\angle AOC = 180°$

So $\overset{\frown}{AC} = \dfrac{\angle AOC}{360°}(2\pi r)$ — Postulate 7.1 Length of an Arc

$= \dfrac{180°}{360°}(2\pi \cdot 8)$ — substituting values

$= \dfrac{1}{2}(16\pi)$

$= 8\pi$

≈ 25.13 cm

$\overset{\frown}{DC} = \dfrac{\angle DOC}{360°}(2\pi r)$ — Postulate 7.1 Length of an Arc

$= \dfrac{25°}{360°}(2\pi \cdot 8)$ — substituting values

$= \dfrac{10\pi}{9}$

≈ 3.49 cm

17. First find $\overset{\frown}{AB}$.

$\overset{\frown}{AB} = \angle AOB$

$= \angle AOC - \angle BOC$

$= 180° - 115°$

$= 65°$

Now use Postulate 7.2 to find Area of Sector AOB.

$A_{\text{Sector } AOB} = \dfrac{\angle AOB}{360°}(\pi r^2)$

$= \dfrac{65°}{360°}(\pi \cdot 8^2)$ — substituting values

$= \dfrac{104}{9}\pi$

$\approx 36.30 \text{ cm}^2$

19. $\angle BCD$ and $\angle BAD$ are inscribed angles that intercept $\overset{\frown}{BD}$.

21. $\angle CEB$ intercepts $\overset{\frown}{CB}$. $\angle CEB$ is not an inscribed angle as vertex E is not on the circle. $\overset{\frown}{AEB} = \overset{\frown}{AC} + \overset{\frown}{CB} = 70° + 100° = 170°$. Since $\overset{\frown}{AEB} \neq 180°$ $\overline{AB}$ is not a diameter of the circle and E is not the center of the circle. Thus $\angle CEB$ is not a central angle.

Therefore, $\angle CEB$ is neither an inscribed angle nor a central angle.

23. $\angle C$ is an inscribed angle that intercepts $\overset{\frown}{BD}$, and thus, $\angle C = \dfrac{1}{2}(80°) = 40°$.

Section 7.1

25. $\angle B$ is an inscribed angle that intercepts $\overset{\frown}{AC}$, and thus,

$$\angle B = \frac{1}{2}(70°) = 35°.$$

27. $\angle C = 40°$ See problem #23.

$\angle B = 35°$ See problem #25.

So $\angle CEB = 180° - 35° - 40°$

$= 105°.$

29. $\overset{\frown}{AD} = 360° - 70° - 100° - 80° = 110°.$

31. $\angle A = \angle D = 30°$ since they are alternate interior angles.

$\angle B = \angle D = 30°$ since they both intercept $\overset{\frown}{AC}$.

$\angle C = \angle B = 30°$ since they are alternate interior angles.

Therefore, $\angle A = \angle B = \angle C = \angle D = 30°$.

33. $\angle A = \angle B = 30°$ see problem #31.

Therefore, $\angle AEB = 180° - 30° - 30° = 120°$ by angle sum of a triangle.

35. We cannot necessarily conclude that $\overset{\frown}{AB} = 120°$ because we do not know that E is the center of the circle.

37. $\angle BAC$ is an inscribed angle that intercepts $\overset{\frown}{BC}$. Thus,

$$\angle BAC = 40° = \frac{1}{2}\overset{\frown}{BC}, \text{ so } \overset{\frown}{BC} = 80° = \angle BOC.$$

39. $\angle B = 90°$ as $\angle B$ is inscribed in a semicircle (Corollary7.3).

Thus ΔABC is a right triangle.

$(AC)^2 = (AB)^2 + (BC)^2$ Pythagorean Theorem

$(AC)^2 = 36^2 + 15^2$ substituting values

$= 1521$

So $AC = 39.$

Hence, radius $OC = \frac{1}{2}(AC)$

$= \frac{1}{2}(39)$

$= 19.5$ in.

41. $\angle B$ is an inscribed angle that intercepts $\overset{\frown}{ADC}$. Thus you have

$$\overset{\frown}{ADC} = 360° - \overset{\frown}{AB} - \overset{\frown}{BC}$$

$$= 360° - 110° - 80° = 170°.$$

Therefore, $\angle B = \frac{1}{2}\overset{\frown}{ADC} = \frac{1}{2}(170°) = 85°.$

Section 7.1

43. $\angle Q = \angle S = 90°$ as $\angle Q$ and $\angle S$ are each inscribed in a semicircle (Corollary 7.3).

Hence ΔPQR and ΔPSR are right triangles.

$(PR)^2 = (PQ)^2 + (QR)^2$ Pythagorean Theorem

$(PR)^2 = 8^2 + 14^2$ substituting values

$= 260$

$PR = \sqrt{260}$

$(PS)^2 = (PR)^2 - (RS)^2$ PythagoreanTheorem

$= 260 - 3^2$ substituting values

$= 251$

Therefore, $PS = \sqrt{251} \approx 15.84$ in.

45. You have $\overset{\frown}{AE} = 120°$, so $\angle AOE = 120°$. Then

$\angle BOA = 180° - 120° = 60°$, so $\overset{\frown}{AB} = 60°$.

$\angle DOE = \angle BOA = 60°$, so $\overset{\frown}{DE} = 60°$.

$\angle DOB = 120°$, so $\overset{\frown}{BCD} = 120°$.

$\overset{\frown}{BC} = \overset{\frown}{BCD} - \overset{\frown}{CD} = 120° - 42° = 78°$

Now you can find the measures of the angles of the pentagon.

$$\angle BCD = \frac{1}{2}(\overset{\frown}{AB} + \overset{\frown}{AE} + \overset{\frown}{DE})$$

$$= \frac{1}{2}(60° + 120° + 60°) = 120°$$

$$\angle ABC = \frac{1}{2}(\overset{\frown}{CD} + \overset{\frown}{DE} + \overset{\frown}{AE})$$

$$= \frac{1}{2}(42° + 60° + 120°) = 111°$$

$\angle DEA = \angle EAB = 90°$ because they are angles inscribed in semicircles.

$$\angle CDE = \frac{1}{2}(\overset{\frown}{AE} + \overset{\frown}{AB} + \overset{\frown}{BC})$$

$$= \frac{1}{2}(120° + 60° + 78°) = 129°$$

47. A complete solution is given in the answer key at the back of the text.

49. A complete solution is given in the answer key at the back of the text.

51. $\overset{\frown}{AC} = \angle AOC$ by Measure of an Arc.

ΔAOB is isosceles as $OA = OB$.

Thus $\angle OAB = \angle ABC$.

51. (continued)

Now $\angle AOC = \angle OAB + \angle ABC$	Exterior Angle Theorem
$= 2\angle ABC$	substitution
Therefore, $\angle ABC = \frac{1}{2}\angle AOC$	Inscribed Angle Theorem
$= \frac{1}{2}\overset{\frown}{AC}.$	substitution

53. A complete solution is given in the answer key at the back of the text.

55. The pendulum swings through an arc of 8°, which is part of a circle. The radius of the circle is 40 inches, the length of the pendulum. The length of the arc traced out by the pendulum is

$$l = \frac{8°}{360°} \cdot 2\pi(40) = \frac{640\pi}{360} = \frac{16\pi}{9} \approx 5.59 \text{ inches.}$$

57. To find the volume of the wedge, find the area of the sector and multiply by the thickness (height) of the wedge.

$$V = \frac{14°}{360°}(\pi \cdot 11^2)4$$

$$= \frac{847\pi}{45}$$

$$\approx 59.13 \text{ inches}^3$$

59. A complete solution is given in the answer key at the back of the text.

Section 7.2

Solutions to odd-numbered textbook problems

1. $\overline{AB}$ and $\overline{CD}$ are chords of a circle with perpendicular bisectors $\overline{GH}$ and $\overline{EF}$ respectively. Thus both $\overline{GH}$ and $\overline{EF}$ contain the center of the circle which implies $\overline{GH}$ and $\overline{EF}$ are diameters of the circle.

   Therefore $GH = EF$. $\overset{\frown}{FGE} = 180°$ as it is the arc of a semicircle.

3. $\angle BEC$ is formed by two intersecting chords. Thus,

   $$\angle BEC = \frac{1}{2}(\overset{\frown}{AD} + \overset{\frown}{BC}) = \frac{1}{2}(125° + 173°) = 149° \text{ (Theorem 7.7).}$$

Section 7.2

5. $\angle CED = \frac{1}{2}(\overset{\frown}{AB} + \overset{\frown}{CD})$ Theorem 7.7

$= \frac{1}{2}(2\angle BDA + 2\angle DAC)$ Theorem 7.1 Inscribed Angle Theorem.

$= \angle BDA + \angle DAC$

$= 40° + 30°$ substituting values

$= 70°.$

7. By Theorem 7.8, $(AF)(FD) = (BF)(EF)$, so $(AF) \cdot 15 = 7(18)$, or $15AF = 126$, and $AF = \frac{126}{15} = 8.4$ in.

9. $(BF)(FE) = (AF)(FD)$ Theorem 7.8

$(BF)(FE) = (5.2)(12)$ substituting values

$BF[3.9(BF)] = 62.4$ substitution

$(BF)^2 = 16$

$BF = 4$ cm

11. $(DE)(EB) = (CE)(EA)$ Theorem 7.8

$(DE)(5) = (2)(4)$ substituting values

$DE = \frac{8}{5}$ cm

$= 1.6$ cm

13. $(DE)(EB) = (AE)(CE)$ Theorem 7.8

$3.5[2(CE)] = (12 - CE)(CE)$

$7(CE) = 12(CE) - (CE)^2$

$(CE)^2 - 5(CE) = 0$

$(CE)(CE - 5) = 0$

$CE = 0$ or $CE = 5$

But $CE = 0$ is not possible as $\overline{CE}$ is in the interior of the circle. Therefore, $CE = 5$ in.

15. $\angle AFE = \frac{1}{2}(\overset{\frown}{AE} + \overset{\frown}{BD})$ Theorem 7.7

$= \frac{1}{2}(\overset{\frown}{AE} + \overset{\frown}{BC} + \overset{\frown}{CD})$ $\overset{\frown}{BD} = \overset{\frown}{BC} + \overset{\frown}{CD}$

$= \frac{1}{2}(95° + 84° + 30°)$ $\overset{\frown}{BC} = 84°$ by Inscribed Angle Theorem 7.1

$= \frac{1}{2}(209°)$

$= 104.5°$

Section 7.2

17. First, find $\overset{\frown}{CD}$.

$$\angle CGD = \frac{1}{2}(\overset{\frown}{AE} + \overset{\frown}{CD}) \quad \text{Theorem 7.7}$$

$$62° = \frac{1}{2}(83° + \overset{\frown}{CD}) \quad \text{substituting values}$$

$$124° = 83° + \overset{\frown}{CD}$$

$$41° = \overset{\frown}{CD}$$

Next, find $\overset{\frown}{BC}$.

$$\angle FGE = \angle CGD = 62° \quad \text{vertical angles}$$

$$\angle BFG = \angle FGE + \angle BEC \quad \text{Exterior Angle Theorem}$$

$$123° = 62° + \angle BEC \quad \text{substituting values}$$

$$\angle BEC = 61°$$

Thus

$$\overset{\frown}{BC} = 2\angle BEC \quad \text{Inscribed Angle Theorem}$$

$$= 2 \cdot 61° \quad \text{substituting values}$$

$$= 122°.$$

Finally, find $\overset{\frown}{ED}$.

$$\angle DFE = 180° - \angle FGE - \angle BEC \quad \text{Angle sum of triangle}$$

$$= 180° - 62° - 61° \quad \text{substituting values}$$

$$= 57°$$

Now,

$$\angle DFE = \frac{1}{2}(\overset{\frown}{AB} + \overset{\frown}{ED}) \quad \text{Theorem 7.7}$$

$$= \frac{1}{2}[\overset{\frown}{AB} + 5(\overset{\frown}{AB})].$$

$$57° = \frac{1}{2}[6(\overset{\frown}{AB})] \quad \text{substituting values}$$

$$57° = 3(\overset{\frown}{AB})$$

$$95° = 5(\overset{\frown}{AB}) \quad \text{multiplying both sides by } \frac{5}{3}$$

Therefore, $95° = \overset{\frown}{ED}$.

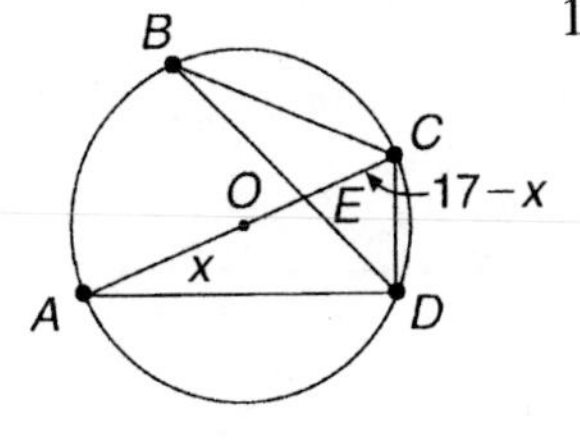

19. Since $\angle ADC$ is inscribed in a semicircle, it is a right angle and ΔADC is a right triangle. Using the Pythagorean Theorem to find AC, you get $15^2 + 8^2 = (AC)^2$, or $289 = (AC)^2$, and thus, $AC = 17$ cm.

Now let $AE = x$ and $CE = 17 - x$.

$$(AE)(CE) = (BE)(DE) \quad \text{Theorem 7.8}$$

$$x(17 - x) = (10)(6)$$

$$17x - x^2 = 60$$

$$0 = x^2 - 17x + 60$$

$$0 = (x - 12)(x - 5)$$

Solving for x, you have $x = 12$ or $x = 5$.

If AE is greater than the radius of $\frac{17}{2}$, then $AE = 12$ cm.

Section 7.2

21. $(AE)(EC) = (BE)(ED)$ Theorem 7.8

$$14(EC) = 12 \cdot 7$$
$$EC = 6 \text{ ft}$$

Now ΔACD is a right triangle as $\angle ACD$ is inscribed in a semicircle.

Use the Pythagorean Theorem to find AD. We are given leg CD and we can find AC.

$$AC = AE + EC$$
$$= 14 + 6$$
$$= 20$$

So, $(AD)^2 = (AC)^2 - (CD)^2$

$$= 20^2 - 12^2$$
$$= 256.$$

Thus, $AD = 16$ ft.

23. Step 1: Use a straight edge to draw two nonparallel chords, $\overline{AB}$ and $\overline{CD}$, in the circle.

    Step 2: Construct the perpendicular bisector of each chord. (See Construction 6.)

    Step 3: By Corollary 7.5, the intersection of the two perpendicular bisectors is the center of the circle.

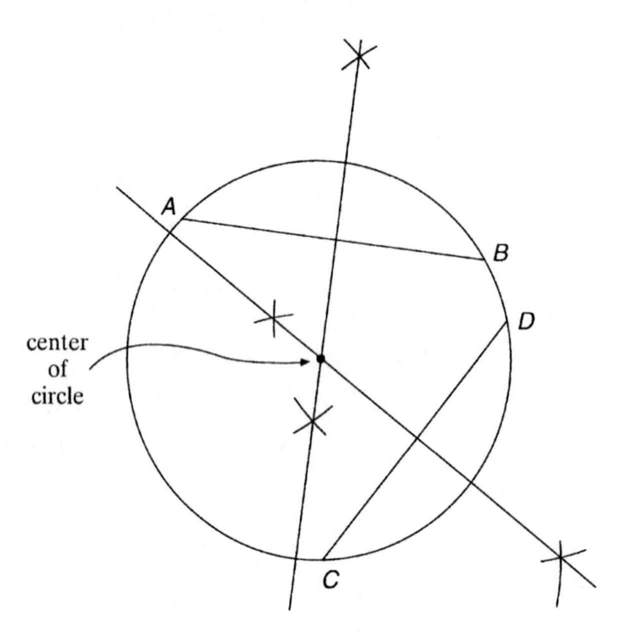

25. Step 1: Construct the perpendicular bisectors of two sides of the cyclic quadrilateral. (See Construction 6.)

    Step 2: By Corollary 7.5, the intersection of the perpendicular bisectors is the center of the circle.

Section 7.2

Note: If the cyclic quadrilateral is a parallelogram then use two nonparallel sides when finding the perpendicular bisectors.

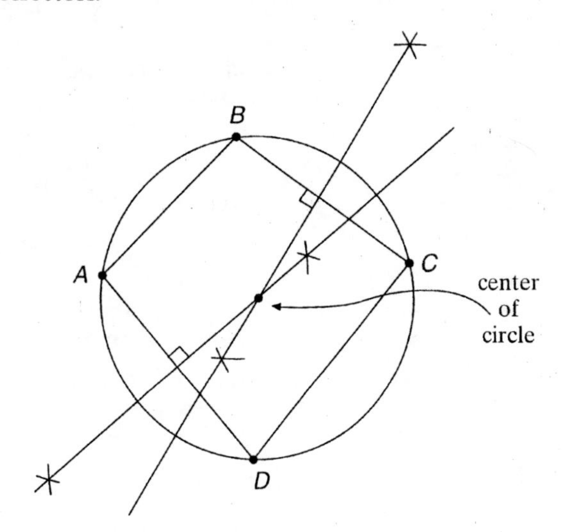

27. A complete solution is given in the answer key at the back of the text.

29. Let O and P be congruent circles and let $\angle AOB = \angle COD$ and $\angle AOB = \angle EPF$.

Then $AO = BO = CO = DO = EP = FP$.

Thus $\Delta AOB \cong \Delta COD$ and $\Delta AOB \cong \Delta EPF$ by SAS.

Therefore, $\overline{AB} \cong \overline{CD}$ and $\overline{AB} \cong \overline{EF}$. That is, in the same or congruent circles, congruent central angles have congruent chords.

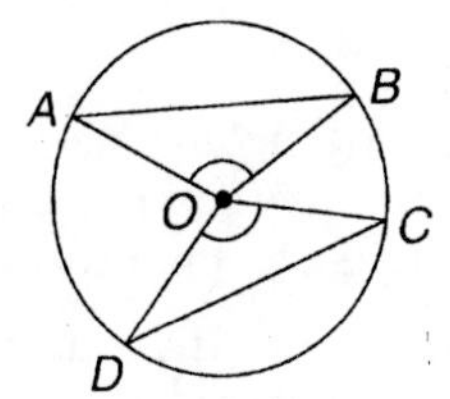

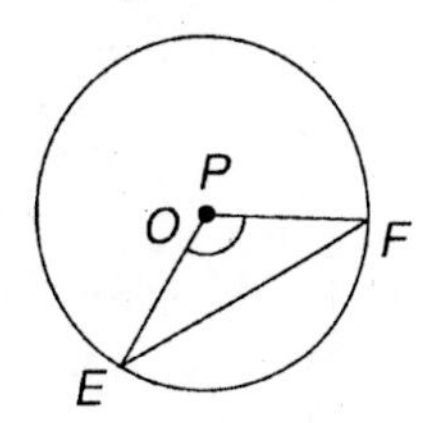

31. A complete solution is given in the answer key at the back of the text.

33. A complete solution is given in the answer key at the back of the text.

35. A complete solution is given in the answer key at the back of the text.

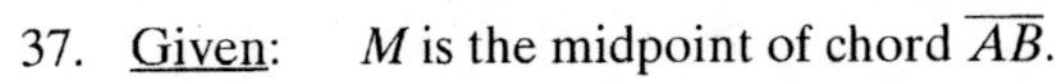

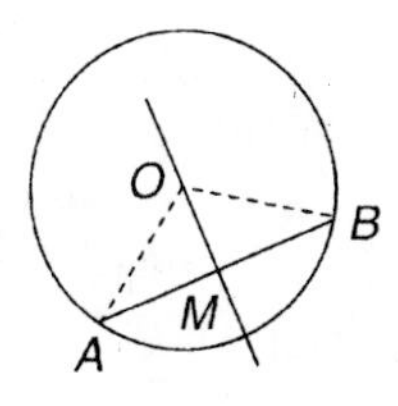

37. Given: M is the midpoint of chord $\overline{AB}$.

Prove: $\overline{OM}$ is the perpendicular bisector of $\overline{AB}$ in circle O.

Proof: $\overline{AM} \cong \overline{BM}$ since M is the midpoint of $\overline{AB}$.

$\overline{OA} \cong \overline{OB}$ because they are radii of the same circle.

$\overline{MO} \cong \overline{MO}$. Thus, $\Delta AMO \cong \Delta BMO$ by SSS. Then $\angle AMO \cong \angle BMO$ by C.P. Since $\angle AMO$ and $\angle BMO$ are also supplementary, they are right angles. Therefore, $\overline{OM}$ is a perpendicular bisector of $\overline{AB}$.

39. A complete solution is given in the answer key at the back of the text.

41. A complete solution is given in the answer key at the back of the text.

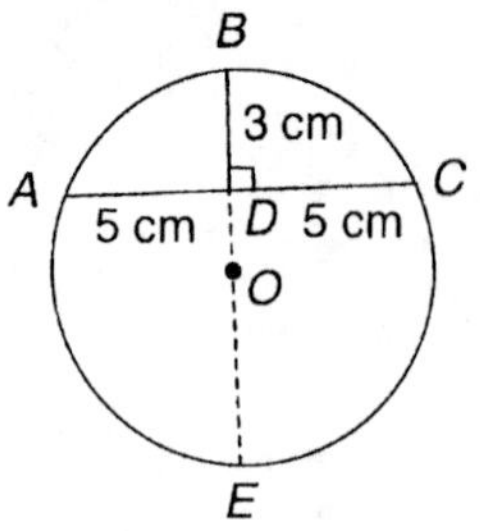

43. If the broken wheel was whole, $\overline{AC}$ and $\overline{BE}$ would be intersecting chords of the circular wheel.

$$(BD)(DE) = (AD)(DC) \qquad \text{Theorem 7.8}$$
$$3(DE) = 5.5$$
$$DE = \frac{25}{3}$$

Since D is the midpoint of $\overline{AC}$ and since $\overline{BD} \perp \overline{AC}$, $\overline{BE}$ is the perpendicular bisector of $\overline{AC}$.

By Theorem 7.1, $\overline{BE}$ goes through the center of the circle which implies $\overline{BE}$ is a diameter of the circle.

Thus,
$$BE = BD + DE$$
$$= 3 + \frac{25}{3} \qquad \text{substituting values}$$
$$= \frac{34}{3}.$$

Therefore, the radius of the wheel is $\frac{1}{2}\left(\frac{34}{3}\right) = \frac{17}{3} \approx 5.67$ cm.

Section 7.3

Tips:

✔ When secants or tangents meet *outside* the circle, the angle formed is half the *difference* of the intercepted arcs, but when secants cross *inside* the circle, the angle formed is half the *sum* of the intercepted arcs.

✔ Do not depend upon a diagram's accuracy. Often figures are not drawn to scale, so work from the information given in the problem instead.

Solutions to odd-numbered textbook problems

1. Since $\angle C$ is formed by two secants that intersect outside the circle, $\angle C = \frac{1}{2}(\overset{\frown}{AE} - \overset{\frown}{BD}) = \frac{1}{2}(90° - 38°) = 26°$.

3.
$$(AC)(BC) = (EC)(DC) \qquad \text{Theorem 7.10}$$
$$10 \cdot 6 = (EC)(EC - 3)$$
$$60 = (EC)^2 - 3(EC)$$
$$(EC)^2 - 3(EC) - 60 = 0$$

Section 7.3

3. (continued)

Use the quadratic formula.

$$EC = \frac{3 \pm \sqrt{3^2 - 4(1)(60)}}{2}$$

$$= \frac{3 \pm \sqrt{249}}{2}$$

Since $EC = \dfrac{3 - \sqrt{249}}{2} < 0$ we reject it and use

$$EC = \frac{3 + \sqrt{249}}{2} \text{ cm} \approx 9.39 \text{ cm}.$$

5. Let $BC = x$ and $AB = x - 8$.

Then $AC = x + x - 8 = 2x - 8$.

By Theorem 7.10 you have $(AC)(BC) = (EC)(DC)$.

Be careful. The letters are arranged differently than in theorem.

$$(2x - 8)(x) = (6 + 9)(9) \qquad \text{sustitution}$$

$$2x^2 - 8x = 135$$

$$2x^2 - 8x - 135 = 0$$

Use the quadratic formula to solve this equation.

$$x = \frac{4 + \sqrt{286}}{2} \approx 10.46 \text{ or } x = \frac{4 - \sqrt{286}}{2} \approx -6.46$$

Since only $x \approx 10.46$ makes sense here, use it and reject $x \approx -6.46$.
Then $AB = x - 8 \approx 10.46 - 8 \approx 2.46$ in.

7. A complete solution is given in the answer key at the back of the text.

9. Let $DE = x$. Since $(AC)(BC) = (EC)(DC)$, you have

$$(10 + 6)(6) = (x + 7)(7)$$

$$96 = 7x + 49$$

$$47 = 7x$$

$$\frac{47}{7} = x \approx 6.71 = DE.$$

11. If $\overparen{PRQ} = 230°$, then $\overparen{QP} = 130°$, as $\overparen{PQR} + \overparen{QP} = 360°$.

$$\angle QPS = \frac{1}{2}\overparen{QP} \qquad \text{Theorem 7.12}$$

$$= \frac{1}{2}(130°)$$

$$= 65°$$

Section 7.3

13. $\angle TPR + \angle RPQ + \angle QPS = 180°$,

$$28° + 44° + \angle QPS = 180° \quad \text{substituting values}$$
$$\angle QPS = 108°$$

Since by Theorem 7.12, $\angle QPS = \frac{1}{2}\overparen{PQ}$, you have $108° = \frac{1}{2}\overparen{PQ}$, and thus $\overparen{PQ} = 216°$.

15. $\angle SUP = \frac{1}{2}(\overparen{PQR} - \overparen{PTS})$ Theorem 7.16

$$24° = \frac{1}{2}(\overparen{PQR} - 40°) \quad \text{substituting values}$$

$$44° = \frac{1}{2}\overparen{PQR}$$

$$88° = \overparen{PQR}$$

17. $\overline{OB}$ is a radius of circle O and tangent to line m at point B, hence $\overline{OB} \perp \overleftrightarrow{CB}$ and $\angle CBO = 90°$.

19. A complete solution is given in the answer key at the back of the text.

21. $\angle AOD$ is a central angle, thus $\angle AOD = \overparen{DA} = 30°$. $\overleftrightarrow{OD}$ contains a diameter of circle O, thus $\overparen{DAE} = 180°$.

Therefore, $\overparen{AE} = 180° - \overparen{AD}$

$$= 180° - 30°$$
$$= 150°.$$

23. ΔOAB is a right triangle since $\overline{AB}$ is a tangent line perpendicular to the radius $\overline{OA}$.

Applying the Pythagorean Theorem, you have

$$(OA)^2 + (AB)^2 = (OB)^2$$
$$4^2 + (AB)^2 = 12^2$$
$$16 + (AB)^2 = 144$$
$$(AB)^2 = 128, \text{so}$$
$$AB = \sqrt{128} = 8\sqrt{2} \approx 11.31 \text{ cm}.$$

25. By Corollary 7.15, $BP = PQ, AP = AR$, and $CQ = CR$.

Since ΔABC is isosceles, $BA = BC$.

so $BP + PA = BQ + QC$

and $PA = QC$.

Thus, $AR = CR$ as $AP = AR$ and $CQ = CR$.

Section 7.3

25. (continued)

Therefore, $AR = CR = 3.5$, and since $AP = AR$, we have $AP = 3.5$ ft.

Now, $QC = AP = 3.5$.

Hence, $BQ = BC - QC$

$= 12 - 3.5$

$= 8.5$ ft.

27. We will give two methods of solving this problem.

Method 1:

By Theorem 7.9, $\angle A = \frac{1}{2}(\overparen{EH} - \overparen{DI})$

$$\angle B = \frac{1}{2}(\overparen{DG} - \overparen{EF})$$

$$\angle C = \frac{1}{2}(\overparen{IF} - \overparen{GH}).$$

So $\angle A = \frac{1}{2}(\overparen{EF} + \overparen{FG} + \overparen{GH} - \overparen{DI})$

$$\angle B = \frac{1}{2}(\overparen{DI} + \overparen{IH} + \overparen{GH} - \overparen{EF})$$

$$\angle C = \frac{1}{2}(\overparen{ID} + \overparen{DE} + \overparen{EF} - \overparen{GH}).$$

Substituting values we have

$$48° = \frac{1}{2}(\overparen{EF} + 60° + \overparen{GH} - \overparen{DI})$$

or

$$36° = \overparen{EF} + \overparen{GH} - \overparen{DI} \qquad \text{(Equation 1)}$$

and

$$72° = \frac{1}{2}(\overparen{DI} + 110° + \overparen{GH} - \overparen{EF})$$

or

$$34° = \overparen{DI} + \overparen{GH} - \overparen{EF} \qquad \text{(Equation 2)}$$

and

$$60° = \frac{1}{2}(\overparen{ID} + 100° + \overparen{EF} - \overparen{GH})$$

or

$$20° = \overparen{ID} + \overparen{EF} - \overparen{GH}. \qquad \text{(Equation 3)}$$

Section 7.3

27. (continued)

Add Equation 1 and Equation 2.

$$\begin{aligned} 36^\circ &= \overset{\frown}{EF} + \overset{\frown}{GH} - \overset{\frown}{ID} \\ 34^\circ &= -\overset{\frown}{EF} + \overset{\frown}{GH} + \overset{\frown}{ID} \\ \hline 70^\circ &= 2\overset{\frown}{GH} \end{aligned}$$

So $\overset{\frown}{GH} = 35^\circ$.

Substitute $\overset{\frown}{GH} = 35^\circ$ into Equation 1 and Equation 3.

$$36^\circ = \overset{\frown}{EF} + 35^\circ - \overset{\frown}{DI} \qquad \text{Equation 1}$$

or

$$1^\circ = \overset{\frown}{EF} - \overset{\frown}{DI}$$

and

$$20^\circ = \overset{\frown}{ID} + \overset{\frown}{EF} - 35^\circ \qquad \text{Equation 3}$$

or

$$55^\circ = \overset{\frown}{ID} + \overset{\frown}{EF}.$$

Thus, add the resulting two equations.

$$\begin{aligned} 1^\circ &= \overset{\frown}{EF} - \overset{\frown}{DI} \\ 55^\circ &= \overset{\frown}{EF} + \overset{\frown}{ID} \\ \hline 56^\circ &= 2\overset{\frown}{EF} \end{aligned}$$

So, $\overset{\frown}{EF} = 28^\circ$.

Hence, $1^\circ = \overset{\frown}{EF} - \overset{\frown}{DI}$

so $\quad 1^\circ = 28^\circ - \overset{\frown}{DI}$

and $\quad \overset{\frown}{DI} = 27^\circ$.

Method 2: It may be easier to solve this problem if you mark the center of the circle, O, and draw radii to D, E, F, G, H, and I. This gives three isosceles triangles: ΔDOE, ΔGOF and ΔHOI. Use the fact that the base angles of an isosceles triangle are congruent to help you determine the measures of the other angles in these triangles.

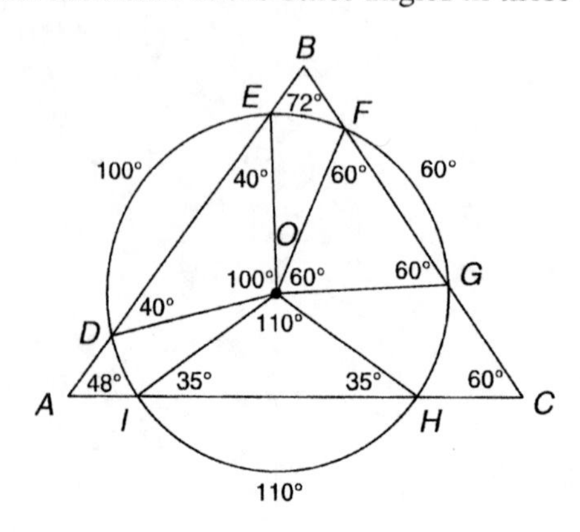

Section 7.3

27. (continued)

Since $\overset{\frown}{DE} = 100°$, $\angle DOE = 100°$. So $\angle EDO + \angle DEO = 180° - 100° = 80°$. Since $\angle EDO \cong \angle DEO$, $\angle EDO = \angle DEO = 40°$. In the same manner you can find that $\angle OFG = \angle FGO = 60°$, and $\angle OIH = \angle OHI = 35°$.

Now look at the quadrilateral $BEOF$.

You know that $\angle B = 72°$, $\angle OEB = 180° - 40° = 140°$, and $\angle BFO = 180° - 60° = 120°$.

Since $\angle B + \angle BFO + \angle FOE + \angle OEB = 360°$,

you have $72° + 120° + \angle FOE + 140° = 360°$.

$$332° + \angle FOE = 360°$$
$$\angle FOE = 28°.$$

$\angle FOE$ is a central angle intercepting $\overset{\frown}{EF}$, so $\overset{\frown}{EF} = 28°$.

Similarly, for quadrilateral $GOHC$, $\angle OHC = 180° - 35° = 145°$ and $\angle CGO = 180° - 60° = 120°$.

Then $\angle C + \angle CGO + \angle GOH + \angle OHC = 360°$

$$60° + 120° + \angle GOH + 145° = 360°$$
$$325° + \angle GOH = 360°$$
$$\angle GOH = 35°.$$

Therefore, $\overset{\frown}{GH} = 35°$.

You can use the same approach to find $\overset{\frown}{DI}$ or you can add the other arcs and subtract from 360°.

$$\begin{aligned}\overset{\frown}{DI} &= 360° - (\overset{\frown}{DE} + \overset{\frown}{EF} + \overset{\frown}{FG} + \overset{\frown}{GH} + \overset{\frown}{HI})\\ &= 360° - (100° + 28° + 60° + 35° + 110°)\\ &= 360° - 333°\\ &= 27°\end{aligned}$$

29. Let x = radius of smaller circle.

Then, $2x$ = measure of side of square, since the smaller circle is inscribed in the square.

The area of the shaded region is the area of the square less the area of the circle, or $(2x)^2 - \pi x^2$.

There are at least two ways to find x.

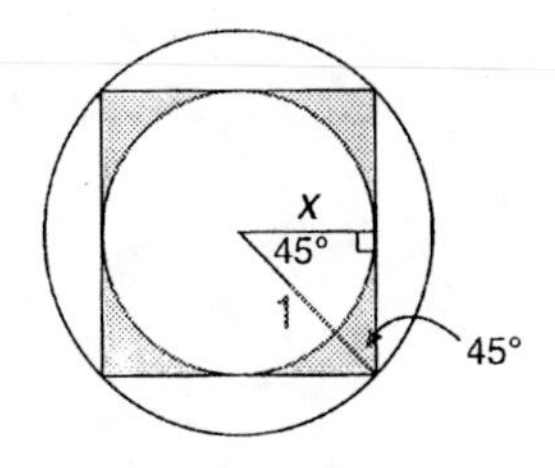

Method 1: Consider the 45°–45° right triangle with hypotenuse of length 1. Then the leg has length $x = \frac{1}{\sqrt{2}}$.

Therefore, the area of the shaded region is

$$\begin{aligned}A &= (2x)^2 - \pi x^2\\ &= 2\left(\frac{1}{\sqrt{2}}\right)^2 - \pi\left(\frac{1}{\sqrt{2}}\right)^2\\ &= 2 - \frac{\pi}{2} \text{ square units.}\end{aligned}$$

Section 7.3

29. (continued)

Method 2: Consider chords $\overline{AC}$ and $\overline{DE}$ of the larger circle which intersect at B.

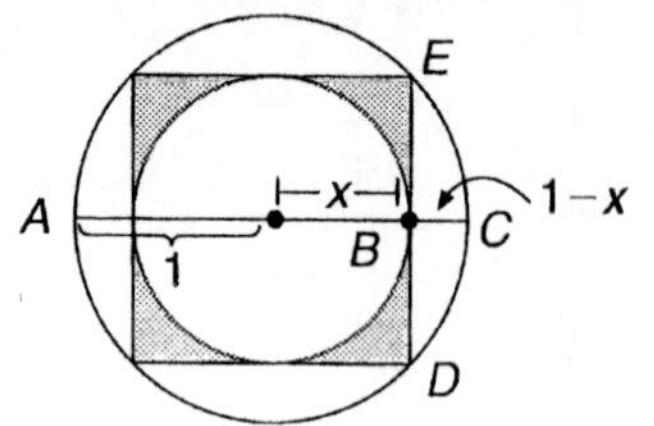

$$(AB)(BC) = (EB)(BD) \qquad \text{Theorem 7.7}$$

or
$$(1 + x)(1 - x) = x^2$$
$$2x^2 - 1 = 0$$
$$x = \pm\sqrt{\frac{1}{2}}$$

We reject $x = -\sqrt{\frac{1}{2}}$ as it does not make sense here and so $x = \sqrt{\frac{1}{2}} = \frac{1}{\sqrt{2}}$.

Therefore, the area of the shaded region is $A = \left(\frac{2}{\sqrt{2}}\right)^2 - \pi\left(\frac{1}{\sqrt{2}}\right)^2$

$$= 2 - \frac{\pi}{2} \text{ square units.}$$

31. A complete solution is given in the answer key at the back of the text.

33. A complete solution is given in the answer key at the back of the text.

35. A complete solution is given in the answer key at the back of the text.

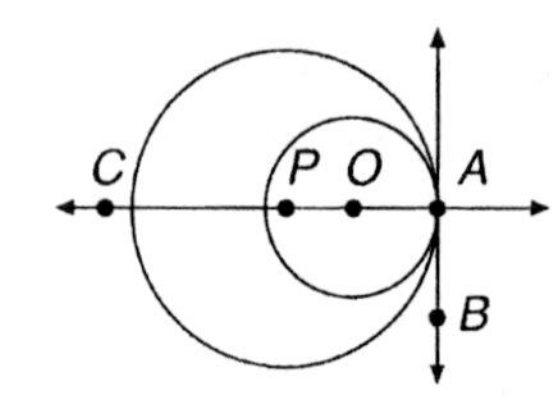

37. Given: The smaller circle O is internally tangent to the larger circle P with A as the point of tangency.

Prove: P, O and A are collinear.

Proof:

Let $\overleftrightarrow{AB}$ be the common tangent line with A the point of tangency. Then $\overleftrightarrow{AB}$ is perpendicular to $\overleftrightarrow{OA}$, and $\overleftrightarrow{AB}$ is perpendicular to $\overleftrightarrow{PA}$. Since $\angle PAB = 90°$ and $\angle OAB = 90°$, P, O and A are collinear.

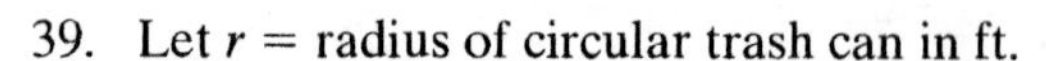

39. Let r = radius of circular trash can in ft.

By Corollary 7.15, $AE = AD$, $CD = CF$, and $BE = BF$.

Now, $AEOD$ is a square with side length r.

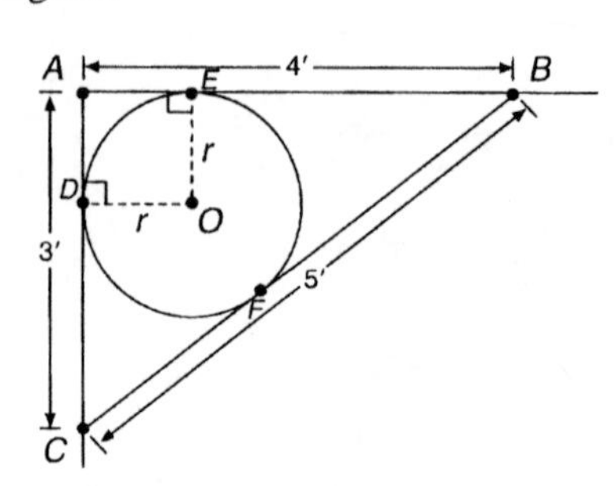

Hence, $CD = 3 - r = CF$

and $BE = 4 - r = BF$.

Thus $BC = BF + FC$

or, $5 = 4 - r + 3 - r$

$5 = 7 - 2r$

$2r = 2$

$r = 1$.

Therefore, the diameter of the circular trash can is 2 ft.

41. Let x = distance from P to center of the earth in miles.

    Now $\Delta ADP \sim \Delta BCP$ by Side Splitting Theorem 6.10.

    Therefore, $\frac{AD}{BC} = \frac{OP}{O'P}$.

    That is, $\frac{870{,}000}{2160} = \frac{93{,}000{,}000 - x}{252{,}700 - x}$

    which yields

    $$(93{,}000{,}000 - x)2160 = (252{,}700 - x)870{,}000$$

    or, $x \approx 21{,}857$

    $\approx 22{,}000$ miles.

    Point P is about $22{,}000 - 3960 = 18{,}040$ miles above the surface of the earth, so no part of the umbra would cover the earth.

Section 7.4

Tips:

✔ Don't forget that a sharp pencil point, good straightedge, and extreme care are essential to producing accurate constructions.

✔ Use a compass that does not slip easily from the radius you set.

Solutions to odd-numbered textbook problems

1. Remember that the center of the circumscribed circle is the point of intersection of the three perpendicular bisectors.

   Step 1: Construct the perpendicular bisectors of any two sides. Remember that to construct the perpendicular bisector of a side you use a radius that is greater than half the length of the side. Draw two arcs of that radius, placing the point of your compass at each endpoint of the side you are bisecting.

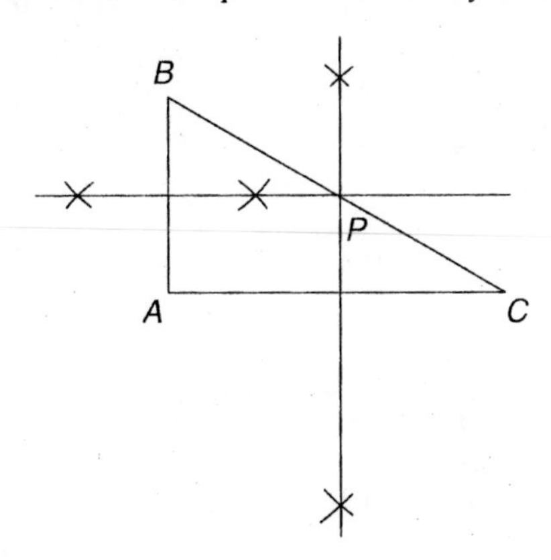

Section 7.4

1. (continued)

Step 2: Now place the point of your compass at P, the point of intersection of the two perpendicular bisectors. Set the radius by placing the tip of the pencil point of the compass at A, B, or C, and draw the circumscribed circle.

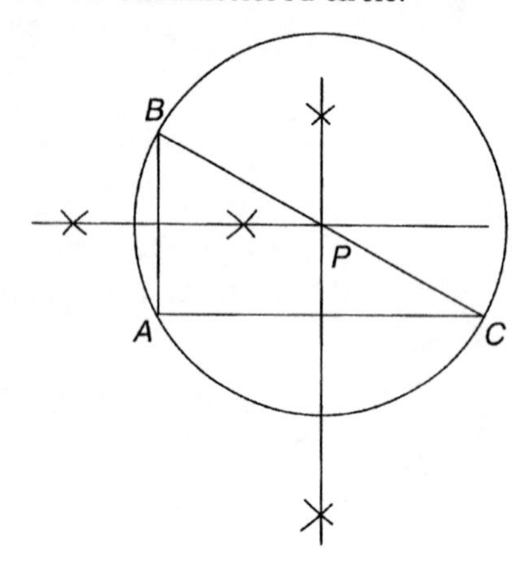

Notice that the circumcenter for right triangle ABC lies on the hypotenuse of the triangle.

3. The center of the circumscribed circle is the point of intersection of the three perpendicular bisectors.

Step 1: Construct the perpendicular bisector of any two sides. (See Construction 6 of Section 4.4.)

Step 2: Place the point of the compass at P, the point of intersection of the two perpendicular bisectors. Set the radius by placing the tip of the pencil point of the compass at G, H, or I, and draw the circumscribed circle.

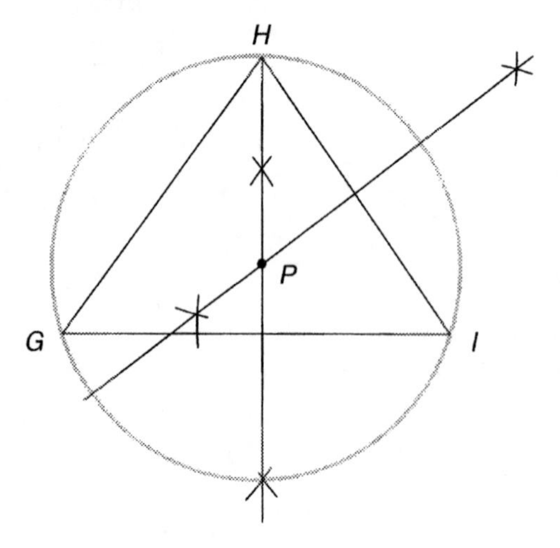

Notice that the circumcenter for isosceles triangle GHI lies on the angle bisector of the vertex angle.

Section 7.4

5. Remember that the orthocenter of a triangle is the point of intersection of the altitudes of the triangle. To find the orthocenter, construct any two altitudes of the triangle. To construct an altitude you construct a perpendicular from a vertex point to the opposite side. (*Hint*: You can extend a side, if necessary, to construct the perpendicular.)

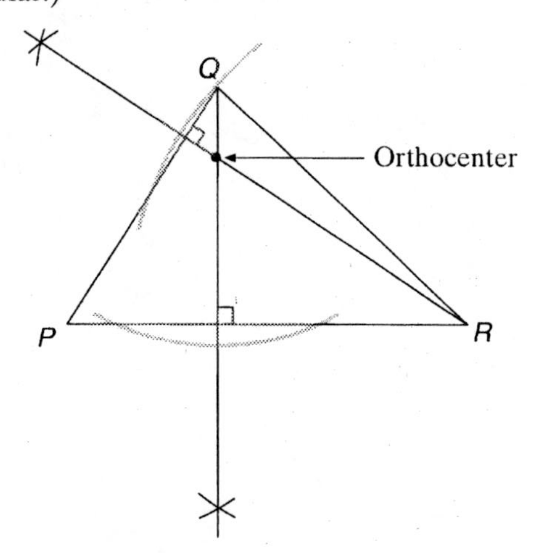

7. Point D will complete the orthocentric set if D is the orthocenter of ΔABC.

   Step 1: Connect points A, B, and C to form a triangle.

   Step 2: To find the orthocenter, construct any two altitudes of ΔABC. To construct an altitude, construct a perpendicular from a vertex point to the opposite side.

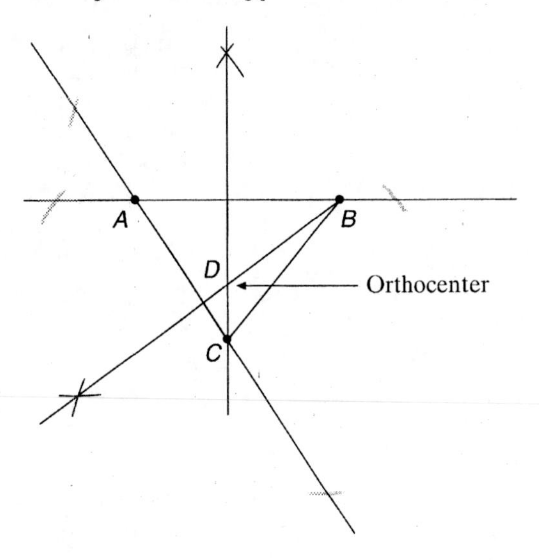

(*Hint*: Extend a side, if necessary, to construct the perpendicular.)

Section 7.4

7. (continued)

Step 3: Copy points D, B, and C and connect the points to form ΔDBC.

Step 4: To find the orthocenter of ΔDBC, construct any two altitudes of ΔDBC.

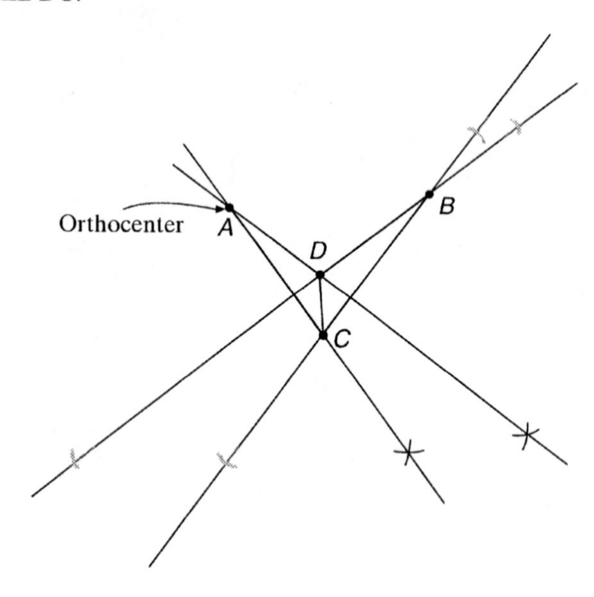

Note: Vertex A of ΔABC is the orthocenter of ΔDBC.

9. Remember that the incenter is the point of intersection of the angle bisectors of the triangle.

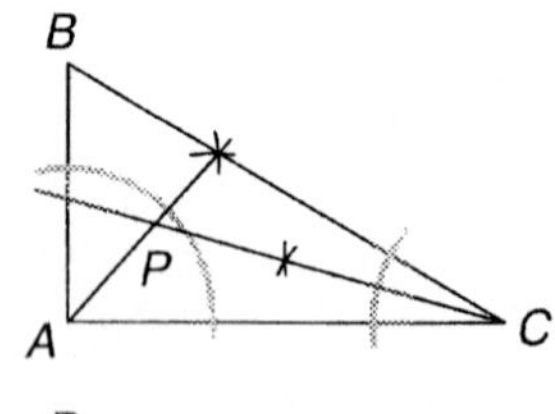

Step 1: To construct an angle bisector, place your compass point at the vertex and draw an arc intersecting both sides of the angle. Then, moving the point of your compass to the points where the arc intersects each of the sides, make two more arcs. This forms an × which when joined with the vertex of the angle, gives the angle bisector.

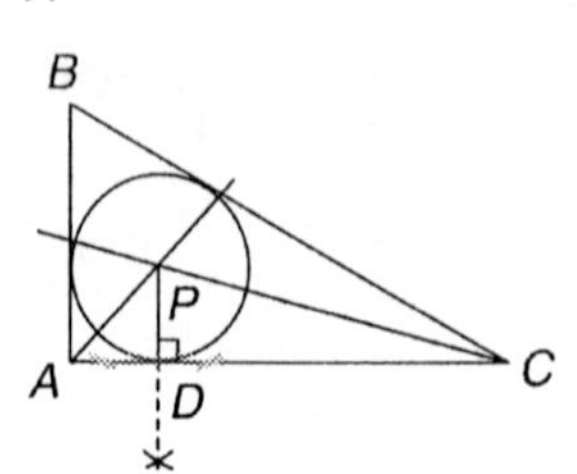

Step 2: Construct a perpendicular from P to $\overline{AC}$. If necessary, you can extend $\overline{AC}$. Label the point of intersection of the perpendicular and $\overline{AC}$ as D. Set the point of your compass at P and the tip of the pencil on D and draw the circle.

11. The center of the inscribed circle is the point of intersection of the three angle bisectors.

Step 1: Construct the angle bisector of any two angles. (See Construction 3 of Section 4.4.) Label as point P the intersection of the angle bisectors.

Section 7.4

11. (continued)

Step 2: Construct the perpendicular from P to $\overline{GI}$. Since ΔGHI is isosceles with vertex angle $\angle H$, the perpendicular from P to $\overline{GI}$ is the same as the angle bisector of $\angle H$. Label the point of intersection of the perpendicular and $\overline{GI}$ as Q. Set the compass at P and the tip of the pencil on Q and draw the circle.

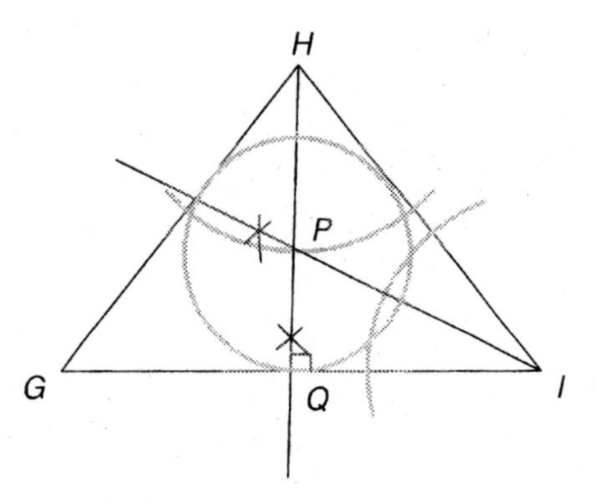

13. The centroid of a triangle is the point of intersection of the medians.

Step 1: Find the medians of ΔABC.

Recall that a median of a triangle is a line segment whose endpoints are a vertex of the triangle and the midpoint of the side opposite that vertex. (Use Construction 6 of Section 4.4 to find the midpoint.)

Step 2: The intersection of any two medians is the centroid, P.

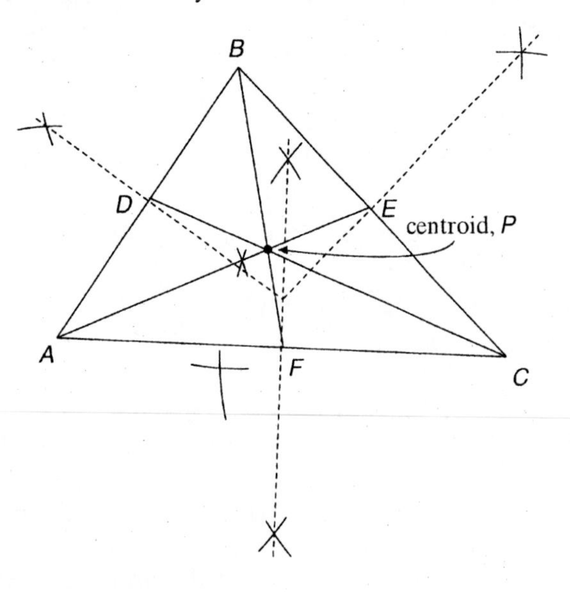

Step 3: Measure to verify

$AP = 2(PE)$,

$BP = 2(PF)$, and

$CP = 2(PD)$.

Section 7.4

15. Step 1: Use a straight edge to construct $\overleftrightarrow{AP}$. Then find the midpoint, M, of $\overline{AP}$.

Step 2: Set compass to radius AM, and with point of compass at P mark off distance and label intersection, E.

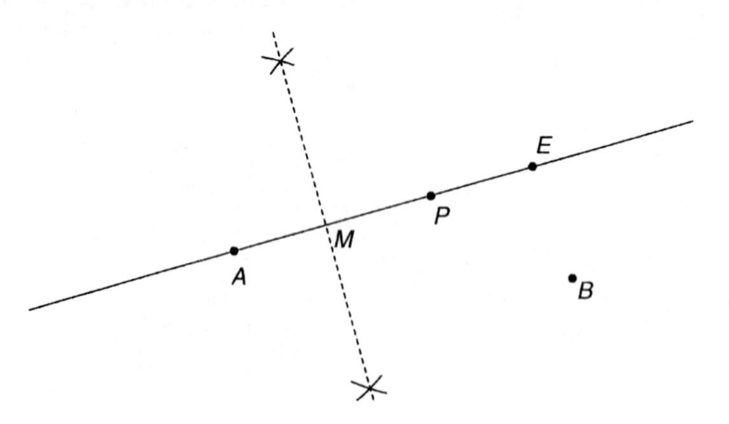

Step 3: Use a straight edge to construct $\overrightarrow{BE}$. Set compass to radius BE, and with point of compass at E mark off distance and label intersection C.

Step 4: Use a straight edge to connect points A, B, and C.

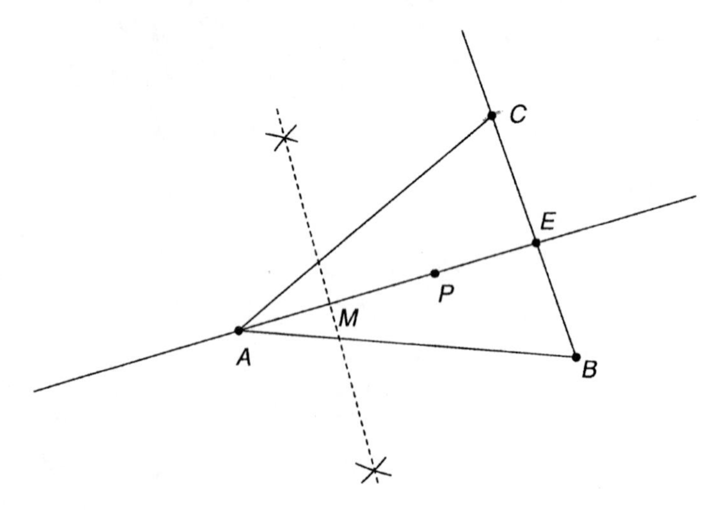

P is the centroid of ΔABC.

17. Follow the procedure of Construction 13 to construct a tangent to a circle at a point on the circle.

19. Follow the procedure of Construction 15 to construct a common internal tangent to two circles.

Section 7.4

21. You can use the fact that an angle inscribed in a semicircle is a right angle to perform this construction.

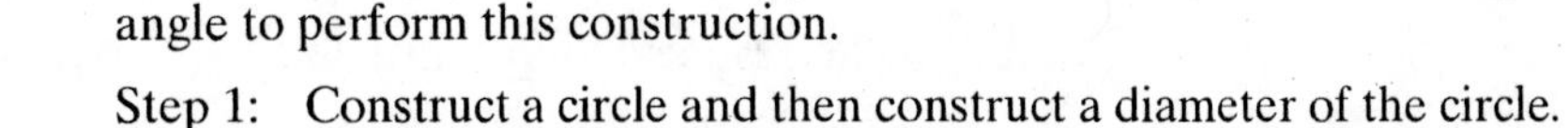

Step 1: Construct a circle and then construct a diameter of the circle.

Step 2: Construct another diameter perpendicular to the first one.

Step 3: Connect the endpoints of the diameters in order around the circle.

Since each of the angles of the quadrilateral formed is a right angle, the quadrilateral is a rectangle. The diagonals of the rectangle are the diameters and are perpendicular, so it must be a rhombus. Therefore, it has congruent sides and four right angles, so the figure constructed must be a square.

23. Step 1: Construct a circle with the center O and then construct a diameter, $\overline{GC}$ of the circle.

Step 2: Construct another diameter, $\overline{AE}$ perpendicular to $\overline{GC}$.

Step 3: Construct two additional diameters, $\overline{BF}$ and $\overline{DH}$ by bisecting two adjacent central angles $\angle AOC$ and $\angle COE$ and extending the angle bisectors. (See Construction 3 of Section 4.4.)

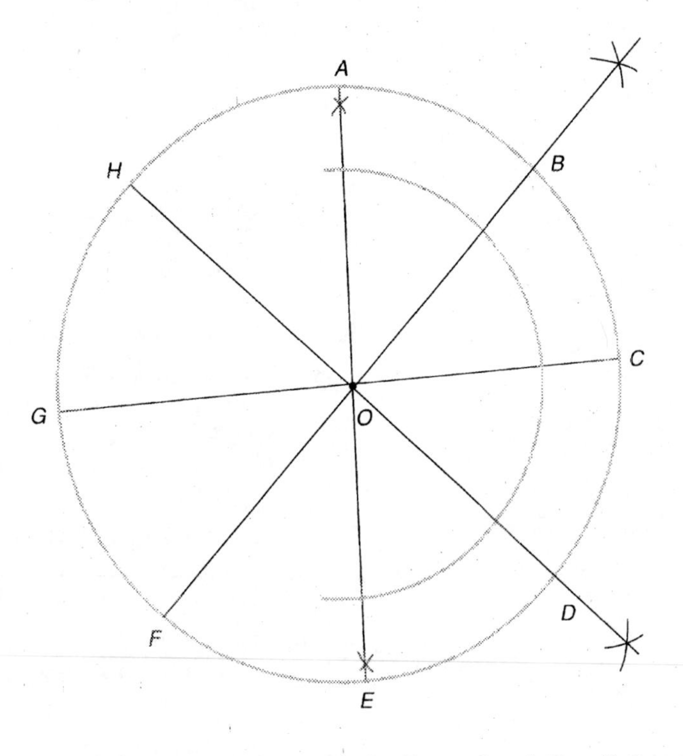

Step 4: Construct a line tangent to circle O at point A. Recall that a radius of a circle is perpendicular to a tangent line at its point of tangency. (Use Construction 4 of Section 4.4.)

Section 7.4

23. (continued)

Step 5: Repeat Step 4 for points B, C, D, E, F and G.

Step 6: Label the points of intersection of the tangent lines M, N, O, P, Q, R, S, and T.

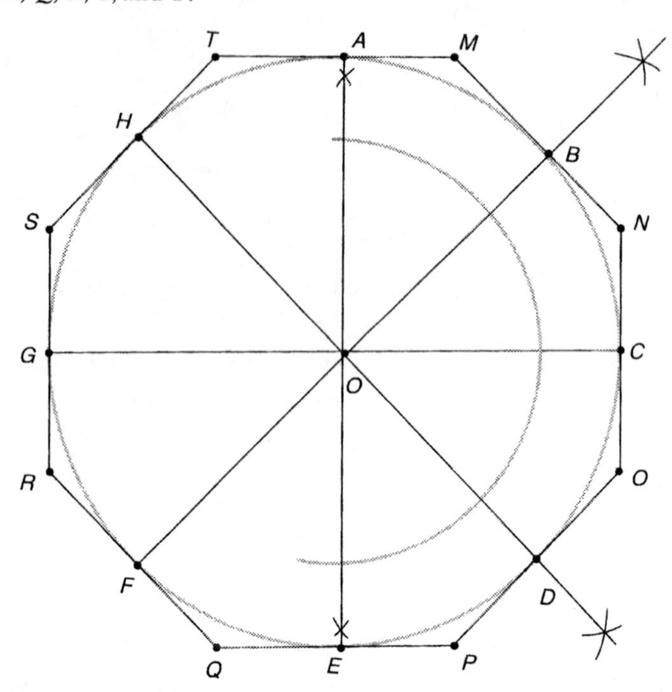

Now, $MNOPQRST$ is an octagon circumscribed about the circle. Notice also that the octagon is regular.

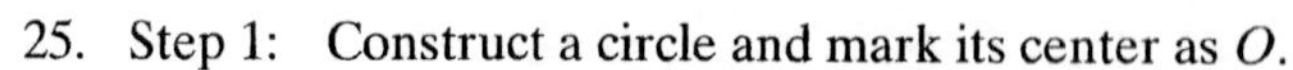

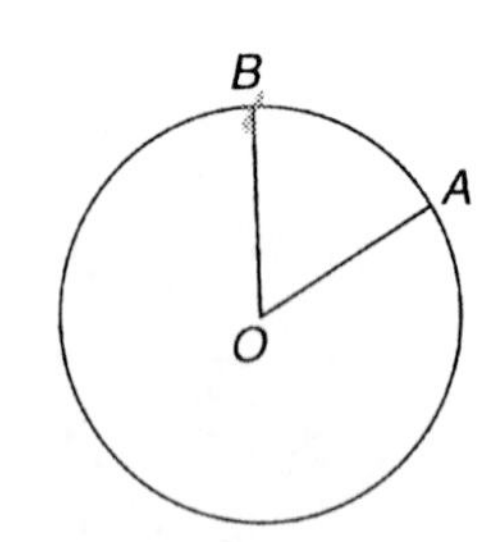

25. Step 1: Construct a circle and mark its center as O.

Step 2: Mark any point A on the circle. Set your compass to the radius of the circle. Place the point of the compass at A and mark off on the circle a distance equal to the radius. Label the point of intersection B.

Step 3: Draw $\overline{OB}$ and $\overline{OA}$ to form $\angle AOB$. If you were to form ΔABO, it would be an equilateral triangle since the same compass radius was used for the length of all the sides. Therefore, you have $\angle AOB = 60°$.

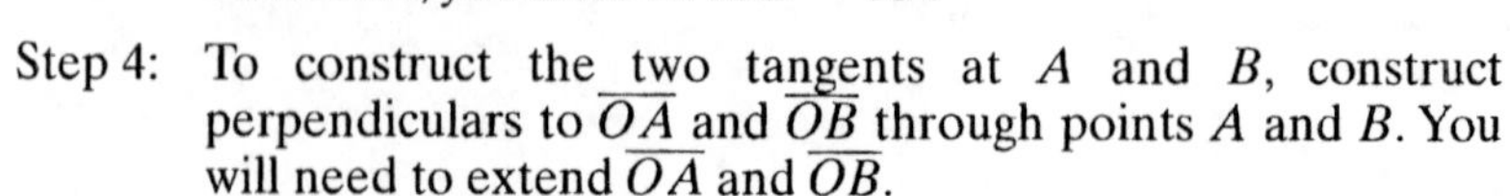

Step 4: To construct the two tangents at A and B, construct perpendiculars to $\overline{OA}$ and $\overline{OB}$ through points A and B. You will need to extend $\overline{OA}$ and $\overline{OB}$.

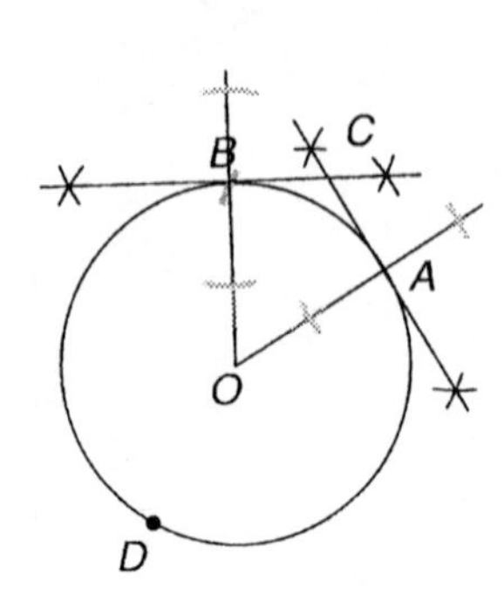

You can use Theorem 7.14 to find the measure of $\angle C$.

$$\angle C = \frac{1}{2}(\overset{\frown}{ADB} - \overset{\frown}{AB})$$

You have $\overset{\frown}{AB} = 60°$ and $\overset{\frown}{ADB} = 360° - 60° = 300°$.

Thus $\angle C = \frac{1}{2}(300° - 60°) = 120°$.

Section 7.4

25. (continued)

For a second way to find $\angle C$, consider the quadrilateral $AOBC$. In it $\angle O = 60°$, and $\angle A = \angle B = 90°$. Then

$$\angle A + \angle B + \angle O + \angle C = 360°$$
$$90° + 90° + 60° + \angle C = 360°$$
$$\angle C = 120°.$$

Quadrilateral $AOBC$ is a kite as $\overline{OA} \cong \overline{OB}$ and $\overline{BC} \cong \overline{AC}$.

27. Step 1: Find the circumcenter of ΔABC. The circumcenter is the intersection of the perpendicular bisectors.

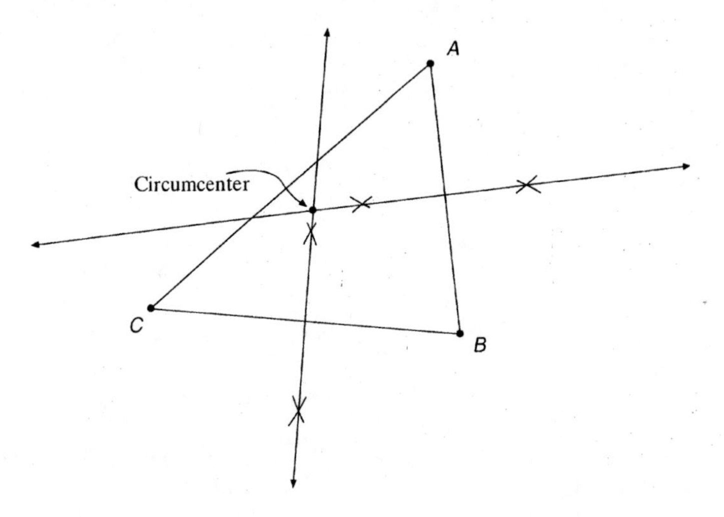

Step 2: Find the orthocenter of ΔABC. The orthocenter is the intersection of the altitudes.

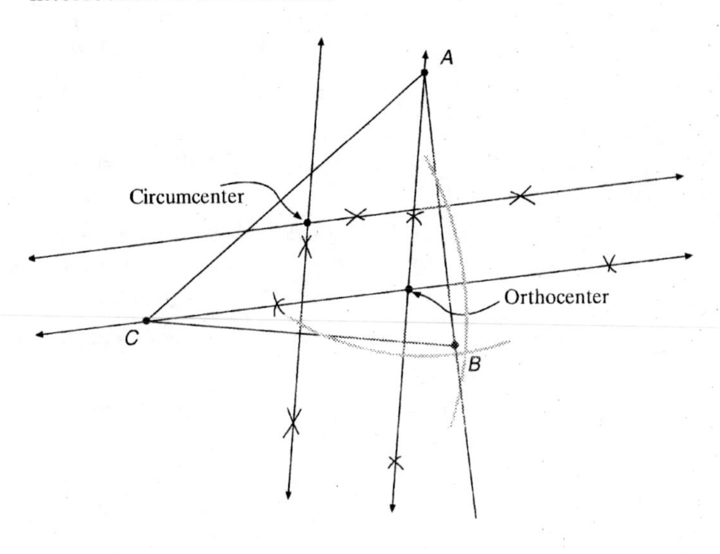

(Use Construction 5 of Section 4.4.)

Section 7.4

27. (continued)

Step 3: Find the centroid of ΔABC. The centroid is the intersection of the medians.

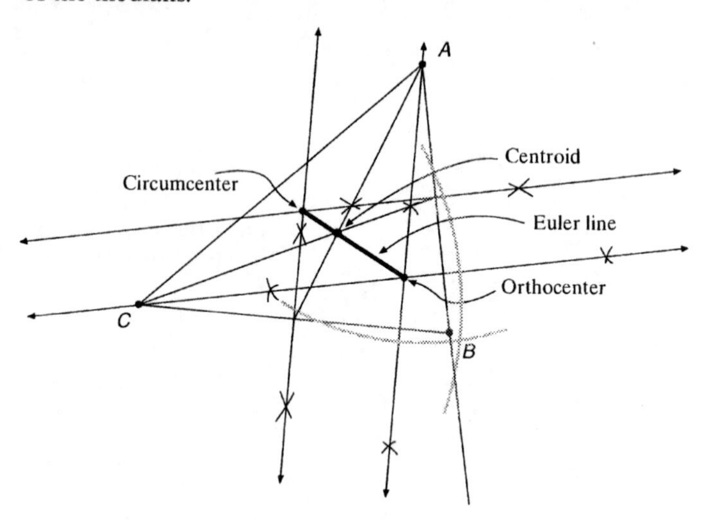

Step 4: Connect the three "centers" of ΔABC to form the Euler line.

29. A complete solution is given in the answer key at the back of the text.

31. A complete solution is given in the answer key at the back of the text.

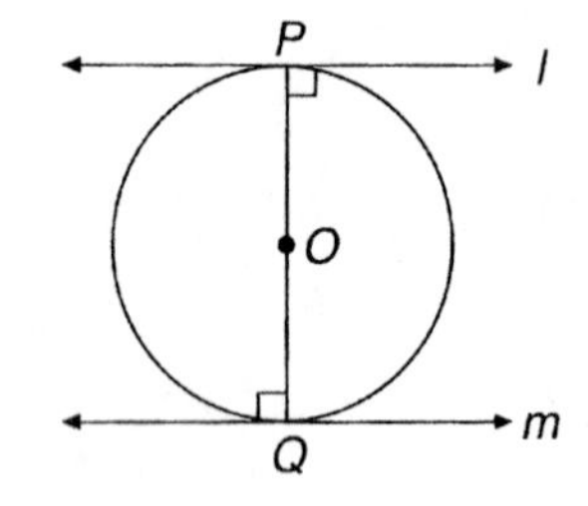

33. Let $\overline{PQ}$ be a diameter of circle O, and let lines l and m be tangent to circle O at points P and Q respectively.

Then, $l \perp \overline{PQ}$ and $m \perp \overline{PQ}$ by Theorem 7.11.

Hence the angles the diameter makes with the tangent lines are right angles.

Thus, we have congruent alternate interior angles.

Therefore, l and m are parallel by Theorem 5.1.

35. A complete solution is given in the answer key at the back of the text.

37. First find AC and FC.

$\angle B = \angle E = \angle D = \angle G = 90°$ and $\Delta ABC \sim \Delta FEC$.

So $$\frac{AB}{AC} = \frac{FE}{FC}$$

$$\frac{7}{15 - FC} = \frac{2}{FC} \qquad \text{substituting values}$$

$$7(FC) = 30 - 2(FC)$$

$$9(FC) = 30$$

$$FC = \frac{30}{9} = \frac{10}{3}$$

and $$AC = 15 - FC = \frac{35}{3}.$$

Section 7.4

37. (continued)

Now find BC and EC.

Since $\Delta ABC \cong \Delta ADC$ and $\Delta FEC \cong \Delta FGC$, we have $BC = DC$ and $EC = GC$.

Use the Pythagorean Theorem.

$$BC = \sqrt{\left(\frac{35}{3}\right) - 7^2}$$
$$= \frac{28}{3}$$
$$EC = \sqrt{\left(\frac{10}{3}\right)^2 - 2^2}$$
$$= \frac{8}{3}$$

Now, find the length of arc $\widehat{BHD}$ and arc $\widehat{EIG}$.

$$\cos(\angle BAC) = \frac{7}{\frac{35}{3}}$$
$$\angle BAC = \cos^{-1}\left(\frac{3}{5}\right)$$

So $$\angle BAD = 2\cos^{-1}\left(\frac{3}{5}\right).$$

Use the length of an Arc, Postulate 7.1.

$$\text{length of arc } \widehat{BHD} = \frac{360 - 2\cos^{-1}\left(\frac{3}{5}\right)}{360}(14\pi)$$

Since $\angle EFG \cong \angle BAD$,

$$\text{length of arc } \widehat{EIG} = \frac{360 - 2\cos^{-1}\left(\frac{3}{5}\right)}{360}(4\pi).$$

Therefore, the length of the belt is

$2(BC) + 2(EC) + \text{arc length } \widehat{BHD} + \text{arc length } \widehat{EIG}$

$$= 2\left(\frac{28}{3}\right) + 2\left(\frac{8}{3}\right) + \frac{360 - 2\cos^{-1}\left(\frac{3}{5}\right)}{360}(14\pi) + \frac{360 - 2\cos^{-1}\left(\frac{3}{5}\right)}{360}(4\pi)$$

$$= 2\left(\frac{36}{3}\right) + \frac{360 - 2\cos^{-1}\left(\frac{3}{5}\right)}{360}(18\pi)$$
$$\approx 24 + 39.86$$
$$\approx 64 \text{ cm}.$$

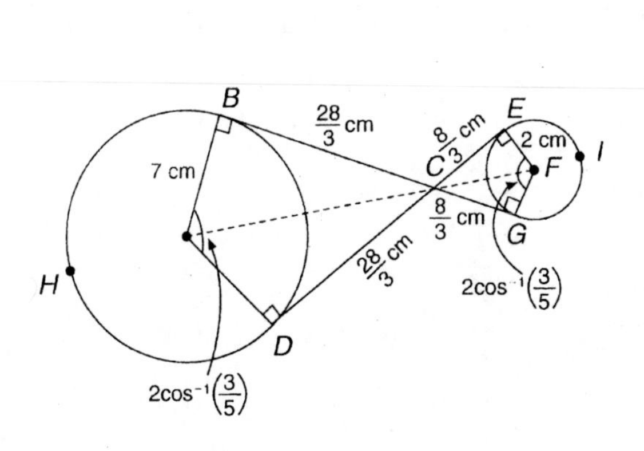

Chapter Review

Additional Problems

1. A complete solution is given in the answer key at the back of the text.

2. A complete solution is given in the answer key at the back of the text.

Solutions to Chapter 7 Review

Section 7.1

1. (a) Inscribed angle since C lies on circle O.
   (b) Central angle since O is the center of the circle.
   (c) $\angle AOB = \overset{\frown}{AB}$
   $= 2\angle ACB$ Inscribed Angle Theorem
   (d) $\overset{\frown}{ABC} = \angle AOC$
   $= 180°$
   (e) $\angle D = 90°$ since $\angle D$ is inscribed in a semicircle.

2. $\overset{\frown}{AD} + \overset{\frown}{AB} = 180°$
   $57° + \overset{\frown}{AB} = 180°$
   $\overset{\frown}{AB} = 123°.$
   Now $\angle AOB = \overset{\frown}{AB}$
   so $3x = 123°$
   and $x = 41°$
   or $\angle BOC = 41°.$
   Therefore, $\angle DOC = 180° - 41°$
   $= 139°.$

3. $\angle A = \frac{1}{2}(\overset{\frown}{BC} + \overset{\frown}{CD})$ Inscribed Angle Theorem
   $= 100°$
   $\angle B = \frac{1}{2}(\overset{\frown}{CD} + \overset{\frown}{AD})$ Inscribed Angle Theorem
   $115° = \frac{1}{2}(145° + \overset{\frown}{AD})$ substituting values
   $230° = 145° + \overset{\frown}{AD}$
   $85° = \overset{\frown}{AD}$
   $\overset{\frown}{AB} = 360° - \overset{\frown}{BC} - \overset{\frown}{CD} - \overset{\frown}{AD}$
   $= 360° - 55° - 145° - 85°$ substituting values
   $= 75°$

Chapter Review

3. (continued)

$$\angle C = \frac{1}{2}(\overset{\frown}{BA} + \overset{\frown}{AD}) \qquad \text{Inscribed Angle Theorem}$$

$$= \frac{1}{2}(75° + 85°) \qquad \text{substituting values}$$

$$= 80°$$

$$\angle D = \frac{1}{2}(\overset{\frown}{AB} + \overset{\frown}{BC}) \qquad \text{Inscribed Angle Theorem}$$

$$= \frac{1}{2}(75° + 55°) \qquad \text{substituting values}$$

$$= 65°$$

To summarize: $\angle A = 100°, \angle C = 80°, \angle D = 65°$

$\overset{\frown}{AB} = 75°$ and $\overset{\frown}{AD} = 85°$.

4. (a) The area of the patio is the area of the sector of a circle. First find the central angle using $l = \frac{x}{360}(2\pi r)$ where x is the central angle, r is the radius of the circle and l is the measure of the arc.

$$54 = \frac{x}{360}(2\pi \cdot 30)$$

$$x = \frac{324}{\pi}$$

Now find the area of the sector.

$$\text{Area of sector} = \frac{x}{360} \cdot \text{area of circle}$$

$$= \frac{\frac{324}{\pi}}{360} \cdot \pi(30)^2$$

$$= 810 \text{ ft}^2$$

(b) Find the central angle.

$$\text{Area of sector} = \frac{x}{360} \cdot \text{area of circle}$$

$$550 = \frac{x}{360}\pi(30)^2$$

$$x = \frac{220}{\pi}$$

Chapter Review

4. (continued)

Now find arc length.

$$l = \frac{x}{360}(2\pi r)$$

$$= \frac{\frac{220}{\pi}}{360}(2\pi \cdot 30)$$

$$= \frac{110}{3} \approx 36.7 \text{ ft}$$

Section 7.2

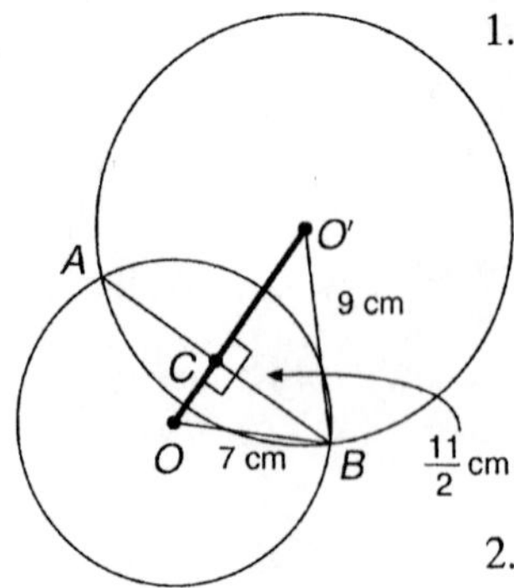

1. By Corollary 7.6, $\overline{OO'}$ is the perpendicular bisector of $\overline{AB}$. Use the Pythagorean Theorem to find OC and $O'C$.

$$OO' = OC + CO'$$

$$= \sqrt{7^2 - \left(\frac{11}{2}\right)^2} + \sqrt{9^2 - \left(\frac{11}{2}\right)^2}$$

$$\approx 11.45 \text{ cm}$$

2. $\angle ACD = \frac{1}{2}\overset{\frown}{AD}$ — Inscribed Angle Theorem

$= \frac{1}{2}(95°)$

$= 47.5°$

$\angle DEC = 180° - \angle BDC - \angle ACD$ — Angle sum of a triangle

$= 180° - 23° - 47.5°$ — substituting values

$= 109.5°$

$\angle AEB = \angle DEC$ — Vertical Angles

$= 109.5°$

3. $(FC)(AF) = (FD)(BF)$ — Theorem 7.8

$5(AF) = 10(BF)$ — substituting values

Thus $5(CG) = 10(BF)$ — since $AF = CG$

and $CG = 2(BF)$.

Now, $(GE)(CG) = (GD)(BG)$ — Theorem 7.8

$10(CG) = 6(4 + BF)$ — substitution

$10[2(BF)] = 24 + 6(BF)$ — substitution

$20(BF) = 24 + 6(BF)$

$14(BF) = 24$

$BF = \frac{12}{7}$.

Chapter Review

4. A complete solution is given in the answer key at the back of the text.

5. A complete solution is given in the answer key at the back of the text.

Section 7.3

1. $\overparen{FDC} = 180° = \overparen{FE} + \overparen{ED} + \overparen{DC}$

$= \overparen{FE} + \overparen{ED} + 25°$

$155° = \overparen{FE} + \overparen{ED}$ Equation 1

$\overparen{FGC} = 180° = \overparen{FG} + \overparen{GB} + \overparen{BC}$

$= \overparen{FG} + 105° + BC$

$75° = \overparen{FG} + \overparen{BC}$

$75° = 2\overparen{FE} + \overparen{BC}$ Equation 2

$\angle BAD = \frac{1}{2}(\overparen{BD} - \overparen{GE})$ Theorem 7.13

$= \frac{1}{2}[\overparen{BC} + \overparen{CD} - (\overparen{EF} + \overparen{FG})]$

$= \frac{1}{2}[\overparen{BC} + \overparen{CD} - 3\overparen{EF}]$ $\overparen{FG} = 2\overparen{EF}$

$35° = \frac{1}{2}[\overparen{BC} + 25° - 3\overparen{EF}]$ substituting values

$45° = \overparen{BC} - 3\overparen{EF}$ Equation 3

Solve the system of Equation 2 and Equation 3.

$$
\begin{array}{r}
3\overparen{EF} - \overparen{BC} = -45° \\
2\overparen{EF} + \overparen{BC} = 75° \\
\hline
5\overparen{EF} = 30° \\
\overparen{EF} = 6°
\end{array}
$$

Substituting $\overparen{EF} = 6°$ into Equation 1 and Equation 3 we obtain

$155° = 6° + \overparen{ED}$

$\overparen{ED} = 149°$

and $45° = \overparen{BC} - 18°$

$\overparen{BC} = 63°.$

Chapter Review

2. Method 1: Use segment relationships to find CE.

   $\angle EBA = 90°$ as it is inscribed in a semicircle. So ΔEBA is a right triangle. Use the Pythagorean Theorem to find AB.

$$(AB)^2 + (BE)^2 = (AE)^2$$
$$(AB)^2 + (8\sqrt{3})^2 = (2\sqrt{73})^2 \quad \text{substituting values}$$
$$(AB)^2 = (2\sqrt{73})^2 - (8\sqrt{3})^2$$
$$= 292 - 192$$
$$= 100$$

   Thus, $AB = 10$.

   Hence, $(CE)(CD) = (CA)(CB)$ by Theorem 7.10

$$(CE)9 = 18.8$$
$$9(CE) = 144$$
$$CE = 16$$

   Method 2: Since $\angle EBA = 90°$ we have $\angle EBC = 90°$. Thus ΔEBC is a right triangle. Use the Pythagorean Theorem to find CE.

$$(CE)^2 = (BE)^2 + (CB)^2$$
$$= (8\sqrt{3})^2 + 8^2$$
$$= 192 + 64$$
$$= 256$$

   Thus, $CE = 16$.

   Therefore, using $CE = 16$ from either Method 1 or Method 2 we have

$$DE = CE - CD$$
$$= 16 - 9$$
$$= 7 \text{ cm.}$$

3. $\angle ACE = \frac{1}{2}\overset{\frown}{CE}$ intersecting tangent and chord.

$$= \frac{1}{2}(170°)$$
$$= 85°$$
$$\angle BCD = 10° - 42.5° - 85°$$
$$= 52.5°$$
$$\overset{\frown}{CD} = 2\angle BCD \quad \text{intersecting tangent and chord}$$
$$= 2(52.5°)$$
$$= 105°$$

Chapter Review

4. (a) ΔBOC is a right triangle since $\overleftrightarrow{BC}$ is tangent to circle O at point C.

$$\angle DBC = \frac{1}{2}\angle ABC$$
$$= \frac{1}{2}(67.38°)$$
$$= 33.69°$$

Thus $\angle DOC = 90° + 33.69°$ Exterior Angle Theorem

$= 123.69°.$

Thus $\overset{\frown}{DC} = 123.69°$ as $\angle DOC$ is a central angle.

Now $\overset{\frown}{EC} = 180° - \overset{\frown}{DC}$

$= 180° - 123.69°$

$= 56.31°.$

Therefore $\overset{\frown}{AE} = 56.31°$ as $\overset{\frown}{EC} = \overset{\frown}{AE}$.

(b) Method 1:

Use the Pythagorean Theorem to find BE.

$$(OB)^2 = (OC)^2 + (BC)^2$$
$$(OE + EB)^2 = (OC)^2 + (BC)^2$$
$$(5 + EB)^2 = 5^2 + (7.5)^2$$
$$5 + EB \pm \sqrt{81.25}$$
$$EB \pm \sqrt{81.25} - 5$$

Reject $EB = -\sqrt{81.25} - 5$ as it does not make sense here and we have

$$EB = \sqrt{81.25} - 5$$
$$= \frac{\sqrt{325}}{2} - \frac{10}{2}$$
$$= \frac{5\sqrt{13} - 10}{2} \approx 4.01 \text{ cm.}$$

Method 2:

Use intersecting tangent and secant segment properties.

$$(BC)(BC) = (BE)(BD)$$
$$(7.5)^2 = (BE)(10 + BE)$$
$$56.25 = (BE)^2 + 10(BE)$$
$$(BE)^2 + 10(BE) - 56.25 = 0$$

Use the quadratic formula to obtain BE.

$$BE = \frac{-10 \pm \sqrt{100 - 4(-56.25)}}{2}$$
$$= \frac{-10 \pm \sqrt{325}}{2}$$
$$BE = \frac{-10 \pm 5\sqrt{13}}{2}$$

4. (continued)

Reject $BE = \dfrac{-10 - 5\sqrt{13}}{2}$ as it does not make sense here

and we have

$$\begin{aligned} BE &= \frac{-10 + 5\sqrt{13}}{2} \\ &= \frac{5\sqrt{13} - 10}{2} \\ &\approx 4.01 \text{ cm.} \end{aligned}$$

5. A complete solution is given in the answer key at the back of the text.

Section 7.4

1. First, find the circumcenter which is the intersection of the perpendicular bisectors.

   Step 1: Find the perpendicular bisector of any two sides of ΔABC. (See Construction 6 of Section 4.4.)

   Step 2: The intersection of the perpendicular bisectors will be the center of the circumscribed circle. The radius of the circle is the distance from the circumcenter to any vertex.

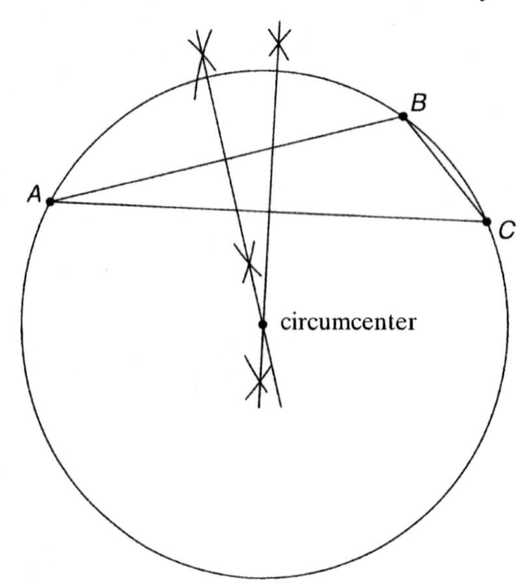

2. (a) The orthocenter is the intersection of the altitudes of ΔABC.

   Step 1: Find the altitudes of any two sides of ΔABC. (See Construction 5 of Section 4.4.)

   Step 2: The intersection of the altitudes is point D, the orthocenter.

Chapter Review 2. (continued)

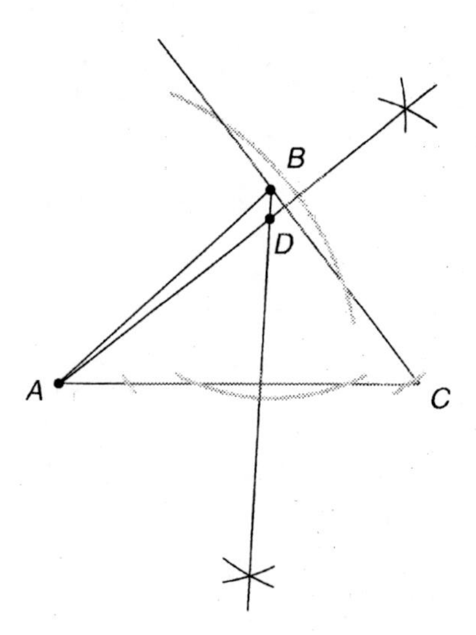

(b) Points A, B, C, and D from an orthocentric set.

Thus,

C is the orthocenter of ΔABD,

B is the orthocenter of ΔACD, and

A is the orthocenter of ΔBCD.

3. Step 1: Construct line through P perpendicular to $\overline{AB}$. (See Construction 5 of Section 4.4.) Label the point of intersection Q.

   Step 2: Set the compass to radius $\overline{PQ}$ and construct circle with center P.

   Step 3: Use Construction 14 of Section 7.4 to construct a tangent to circle P from point A.

   Step 4: Repeat Step 3, only this time construct a tangent to circle P from point B.

   Step 5: The intersection of the tangent lines is point C of ΔABC which has P as its incenter.

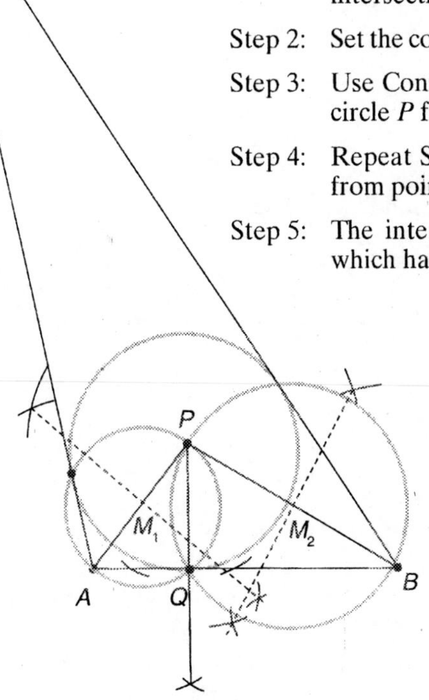

Chapter Review

4. The centroid is the intersection of the medians of ΔABC.

   Step 1: Construct the medians of any two sides of ΔABC. Use Construction 6 of Section 4.4 to find the midpoint.

   Step 2: Find the intersection of the medians. This is the centroid.

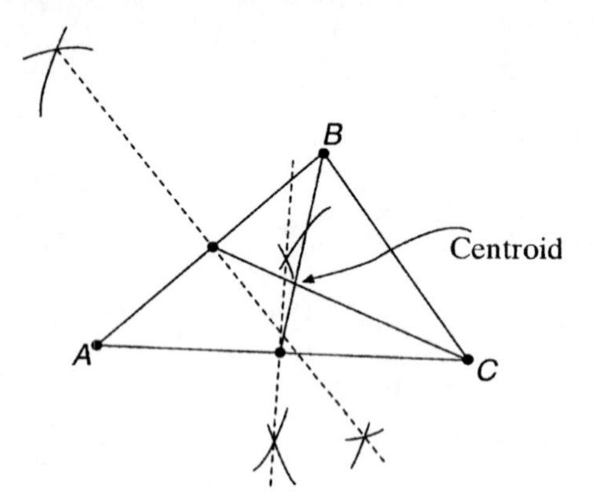

5. Step 1: Follow the procedure given in Problem 3 of Section 5.5 to construct a square.

   Step 2: Construct the perpendicular bisectors of any two adjacent sides of the square. (See Construction 6 of Section 4.4.)

   Step 3: The intersection of the perpendicular bisectors, O, is the center of the circle.

   Step 4: Set the compass to radius OA where A is the intersection of a perpendicular bisector and a side of the square. Construct circle O with radius OA.

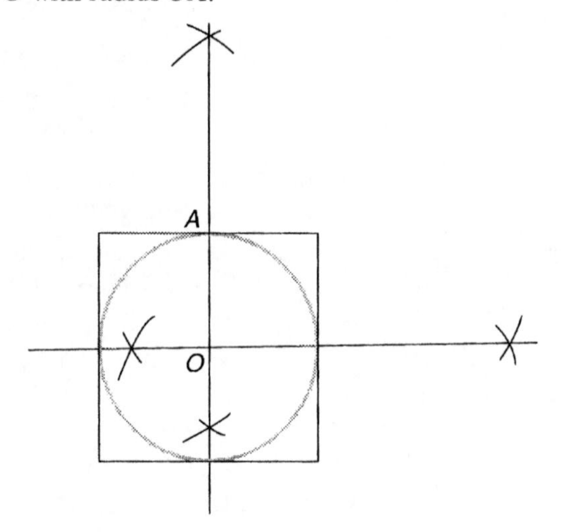

Circle O is inscribed in the square.

6. (a) Follow the procedure given in Construction 16 to construct a common external tangent to circle O and circle O'.

   (b) Follow the procedure given in Construction 15 to construct a common internal tangent to circle O and circle O'.

Chapter Test

Solutions to Chapter 7 Test

1. T — An angle is inscribed in a semicircle if and only if it is a right angle. Thus, the hypotenuse of the triangle must be a diameter if the right angle is inscribed in a semicircle.

2. F — Since the incenter of a triangle is the center of the inscribed circle, it must be inside the triangle.

3. T — See the examples in problems 1–3 in Section 7.4.

4. F — The longest possible chord is the diameter, which has length 4 inches.

5. F — The measure of an inscribed angle is half the measure of its intercepted arc.

6. T — The measure of an arc of a circle equals the measure of the central angle that intercepts the arc.

7. T — If x = the measure of the central angle, then you have $\frac{2\pi}{\pi(4)^2} = \frac{x}{360}$.

   Thus, $16\pi x = 720\pi$, and $16x = 720$ or $x = 45°$.

8. T — Tangent lines are perpendicular to the radius at the point of tangency.

9. T — See Corollary 7.6

10. F — The measure of $\angle POQ$ is less than the measure of $\angle AOB$.

11. (a) $\overleftrightarrow{AC}$ is the only secant line shown.
    (b) $\angle ECD$, $\angle BCE$, and $\angle BCD$ are inscribed angles.
    (c) $\overparen{ECD}$, $\overparen{DEB}$, and $\overparen{BDE}$ are major arcs.
    (d) $\angle OEA$ is a right angle.
    (e) $\overline{BC}$, $\overline{CD}$, and $\overline{CE}$ are all chords.

12. Since $\angle BAC = \frac{1}{2}\overparen{BC}$, you have $35° = \frac{1}{2}\overparen{BC}$, or $\overparen{BC} = 70°$. Then $\overparen{AB} = \overparen{AC} - \overparen{BC} = 126° - 70° = 56°$.

13. You have $\overparen{CD} = 360° - (\overparen{AD} + \overparen{AC})$

    $= 360° - (140° + 126°)$

    $= 94°$.

    Thus, $\angle CAD = \frac{1}{2}\overparen{CD} = \frac{1}{2}(94°) = 47°$.

Chapter Test

14. By Theorem 7.7, $\angle BEC = \frac{1}{2}(\overparen{AD} + \overparen{BC})$.

Since $\angle BAC = 35°$, you have $\overparen{BC} = 70°$.

Then $\angle BEC = \frac{1}{2}(140° + 70°) = 105°$.

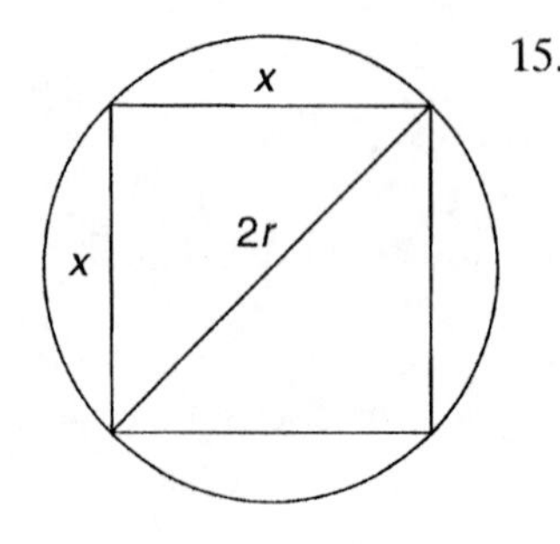

15. (a) The diagonal of the square is $2r$, where r is the radius of the circle. If the sides of the square are each x, then the area of the square is x^2. Applying the Pythagorean Theorem to the right triangle formed by two sides of the square and its diagonal gives you

$$x^2 + x^2 = (2r)^2$$
$$2x^2 = 4r^2$$
$$0.5x^2 = r^2$$

The area of the circle is πr^2. Substituting for r^2 gives you $\pi(0.5x^2) = 0.5\pi x^2$. To find the percentage of area that the square occupies, form a ratio.

$$\frac{\text{area of the square}}{\text{area of the circle}} = \frac{x^2}{0.5\pi x^2}$$
$$= \frac{1}{0.5\pi} \approx 0.6366 \approx 63.7\%.$$

Thus, the square plug fills 63.7% of the circular hole.

(b) If the circle has radius x, the sides of the square are each $2x$. Then the percentage of the hole that the plug fills is given by

$$\frac{\text{area of the circle}}{\text{area of the square}} = \frac{\pi x^2}{(2x)^2}$$
$$= \frac{\pi x^2}{4x^2} = \frac{\pi}{4} \approx 0.785 = 78.5\%.$$

(c) The circular plug fits best.

16. (a) Use Theorem 7.9.

$$\angle ADB = \frac{1}{2}(\overparen{AB} - \overparen{CE})$$
$$8° = \frac{1}{2}(\overparen{AB} - 14°)$$
$$16° = \overparen{AB} - 14°$$
$$30° = \overparen{AB}$$

(b) First find $\overparen{EF}$.

$$\overparen{EF} = 360° - (\overparen{FG} + \overparen{AG} + \overparen{AB} + \overparen{BC} + \overparen{CE})$$
$$= 360° - (95° + 87° + 30° + 115° + 14°)$$
$$= 19°.$$

Chapter Test

15. (continued)

$$\text{Then } \angle EDF = \frac{1}{2}(\widehat{AG} - \widehat{EF})$$
$$= \frac{1}{2}(87^\circ - 19^\circ)$$
$$= 34^\circ.$$

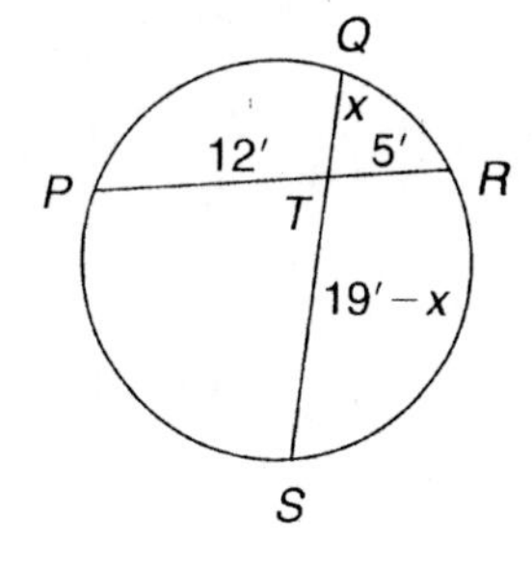

17. By Theorem 7.8, $(PT)(TR) = (TS)(QT)$.
If x represents QT, then $TS = 19 - x$. Substitute for QT and TS.

$$(12)(5) = (19 - x)(x)$$
$$60 = 19x - x^2$$
$$x^2 - 19x + 60 = 0$$
$$(x - 15)(x - 4) = 0$$
$$x = 4 \text{ or } x = 15$$

If $TS > QT$, then $x = 4$ is the correct solution. This means that $TS = 19 - 4 = 15$ ft and $QT = 4$ ft.

18. $\angle OAB = 90^\circ$ because tangent lines are perpendicular to the radius at the point of tangency. Therefore, ΔAOB is a right triangle. Let x be the length of the radius. Then $OA = x$ and $OB = x + 3$. Use the Pythagorean Theorem.

$$x^2 + 7^2 = (x + 3)^2$$
$$x^2 + 49 = x^2 + 6x + 9$$
$$49 = 6x + 9$$
$$40 = 6x$$
$$\frac{40}{6} = x \text{ or } x = \frac{20}{3}$$

Then $DE = 2x = 2 \cdot \frac{20}{3} = \frac{40}{3} \approx 13.33$.

19. See the solution to problem #9 in Section 7.4.

20. A complete description of this construction has been given in Section 7.4. See Construction 14.

21. A complete solution is given in the answer key at the back of the text.

22. A complete solution is given in the answer key at the back of the text.

Chapter Test

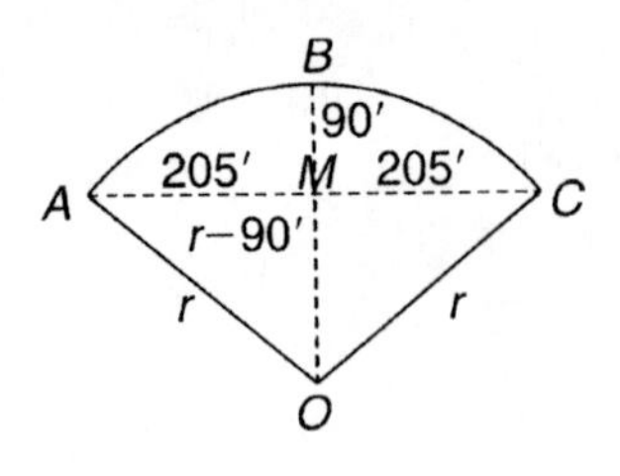

23. Extend $\overline{BM}$ so that it passes through the center O of the circle containing $\overset{\frown}{AC}$. Then draw $\overline{OA}$ and $\overline{OC}$. Let r be the radius of the circle. Then you have $OB = r$ and $OM = r - 90$. Since $\overline{BO}$ is the perpendicular bisector of chord $\overline{AC}$, ΔAMO is a right triangle. Apply the Pythagorean Theorem to ΔAMO.

$$(205)^2 + (r - 90)^2 = r^2$$
$$42025 + r^2 - 180r + 8100 = r^2$$
$$50125 - 180r = 0$$
$$50125 = 180r$$
$$278 \text{ ft} \approx r$$

The radius of curvature is approximately 278 ft.

24. By Postulate 7.2, the area of the field is given by

$$\text{Area} = \frac{95^\circ}{360^\circ} \cdot \pi(420)^2 \approx 146{,}200 \text{ ft}^2.$$

8

Coordinate Geometry

Section 8.1

Tips:

- ✔ An easy way to remember the distance formula is to think of it as an application of the PythagoreanTheorem.
- ✔ The coordinates of the midpoint of a segment are the average of the x-coordinates and the average of the y-coordinates.

Solutions to odd-numbered textbook problems

1. A complete solution is given in the answer key at the back of the text.

3. Use the distance formula, Theorem 8.1.

   (a) $\sqrt{(3-0)^2+(4-0)^2} = \sqrt{9+16} = \sqrt{25} = 5$

   (b) $\sqrt{[3-(-1)]^2+(5-2)^2} = \sqrt{4^2+3^2} = \sqrt{25} = 5$

   (c) $\sqrt{[-5-(-1)]^2+[-3-(-3)]^2} = \sqrt{(-4)^2+0^2} = \sqrt{16} = 4$

   (d) $\sqrt{(3-3)^2+[5-(-4)]^2} = \sqrt{0^2+9^2} = \sqrt{81} = 9$

5. Theorem 5.17 states that if the opposite sides of a quadrilateral are congruent, then the quadrilateral is a parallelogram. To show that $ABCD$ is a parallelogram, you need only show that opposite sides are of equal length.

   $AB = \sqrt{[2-(-1)]^2+(4-1)^2} = \sqrt{3^2+3^2} = \sqrt{18} = 3\sqrt{2}$

   $BC = \sqrt{(6-2)^2+(1-4)^2} = \sqrt{4^2+(-3)^2} = \sqrt{25} = 5$

   $CD = \sqrt{(3-6)^2+(-2-1)^2} = \sqrt{(-3)^2+(-3)^2} = \sqrt{18} = 3\sqrt{2}$

   $DA = \sqrt{(-1-3)^2+[1-(-2)]^2} = \sqrt{(-4)^2+3^2} = \sqrt{25} = 5$

   Therefore, $AB = CD$ and $BC = DA$. The quadrilateral is a parallelogram.

7. If the sum of the two shorter distances equals the longest distance, the points are collinear.

   (a) $PQ = \sqrt{[-2-(-1)]^2+(3-4)^2} = \sqrt{(-1)^2+(-1)^2}$
   $= \sqrt{1+1} = \sqrt{2}$

   $QR = \sqrt{[-4-(-2)]^2+(1-3)^2} = \sqrt{(-2)^2+(-2)^2}$
   $= \sqrt{4+4} = \sqrt{8} = \sqrt{4 \cdot 2} = 2\sqrt{2}$

   $PR = \sqrt{[-4-(-1)]^2+(1-4)^2} = \sqrt{(-3)^2+(-3)^2}$
   $= \sqrt{9+9} = \sqrt{18} = \sqrt{9 \cdot 2} = 3\sqrt{2}$

   Since $\sqrt{2} + 2\sqrt{2} = 3\sqrt{2}$, points P, Q, and R are collinear.

Section 8.1

7. (continued)

(b) $PQ = \sqrt{[3-(-2)]^2 + (4-1)^2} = \sqrt{5^2 + 3^2}$
$= \sqrt{25 + 9} = \sqrt{33} \approx 5.745$
$QR = \sqrt{(12-3)^2 + (10-4)^2} = \sqrt{9^2 + 6^2}$
$= \sqrt{81 + 36} = \sqrt{117} \approx 10.817$
$PR = \sqrt{(12-2)^2 + (10-1)^2} = \sqrt{14^2 + 9^2}$
$= \sqrt{196 + 81} = \sqrt{277} \approx 16.643$
Since $\sqrt{33} + \sqrt{117} \neq \sqrt{277}$, P, Q, and R are not collinear.

(c) $PQ = \sqrt{[2-(-2)]^2 + [-1-(-3)]^2} = \sqrt{4^2 + 2^2}$
$= \sqrt{16 + 4} = \sqrt{20} = \sqrt{4 \cdot 5} = 2\sqrt{5}$
$QR = \sqrt{(10-2)^2 + [3-(-1)]^2} = \sqrt{8^2 + 4^2}$
$= \sqrt{64 + 16} = \sqrt{80} = \sqrt{16 \cdot 5} = 4\sqrt{5}$
$PR = \sqrt{[10-(-2)]^2 + [3-(-3)]^2} = \sqrt{12^2 + 6^2}$
$= \sqrt{144 + 36} = \sqrt{180} = \sqrt{36 \cdot 5} = 6\sqrt{5}$
Since $2\sqrt{5} + 4\sqrt{5} = 6\sqrt{5}$, points P, Q, and R are collinear.

9. Use the distance formula with points $S(5, 2)$ and $T(2n, n)$.

$$\sqrt{(5-2n)^2 + (2-n)^2} = \sqrt{2}$$
$$(5-2n)^2 + (2-n)^2 = 2$$
$$25 - 20n + 4n^2 + 4 - 4n + n^2 = 2$$
$$5n^2 - 24n + 27 = 0$$
$$(5n-9)(n-3) = 0$$
$$5n - 9 = 0 \text{ or } n - 3 = 0$$
$$n = \frac{9}{5} \qquad n = 3$$

Therefore, $n = 3$ or $n = \frac{9}{5}$.

11. If the sum of the two shorter distances equals the longest distance, the points are collinear. Let C be the point given in each part of this problem.

(a) $AB = \sqrt{[1-(-3)]^2 + (3-0)^2} = \sqrt{16 + 9} = 5$
$BC = \sqrt{(6-1)^2 + (6-3)^2} = \sqrt{25 + 9} = \sqrt{34}$
$AC = \sqrt{[6-(-3)]^2 + (6-0)^2} = \sqrt{81 + 36} = \sqrt{117} = 3\sqrt{13}$
Since $5 + \sqrt{34} \neq 3\sqrt{13}$, points A, B, and C are not collinear.

(b) $AC = \sqrt{[0-(-3)]^2 + (2-0)^2} = \sqrt{9 + 4} = \sqrt{13}$
$CB = \sqrt{(1-0)^2 + (3-2)^2} = \sqrt{1 + 1} = \sqrt{2}$
$AB = \sqrt{[1-(-3)]^2 + (3-0)^2} = \sqrt{16 + 9} = \sqrt{25} = 5$
Since $\sqrt{13} + \sqrt{2} \neq 5$, points A, B, and C are not collinear.

Section 8.1

11. (continued)

(c) $CA = \sqrt{[-3-(-7)]^2 + [0-(-3)]^2} = \sqrt{16+9} = \sqrt{25} = 5$
$AB = \sqrt{[1-(-3)]^2 + (3-0)^2} = \sqrt{16+9} = \sqrt{25} = 5$
$CB = \sqrt{[1-(-7)]^2 + [3-(-3)]^2} = \sqrt{64+36} = \sqrt{100} = 10$
Since $5 + 5 = 10$, points A, B, and C are collinear.

(d) $AB = \sqrt{[1-(-3)]^2 + (3-0)^2} = \sqrt{16+9} = \sqrt{25} = 5$
$BC = \sqrt{(9-1)^2 + (9-3)^2} = \sqrt{64+36} = \sqrt{100} = 10$
$AC = \sqrt{[9-(-3)]^2 + (9-0)^2} = \sqrt{144+81} = \sqrt{225} = 15$
Since $5 + 10 = 15$, points A, B, and C are collinear.

13. (a) Use the midpoint formula.
$\left(\frac{0+(-3)}{2}, \frac{2+2}{2}\right) = \left(\frac{-3}{2}, 2\right)$

(b) $\left(\frac{-5+3}{2}, \frac{-1+5}{2}\right) = (-1, 2)$

(c) $\left(\frac{-2+(-3)}{2}, \frac{3+6}{2}\right) = \left(\frac{-5}{2}, \frac{9}{2}\right)$

(d) $\left(\frac{3+3}{2}, \frac{-5+7}{2}\right) = (3, 1)$

(e) $\left(\frac{1+3}{2}, \frac{5+9}{2}\right) = (2, 7)$

(f) $\left(\frac{6+(-3)}{2}, \frac{-2+5}{2}\right) = \left(\frac{3}{2}, \frac{3}{2}\right)$

15. (a) Find the lengths of the sides of the triangle.
$AB = 5, BC = 5\ AC = \sqrt{(5-0)^2 + (-5-0)^2} = \sqrt{50}$
Since $AB = BC$, the triangle is isosceles.
Now $(AB)^2 + (BC)^2 = 5^2 + 5^2$
$= 50$
$= (AC)^2.$
So by the Pythagorean Theorem, the triangle is a right triangle.
Therefore, ΔABC is a right isosceles triangle.

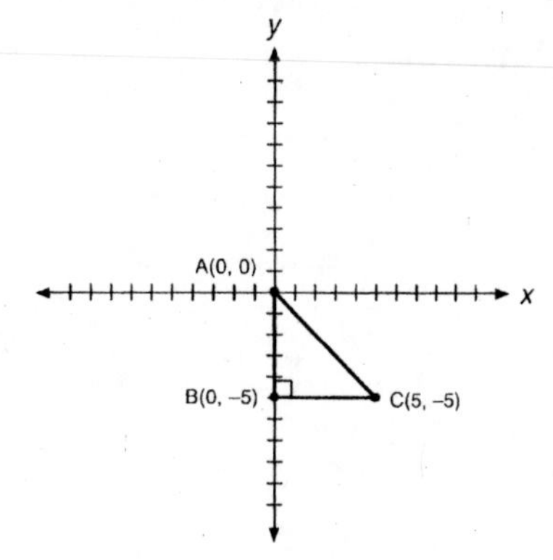

Section 8.1

15. (continued)

(b) $AB = \sqrt{[1-(-3)]^2 + (3-1)^2} = \sqrt{4^2 + 2^2} = \sqrt{20} = 2\sqrt{5}$

$BC = \sqrt{(5-1)^2 + (-5-3)^2} = \sqrt{4^2 + (-8)^2} = \sqrt{80} = 4\sqrt{5}$

$AC = \sqrt{[5-(-3)]^2 + (-5-1)^2} = \sqrt{8^2 + (-6)^2} = \sqrt{100} = 10$

All sides of the triangle are of different lengths; thus, ΔABC is scalene.

Now $(AB)^2 + (BC)^2 = 20 + 80$

$= 100$

$= (AC)^2.$

So ΔABC is a right triangle, by the Pythagorean Theorem.

Therefore, ΔABC is a right scalene triangle.

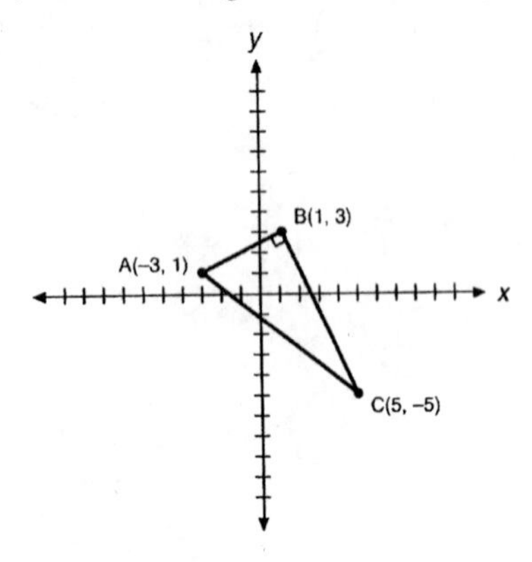

(c) $AB = \sqrt{[2-(-2)]^2 + [2-(-1)]^2} = \sqrt{4^2 + 3^2} = \sqrt{25} = 5$

$BC = \sqrt{(6-2)^2 + (-1-2)^2} = \sqrt{4^2 + (-3)^2} = \sqrt{25} = 5$

$AC = 8$

Since $AB = BC$, ΔABC is isosceles.

$(AB)^2 + (BC)^2 = 5^2 + 5^2$

$= 50$

< 64

So $(AB)^2 + (BC)^2 < (AC)^2$, hence $\angle B > 90°$.

Therefore, ΔABC is an obtuse isosceles triangle.

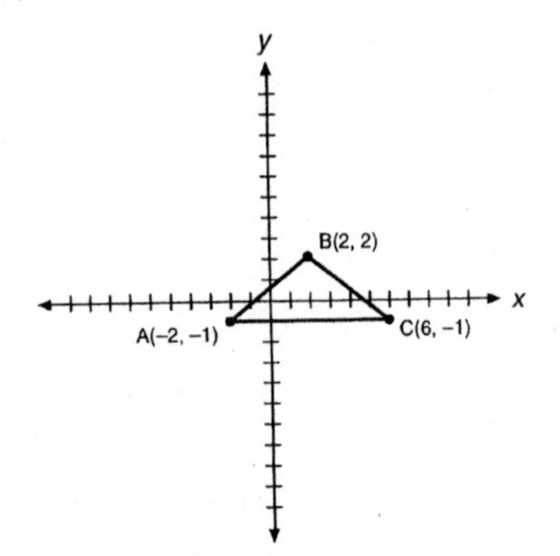

Section 8.1

15. (continued)

(d) $AB = \sqrt{[-1-(-4)]^2 + [3-(-2)]^2} = \sqrt{3^2+5^2} = \sqrt{34}$

$BC = \sqrt{[4-(-1)]^2 + (-2-3)^2} = \sqrt{5^2+(-5)^2} = \sqrt{50}$

$AC = 8$

ΔABC is scalene since $AB \neq BC \neq AC$.

$$\begin{aligned}(AB)^2 + (BC)^2 &= 34 + 50\\ &= 84\\ &> 64\end{aligned}$$

So $(AB)^2 + (BC)^2 > (AC)^2$, hence $\angle B < 90°$.

Therefore, ΔABC is an acute scalene triangle.

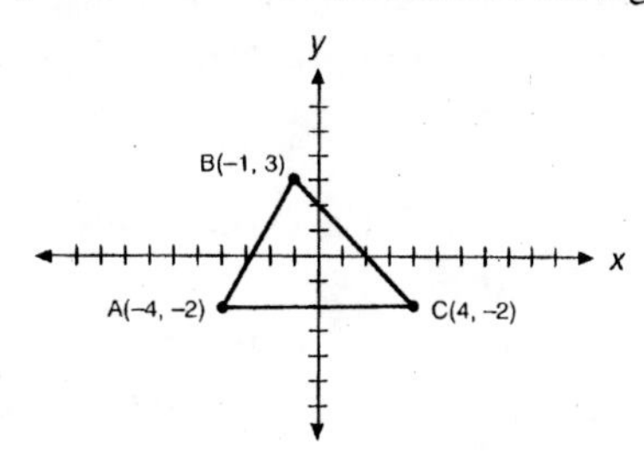

17. (a) Given points $A(-2,3)$, $B(5,5)$, and $C(7,-2)$,

$AB = \sqrt{[5-(-2)]^2 + (5-3)^2} = \sqrt{7^2+2^2} = \sqrt{53}$

$BC = \sqrt{(7-5)^2 + (-2-5)^2} = \sqrt{2^2+(-7)^2} = \sqrt{53}$

$AC = \sqrt{[7-(-2)]^2 + (-2-3)^2} = \sqrt{9^2+(-5)^2} = \sqrt{106}.$

$$\begin{aligned}\text{Now } (AB)^2 + (BC)^2 &= 53 + 53\\ &= 106\\ &= (AC)^2.\end{aligned}$$

Therefore, ΔABC is a right triangle.

(b) Given points $A(0,-6)$, $B(1,0)$, and $C(6,-7)$,

$AB = \sqrt{(1-0)^2 + [0-(-6)]^2} = \sqrt{1^2+16^2} = \sqrt{37}$

$BC = \sqrt{(6-1)^2 + (-7-0)^2} = \sqrt{5^2+(-7)^2} = \sqrt{74}$

$AC = \sqrt{(6-0)^2 + [-7-(-6)]^2} = \sqrt{6^2+(-1)^2} = \sqrt{37}$

$$\begin{aligned}\text{Now } (AB)^2 + (AC)^2 &= 37 + 37\\ &= 74\\ &= (BC)^2.\end{aligned}$$

Therefore, ΔABC is a right triangle.

(c) Given points $A(-7,-5)$, $B(-4,8)$, and $C(3,5)$,

$AB = \sqrt{[-4-(-7)]^2 + [8-(-5)]^2} = \sqrt{3^2+13^2} = \sqrt{178}$

$BC = \sqrt{[3-(-4)]^2 + (5-8)^2} = \sqrt{7^2+(-3)^2} = \sqrt{58}$

$AC = \sqrt{[3-(-7)]^2 + [5-(-5)]^2} = \sqrt{10^2+10^2} = \sqrt{200}$

$$\begin{aligned}\text{Now } (AB)^2 + (BC)^2 &= 178 + 58\\ &= 236\\ &\neq 200.\end{aligned}$$

Since $(AB)^2 + (BC)^2 \neq (AC)^2$, ΔABC is not a right triangle.

Section 8.1

19. The diagonals of a rectangle are congruent.

The diagonals of quadrilateral $PQRS$ are $\overline{PR}$ and $\overline{QS}$.

$PR = \sqrt{(2-1)^2 + (5-1)^2} = \sqrt{1^2 + 4^2} = \sqrt{17}$

$QS = \sqrt{(-1-4)^2 + (4-2)^2} = \sqrt{(-5)^2 + 2^2} = \sqrt{29}$

Therefore, $PR \neq QS$ so $\overline{PR}$ and $\overline{QS}$ are not congruent.

Hence, $PQRS$ is not a rectangle.

21. If ΔABC is an equilateral triangle, $AB = BC = CA$. You know that $AB = \sqrt{(0-0)^2 + (10-0)^2} = 10$. By Theorem, 4.9, C is on the perpendicular bisector of $\overline{AB}$, and $\overline{AB}$ is on the y-axis. Thus, the y-coordinate will be the same as the y-coordinate of the midpoint of $\overline{AB}$. So you have $y = \dfrac{0 + 10}{2} = 5$.

If C has coordinates $(x, 5)$, you can find x by using the distance formula.

$$CA = 10 = \sqrt{(x-0)^2 + (5-0)^2}$$
$$10 = \sqrt{x^2 + 25}$$
$$100 = x^2 + 25$$
$$75 = x^2$$
$$\pm\sqrt{75} = \pm\sqrt{25 \cdot 3} = \pm 5\sqrt{3} = x$$

Thus, $x = \pm 5\sqrt{3} \approx \pm 8.66$.

There are two possibilities for the third vertex C:

$(5\sqrt{3}, 5)$ or $(-5\sqrt{3}, 5)$.

23. Find the lengths of the sides of quadrilateral $ABCD$ and $A'B'C'D'$.

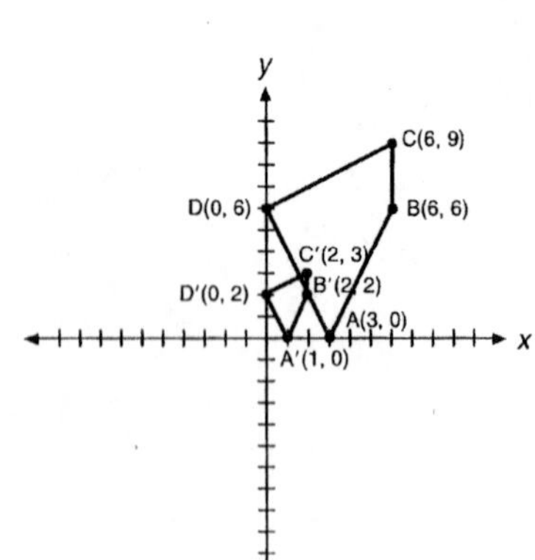

$$AB = \sqrt{(6-3)^2 + (6-0)^2} = \sqrt{3^2 + 6^2} = \sqrt{45} = 3\sqrt{5}$$
$$BC = 3$$
$$CD = \sqrt{(0-6)^2 + (6-9)^2} = \sqrt{(-6)^2 + (-3)^2} = \sqrt{45} = 3\sqrt{5}$$
$$DA = \sqrt{(3-0)^2 + (0-6)^2} = \sqrt{3^2 + (-6)^2} = \sqrt{45} = 3\sqrt{5}$$

$$A'B' = \sqrt{(2-1)^2 + (2-0)^2} = \sqrt{1^2 + 2^2} = \sqrt{5} = \frac{1}{3}AB$$
$$B'C' = 1 = \frac{1}{3}BC$$
$$C'D' = \sqrt{(0-2)^2 + (2-3)^2} = \sqrt{(-2)^2 + (-1)^2} = \sqrt{5} = \frac{1}{3}CD$$
$$D'A' = \sqrt{(1-0)^2 + (0-2)^2} = \sqrt{1^2 + (-2)^2} = \sqrt{5} = \frac{1}{3}DA$$

The sides of $A'B'C'D'$ are one-third as long as the sides of $ABCD$.

Section 8.1

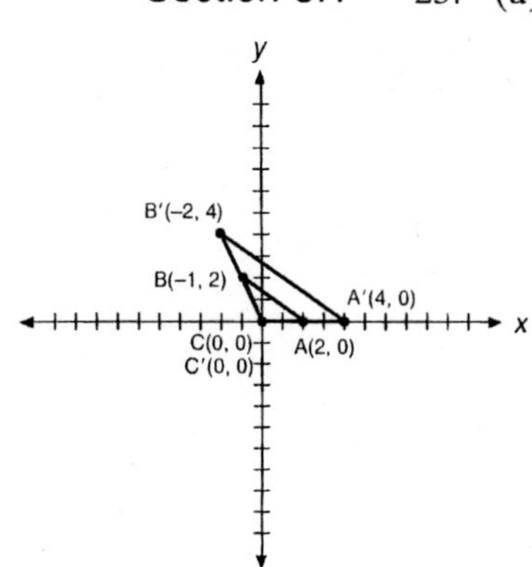

25. (a) Find the lengths of the sides of ΔABC and $\Delta A'B'C'$.

$AB = \sqrt{(-1 - 2)^2 + (2 - 0)^2} = \sqrt{(-3)^2 + 2^2} = \sqrt{13}$

$BC = \sqrt{[0 - (-1)]^2 + (0 - 2)^2} = \sqrt{1^2 + (-2)^2} = \sqrt{5}$

$AC = 2$

$A'B' = \sqrt{(-2 - 4)^2 + (4 - 0)^2} = \sqrt{(-6)^2 + 4^2} = \sqrt{52} = 2\sqrt{13} = 2(AB)$

$B'C' = \sqrt{[0 - (-2)]^2 + (0 - 4)^2} = \sqrt{2^2 + (-4)^2} = \sqrt{20} = 2\sqrt{5} = 2(BC)$

$A'C' = 4 = 2(AC)$

The sides of $\Delta A'B'C'$ are twice as long as the sides of ΔABC.

(b) Find the area of ΔABC, and $\Delta A'B'C'$.

$$A_{\Delta ABC} = \frac{1}{2}bh = \frac{1}{2}(2)(2) = 2$$

$$A_{\Delta A'B'C'} = \frac{1}{2}bh = \frac{1}{2}(4)(4) = 8 = 4A_{\Delta ABC}$$

The area of $\Delta A'B'C'$ is four times the area of ΔABC.

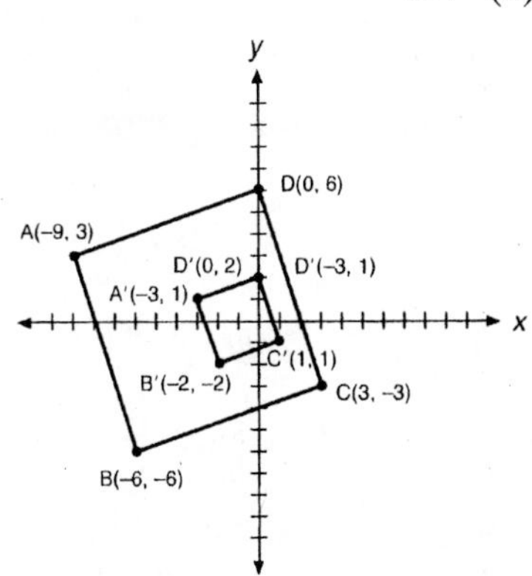

27. (a) Find AB, BC, and diagonals AC and BD

$AB = \sqrt{[-6 - (-9)]^2 + (-6 - 3)^2} = \sqrt{3^2 + (-9)^2} = \sqrt{90} = 3\sqrt{10}$

$BC = \sqrt{[3 - (-6)]^2 + [-3 - (-6)]^2} = \sqrt{9^2 + 3^2} = \sqrt{90} = 3\sqrt{10}$

$AC = \sqrt{[3 - (-9)]^2 + (-3 - 3)^2} = \sqrt{12^2 + (-6)^2} = \sqrt{180} = 6\sqrt{5}$

$BD = \sqrt{[0 - (-6)]^2 + [6 - (-6)]^2} = \sqrt{6^2 + 12^2} = \sqrt{180} = 6\sqrt{5}$

Since the diagonals are congruent and since two adjacent sides are congruent we conclude that $ABCD$ is a square.

(b) Area $= (3\sqrt{10})^2 = 90$ square units

(c) Verify that quadrilateral $A'B'C'D'$ is a square.

$A'B' = \sqrt{[-2 - (-3)]^2 + (-2 - 1)^2} = \sqrt{1 + (-3)^2} = \sqrt{10}$

$B'C' = \sqrt{[1 - (-2)]^2 + [-1 - (-2)]^2} = \sqrt{3^2 + 1^2} = \sqrt{10}$

$A'C' = \sqrt{[1 - (-3)]^2 + (-1 - 1)^2} = \sqrt{4^2 + (-2)^2} = \sqrt{20} = 2\sqrt{5}$

$B'D' = \sqrt{[0 - (-2)]^2 + [2 - (-2)]^2} = \sqrt{2^2 + 4^2} = \sqrt{20} = 2\sqrt{5}$

The diagonals are congruent and two adjacent sides are congruent, thus $A'B'C'D'$ is a square.

$$\text{Area} = (\sqrt{10})^2 = 10 \text{ square units}$$

The area of $A'B'C'D'$ is one-ninth the area of $ABCD$.

Section 8.1

29. The circle consists of every point whose distance from the center O is 13. Let $P(x, y)$ be the coordinates of a point on the circle, then $OP = \sqrt{(x-0)^2 + (y-0)^2} = 13.$

or $$x^2 + y^2 = 13^2.$$

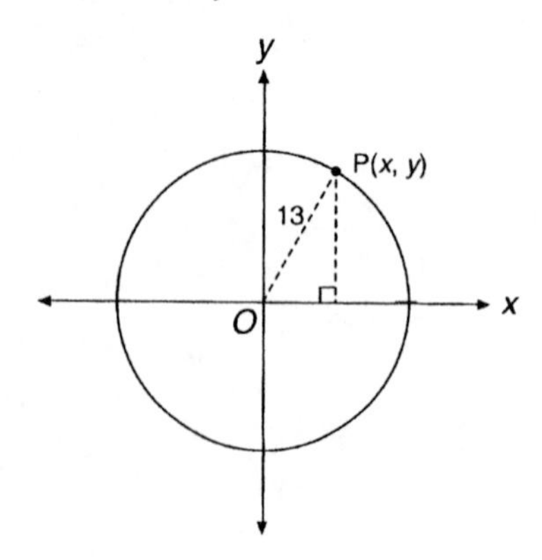

There are an infinite number of points that satisfy $x^2 + y^2 = 13$, let us find those with integer values. The following points are the most obvious:

$(13, 0), (-13, 0), (0, 13)$ and $(0, -13)$.

Notice that for any point on the circle that does not lie on the x-axis or y-axis, a right triangle can be formed in the circle with hypotenuse 13.

Thus, to find the other integer valued coordinates use the common 5 –12–13 right triangle ratio.

The other points are: (5, 12), (5, −12), (−5, 12), (−5, −12), (12, 5), (−12, 5), (12, −5), and (−12, −5).

31. A complete solution is given in the answer key at the back of the text.

33. See problem #32 for the location of each octant. All octants above the xy-plane, octants 1, 2, 3, and 4, have positive z-coordinates. All octants below the xy-plane, i.e., octants 5, 6, 7, and 8, have negative z-coordinates. Octants 1, 4, 5, and 8 have positive x-coordinates. Similarly, octants 1, 2, 5, and 6 have positive y-coordinates. Therefore, the octants have coordinates (x, y, z) as follows: Octant 1 $(+, +, +)$, Octant 2 $(-, +, +)$, Octant 3 $(-, -, +)$, Octant 4 $(+, -, +)$, Octant 5 $(+, +, -)$, Octant 6 $(-, +, -)$, Octant 7 $(-, -, -)$, Octant 8 $(+, -, -)$.

35. If $P(x_1, y_1, z_1)$ and $Q(x_2, y_2, z_2)$ are two points in three-dimensional space, then $PQ = \sqrt{(x_2 - x_1)^2 + (y_2 - y_1)^2 + (z_2 - z_1)^2}$.

(a) For $P(1, 2, 3)$ and $Q(2, 3, 1)$,

$$PQ = \sqrt{(2-1)^2 + (3-2)^2 + (1-3)^2} = \sqrt{1^2 + 1^2 + (-2)^2} = \sqrt{6}.$$

(b) For $P(-1, 0, 5)$ and $Q(6, 2, -1)$,

$$PQ = \sqrt{[6-(-1)]^2 + (2-0)^2 + (-1-5)^2} = \sqrt{7^2 + 2^2 + (-6)^2} = \sqrt{89}.$$

Section 8.1

37. (a)

$\angle A = 30°$

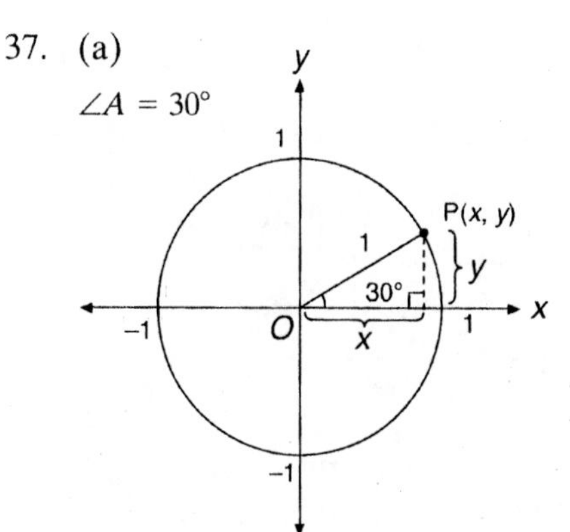

Form a right triangle with $\angle A = 30°$. Now we have a 30°–60° right triangle so $x = \frac{\sqrt{3}}{2}$ and $y = \frac{1}{2}$. Thus point P has coordinates $\left(\frac{\sqrt{3}}{2}, \frac{1}{2}\right)$.

(b)

$\angle A = 90°$

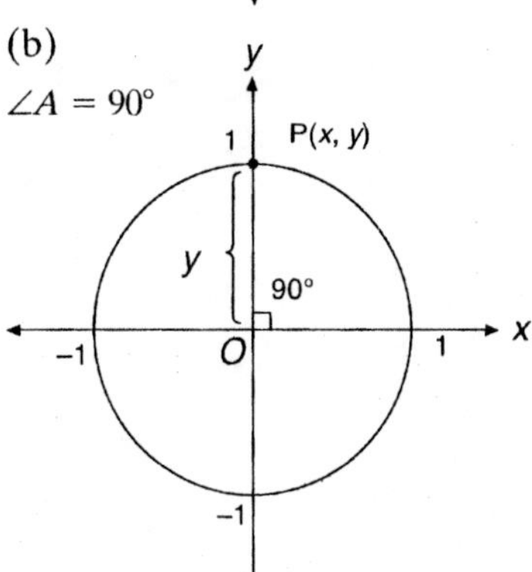

The coordinates of P, which lies on the y-axis, are $(0, 1)$.

(c)

$\angle A = 135°$

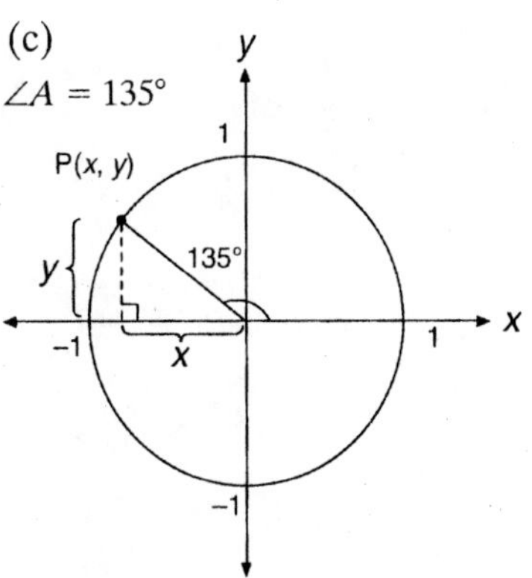

Form a right triangle with hypotenuse OP and one leg on the x-axis. This is a 45°–45° right triangle, with each leg of the length $\frac{\sqrt{2}}{2}$.

Since the x-coordinate of point P lies in quadrant II, it is a negative value.

Therefore, the coordinates of point P are $\left(-\frac{\sqrt{2}}{2}, \frac{\sqrt{2}}{2}\right)$.

(d)

$\angle A = 0°$

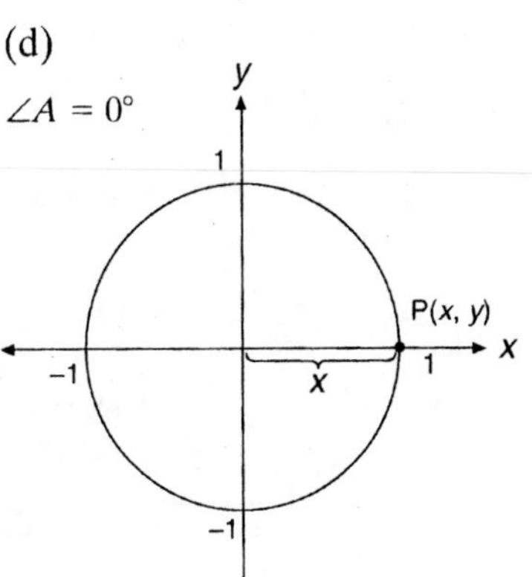

The coordinates of P, which lies on the x-axis, are $(1, 0)$.

Section 8.1

39. (a) If $\angle A = 30°$, then the coordinates of the point $P(x, y)$ on the unit circle are $(x, y) = (\cos 30°, \sin 30°) = \left(\frac{\sqrt{3}}{2}, \frac{1}{2}\right)$.

(b) If $\angle A = 90°$, then the coordinates of the point $P(x, y)$ on the unit circle are $(x, y) = (\cos 90°, \sin 90°) = (0, 1)$.

(c) If $\angle A = 135°$, then the coordinates of the point $P(x, y)$ on the unit circle are $(x, y) = (\cos 135°, \sin 135°) = \left(\frac{-\sqrt{2}}{2}, \frac{\sqrt{2}}{2}\right)$.

(d) If $\angle A = 0°$, then the coordinates of the point $P(x, y)$ on the unit circle are $(x, y) = (\cos 0°, \sin 0°) = (1, 0)$.

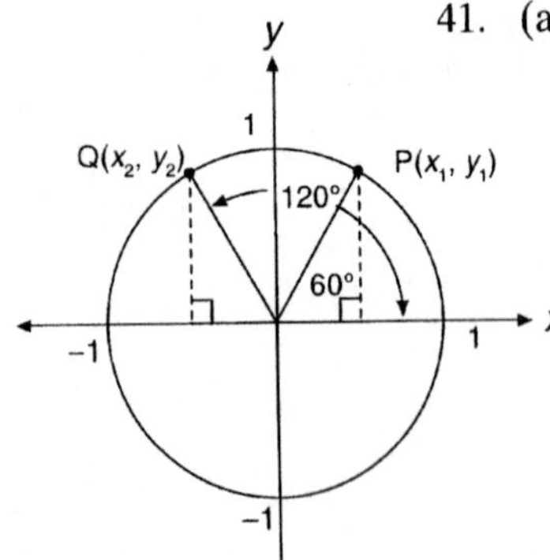

41. (a) Find the y-values of points P and Q.

$$y_1 = \sin 60° = \frac{\sqrt{3}}{2}$$

$$y_2 = \sin 120° = \frac{\sqrt{3}}{2}$$

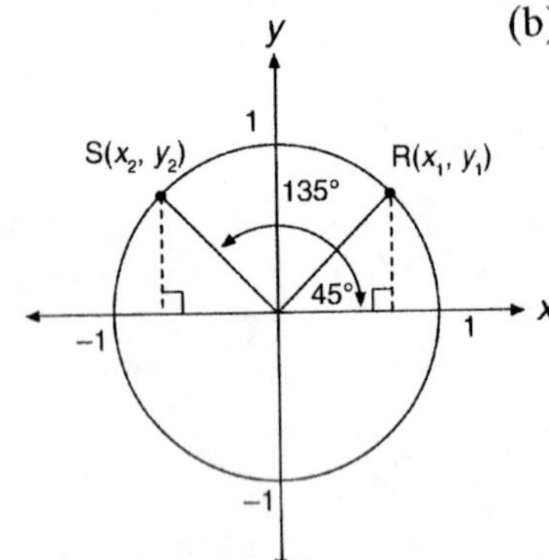

(b) Find the y-values of points R and S.

$$y_1 = \sin 45° = \frac{\sqrt{2}}{2}$$

$$y_2 = \sin 45° = \frac{\sqrt{2}}{2}$$

(c) $\sin A = \sin (180° - A)$

The points associated with these angles are symmetric with respect to the y-axis.

43. (a) Law of Cosines: $c^2 = a^2 + b^2 - 2ab \cos C$

$c^2 = (10.1)^2 + 12^2 - 2(10.1)(12) \cos 115°$

$c^2 = 246 - 242.4 \cos 115°$

$c = \sqrt{246 - 242.4 \cos 115°} \approx 18.7$

Section 8.1

43. (continued)

(b) Law of Cosines: $a^2 = b^2 + c^2 - 2bc \cos A$

$$a^2 = (18.2)^2 + (25.3)^2 - 2(18.2)(25.3) \cos 170°$$
$$= 971.33 - 920.92 \cos 170°$$
$$a = \sqrt{971.33 - 920.92 \cos 170°} \approx 43.3$$

45. (a) A complete solution is given in the answer key at the back of the text.

(b) Let $(x_1, y_1) = A(2, 3), (x_2, y_2) = B(4, 6)$ and $(x_3, y_3) = C(9, 8)$.

Use the area formula in part (a).

$$A = \frac{1}{2}[2(8 - 6) + 4(3 - 8) + 9(6 - 3)]$$
$$= \frac{1}{2}[4 - 20 + 27]$$
$$= \frac{11}{2} = 5.5 \text{ square units}$$

The area formula developed in the "Geometry Investigation" is

$$A = \frac{1}{2}(x_2 - x_1)(y_1 + y_2) + \frac{1}{2}(x_3 - x_2)(y_2 + y_3) - \frac{1}{2}(x_3 - x_1)(y_1 + y_3)$$
$$= \frac{1}{2}(4 - 2)(3 + 6) + \frac{1}{2}(9 - 4)(6 + 8) - \frac{1}{2}(9 - 2)(3 + 8)$$
$$= \frac{1}{2}(2 \cdot 9 + 5 \cdot 14 - 7 \cdot 11)$$
$$= \frac{1}{2}(18 + 70 - 77)$$
$$= \frac{11}{2} = 5.5 \text{ square units.}$$

The areas are the same.

(c) Let $(x_1, y_1) = A(-5, -4)$

$(x_2, y_2) = B(4, -4)$

$(x_3, y_3) = C(-5, 9)$.

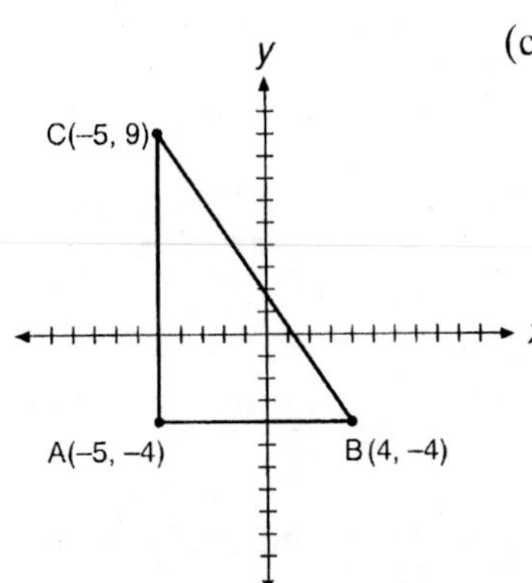

Use the formula from part (a).

$$A = \frac{1}{2}\{(-5)[9 - (-4)] + 4(-4 - 9) + (-5)[-4 - (-4)]\}$$
$$= \frac{1}{2}[(-5)(13) + 4(-13) + (-5)(0)]$$
$$= \frac{1}{2}[-65 - 52]$$
$$= \frac{-117}{2} = -58.5$$

Section 8.1 45. (continued)

Using the area formula $A = \frac{1}{2}bh$ we obtain

$$A = \frac{1}{2}(9)(13)$$
$$= \frac{117}{2} = 58.5 \text{ square units.}$$

The formula from part (a) gives an answer that is opposite the area of the triangle.

The results differ because of the labeling of points A, B, and C. If we labeled the coordinates $A(-5, -4)$, $B(-5, 9)$, and $C(4, -4)$ the formula from part (a) would result in 58.5.

The triangle given in "Geometry Investigation" at the beginning of the chapter had each y-value as positive. If any or all of the triangle lies below the x-axis, we can add a positive constant k to each y-value to translate ΔABC above the x-axis. The following shows that the formula given in part (a) still holds.

$$\frac{1}{2}\{x_1[y_3 + k - (y_2 + k)] + x_2[y_1 + k - (y_2 + k)] + x_3[y_2 + k - (y_1 + k)]\}$$

$$= \frac{1}{2}[x_1(y_3 + k - y_2 - k) + x_2(y_1 + k - y_2 - k) + x_3(y_2 + k - y_1 - k)]$$

$$= \frac{1}{2}[x_1(y_3 - y_2) + x_2(y_1 - y_2) + x_3(y_2 - y_1)]$$

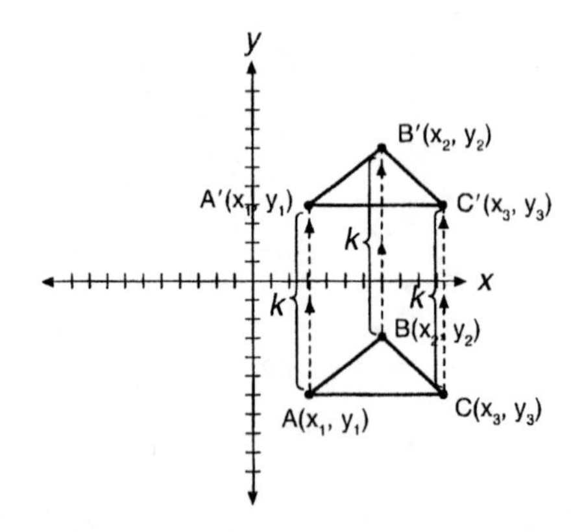

Let ΔABC be any triangle. If any or all of ΔABC lies below the x-axis add a positive constant k to each y-value to translate ΔABC to a location above the x-axis. The area of this new triangle will be the same as the area of ΔABC.

Let $P(x_1, y_1)$ and $Q(x_2, y_2)$ and $R(x_3, y_3)$ be vertices of this new triangle where $x_1 \leq x_2 \leq x_3$.

Suppose $y_2 \geq \frac{y_1 + y_3}{2}$, then ΔPQR is of the same special form as in the "Geometry Investigation" at the beginning of the chapter.

Section 8.1 45. (continued)

Area of ΔPQR = area of $DPQE$ + area of $EQRF$ – area of $DPRF$

$$= \frac{1}{2}[(x_2 - x_1)(y_1 + y_2) + (x_3 - x_2)(y_2 + y_3) - (x_3 - x_1)(y_1 + y_2)]$$

$$= \frac{1}{2}[x_1(y_3 - y_2) + x_2(y_1 - y_3) + x_3(y_2 - y_1)] \text{ by part (a).}$$

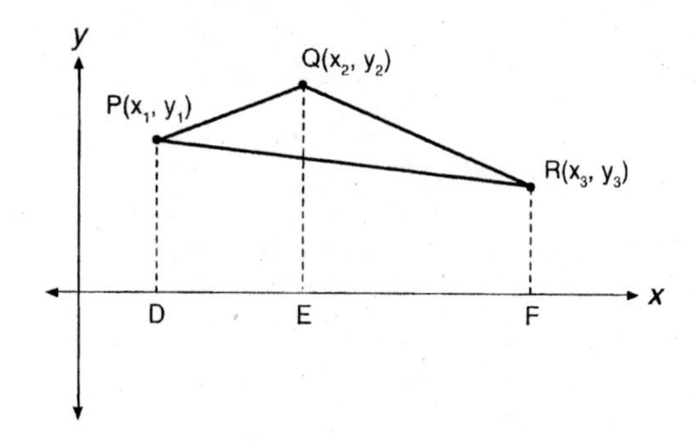

Suppose $y_2 < \frac{y_1 + y_3}{2}$.

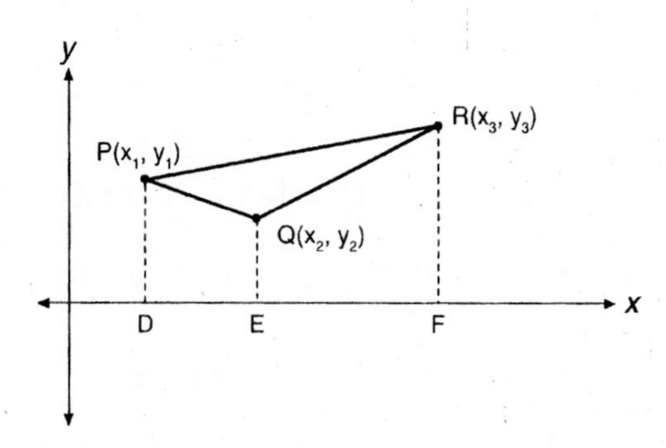

Now, Area of ΔPQR = Area of $DPRF$ – Area of $DPQE$ – Area of $EQRF$

$$= \frac{1}{2}[(x_3 - x_1)(y_1 + y_3) - (x_2 - x_1)(y_1 + y_2) - (x_3 - x_2)(y_2 + y_3)]$$

$$= -\frac{1}{2}[(x_2 - x_1)(y_1 + y_2) + (x_3 - x_2)(y_2 + y_3) - (x_3 - x_1)(y_1 + y_3)]$$

$$= -\frac{1}{2}[x_1(y_3 - y_2) + x_2(y_1 - y_3) + x_3(y_2 - y_1)] \text{ by part (a).}$$

Therefore the area of any triangle is either the value obtained using the formula in part (a) or its opposite.

Hence Area of ΔABC

$$= \frac{1}{2}\left|x_1(y_3 - y_2) + x_2(y_1 - y_3) + x_3(y_2 - y_1)\right|.$$

Section 8.1

47. To verify that P, M, and Q are collinear, show $PM + MQ = PQ$.

$$PM = \sqrt{\left(\frac{x_1 + x_2}{2} - x_1\right)^2 + \left(\frac{y_1 + y_2}{2} - y_1\right)^2}$$

$$= \sqrt{\left(\frac{x_2 - x_1}{2}\right)^2 + \left(\frac{y_2 - y_1}{2}\right)^2}$$

$$MQ = \sqrt{\left(x_2 - \frac{x_1 + x_2}{2}\right)^2 + \left(y_2 - \frac{y_1 + y_2}{2}\right)^2}$$

$$= \sqrt{\left(\frac{x_2 - x_1}{2}\right)^2 + \left(\frac{y_2 - y_1}{2}\right)^2}$$

Hence $PM = MQ$.

Now $PM + MQ = 2\,PM$

$$= 2\sqrt{\left(\frac{x_2 - x_1}{2}\right)^2 + \left(\frac{y_2 - y_1}{2}\right)^2}$$

$$= 2\sqrt{\frac{(x_2 - x_1)^2 + (y_2 - y_1)^2}{4}}$$

$$= \frac{2}{2}\sqrt{(x_2 - x_1)^2 + (y_2 - y_1)^2}$$

$$= \sqrt{(x_2 - x_1)^2 + (y_2 - y_1)^2}$$

$$= PQ.$$

Therefore, $PM + MQ = PQ$.

49. A complete solution is given in the answer key at the back of the text.

51. Label the outermost trees $T_1 = (0, 5)$ and $T_5 = (14, 0)$.

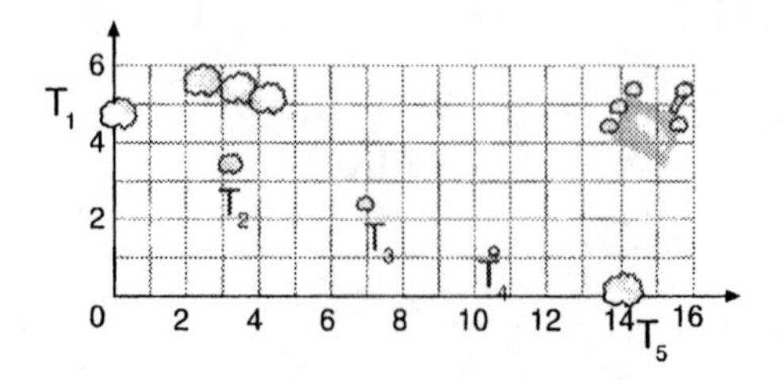

Then the coordinates of T_3 will be the midpoint of T_1 and T_5.

$$T_3 = \left(\frac{0 + 14}{2}, \frac{5 + 0}{2}\right) = (7, 2.5)$$

Next, find the midpoint of T_1 and T_3; label this tree T_2.

$$T_2 = \left(\frac{0 + 7}{2}, \frac{5 + 2.5}{2}\right) = (3.5, 3.75)$$

Finally, find the midpoint of T_3 and T_5; label this tree T_4.

$$T_4 = \left(\frac{7 + 14}{2}, \frac{2.5 + 0}{2}\right) = (10.5, 1.25)$$

Section 8.2

53. When the person has rowed for 30 minutes, the distances are given in the diagram. To find c, the distance from the ship, use the Law of Cosines.

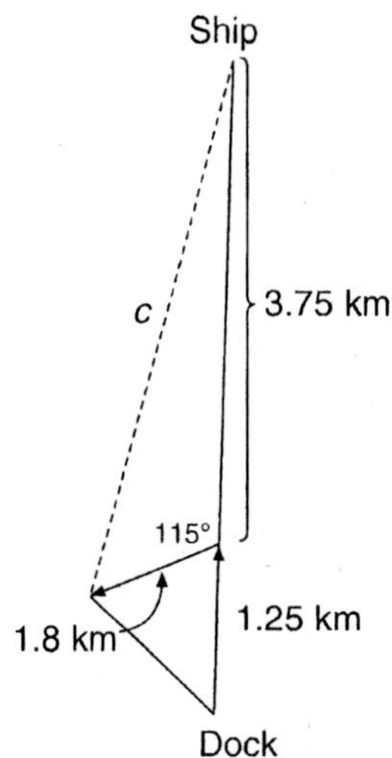

$$c^2 = a^2 + b^2 - 2ab\cos C$$

$$c^2 = (1.8)^2 + (3.75)^2 - 2(1.8)(3.75)\cos 115°$$

$$= 17.3025 - 13.5\cos 115°$$

$$c = \sqrt{17.3025 - 13.5\cos 115°}$$

$$\approx 4.8 \text{ km}$$

Section 8.2

Tips:

✔ When using the slope formula, you can subtract the coordinates of the points in either order; the slope will be the same. Just be consistent so that the first x-coordinate in the denominator and the first y-coordinate in the numerator are from the same point.

✔ One of the most common mistakes in using the slope formula is to forget that the y-coordinates are on the *top* of the fraction. It may help to remember slope as "rise over run".

✔ The letter m is generally used to represent slope.

✔ Given *any* two points on a line, the slope of the line segment between them is the same as the slope of the line. That is, any two points on a line can be used to find its slope.

✔ Remember that the slope of a vertical line is undefined, but the slope of a horizontal line is zero.

Solutions to odd-numbered textbook problems

1. Let $P(x_1, y_1) = (-4, 5)$ and $Q(x_2, y_2) = (6, -3)$.

$$\text{Slope of } \overline{PQ} = \frac{-3 - 5}{6 - (-4)} = \frac{-8}{10} = \frac{-4}{5}.$$

Now let $P(x_1, y_1) = (6, -3)$ and $Q(x_2, y_2) = (-4, 5)$.

$$\text{Slope of } \overline{PQ} = \frac{5 - (-3)}{-4 - 6} = \frac{8}{-10} = \frac{-4}{5}$$

The slopes are the same.

Section 8.2

3. Use the slope formula.

(a) $m = \frac{3 - 2}{5 - 3} = \frac{1}{2}$

(b) $m = \frac{-3 - 1}{-5 - (-2)} = \frac{-4}{-3} = \frac{4}{3}$

(c) $m = \frac{-5 - (-5)}{-6 - 3} = \frac{0}{-9} = 0$

These two points are on a horizontal line. The slope of any horizontal line is 0.

(d) $m = \frac{2 - (-1)}{4 - 4} = \frac{3}{0}$, which is undefined.

These two points are on a vertical line. Any vertical line has undefined slope. Sometimes you say that a vertical line has *no* slope, which means that the slope is undefined, not that it is zero.

5. (a) The slope of line l is positive since line l rises from left to right.
   (b) The slope of line l is zero since line l is horizontal.

7. (a) The line falls from left to right since the slope is negative.
   (b) The line is horizontal since the slope is zero.
   (c) The line rises from left to right since the slope is positive.
   (d) The line is vertical since the slope is undefined.

9. Slope of $\overline{AB} = \frac{2 - (-2)}{1 - 3} = \frac{4}{-2} = -2$

Slope of $\overline{BC} = \frac{10 - 2}{-3 - 1} = \frac{8}{-4} = -2$

Slope of $\overline{AC} = \frac{10 - (-2)}{-3 - 3} = \frac{12}{-6} = -2$

The three points A, B, and C are collinear since the slopes of $\overline{AB}$, $\overline{BC}$ and $\overline{AC}$ are the same.

11. Slope of $\overline{AB} = \frac{-4 - (-11)}{4 - (-8)} = \frac{7}{12}$

Slope of $\overline{BC} = \frac{-1 - (-4)}{12 - 4} = \frac{3}{8}$

Slope of $\overline{CD} = \frac{-8 - (-1)}{-2 - 12} = \frac{-7}{-14} = \frac{7}{14} = \frac{1}{2}$

Slope of $\overline{AC} = \frac{-1 - (-11)}{12 - (-8)} = \frac{10}{20} = \frac{1}{2}$

Slope of $\overline{AD} = \frac{-8 - (-11)}{-2 - (-8)} = \frac{3}{6} = \frac{1}{2}$

A, C, and D are collinear since the slopes of $\overline{AC}$, $\overline{CD}$, and $\overline{AD}$ are the same.

Section 8.2

13. $\overline{AB} \parallel \overline{PQ}$ if the slope of $\overline{AB}$ equals the slope of $\overline{PQ}$.

(a) Slope of $\overline{AB} = \dfrac{5-0}{4-(-1)} = \dfrac{5}{5} = 1$

Slope of $\overline{PQ} = \dfrac{4-9}{-2-3} = \dfrac{-5}{-5} = 1$

$\overline{AB} \parallel \overline{PQ}$ since the slopes of $\overline{AB}$ and $\overline{PQ}$ are equal.

(b) Slope of $\overline{AB} = \dfrac{8-4}{6-0} = \dfrac{4}{6} = \dfrac{2}{3}$

Slope of $\overline{PQ} = \dfrac{-2-(-6)}{2-(-4)} = \dfrac{4}{6} = \dfrac{2}{3}$

$\overline{AB} \parallel \overline{PQ}$ since the slopes of $\overline{AB}$ and $\overline{PQ}$ are equal.

15. Parallel lines have the same slope. To find the slope of a line parallel to $\overline{PQ}$ we need to find the slope of $\overline{PQ}$.

(a) Slope of $\overline{PQ} = \dfrac{-7-2}{5-3} = \dfrac{-9}{2}$

The slope of a line parallel to $\overline{PQ}$ is $-\dfrac{9}{2}$.

(b) Slope of $\overline{PQ} = \dfrac{9-1}{-1-(-3)} = \dfrac{8}{2} = 4$

The slope of a line parallel to $\overline{PQ}$ is 4.

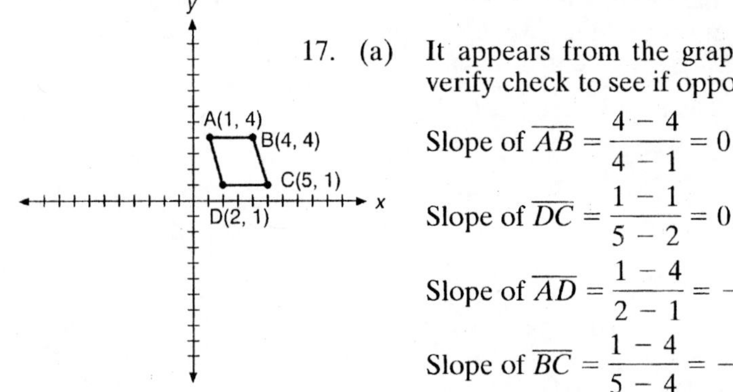

17. (a) It appears from the graph that $ABCD$ is a parallelogram. To verify check to see if opposite sides are parallel.

Slope of $\overline{AB} = \dfrac{4-4}{4-1} = 0$

Slope of $\overline{DC} = \dfrac{1-1}{5-2} = 0$

Slope of $\overline{AD} = \dfrac{1-4}{2-1} = -3$

Slope of $\overline{BC} = \dfrac{1-4}{5-4} = -3$

$ABCD$ is a parallelogram since opposite sides are parallel.

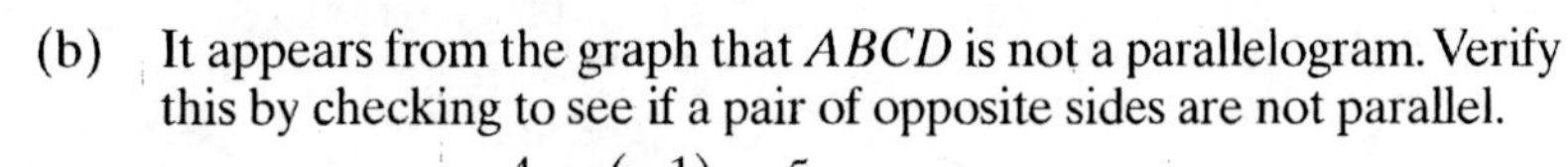

(b) It appears from the graph that $ABCD$ is not a parallelogram. Verify this by checking to see if a pair of opposite sides are not parallel.

Slope of $\overline{AB} = \dfrac{4-(-1)}{6-1} = \dfrac{5}{5} = 1$

Slope of $\overline{BC} = \dfrac{-1-4}{6-6} = \dfrac{-5}{0}$ undefined

Slope of $\overline{CD} = \dfrac{-1-(-4)}{6-1} = \dfrac{3}{5}$

Slope of $\overline{AD} = \dfrac{-4-(-1)}{1-1} = \dfrac{-3}{0}$ undefined

$\overline{AD} \parallel \overline{BC}$ but $\overline{AB} \nparallel \overline{CD}$.

$ABCD$ is not a parallelogram since a pair of opposite sides are not parallel.

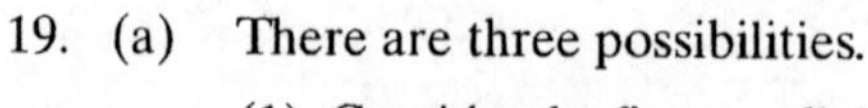

Section 8.2 19. (a) There are three possibilities.

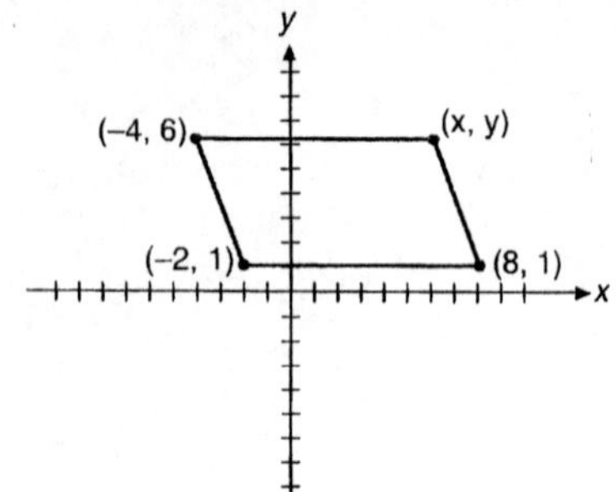

(1) Consider the first parallelogram. To get from $(-2,1)$ to $(-4,6)$, you move 2 units left and 5 units up. Similarly, (x, y) will be 2 units left and 5 units up from $(8, 1)$. Therefore, you have $x = 8 - 2 = 6$ and $y = 1 + 5 - 6$. Thus, the fourth vertex is $(6,6)$.

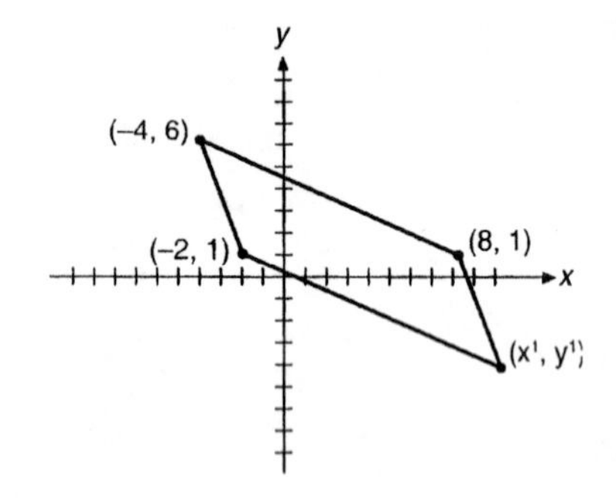

(2) If instead you connect the vertices of the quadrilateral as shown in the second figure, you move 2 units right and 5 units down to get from $(-4,6)$ to $(-2,1)$ or from $(8,1)$ to (x', y'). Therefore, $x' = 8 + 2 = 10$ and $y' = 1 - 5 = -4$. The fourth vertex is $(10, -4)$.

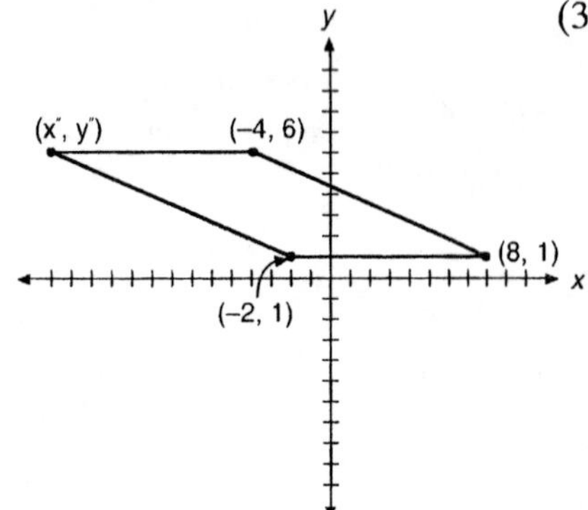

(3) If the vertices of the quadrilateral are connected as shown in the third figure, you move 12 units left and 5 units up to get from $(8, 1)$ to $(-4, 6)$ or from $(-2, 1)$ to (x'', y''). Therefore, $x'' = -2 - 12 = -14$ and $y'' = 1 + 5 = 6$.

The fourth vertex is $(-14, 6)$.

(b) There are three possibilities here too.

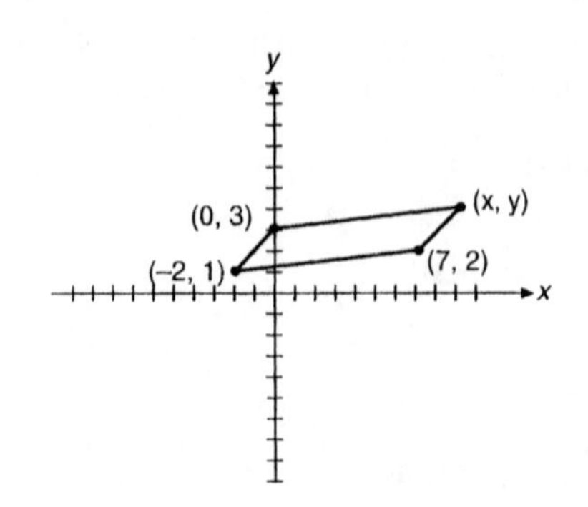

(1) Consider the first parallelogram. Moving from $(-2, 1)$ to $(0, 3)$ or from $(7, 2)$ to (x, y), you move right 2 and up 2. Therefore, $x = 7 + 2 = 9$ and $y = 2 + 2 = 4$, making the fourth vertex $(9, 4)$.

Section 8.2 19. (continued)

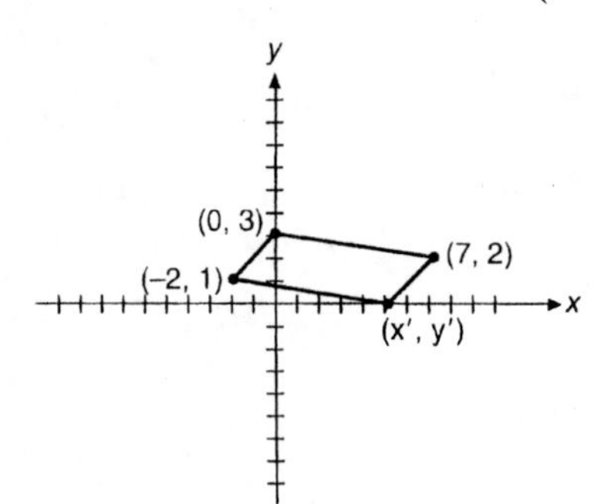

(2) In the second parallelogram, moving from (0, 3) to (−2, 1) or from (7, 2) to (x', y'), you move left 2 and down 2. Then you have $x' = 7 - 2 = 5$ and $y' = 2 - 2 = 0$, yielding (5, 0) as the fourth vertex.

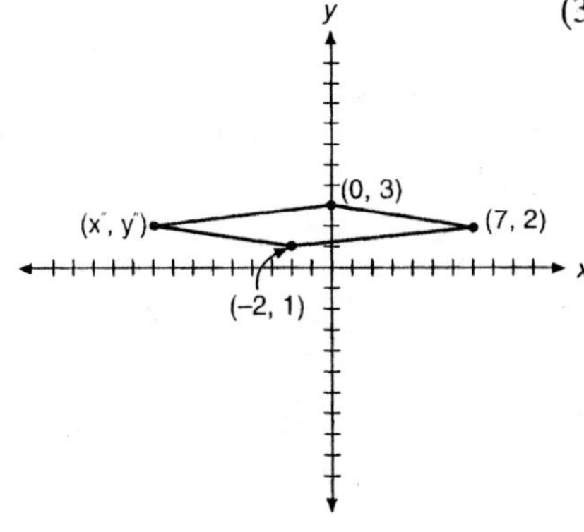

(3) In the third parallelogram, moving from (7, 2) to (0, 3) or from (−2, 1) to (x'', y''), you move left 7 and up 1. Therefore $x'' = -2 - 7 = -9$ and $y'' = 1 + 1 = 2$, giving (−9, 2) as the fourth vertex.

21. $ABCD$ is a parallelogram if the slopes of opposite sides are equal.

$$\text{Slope of } \overline{AB} = \frac{4-1}{2-(-1)} = \frac{3}{3} = 1$$

$$\text{Slope of } \overline{CD} = \frac{-2-1}{3-6} = \frac{-3}{-3} = 1$$

$$\text{Slope of } \overline{BC} = \frac{1-4}{6-2} = \frac{-3}{4}$$

$$\text{Slope of } \overline{AD} = \frac{-2-1}{3-(-1)} = \frac{-3}{4}$$

$ABCD$ is a parallelogram since slope of $\overline{AB}$ = slope of $\overline{CD}$ and slope of $\overline{BC}$ = slope of $\overline{AD}$.

23. The product of the slopes of perpendicular lines is −1, or the slope of one line is zero and the slope of the other line is undefined.

(a) $\text{Slope of } \overline{AB} = \dfrac{5-6}{2-1} = -1$

Let m be the slope of a line perpendicular to $\overline{AB}$

then $-1 \cdot m = -1$

and $m = 1$.

Section 8.2 23. (continued)

(b) Slope of $\overline{AB} = \frac{2-2}{-5-4} = \frac{0}{-9} = 0$

$\overline{AB}$ is a horizontal line since the slope is zero. Any line perpendicular to $\overline{AB}$ is a vertical line. The slope of a vertical line is undefined.

25. (a) Slope of $\overline{AB} = \frac{9-3}{3-3} = \frac{6}{0}$ undefined

Slope of $\overline{BC} = \frac{9-9}{12-3} = \frac{0}{9} = 0$

Slope of $\overline{CA} = \frac{3-9}{3-12} = \frac{-6}{-9} = \frac{2}{3}$

$\overline{AB}$ and $\overline{BC}$ are perpendicular since $\overline{AB}$ is a vertical line and $\overline{BC}$ is a horizontal line.

(b) Slope of $\overline{AB} = \frac{1-8}{7-2} = \frac{-7}{5}$

Slope of $\overline{BC} = \frac{6-1}{14-7} = \frac{5}{7}$

Slope of $\overline{CA} = \frac{8-6}{2-14} = \frac{2}{-12} = \frac{1}{-6}$

Now $\left(\frac{-7}{5}\right)\left(\frac{5}{7}\right) = -1$, so $\overline{AB}$ and $\overline{BC}$ are perpendicular.

27. Use the slope formula.

$$m = \frac{-3-4}{a-(-11)} = \frac{-7}{a+11}$$

Since $m = -\frac{1}{3} = \frac{-1}{3}$, you have $\frac{-7}{a+11} = \frac{-1}{3}$.

Cross-multiplying, $-a - 11 = -21$

$$-a = -10$$

$$a = 10.$$

29. The diagonals of a rhombus are perpendicular.

Slope of $\overline{AC} = \frac{-2-(-1)}{7-(-4)} = \frac{-1}{11}$.

Let m be the slope of the other diagonal, then

$$\frac{-1}{11}m = -1$$

and $m = 11$.

31. If the points are the vertices of a right triangle, two of the sides will be perpendicular and the product of their slopes will be -1 (See Theorem 8.5). To determine if a triangle is a right triangle, first find the slopes of each of the sides.

Section 8.2

31. (continued)

(a) Let $A(2, 2)$, $B(8, 6)$, and $C(4, 8)$ be the coordinates of the vertices of ΔABC.

$$\text{Slope of } \overline{AB} = \frac{6-2}{8-2} = \frac{4}{6} = \frac{2}{3}$$

$$\text{Slope of } \overline{BC} = \frac{8-6}{4-8} = \frac{2}{-4} = \frac{1}{2}$$

$$\text{Slope of } \overline{CA} = \frac{2-8}{2-4} = \frac{-6}{-2} = 3$$

Since no pair of the slopes has a product of -1, there is no right angle and no right triangle.

(b) Let $A(1, -1)$, $B(9, -4)$, and $C(10, 11)$ be the coordinates of the vertices of ΔABC.

$$\text{Slope of } \overline{AB} = \frac{-4-(-1)}{9-1} = \frac{-3}{8}$$

$$\text{Slope of } \overline{BC} = \frac{11-(-4)}{10-9} = \frac{15}{1} = 15$$

$$\text{Slope of } \overline{CA} = \frac{-1-11}{1-10} = \frac{-12}{-9} = \frac{4}{3}$$

Since no two slopes multiply to -1, this is not a right triangle.

(c) Let $A(8, 1)$, $B(6, 3)$, and $C(1, -6)$ be the coordinates of the vertices of ΔABC.

$$\text{Slope of } \overline{AB} = \frac{3-1}{6-8} = \frac{2}{-2} = -1$$

$$\text{Slope of } \overline{BC} = \frac{-6-3}{1-6} = \frac{-9}{-5} = \frac{9}{5}$$

$$\text{Slope of } \overline{CA} = \frac{1-(-6)}{8-1} = \frac{7}{7} = 1$$

Since (Slope of $\overline{AB}$) $\cdot$ (Slope of $\overline{CA}$) $= (-1)(1) = -1$, $\overline{AB}$ and $\overline{CA}$ are perpendicular. Therefore ΔABC is a right triangle.

33. It might be helpful to graph the two kites.

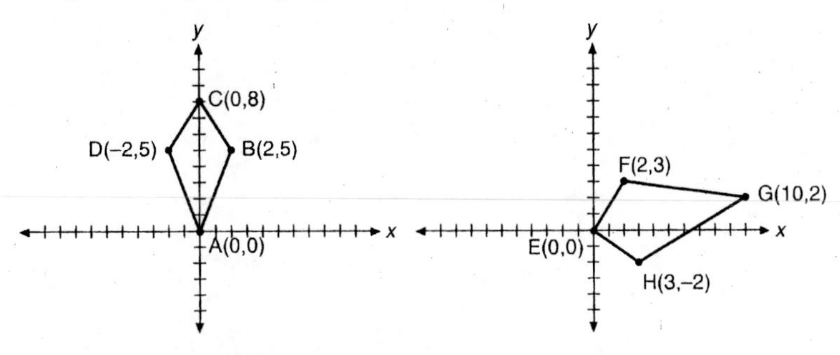

(a) Find the lengths of the diagonals.

(i) $AC = 8, BD = 4; AC \neq BD$

(ii) $EG = \sqrt{(10-0)^2 + (2-0)^2} = \sqrt{104}$

$FH = \sqrt{(2-3)^2 + [3-(-2)]^2} = \sqrt{(-1)^2 + 5^2} = \sqrt{26}$

$EG \neq FH$

Therefore, the diagonals are not congruent.

Section 8.2 33. (continued)

(b) Find the slopes of the diagonals.

(i) Slope of $\overline{AC} = \frac{8-0}{0-0} = \frac{8}{0}$ undefined; $\overline{AC}$ is a vertical line.

Slope of $\overline{BD} = \frac{5-5}{2-(-2)} = \frac{0}{4} = 0$; $\overline{BD}$ is a horizontal line.

$\overline{AC} \perp \overline{BD}$

(ii) Slope of $\overline{EG} = \frac{2-0}{10-0} = \frac{2}{10} = \frac{1}{5}$

Slope of $\overline{FH} = \frac{3-(-2)}{2-3} = \frac{5}{-1} = -5$

$\left(\frac{1}{5}\right)(-5) = -1$ so $\overline{EG} \perp \overline{FH}$.

The diagonals are perpendicular.

(c) Find the midpoint of the diagonals. If they are the same then the diagonals bisect each other.

(i) Midpoint of $\overline{AC} = \left(\frac{0+0}{2}, \frac{0+8}{2}\right) = (0, 4)$

Midpoint of $\overline{BD} = \left(\frac{-2+2}{2}, \frac{5+5}{2}\right) = (0, 5)$

$(0, 4) \neq (0, 5)$ so the diagonals do not bisect each other.

(ii) Midpoint of $\overline{EG} = \left(\frac{0+10}{2}, \frac{0+2}{2}\right) = (5, 1)$

Midpoint of $\overline{FH} = \left(\frac{2+3}{2}, \frac{3+-2}{2}\right) = \left(\frac{5}{2}, \frac{1}{2}\right)$

$(5, 1) \neq \left(\frac{5}{2}, \frac{1}{2}\right)$ so the diagonals do not bisect each other.

The diagonals of the kites do not bisect each other.

(d) If adjacent sides of a kite are perpendicular then a right angle is formed.

(i) Slope of $\overline{AB} = \frac{5-0}{2-0} = \frac{5}{2}$

Slope of $\overline{BC} = \frac{8-5}{0-2} = \frac{3}{-2}$

Slope of $\overline{CD} = \frac{5-8}{-2-0} = \frac{3}{2}$

Slope of $\overline{AD} = \frac{5-0}{-2-0} = \frac{-5}{2}$

The product of any two slopes is not -1, therefore no right angles are formed.

Section 8.2

33. (continued)

(ii) Slope of $\overline{EF} = \frac{3-0}{2-0} = \frac{3}{2}$

Slope of $\overline{FG} = \frac{2-3}{10-2} = \frac{1}{8}$

Slope of $\overline{GH} = \frac{-2-2}{3-10} = \frac{4}{7}$

Slope of $\overline{EH} = \frac{-2-0}{3-0} = \frac{-2}{3}$

Only one right angle is formed as $\overline{EF} \perp \overline{EH}$.

The kites do not have two right angles.

35. A complete solution is given in the answer key at the back of the text.

37. $ABCD$ is a rhombus if opposite sides are parallel and diagonals are perpendicular.

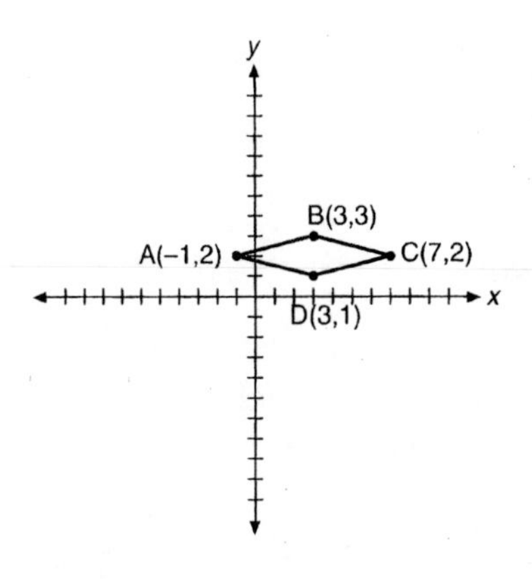

Find the slopes of the sides of $ABCD$.

Slope of $\overline{AB} = \frac{3-2}{3-(-1)} = \frac{1}{4}$

Slope of $\overline{BC} = \frac{2-3}{7-3} = \frac{-1}{4}$

Slope of $\overline{CD} = \frac{1-2}{3-7} = \frac{1}{4}$

Slope of $\overline{DA} = \frac{2-1}{-1-3} = \frac{-1}{4}$

Opposite sides of $ABCD$ are parallel.

Find the slopes of diagonals $\overline{AC}$ and $\overline{BD}$.

Slope of $\overline{AC} = \frac{2-2}{7-(-1)} = \frac{0}{8} = 0$

Slope of $\overline{BD} = \frac{1-3}{3-3} = \frac{-2}{0}$ undefined

Diagonal $\overline{AC}$ is horizontal and diagonal $\overline{BD}$ is vertical, so $\overline{AC} \perp \overline{BD}$.

$ABCD$ is a rhombus since its opposite sides are parallel and its diagonals are perpendicular.

Section 8.2

39. A complete solution is given in the answer key at the back of the text.

41. Set up the proportion $\dfrac{\text{vertical drop}}{\text{gutter length}} = \dfrac{\text{vertical drop}}{\text{gutter length}}$.

(a) Let x = vertical drop in inches.

1 foot = 12 inches and 25 feet = 300 inches

$$\frac{\frac{1}{16}}{12} = \frac{x}{300}$$

$$12x = \frac{1}{16}(300)$$

$$x = \frac{25}{16} \approx 1.56$$

Hence the vertical drop is approximately 1.56 inches.

(b) Let x = gutter length in inches.

$$\frac{\frac{1}{16}}{12} = \frac{\frac{5}{8}}{x}$$

$$\frac{1}{16}x = 12\left(\frac{5}{8}\right)$$

$$x = 120$$

Now, 120 inches $\times \dfrac{1 \text{ foot}}{12 \text{ inches}} = 10$ feet.

Hence, the gutter length is 10 feet.

43.

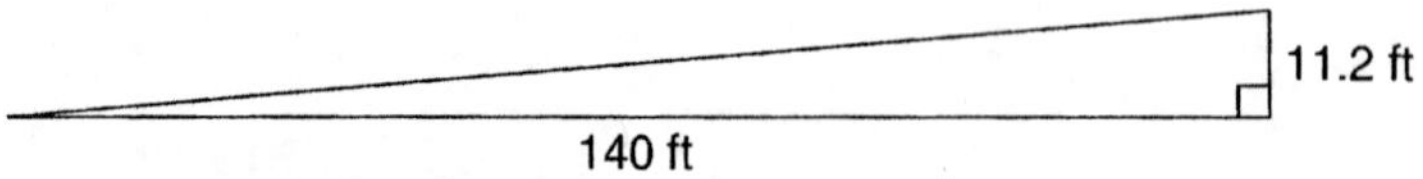

11.2 ft

140 ft

The slope of the road is rise over run.

So, $\dfrac{\text{rise}}{\text{run}} = \dfrac{11.2}{140} = 0.08$.

The angle of inclination $= \tan^{-1}\left(\dfrac{11.2}{140}\right)$

$\approx 4.6°$.

45.

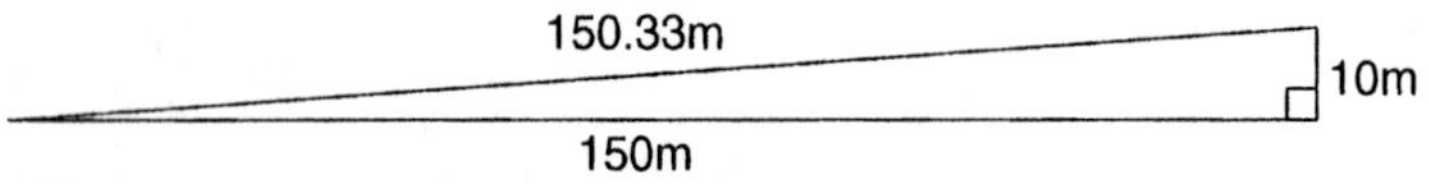

150.33m

10m

150m

The percent grade of the ramp is the slope of the ramp, or

$\dfrac{10}{150} = 0.0\overline{6} \approx 6.67\%$.

Section 8.3

Section 8.3

Tips:

✔ If a line is vertical, its equation has the form $x = a$, where a is the x-coordinate common to all points on the line, including the x-intercept. If you are given two points with the same x-coordinate a, the line containing them is vertical and has the equation $x = a$.

✔ If a line is horizontal, its equation has the form $y = b$, where b is the y-coordinate common to all points on the line, including the y-intercept. If you are given two points with the same y-coordinate b, the equation of the line containing them is $y = b$, and the line is horizontal.

✔ Use the point-slope form to determine the equation of a line, but write your answer in slope-intercept form.

Solutions to odd-numbered textbook problems

1. To show that P lies on the line, show that the coordinates of P satisfy the equation of the line. That is, substitute the x-coordinate and y-coordinate of P for x and y in the equation and show that the equation holds.

   (a) Substitute $P(1, 5)$ into $y = 7x - 2$.

   $$5 = 7(1) - 2$$
   $$5 = 5$$

   Thus, P lies on the line with equation $y = 7x - 2$.

   (b) Substitute $P(7.5, -3)$ into $-2x = 6y + 3$.

   $$-2(7.5) = 6(-3) + 3$$
   $$-15 = -18 + 3$$
   $$-15 = -15$$

   Thus, P lies the line $-2x = 6y + 3$.

3. To find the coordinates of P that satisfy the equation of the line, choose any value for one of the coordinates and substitute that value into the equation. Then, solve the equation for the other variable to find the other coordinate.

   (a) (i) Let $x = 2$. Then $2 - y = 4$

   $$-y = 2$$
   $$y = -2$$

   Thus, $P(2, -2)$ satisfies the equation $x - y = 4$.

   (ii) Let $x = 0$. Then $0 - y = 4$

   $$-y = 4$$
   $$y = -4$$

   Thus, $P(0, -4)$ satisfies the equation $x - y = 4$.

Section 8.3

3. (continued)

(iii) Let $x = -\frac{1}{3}$. Then $-\frac{1}{3} - y = 4$

$$-y = 4\frac{1}{3}$$

$$y = -4\frac{1}{3}.$$

Thus, $P\left(-\frac{1}{3}, -4\frac{1}{3}\right)$ satisfies the equation $x - y = 4$.

(b) In part (a) values were chosen for the x-coordinate.
In part (b) values will be chosen for the y-coordinate.

(i) Let $y = 0$. Then $2x - 3(0) = 6$

$$2x = 6$$

$$x = 3.$$

Thus, $P(3, 0)$ satisfies the equation $2x - 3y = 6$.

(ii) Let $y = -2$. Then $2x - 3(-2) = 6$

$$2x = 0$$

$$x = 0.$$

Thus, $P(0, -2)$ satisfies the equation $2x - 3y = 6$.

(iii) Let $y = 2$. Then $2x - 3(2) = 6$

$$2x = 12$$

$$x = 6.$$

Thus, $P(6, 2)$ satisfies the equation $2x - 3y = 6$.

5. In the equation $y = mx + b$, m is the slope and b is the y-intercept.

(a) $2 + y = -3x$

$$y = -3x - 2$$

The slope is -3 and the y-intercept is -2.

(b) $-2x + y = -5$

$$y = 2x - 5$$

The slope is 2 and the y-intercept is -5.

(c) $2y = 5x + 6$

$$y = \frac{5}{2}x + 3$$

The slope is $\frac{5}{2}$ and the y-intercept is 3.

(d) $2x - 7y = 8$

$$-7y = -2x + 8$$

$$y = \frac{2}{7}x - \frac{8}{7}$$

The slope is $\frac{2}{7}$ and the y-intercept is $-\frac{8}{7}$.

Section 8.3

7. When you are given the slope m and the y-intercept b of a line, substitute them into the slope-intercept form of a line, $y = mx + b$, to write the equation of the line.

   (a) $m = 3$ and $b = 7$, so $y = 3x + 7$ is the equation of the line.

   (b) $m = -1$ and $b = -3$, so $y = -1x - 3$ or $y = -x - 3$ is the equation of the line.

9. The slope-intercept equation of a line is $y = mx + b$.

   (a) Find the slope of the line containing $(6, 3)$ and $(0, 2)$.

   $$m = \frac{2 - 3}{0 - 6} = \frac{-1}{-6} = \frac{1}{6}$$

   Now use the point-slope equation, $y - y_1 = m(x - x_1)$ using one of the given points. That is, $(x_1, y_1) = (6, 3)$.

   $$y - 3 = \frac{1}{6}(x - 6)$$

   $$y - 3 = \frac{1}{6}x - 1$$

   $$y = \frac{1}{6}x + 2 \qquad \text{slope-intercept form of the line}$$

   Instead of using the point $(6, 3)$, the point $(0, 2)$ could have been used to obtain the slope-intercept equation.

   $$y - 2 = \frac{1}{6}(x - 0)$$

   $$y - 2 = \frac{1}{6}x$$

   $$y = \frac{1}{6}x + 2 \qquad \text{slope-intercept form of the line}$$

   (b) Find the slope of the line containing $(-4, 8)$ and $(3, -6)$.

   $$m = \frac{-6 - 8}{3 - (-4)} = \frac{-14}{7} = -2$$

   Use the point $(x_1, y_1) = (-4, 8)$ and $m = -2$ in the point-slope equation. Then solve for y.

   $$y - 8 = -2[x - (-4)]$$
   $$y - 8 = -2(x + 4)$$
   $$y - 8 = -2x - 8$$
   $$y = -2x \qquad \text{slope-intercept form of the line}$$

11. (a) The slope of $y = 5x$ is 5. Since 5 is positive, the line $y = 5x$ rises to the right.

   (b) The slope of $y = -\frac{1}{3}x - 1$ is $-\frac{1}{3}$. Since $-\frac{1}{3}$ is negative, the line $y = -\frac{1}{3}x - 1$ falls to the right.

   (c) The line $y = -10$ can be written as $y = 0x - 10$. The slope of this line is 0. Therefore, the line $y = -10$ is horizontal.

Section 8.3

13. (a) A vertical line has an undefined slope and is of the form $x = a$. Since the line contains $(0, 0)$, the equation is $x = 0$.

    (b) A horizontal line has a slope of 0 and is of the form $y = b$, where b is the y-intercept. Since the line contains $(2, 5)$, the equation is $y = 5$.

    (c) The vertical line has no slope and contains $(-9, 3)$. The equation is $x = -9$.

    (d) The horizontal line has a slope of 0 and contains $(-10, 7)$. The equation is $y = 7$.

15. (a) Divide by 3 to write the equation in slope-intercept form.

$$3y = 4x + 3$$

$$y = \frac{4}{3}x + 1$$

    (b) The constant term in the slope-intercept form of the equation is the y-intercept, so the y-intercept here is 1.

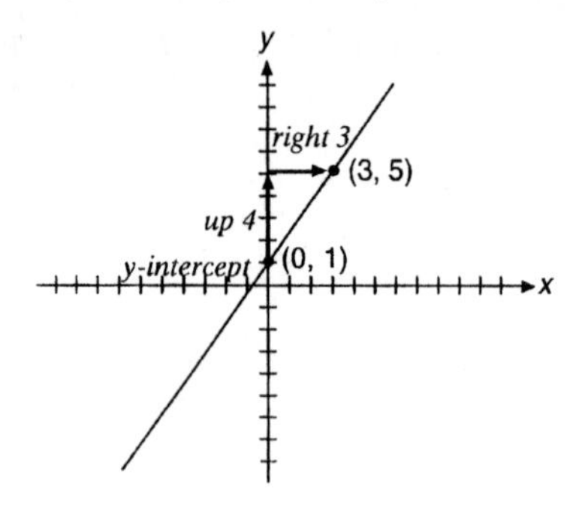

    (c) The slope is $\frac{4}{3}$, the coefficient of x in the slope-intercept form of the equation. The slope represents a rise of 4 and a run of 3. Starting at the y-intercept on the graph, move up 4 units and right 3 units to plot another point. Then draw the line through the two points.

17. (a) For $y = 2x - 1$, the slope is 2 and the y-intercept is -1. Hence, $(0, -1)$ lies on the line. Also the slope represents a rise of 2 and a run of 1. Starting at the y-intercept on the graph, move up 2 units and right 1 unit to plot another point. This point has coordinates $(1, 1)$. Draw a line through the two points.

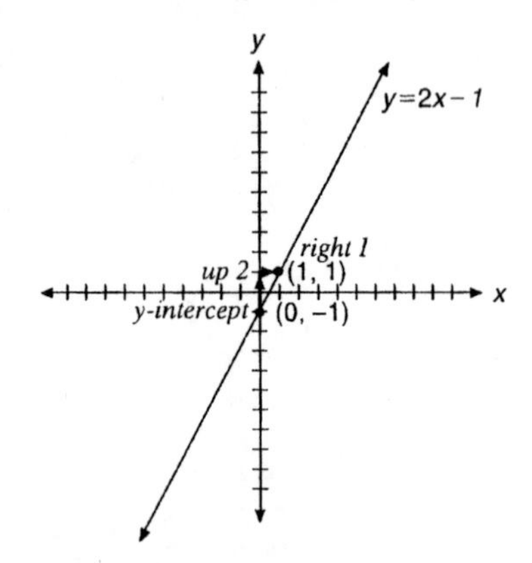

Section 8.3 17. (continued)

(b) The line $x = -2$ is a vertical line where every point on the line has an x-coordinate of -2.

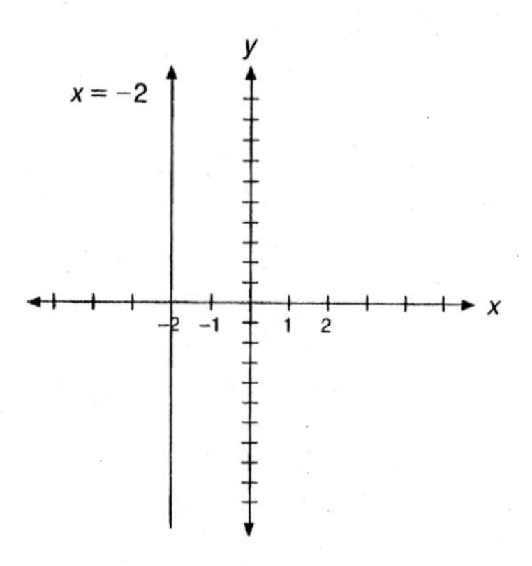

(c) Solve $2x = -6y$ for y.

$$y = \frac{2}{-6}x$$

$$y = -\frac{1}{3}x.$$

The y-intercept is 0 and the slope is $-\frac{1}{3}$. Hence $(0, 0)$ lies on the line. Also the slope represents a rise of -1 and a run of 3. Starting at the y-intercept on the graph, move down 1 unit and right 3 units to plot another point. This point has coordinates $(3, -1)$. Draw a line through the two points.

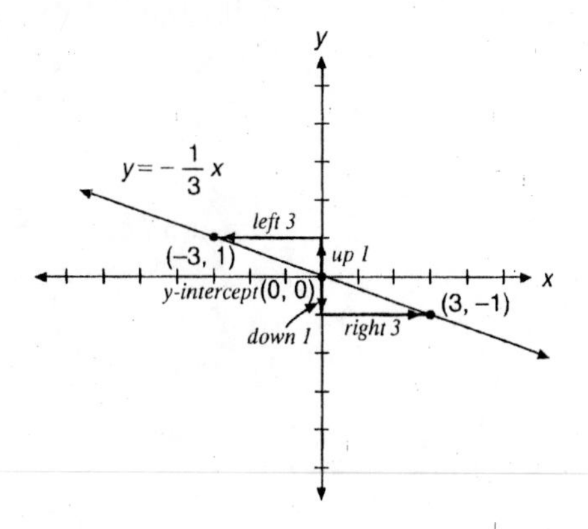

Note: We could have thought of the slope as a rise of 1 and a run of -3. This would result in moving up 1 and left 3 from the y-intercept. The coordinates of this point are $(-3, 1)$ and it, too, lies on the line $y = -\frac{1}{3}x$.

Section 8.3 19.

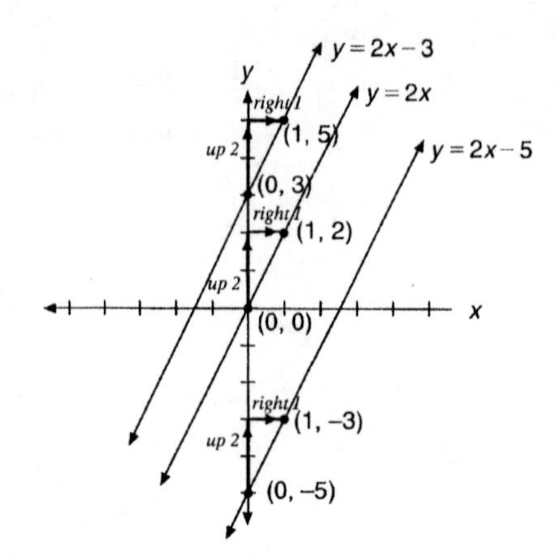

(a) (i) The line $y = 2x$ has a slope of 2 and y-intercept of 0.

(ii) The line $y = 2x + 3$ has a slope of 2 and y-intercept of 3.

(iii) The line $y = 2x - 5$ has a slope of 2 and y-intercept of -5.

(b) The three lines have the same slope. The lines are parallel.

(c) The lines all have the same slope of c and so are parallel.

21. Since parallel lines have the same slope, you must first find the slope of the given line. That is done most easily by solving the equation for y. Once it is in the $y = mx + b$ form, the slope is the coefficient of x. To write the equation of the desired line, substitute the slope and the given point into the point-slope form: $y - y_1 = m(x - x_1)$. Simplify the result and you are done.

(a) You are given $(5, -1)$ and $y = 2x - 3$, so $m = 2$.

Substitute -1 for y_1, 5 for x_1, and 2 for m.

$$y - (-1) = 2(x - 5)$$
$$y + 1 = 2x - 10$$
$$y = 2x - 11$$

The equation of the line through $(5, -1)$ that is parallel to $y = 2x - 3$ is $y = 2x - 11$.

(b) You are given $(1, -2)$ and $y = -x - 2$, so $m = -1$, $x_1 = 1$, and $y_1 = -2$.

$$y - (-2) = -1\,(x - 1)$$
$$y + 2 = -x + 1$$
$$y = -x - 1$$

The equation of the line through $(1, -2)$ and parallel to $y = -x - 2$ is $y = -x - 1$.

(c) You are given $(2, -5)$ and $3x + 5y = 1$.

First solve for y.

$$5y = -3x + 1$$
$$y = -\frac{3}{5}x + \frac{1}{5}, \text{ so } m = -\frac{3}{5}.$$

Section 8.3 21. (continued)

Substitute $m = \frac{3}{5}$, $x_1 = 2$ and $y_1 = -5$ into the point-slope form.

$$y - (-5) = -\frac{3}{5}(x - 2)$$

$$y + 5 = -\frac{3}{5}x + \frac{6}{5}$$

$$y = -\frac{3}{5}x + \frac{6}{5} - 5$$

$$y = -\frac{3}{5}x + \frac{6}{5} - \frac{25}{5}$$

$$y = -\frac{3}{5}x - \frac{19}{5}$$

The equation of the line through $(2, -5)$ and parallel to $3x + 5y = 1$ is $y = -\frac{3}{5}x - \frac{19}{5}$.

(d) You are given $(-1, 0)$ and $3x + 2y = 6$.

First solve for y.

$$2y = -3x + 6$$

$$y = -\frac{3}{2}x + 3, \text{ so } m = -\frac{3}{2}.$$

Substitute $m = -\frac{3}{2}$, $x_1 = -1$, and $y_1 = 0$ into the point-slope form.

$$y - 0 = -\frac{3}{2}[x - (-1)]$$

$$y = -\frac{3}{2}(x + 1)$$

$$y = -\frac{3}{2}x - \frac{3}{2}$$

The equation of the line through $(-1, 0)$ and parallel to $3x + 2y = 6$ is $y = -\frac{3}{2}x - \frac{3}{2}$.

23. The perpendicular bisector contains the midpoint of the segment and has a slope perpendicular to the slope of the segment.

Given $A(-3, -1)$ and $B(6, 2)$.

Midpoint of $\overline{AB} = \left(\frac{-3 + 6}{2}, \frac{-1 + 2}{2}\right) = \left(\frac{3}{2}, \frac{1}{2}\right)$

Slope of $\overline{AB} = \frac{2 - (-1)}{6 - (-3)} = \frac{3}{9} = \frac{1}{3}$

Section 8.3

23. (continued)

Thus, the perpendicular bisector of $\overline{AB}$ has a slope of -3 and contains $\left(\frac{3}{2}, \frac{1}{2}\right)$.

$$y - \frac{1}{2} = -3\left(x - \frac{3}{2}\right) \qquad \text{point-slope form of line}$$

$$y - \frac{1}{2} = -3x + \frac{9}{2}$$

$$y = -3x + \frac{10}{2}$$

$$y = -3x + 5 \qquad \text{slope-intercept form of line}$$

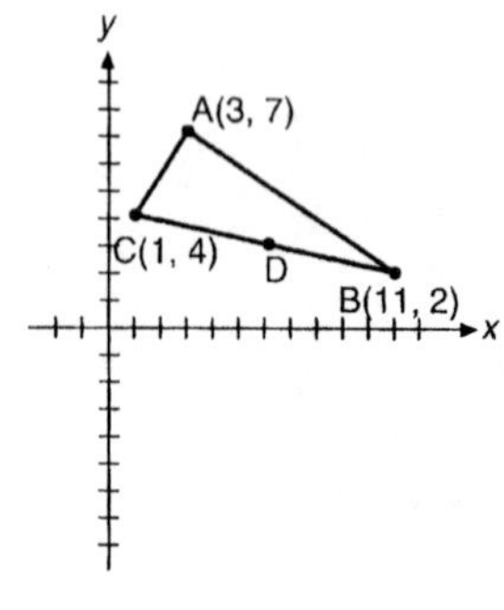

25. The median $\overline{AD}$ joins A to the midpoint of the opposite side, $\overline{BC}$. First you must find the midpoint D.

$$D = \left(\frac{1 + 11}{2}, \frac{4 + 2}{2}\right) = (6, 3)$$

The slope of $\overline{AD}$ is $m = \dfrac{3 - 7}{6 - 3} = \dfrac{-4}{3}$.

Using the point slope form and the point A, write the equation of the line containing $\overline{AD}$.

$$y - 7 = -\frac{4}{3}(x - 3)$$

$$y - 7 = -\frac{4}{3}x + 4$$

$$y = -\frac{4}{3}x + 11$$

27. (a) You are given $y = 2x - 4$ and $y = -5x + 17$.

Substituting $2x - 4$ for y in the second equation, you have

$$2x - 4 = -5x + 17$$

$$2x = -5x + 21$$

$$7x = 21$$

$$x = 3.$$

Substitute $x = 3$ into either equation to find y.

Using the first equation,

$$y = 2 \cdot 3 - 4 = 2.$$

The point (3, 2) is a solution of the system. (You should substitute (3, 2) into both of the original equations to verify that it is a solution.)

Section 8.3

27. (continued)

(b) You are given $4x + y = 8$ and $5x + 3y = 3$. It is most convenient to solve for y in the first equation.

$$4x + y = 8, \text{ so } y = 8 - 4x.$$

Substitute $8 - 4x$ in for y in the second equation, and solve for x.

$$5x + 3(8 - 4x) = 3$$
$$5x + 24 - 12x = 3$$
$$-7x + 24 = 3$$
$$-7x = -21$$
$$x = 3$$

Substituting $x = 3$ into $y = 8 - 4x$ yields

$$y = 8 - 4 \cdot 3 = -4.$$

The solution is $(3, -4)$.

(c) You are given $x - 2y = 1$ and $x = 2y + 3$. Substitute $2y + 3$ in place of x in the first equation.

$$(2y + 3) - 2y = 1$$
$$3 = 1$$

Uh, oh! The resulting equation is a contradiction, which means there is no solution. The lines are parallel and do not intersect.

(d) You are given $x - y = 5$ and $2x - 4y = 7$. Solve the first equation for x.

$$x - y = 5, \text{ so } x = 5 + y$$

Substitute $5 + y$ for x in the second equation.

$$2(5 + y) - 4y = 7$$
$$10 + 2y - 4y = 7$$
$$10 - 2y = 7$$
$$-2y = -3$$
$$y = \frac{3}{2} \text{ or } 1.5$$

Then $x = 5 + y$, so $x = 5 + 1.5 = 6.5 = 6\frac{1}{2} = \frac{13}{2}$.

The solution is $\left(\frac{13}{2}, \frac{3}{2}\right)$.

Section 8.3

29. (a) Both lines have the same slope of 2, but different y-intercepts. Hence, there is no solution as the lines are parallel.

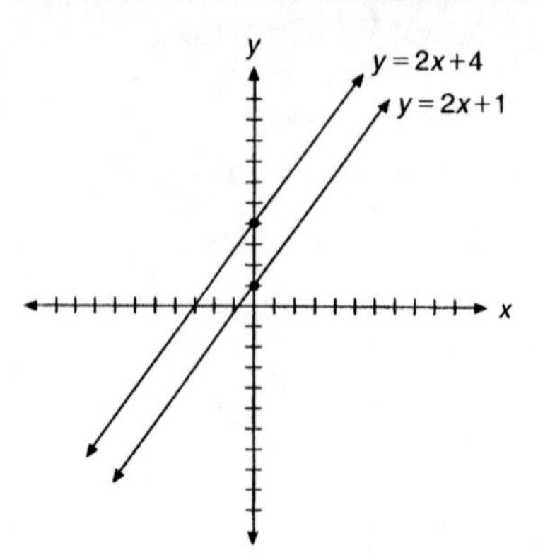

(b) Rewrite both lines in slope-intercept form

$$x + 2y = 4 \qquad\qquad x + 2y = -2$$

$$2y = -x + 4 \qquad\qquad 2y = -x - 2$$

$$y = -\frac{1}{2}x + 2 \qquad\qquad y = -\frac{1}{2}x - 1$$

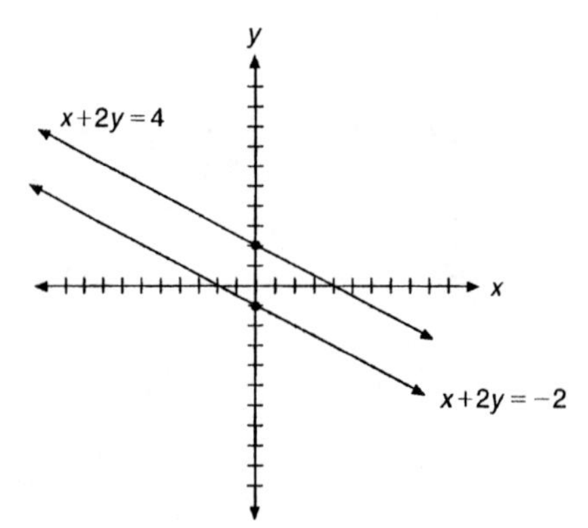

Both lines have the same slope of $-\frac{1}{2}$, but different y-intercepts.

Hence there is no solution as the lines are parallel.

(c) Rewrite both lines in slope-intercept form.

$$-2x + 3y = 9 \qquad\qquad x + y = -2$$

$$3y = 2x + 9 \qquad\qquad y = -x - 2$$

$$y = \frac{2}{3}x + 3$$

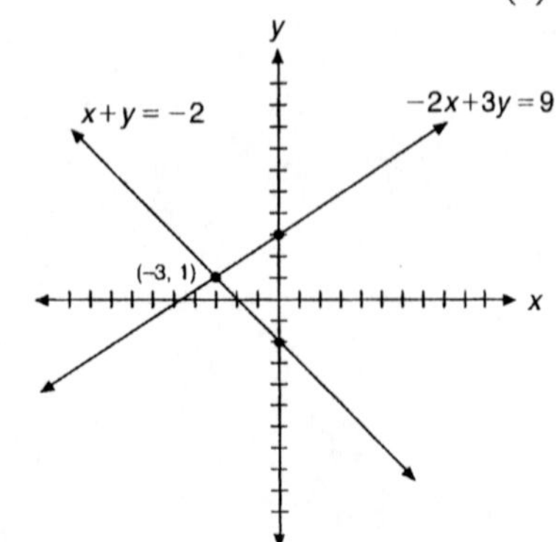

The point of intersection of the two lines is $(-3, 1)$.

Section 8.3

29. (continued)

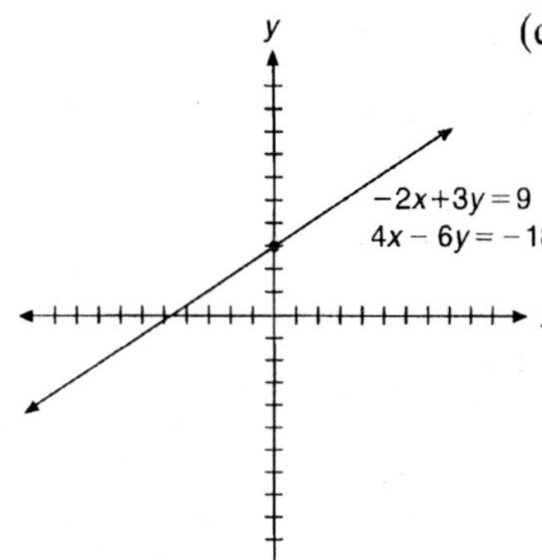

(d) Rewrite both lines in slope-intercept form.

$$-2x + 3y = 9 \qquad\qquad 4x - 6y = -18$$

$$y = \frac{2}{3}x + 3 \qquad\qquad -6y = -4x - 18$$

$$y = \frac{2}{3}x + 3$$

Both equations represent the same line as the slopes are the same and the y-intercepts are the same. Hence both equations represent the same line and all points on the line $-2x + 3y = 9$ are solutions to the system of equations.

31. Adding the equations together yields

$$\begin{array}{rl} 2x + y &= 7 \\ + \; 3x - y &= 3 \\ \hline 5x &= 10 \\ x &= 2. \end{array}$$

Substituting $x = 2$ into the first equation, you have

$$2(2) + y = 7$$
$$4 + y = 7$$
$$y = 3.$$

The solution is $(2, 3)$.

Note: You could have substituted $x = 2$ into either equation to find y.

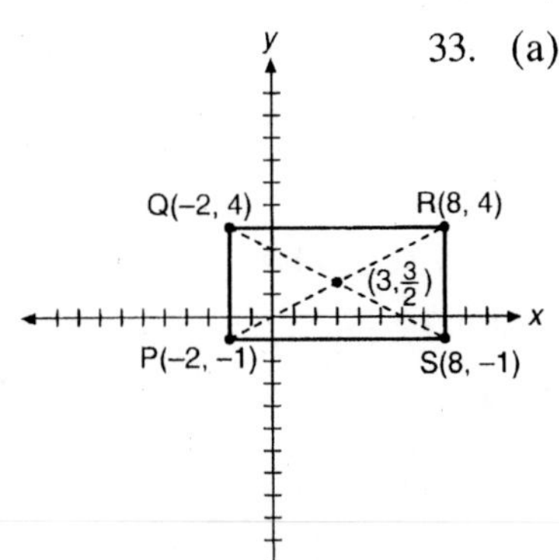

33. (a) Find the equation of diagonals $\overline{PR}$ and $\overline{QS}$.

Slope of $\overline{PR} = \dfrac{4 - (-1)}{8 - (-2)} = \dfrac{5}{10} = \dfrac{1}{2}$

Equation of diagonal $\overline{PR}$:

$$y - 4 = \frac{1}{2}(x - 8)$$

$$y = \frac{1}{2}x.$$

Slope of $\overline{QS} = \dfrac{-1 - 4}{8 - (-2)} = \dfrac{-5}{10} = -\dfrac{1}{2}$

Equation of diagonal $\overline{QS}$:

$$y - (-1) = \frac{-1}{2}(x - 8)$$

$$y + 1 = \frac{-1}{2}x + 4$$

$$y = -\frac{1}{2}x + 3.$$

Section 8.3

33. (continued)

(b) Substitute $y = \frac{1}{2}x$ into the equation $y = -\frac{1}{2}x + 3$.

$$\frac{1}{2}x = \frac{-1}{2}x + 3$$
$$x = 3$$

Now substitute $x = 3$ into $y = \frac{1}{2}x$ to obtain $y = \frac{3}{2}$.

Therefore, the point of intersection of the diagonals is $\left(3, \frac{3}{2}\right)$.

35. (a) Substitute $P(3, 4)$ into the equation $x^2 + y^2 = 25$.

$$3^2 + 4^2 = 25$$
$$9 + 16 = 25$$
$$25 = 25$$

Thus, P lies on the circle with equation $x^2 + y^2 = 25$.

(b) Substitute $P(-3, 5)$ into the equation $x^2 + y^2 = 34$.

$$(-3)^2 + 5^2 = 34$$
$$9 + 25 = 34$$
$$34 = 34$$

Thus, P lies on the circle with equation $x^2 + y^2 = 34$.

37. The circle with center (h, k) and radius r has the equation $(x - h)^2 + (y - k)^2 = r^2$

(a) Rewrite $(x + 7)^2 + (y - 1)^2 = 121$ in the form

$$[x - (-7)]^2 + (y - 1)^2 = 11^2$$

Thus, the circle has center $(-7, 1)$ and radius 11.

(b) Rewrite $(x - 5)^2 + (y - 8)^2 = 27$ in the form

$$(x - 5)^2 + (y - 8)^2 = (3\sqrt{3})^2.$$

Thus, the circle has center $(5, 8)$ and radius $3\sqrt{3}$.

39. The equation of a circle with center (h, k) and radius r is $(x - h)^2 + (y - k)^2 = r^2$. Use this form to write each equation below.

(a) You have $h = -1$, $k = -2$, and $r = \sqrt{5}$, so the equation of the circle will be

$$[x - (-1)]^2 + [y - (-2)]^2 = (\sqrt{5})^2$$
$$(x + 1)^2 + (y + 2)^2 = 5.$$

Section 8.3

39. (continued)

(b) The radius is the distance between the center and the given point on the circle.

$$r = \sqrt{(-2-2)^2 + [1-(-4)]^2} = \sqrt{(-4)^2 + (5)^2}$$
$$= \sqrt{16+25} = \sqrt{41}$$

Thus, you have $h = 2, k = -4$, and $r = \sqrt{41}$. The equation of the circle is

$$(x-2)^2 + [y-(-4)]^2 = (\sqrt{41})^2$$
$$(x-2)^2 + (y+4)^2 = 41.$$

(c) The radius is half the diameter, which you can find as the distance between the two given points. The center is the midpoint of the diameter.

$$r = \frac{1}{2}\sqrt{[3-(-1)]^2 + (-2-6)^2} = \frac{1}{2}\sqrt{(4)^2 + (-8)^2}$$
$$= \frac{1}{2}\sqrt{16+64} = \frac{1}{2}\sqrt{80}$$

The center is $\left(\frac{-1+3}{2}, \frac{6+(-2)}{2}\right) = (1,2)$.

The equation of the circle is

$$(x-1)^2 + (y-2)^2 = \left(\frac{1}{2}\sqrt{80}\right)^2$$
$$(x-1)^2 + (y-2)^2 = \frac{1}{4}\cdot 80$$
$$(x-1)^2 + (y-2)^2 = 20.$$

41. The centroid is the intersection of the medians of a triangle. Find any two medians of the triangle with vertices $A(-5,-1)$, $B(3,3)$, and $C(5,-5)$.

(i) Find the midpoint, D, of $\overline{BC}$.

$$D = \left(\frac{3+5}{2}, \frac{3+(-5)}{2}\right) = (4,-1)$$

Find the slope of $\overline{AD}$.

$$\text{Slope of } \overline{AD} = \frac{-1-(-1)}{4-(-5)} = \frac{0}{9} = 0$$

So the equation of the line containing median $\overline{AD}$ is

$$y-(-1) = 0(x-4)$$
$$y = -1. \qquad \text{Equation 1}$$

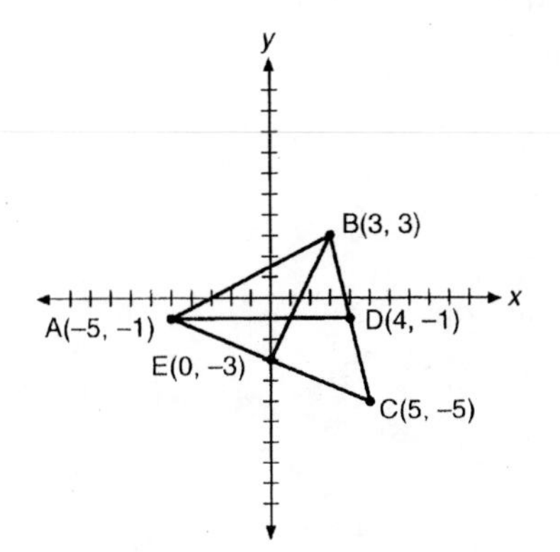

Section 8.3

41. (continued)

(ii) Find the midpoint, E, of $\overline{AC}$.

$$E = \left(\frac{-5 + 5}{2}, \frac{-1 + (-5)}{2}\right) = (0, -3)$$

Find the slope of $\overline{BE}$.

$$\text{Slope of } \overline{BE} = \frac{-3 - 3}{0 - 3} = \frac{-6}{-3} = 2$$

So the equation of the line containing median $\overline{BE}$ is

$$y - (-3) = 2(x - 0)$$

$$y = 2x - 3. \qquad \text{Equation 2}$$

(iii) Solve simultaneously Equations 1 and 2 containing medians $\overline{AD}$ and $\overline{BE}$, respectively. Here we are using substitution.

$$2x - 3 = -1$$

$$2x = 2$$

$$x = 1$$

Therefore, the coordinates of the centroid are $(1, -1)$.

43. The circumcenter, P, of a triangle is the intersection of the perpendicular bisectors of its sides. To find the coordinates of P you must first find the equations of any two of the perpendicular bisectors.

(i) Find the midpoint, D, and the slope of side $\overline{AB}$, and use them to write the equation of the line which contains $\overline{DP}$.

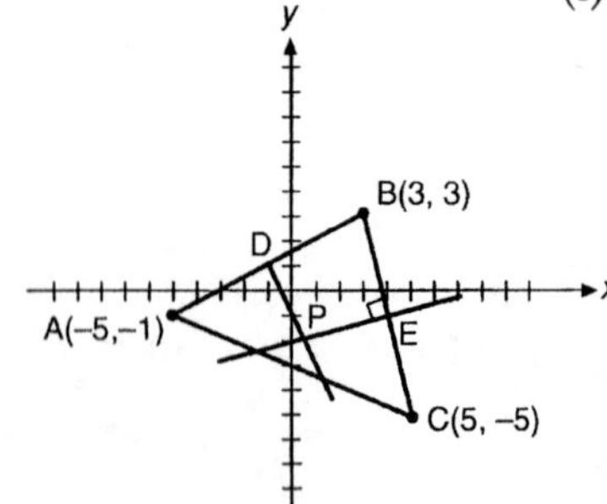

$$D = \left(\frac{-5 + 3}{2}, \frac{-1 + 3}{2}\right) = (-1, 1)$$

$$\text{Slope of } \overline{AB} = \frac{3 - (-1)}{3 - (-5)} = \frac{4}{8} = \frac{1}{2}$$

Now you know that the slope of a line perpendicular to $\overline{AB}$ is $m_{\perp} = -\frac{2}{1} = -2$ (inverting the slope of $\overline{AB}$ and changing its sign).

The equation of the line containng $\overline{DP}$ is

$$y - 1 = -2[x - (-1)]$$

$$y - 1 = -2(x + 1)$$

$$y - 1 = -2x - 2$$

$$y = -2x - 1.$$

(ii) Now find the midpoint E and the slope of side $\overline{BC}$.

$$E = \left(\frac{5 + 3}{2}, \frac{-5 + 3}{2}\right) = (4, -1)$$

$$\text{Slope of } \overline{BC} = \frac{-5 - 3}{5 - 3} = \frac{-8}{2} = -4, \text{ so } m_{\perp} = \frac{1}{4}.$$

Section 8.3 43. (continued)

The equation of the line containing $\overline{EP}$ is

$$y - (-1) = \frac{1}{4}(x - 4)$$

$$y + 1 = \frac{1}{4}x - 1$$

$$y = \frac{1}{4}x - 2.$$

(iii) To find P, determine the point of intersection of these lines by solving simultaneously the equations of the lines containing $\overline{AB}$ and $\overline{BC}$.

If $y = -2x - 1$ and $y = \frac{1}{4}x - 2$, then substituting for y, you have

$$-2x - 1 = \frac{1}{4}x - 2.$$

Multiply by 4 to eliminate the fractions.

$$-8x - 4 = x - 8$$

$$-8x = x - 4$$

$$-9x = -4$$

$$x = \frac{4}{9}.$$

If $x = \frac{4}{9}$, then find y by substituting $\frac{4}{9}$ for x in either equation. Using the first equation, you have

$$y = -2 \cdot \frac{4}{9} - 1 = \frac{-8}{9} - 1 = \frac{-8}{9} - \frac{9}{9} = \frac{-17}{9}.$$

The circumcenter is $\left(\frac{4}{9}, \frac{-17}{9}\right)$.

45. (a)

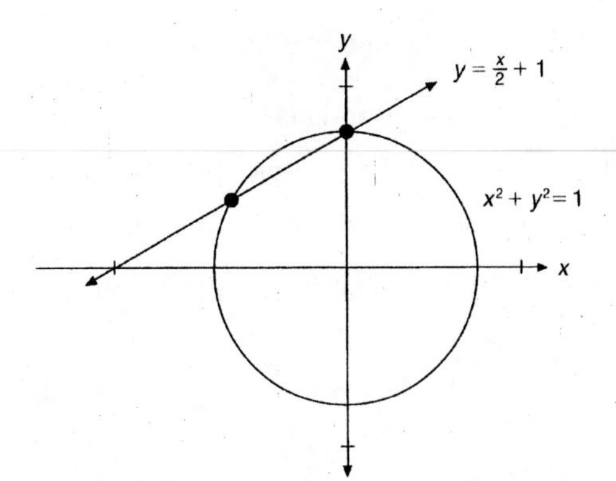

From the graph it appears there are two solutions.

Section 8.3 45. (continued)

(b) Substitute $y = \frac{x}{2} + 1$ into $x^2 + y^2 = 1$.

$$x^2 + \left(\frac{x}{2} + 1\right)^2 = 1$$

$$x^2 + \frac{x^2}{4} + x + 1 = 1$$

$$\frac{5x^2}{4} + x = 0$$

$$x\left(\frac{5x}{4} + 1\right) = 0$$

$$x = 0 \text{ or } \frac{5x}{4} + 1 = 0$$

$$\frac{5x}{4} = -1$$

$$x = \frac{-4}{5}$$

When $x = 0, y = \frac{0}{2} + 1$

$y = 1.$

When $x = \frac{-4}{5}, y = \frac{1}{2}\left(\frac{-4}{5}\right) + 1$

$= -\frac{2}{5} + 1$

$= \frac{3}{5}.$

Therefore, the solutions are $(0, 1)$ and $\left(\frac{-4}{5}, \frac{3}{5}\right)$.

47. The triangle will have the origin, the x-intercept and y-intercept of the line as its vertices. Since it will be a right triangle, the base could be on the x-axis and the height on the y-axis. First find the intercepts of the line.

To find the y-intercept, substitute 0 for x and solve for y.

$2 \cdot 0 + 3y = 6$, so $3y = 6$, and $y = 2$.

To find the x-intercept, substitute 0 for y and solve for x.

$2x + 3 \cdot 0 = 6$, so $2x = 6$, and $x = 3$.

Thus, the base is 3 and the height is 2. You can now determine the area.

$$A = \frac{1}{2}bh = \frac{1}{2} \cdot 3 \cdot 2 = 3 \text{ square units}$$

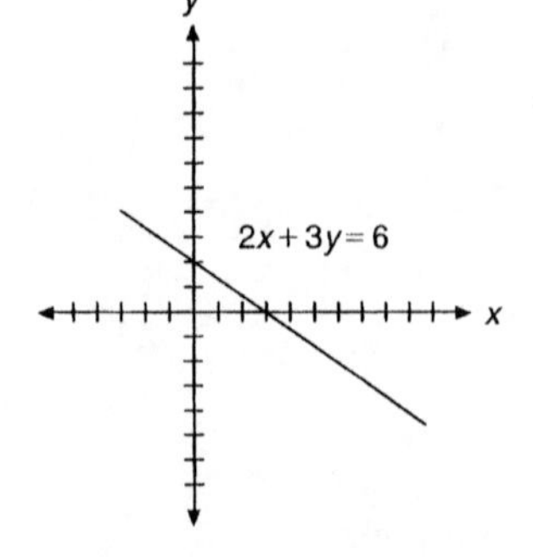

Section 8.3

49. Let O be a circle with center (h, k), let (x, y) be a point on the circle, and let r be the distance from (x, y) to (h, k).

Then $\sqrt{(x-h)^2 + (y-k)^2} = r$ by the distance formula.

Therefore, $(x-h)^2 + (y-k)^2 = r^2$.

51. A complete solution is given in the answer key at the back of the text.

53. The total cost can be calculated as

Cost = fixed fee + (cost per mile) · (number of miles).

(a) Cost for 10 miles $= 0.60 + 0.50(10) =$ \$5.60

Cost for 25 miles $= 0.60 + 0.50(25) =$ \$13.10

(b) Cost = fixed fee + (cost per mile) · (number of miles)

$$y = 0.60 + 0.50x, \text{ or } y = 0.5x + 0.6$$

55. (a) The total cost can be calculated as

Cost = (length of fence in feet) · (cost per linear foot)

Let x = length of one side of fencing in feet.

Then, $y = (2x + 60)(7.25)$

or $y = 14.5x + 435$.

(b) The slope is 14.5. For each foot of fencing added, the cost increases by \$14.50.

(c) The y-intercept is 435. This is a fixed cost, which means the owner will pay this cost of \$435 even if the side fence has length zero feet.

(d) Substitute \$1700 in for y.

$$14.5x + 435 = 1700$$
$$14.5x = 1265$$
$$x = \frac{1265}{14.5} \approx 87 \text{ feet}$$

The maximum length for the side of the fenced area is approximately 87 feet.

57. (a) The cost can be calculated as follows.

Total cost = 400 + 100 + \$2 · (number of people).

If x = number of people, then cost $= 500 + 2x$.

Total revenue = (price of a ticket) · (number of people), so revenue $= 7x$. If the organization breaks even, then the total cost is the same as the revenue.

$$\text{Cost} = \text{Revenue}$$
$$7x = 500 + 2x$$
$$5x = 500$$
$$x = 100 \text{ people}$$

One hundred people are needed to break even when tickets are \$7 each.

57. (continued)

(b) If tickets sell for \$6 each, the revenue is $6x$, so cost equals revenue when

$$\begin{aligned} 6x &= 500 + 2x \\ 4x &= 500 \\ x &= 125 \text{ people.} \end{aligned}$$

To break even when tickets are \$6 each, 125 people are needed.

(c) If the organization wants a profit, then the revenue must equal cost plus profit. That is,

$$\begin{aligned} \text{Revenue} &= \text{Total cost} + \text{Profit} \\ 6x &= 500 + 2x + 400 \\ 6x &= 900 + 2x \\ 4x &= 900 \\ x &= 225 \text{ people.} \end{aligned}$$

To make a \$400 profit on \$6 tickets, 225 people need to attend the dance.

Section 8.4

Tips

✔ Whenever possible, place vertices at the origin or on the axes when setting up a problem. In that way, some coordinates are 0 and easier to work with.

✔ Make your drawing as general as possible so that lengths of sides, etc., are not specific numerical values but are represented by letters.

✔ Be sure not to choose coordinates in such a way that you inadvertently create a figure with special properties that were not given in the problem. For example, you may make a right triangle by your choice of coordinates.

Solutions to odd-numbered textbook problems

1. Since the two bases of a trapezoid are parallel, the y-coordinate of C must be b. Thus, its coordinates could be (c, b). Similarly, the coordinates of D could be $(d, 0)$. To show that the figure is a trapezoid you must show that exactly two sides are parallel. Use slope to verify this. The vertices of the trapezoid are $A(0, 0)$, $B(a, b)$, $C(c, b)$, and $D(d, 0)$.

$$\text{Slope of } \overline{AB} = \frac{b-0}{a-0} = \frac{b}{a}, \text{ Slope of } \overline{BC} = \frac{b-b}{c-a} = \frac{0-0}{c-a} = 0,$$

$$\text{Slope of } \overline{CD} = \frac{d-b}{0-c} = \frac{d-b}{-c}, \text{ and Slope of } \overline{DA} = \frac{0-0}{d-0} = 0.$$

Section 8.4

1. (continued)

   Since the slope of $\overline{BC}$ equals the slope of $\overline{DA}$, $\overline{BC}$ and $\overline{DA}$ are parallel. The other two sides do not have the same slope, so they are not parallel. Therefore $ABCD$ is a trapezoid. To show that it is not isosceles find AB and CD.

   $$AB = \sqrt{(b-0)^2 + (a-0)^2} = \sqrt{a^2 + b^2}$$
   $$CD = \sqrt{(d-c)^2 + (0-b)^2} = \sqrt{(d-c)^2 + b^2}$$

   $AB \neq CD$, so $ABCD$ is not isosceles.

3. A complete solution is given in the answer key at the back of the text.

5. A complete solution is given in the answer key at the back of the text.

7. A complete solution is given in the answer key at the back of the text.

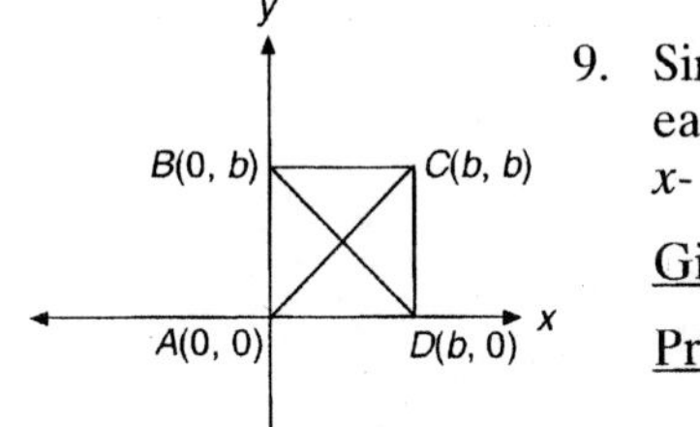

9. Since $ABCD$ is a square, you can place one vertex at the origin, one on each axis the same distance from the origin and the third using equal x- and y-coordinates. (See diagram.)

   Given: Square $ABCD$ with coordinates as shown

   Prove: $\overline{AC} \perp \overline{BD}$.

   Proof: Slope of $\overline{AC} = \dfrac{b-0}{b-0} = \dfrac{b}{b} = 1$, and

   Slope of $\overline{BD} = \dfrac{0-b}{b-0} = \dfrac{-b}{b} = -1$

   (Slope of $\overline{AC}$) (Slope of $\overline{BD}$) $= 1 \cdot (-1) = -1$, so $\overline{AC} \perp \overline{BD}$ (by Theorem 8.5).

11. A complete solution is given in the answer key at the back of the text.

13. A complete solution is given in the answer key at the back of the text.

15. Let $ABCD$ be an isosceles trapezoid with vertices $A(0, 0)$, $B(a, b)$, $C(c, b)$ and $D(c + a, 0)$.

    To verify that $ABCD$ is an isosceles trapezoid, show $AB = CD$.

    $AB = \sqrt{(a-0)^2 + (b-0)^2} = \sqrt{a^2 + b^2}$

    $CD = \sqrt{[c-(c+a)]^2 + (b-0)^2} = \sqrt{(-a)^2 + b^2} = \sqrt{a^2 + b^2}$

    Thus $AB = CD$.

    Now, $AC = \sqrt{(c-0)^2 + (b-0)^2} = \sqrt{c^2 + b^2}$

    and $BD = \sqrt{(c + a - a)^2 + (0-b)^2} = \sqrt{c^2 + (-b)^2} = \sqrt{c^2 + b^2}$.

    Therefore, $AC = BD$.

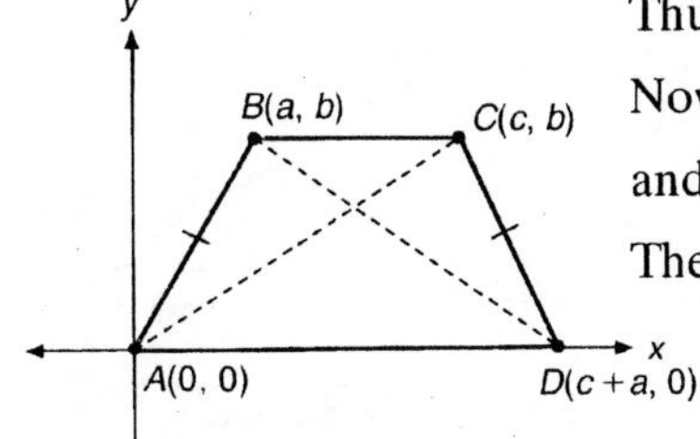

Section 8.4

17. Place one vertex of the rectangle at the origin. Place another vertex on the y-axis, and a third on the x-axis. The fourth vertex should have he same y-coordinate as the vertex on the y-axis and the same x-coordinate as the vertex on the x-axis. (See diagram.)

<u>Given</u>: $ABCD$ is a rectangle and M, N, O and P are midpoints of the sides.

<u>Prove</u>: $MNOP$ is a rhombus.

<u>Proof</u>:

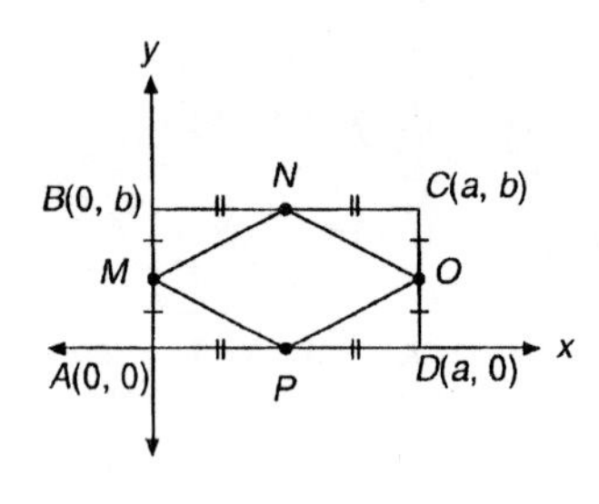

Find the coordinates of the midpoints of the sides of $ABCD$.

$$M = \left(\frac{0+0}{2}, \frac{0+b}{2}\right) = \left(0, \frac{b}{2}\right),$$

$$N = \left(\frac{0+a}{2}, \frac{b+b}{2}\right) = \left(\frac{a}{2}, b\right),$$

$$O = \left(\frac{a+a}{2}, \frac{b+0}{2}\right) = \left(a, \frac{b}{2}\right), \text{ and}$$

$$P = \left(\frac{0+a}{2}, \frac{0+0}{2}\right) = \left(\frac{a}{2}, 0\right).$$

Now determine the length of each side of $MNOP$.

$$MN = \sqrt{\left(\frac{a}{2} - 0\right)^2 + \left(b - \frac{b}{2}\right)^2} = \sqrt{\left(\frac{a}{2}\right)^2 + \left(\frac{b}{2}\right)^2} = \sqrt{\frac{a^2}{4} + \frac{b^2}{4}}$$

$$NO = \sqrt{\left(a - \frac{a}{2}\right)^2 + \left(\frac{b}{2} - b\right)^2} = \sqrt{\left(\frac{a}{2}\right)^2 + \left(\frac{-b}{2}\right)^2} = \sqrt{\frac{a^2}{4} + \frac{b^2}{4}}$$

$$OP = \sqrt{\left(\frac{a}{2} - a\right)^2 + \left(0 - \frac{b}{2}\right)^2} = \sqrt{\left(\frac{-a}{2}\right)^2 + \left(\frac{-b}{2}\right)^2} = \sqrt{\frac{a^2}{4} + \frac{b^2}{4}}$$

$$PM = \sqrt{\left(0 - \frac{a}{2}\right)^2 + \left(\frac{b}{2} - 0\right)^2} = \sqrt{\left(\frac{-a}{2}\right)^2 + \left(\frac{b}{2}\right)^2} = \sqrt{\frac{a^2}{4} + \frac{b^2}{4}}$$

Since $MN = NO = OP = PM$, $MNOP$ is a rhombus.

19. A complete solution is given in the answer key at the back of the text.

Section 8.4

21. Place one vertex of the isosceles triangle at the origin. If you assign to B the coordinates (a, b), then since B is on the perpendicular bisector of $\overline{AC}$, the coordinates of C will be $(2a, 0)$. (See diagram.)

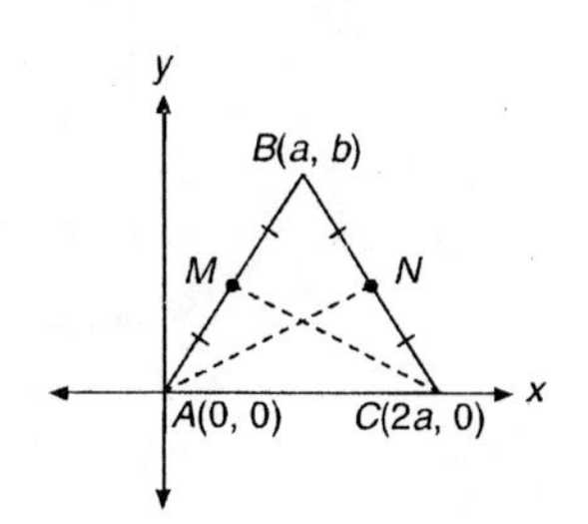

Given: M and N are midpoints of the congruent sides of isosceles triangle ABC, where $\overline{AB} \cong \overline{BC}$.

Prove: $AN = CM$

Proof: First find the coordinates of M and N.

$$M = \left(\frac{0 + a}{2}, \frac{0 + b}{2}\right) = \left(\frac{a}{2}, \frac{b}{2}\right)$$

$$N = \left(\frac{a + 2a}{2}, \frac{b + 0}{2}\right) = \left(\frac{3a}{2}, \frac{b}{2}\right)$$

Now find the lengths of $\overline{AN}$ and $\overline{CM}$.

$$AN = \sqrt{\left(\frac{3a}{2} - 0\right)^2 + \left(\frac{b}{2}\right)^2} = \sqrt{\left(\frac{3a}{2}\right)^2 + \left(\frac{b}{a}\right)^2}$$

$$= \sqrt{\frac{9a^2}{4} + \frac{b^2}{4}}$$

$$CM = \sqrt{\left(2a - \frac{a}{2}\right)^2 + \left(0 - \frac{b}{2}\right)^2}$$

$$= \sqrt{\left(\frac{3a}{2}\right)^2 + \left(-\frac{b}{2}\right)^2}$$

$$= \sqrt{\frac{9a^2}{4} + \frac{b^2}{4}}$$

Therefore, $AN = CM$.

23. A complete solution is given in the answer key at the back of the text.

25. A complete solution is given in the answer key at the back of the text.

27. Prove: The medians of a triangle are concurrent.

Proof: Let ΔABC have coordinates $A(0, 0)$, $B(a, b)$, and $C(c, 0)$. Let M, N, and P be midpoints of $\overline{AB}$, $\overline{BC}$, and $\overline{AC}$.

Use the midpoint formula to find the coordinates of M, N, and P.

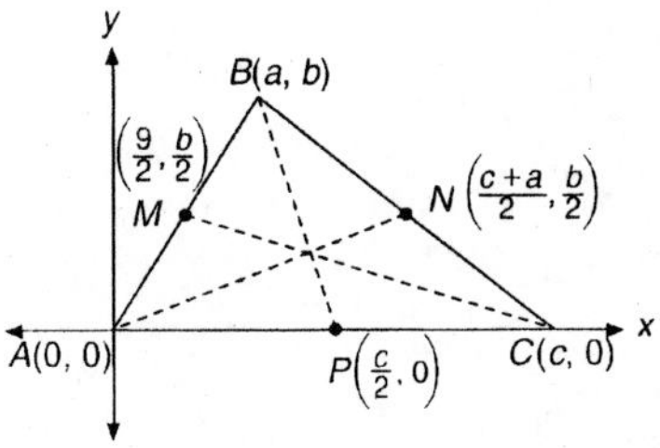

$$\text{Midpoint of } \overline{AB} = \left(\frac{0 + a}{2}, \frac{0 + b}{2}\right) = \left(\frac{a}{2}, \frac{b}{2}\right) = M$$

$$\text{Midpoint of } \overline{BC} = \left(\frac{c + a}{2}, \frac{0 + b}{2}\right) = \left(\frac{c + a}{2}, \frac{b}{2}\right) = N$$

$$\text{Midpoint of } \overline{AC} = \left(\frac{0 + c}{2}, \frac{0 + 0}{2}\right) = \left(\frac{c}{2}, 0\right) = P$$

Find the equations of the lines containing the medians.

Section 8.4 27. (continued)

(i) Find line containing median $\overline{MC}$.

$$\text{Slope of } \overline{MC} = \frac{0 - \frac{b}{2}}{c - \frac{a}{2}} = \frac{-\frac{b}{2}}{\frac{2c - a}{2}} = \frac{-b}{2c - a}$$

$$y - 0 = \frac{-b}{2c - a}(x - c)$$

$$= \frac{-b}{2c - a}x = \frac{bc}{2c - a} \qquad \text{(Line 1)}$$

(ii) Find line containing median $\overline{NA}$.

$$\text{Slope of } \overline{NA} = \frac{0 - \frac{b}{2}}{0 - \left(\frac{c + a}{2}\right)} = \frac{\frac{-b}{2}}{\frac{-(c + a)}{2}} = \frac{b}{c + a}$$

$$y - 0 = \frac{b}{c + a}(x - 0)$$

$$y = \frac{b}{c + a}x \qquad \text{(Line 2)}$$

(iii) Find line containing median $\overline{PB}$.

$$\text{Slope of } \overline{PB} = \frac{b - 0}{a - \frac{c}{2}} = \frac{b}{\frac{2a - c}{2}} = \frac{2b}{2a - c}$$

$$y - 0 = \frac{2b}{2a - c}\left(x - \frac{c}{2}\right)$$

$$y = \frac{2b}{2a - c}x - \frac{bc}{2a - c} \qquad \text{(Line 3)}$$

Now show that the lines containing the medians intersect in the same point.

Find coordinates of the intersection of Line 1 and Line 2.

$$\frac{-b}{2c - a}x + \frac{bc}{2c - a} = \frac{b}{c + a}x$$

$$\left(\frac{b}{c + a} + \frac{b}{2c - a}\right)x = \frac{bc}{2c - a}$$

$$\frac{b(2c - a) + b(c + a)}{(c + a)(2c - a)}x = \frac{bc}{2c - a}$$

$$\frac{2bc - ab + bc + ab}{(c + a)(2c - a)}x = \frac{bc}{2c - a}$$

$$\frac{3bc}{(c + a)(2c - a)}x = \frac{bc}{2c - a}$$

$$x = \frac{a + c}{3}$$

Additional Problems

27. (continued)

Substitute this x-coordinate back into either Line 1 or Line 2. Let's use Line 2.

$$y = \frac{b}{c+a}\left(\frac{a+c}{3}\right)$$
$$= \frac{b}{3}$$

The coordinates of the intersection point of Line 1 and Line 2 are $\left(\frac{a+c}{3}, \frac{b}{3}\right)$.

To show that all three medians intersect in the same point, show that the intersection point, $\left(\frac{a+c}{3}, \frac{b}{3}\right)$, of Line 1 and Line 2 also lies on Line 3. That is, verify that $\left(\frac{a+c}{3}, \frac{b}{3}\right)$ lies on

$$y = \frac{2b}{2a-c}x - \frac{bc}{2a-c}$$

$$\frac{2b}{2a-c}\left(\frac{a+c}{3}\right) - \frac{bc}{2a-c} = \frac{2ab+2bc-3bc}{3(2a-c)}$$
$$= \frac{2ab-bc}{3(2a-c)}$$
$$= \frac{b(2a-c)}{3(2a-c)}$$
$$= \frac{b}{3}$$

Thus, $\left(\frac{a+c}{3}, \frac{b}{3}\right)$ also lies on Line 3.

Therefore, the intersection of the three medians are concurrent.

This point is called the centroid.

Additional Problems

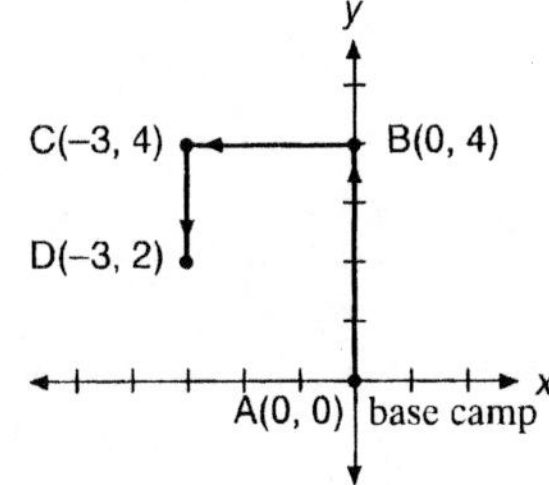

1. Place base camp on the coordinate system with coordinates $A(0,0)$.

   After, the explorer travels 4 km north, his coordinates are $B(0,4)$.

   After he travels 3 km west, his coordinates are $C(-3,4)$. Finally, after traveling 2 km south, his coordinates are $D(-3,2)$.

   This distance the explorer is from base camp is the distance between $A(0,0)$ and $D(-3,2)$.

   $AD = \sqrt{(-3-0)^2 + (2-0)^2} = \sqrt{9+4} = \sqrt{13} \approx 3.6$ km

Chapter Review

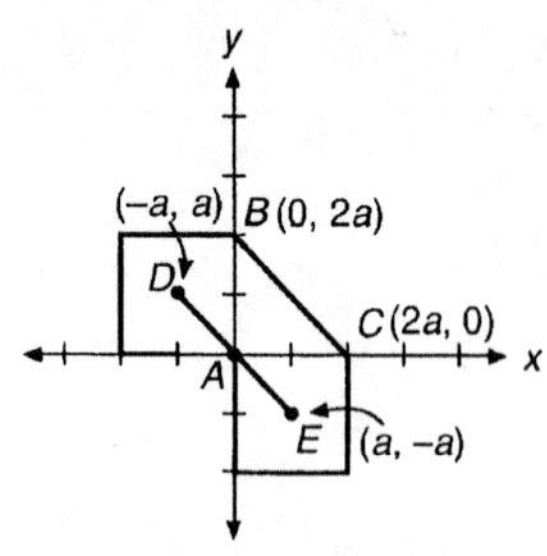

2. Place the figure on the coordinate system with $B(0, 2a)$ and $C(2a, 0)$ Then we have $D(-a, a)$ and $E(a, -a)$.

$$\text{Slope of } \overline{BC} = \frac{0 - 2a}{2a - 0} = -1$$

$$\text{Slope of } \overline{DE} = \frac{-a - a}{a - (-a)} = \frac{-2a}{2a} = -1$$

Therefore, $\overline{BC} \parallel \overline{DE}$.

Solutions to Chapter 8 Review

Section 8.1

1.

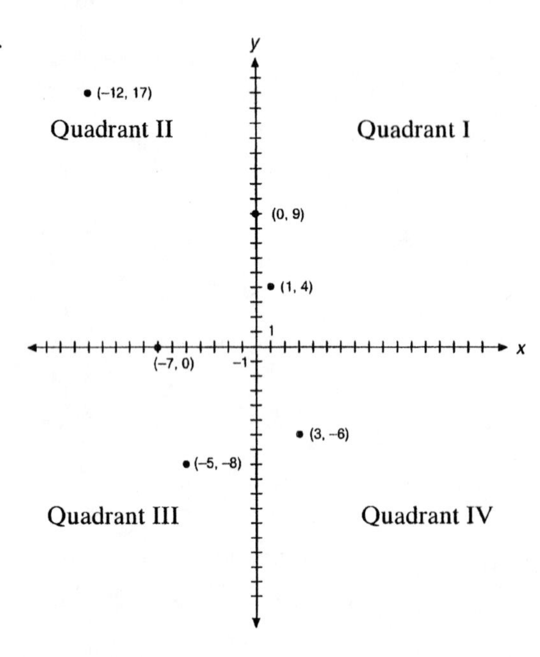

(a) Quadrant III
(b) Quadrant IV
(c) y-axis
(d) Quadrant II
(e) x-axis
(f) Quadrant I

Chapter Review

2.

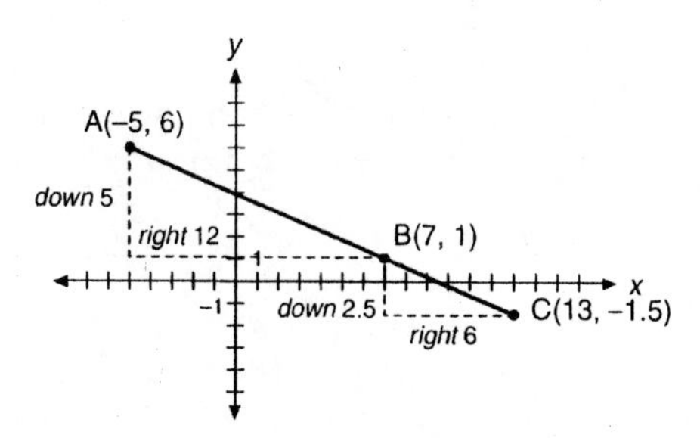

Since A, B, and C are collinear. The slopes of $\overline{AB}$, $\overline{BC}$ and $\overline{AC}$ are the same.

The slope of $\overline{AB} = \dfrac{1 - 6}{7 - (-5)} = \dfrac{-5}{12}$. That is, to get to point B from point A move down 5 and right 12.

Since $AB = 2BC$, the coordinates of point C can be found by starting at point B and moving down 2.5 and right 6.

$$\left(\text{Note: } \frac{-2.5}{6} = \frac{\frac{1}{2}(-5)}{\frac{1}{2}(12)}\right)$$

Therefore, the coordinates of C are (13, −1.5).

3.

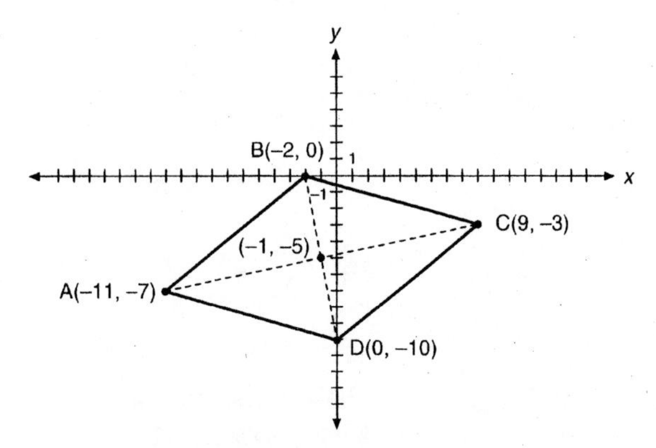

(a) $AB = \sqrt{[-2 - (-11)]^2 + [0 - (-7)]^2} = \sqrt{9^2 + 7^2} = \sqrt{130}$
$BC = \sqrt{[9 - (-2)]^2 + (-3 - 0)^2} = \sqrt{11^2 + (-3)^2} = \sqrt{130}$
$CD = \sqrt{(9 - 0)^2 + [-3 - (-10)]^2} = \sqrt{9^2 + 7^2} = \sqrt{130}$
$DA = \sqrt{(-11 - 0)^2 + [-7 - (-10)]^2} = \sqrt{(-11)^2 + 3^2} = \sqrt{130}$
$ABCD$ is a rhombus as all sides are congruent.

Chapter Review

3. (continued)

   (b) The diagonals of a rhombus are perpendicular bisectors of each other. Thus, the midpoint of $\overline{AC}$ equals the midpoint of $\overline{BD}$.

   $$\text{Midpoint of } \overline{AC} = \left(\frac{-11 + 9}{2}, \frac{-7 + (-3)}{2}\right) = (-1, -5)$$

   $$\text{Midpoint of } \overline{BD} = \left(\frac{2 + 0}{2}, \frac{0 + (-10)}{2}\right) = (-1, -5)$$

   The coordinates of the midpoint of the diagonals of $ABCD$ are $(-1, -5)$.

4. $AB = \sqrt{[-2 - (-4)]^2 + (6 - 2)^2} = \sqrt{2^2 + 4^2} = \sqrt{20} = 2\sqrt{5}$
   $BC = \sqrt{[1 - (-2)]^2 + (12 - 6)^2} = \sqrt{3^2 + 6^2} = \sqrt{45} = 3\sqrt{5}$
   $AC = \sqrt{[1 - (-4)]^2 + (12 - 2)^2} = \sqrt{5^2 + 10^2} = \sqrt{125} = 5\sqrt{5}$
   Since $2\sqrt{5} + 3\sqrt{5} = 5\sqrt{5}$ we have $AB + BC = AC$ and A, B, and C are collinear.

5. A complete solution is given in the answer key at the back of the text.

Section 8.2

1. The slope of the line through (x_1, y_1) and (x_2, y_2) is $\frac{y_2 - y_1}{x_2 - x_1}$.

   Thus, $\frac{4}{5} = \frac{12 - 7}{-3 - x}$. Now, solve for x.

   $$\begin{aligned} 4(-3 - x) &= 5(12 - 7) \\ -12 - 4x &= 25 \\ -4x &= 37 \\ x &= \frac{-37}{4} \end{aligned}$$

2. Parallel lines have the same slope and either the product of the slopes of perpendicular lines is -1 or one line has zero slope and the slope of the other line is undefined.

   (a) Slope of parallel line is 9.

   Slope of perpendicular line is $-\frac{1}{9}$.

   (b) Slope of parallel line is 0.

   Slope of perpendicular line is undefined.

   (c) Slope of parallel line is $\frac{2}{7}$.

   Slope of perpendicular line is $-\frac{7}{2}$.

Chapter Review

2. (continued)

   (d) Slope of parallel line is $-\frac{5}{3}$.

   Slope of perpendicular line is $\frac{3}{5}$.

   (e) Slope of parallel line is undefined.

   Slope of perpendicular line is 0.

3.

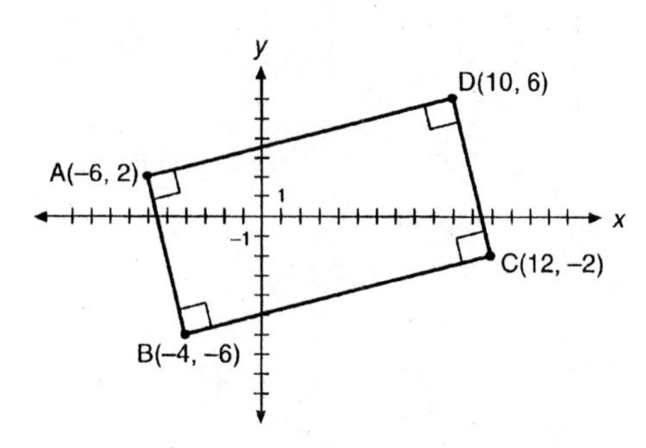

Find the slopes of the sides of quadrilateral $ABCD$

$$\text{Slope of } \overline{AB} = \frac{-6-2}{-4-(-6)} = \frac{-8}{2} = -4$$

$$\text{Slope of } \overline{BC} = \frac{-2-(-6)}{12-(-4)} = \frac{4}{16} = \frac{1}{4}$$

$$\text{Slope of } \overline{CD} = \frac{6-(-2)}{10-12} = \frac{8}{-2} = -4$$

$$\text{Slope of } \overline{DA} = \frac{2-6}{-6-10} = \frac{-4}{-16} = \frac{1}{4}$$

$ABCD$ is a rectangle as opposite sides are parallel and adjacent sides are perpendicular.

4. (a)

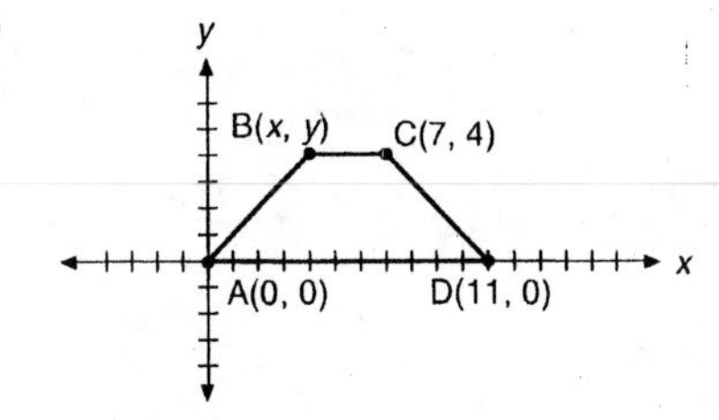

The slope of $\overline{AD} = 0$ so the slope of $\overline{BC}$ must also be zero.

Thus the y-coordinate of point B must be 4.

Chapter Review

4. (continued)

To find the x-coordinate, set $AB = CD$ and solve for x.

$AB = \sqrt{(x-0)^2 + (4-0)^2} = \sqrt{x^2 + 4^2}$

$CD = \sqrt{(11-7)^2 + (0-4)^2} = \sqrt{4^2 + (-4)^2} = \sqrt{32}$

$$\sqrt{x^2 + 4^2} = \sqrt{32}$$

$$x^2 + 4^2 = 32$$

$$x^2 = 16$$

$$x = 4 \text{ or } x = -4$$

Reject $x = -4$ as it does not make sense here because B needs to be in Quadrant I.

Thus the coordinates of B are $(4, 4)$.

(b)

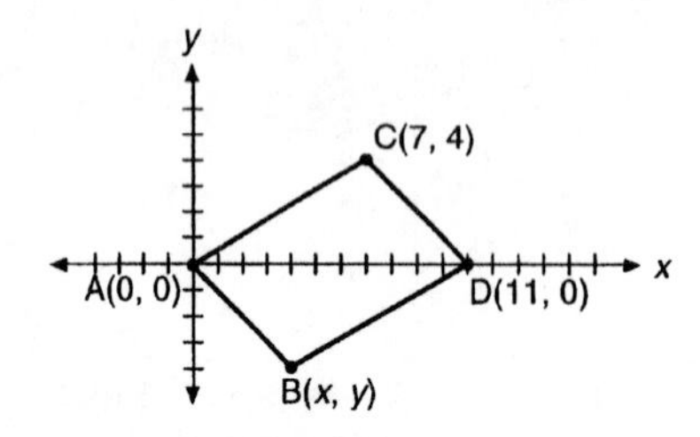

The slopes of opposite sides of a parallelogram are the same.

(i) Slope of $\overline{AB}$ = Slope of $\overline{CD}$

Slope of $\overline{AB} = \dfrac{y-0}{x-0} = \dfrac{y}{x}$

Slope of $\overline{CD} = \dfrac{0-4}{11-7} = \dfrac{-4}{4} = -1$

Therefore, $\dfrac{y}{x} = -1$ and $y = -x$.

(ii) Slope of $\overline{AC}$ = Slope of $\overline{BD}$

Slope of $\overline{AC} = \dfrac{4-0}{7-0} = \dfrac{4}{7}$

Slope of $\overline{BD} = \dfrac{0-y}{11-x} = \dfrac{-y}{11-x}$

Therefore, $\dfrac{-y}{11-x} = \dfrac{4}{7}$

$$-7y = 4(11-x)$$

$$y = -\frac{4}{7}(11-x)$$

$$y = \frac{4}{7}x - \frac{44}{7}.$$

Chapter Review

4. (continued)

Now we have two equations $y = -x$ and $y = \frac{4}{7}x - \frac{44}{7}$.

Solve these two equations using substitution.

$$-x = \frac{4}{7}x - \frac{44}{7}$$

$$\frac{11}{7}x = \frac{44}{7}$$

$$x = 4$$

If $x = 4$ then $y = -4$.

Thus, the coordinates of point B are $(4, -4)$.

5. A complete solution is given in the answer key at the back of the text.

Section 8.3

1. An equation in the form $y = mx + b$ has slope $= m$ and y-intercept $= b$. See answer key in the back of the text for the graphs of the lines.

(a) $y = -\frac{3}{4}x + 6$; slope $= -\frac{3}{4}$, y-intercept $= 6$.

(b)
$$x = 4y - 12$$
$$x + 12 = 4y$$
$$y = \frac{1}{4}x + 3;\ \text{slope} = \frac{1}{4},\ y\text{-intercept} = 3.$$

(c)
$$\frac{y}{5} = 2 - x$$
$$y = 10 - 5x$$
$$y = -5x + 10;\ \text{slope} = -5,\ y\text{-intercept} = 10.$$

(d)
$$-3x + 3y = 21$$
$$3y = 3x + 21$$
$$y = x + 7;\ \text{slope} = 1,\ y\text{-intercept} = 7.$$

2. (a) A horizontal line is of the form $y = b$. Since this line contains $(-12, 5)$, the equation is $y = 5$.

(b) A vertical line is of the form $x = a$. Since this line contains $(4, -7)$, the equation is $x = 4$.

(c) Find the slope of the line $2x - 3y = 12$.

$$2x - 3y = 12$$
$$-3y = -2x + 12$$
$$y = \frac{2}{3}x - 4$$

Chapter Review

2. (continued)

The slope of the line $y = \frac{2}{3}x - 4$ is $\frac{2}{3}$.

Thus, the slope of a line perpendicular is $-\frac{3}{2}$.

Use the point-slope formula of a line with point $(2, -4)$ and slope $-\frac{3}{2}$. Then rewrite the line in slope-intercept form.

$$y - -4 = -\frac{3}{2}(x - 2) \quad \text{point-slope form}$$

$$y + 4 = -\frac{3}{2}x + 3$$

$$y = -\frac{3}{2}x - 1. \quad \text{slope-intercept form}$$

3. (a) Solve $x + y = -3$ for y.

$$y = -x - 3$$

Substitute $-x - 3$ in for y in $-5x + 4y = 24$.

$$-5x + 4(-x - 3) = 24$$

$$-5x - 4x - 12 = 24$$

$$-9x = 36$$

$$x = -4$$

Now substitute $x = -4$ into either one of the original equations.

$$-4 + y = -3$$

$$y = 1$$

The solution is $(-4, 1)$.

(b) Solve $-\frac{3}{2}x + y = 3$ for y.

$$y = \frac{3}{2}x + 3.$$

Substitute $\frac{3}{2}x + 3$ in for y in $2y = 3x - 4$.

$$2\left(\frac{3}{2}x + 3\right) = 3x - 4$$

$$3x + 6 = 3x - 4$$

$$6 = 4$$

But $6 = 4$ is a false statement.

Thus, there is no solution. The lines are parallel.

Chapter Review

3. (continued)

This conclusion can also be arrived at by solving each equation for y and noticing that both lines have the same slope, but different y-intercepts.

$$-\frac{3}{2}x + y = 3$$

$$y = \frac{3}{2}x + 3;\ \text{slope} = \frac{3}{2},\ y\text{-intercept} = 3$$

$$2y = 3x - 4$$

$$y = \frac{3}{2}x - 2;\ \text{slope} = \frac{3}{2},\ y\text{-intercept} = -2.$$

4. The perpendicular bisector of $\overline{AB}$ has a slope perpendicular to the slope of $\overline{AB}$ and contains the midpoint of $\overline{AB}$.

(i) Slope of $\overline{AB} = \frac{5-0}{2-0} = \frac{5}{2}$

Thus, the slope of the perpendicular bisector of $\overline{AB}$ is $-\frac{2}{5}$.

(ii) Midpoint of $\overline{AB} = \left(\frac{0+2}{2}, \frac{0+5}{2}\right) = \left(1, \frac{5}{2}\right)$

(iii) Use the slope and the midpoint to find an equation of the perpendicular bisector.

$$y - \frac{5}{2} = \frac{-2}{5}(x - 1) \qquad \text{point-slope form}$$

$$y - \frac{5}{2} = \frac{-2}{5}x + \frac{2}{5}$$

$$y = \frac{-2}{5}x + \frac{4}{10} + \frac{25}{10}$$

$$y = \frac{-2}{5}x + \frac{29}{10} \qquad \text{slope-intercept form}$$

5. The equation of a circle with center (h, k) and radius r is $(x - h)^2 + (y - k)^2 = r^2$.

The radius of the circle is the distance from a point on the circle, $(-5, 3)$, to the center of the circle $(2, 4)$.

$$r = \sqrt{(-5-2)^2 + (3-4)^2} = \sqrt{(-7)^2 + (-1)^2} = \sqrt{50} = 5\sqrt{2}$$

Therefore, the equation of the circle is

$$(x-2)^2 + (y-4)^2 = 50.$$

Chapter Test **Section 8.4**

1. The diagonals of a rhombus are perpendicular bisectors of each other. Thus points B, C, and D must lie on the axes. Therefore, the appropriate coordinates of the other vertices are $B(0, b)$, $C(a, 0)$ and $D(0, -b)$.

2. We can verify that the sides of $ABCD$ are congruent.
$AB = \sqrt{[0 - (-a)]^2 + (b - 0)^2} = \sqrt{a^2 + b^2}$
$BC = \sqrt{(a - 0)^2 + (0 - b)^2} = \sqrt{a^2 + (-b)^2} = \sqrt{a^2 + b^2}$
$CD = \sqrt{(0 - a)^2 + (-b - 0)^2} = \sqrt{(-a)^2 + (-b)^2} = \sqrt{a^2 + b^2}$
$DA = \sqrt{(-a - 0)^2 + [0 - (-b)]^2} = \sqrt{(-a)^2 + b^2} = \sqrt{a^2 + b^2}$
Therefore, since $\overline{AB} \cong \overline{BC} \cong \overline{CD} \cong \overline{DA}$, $ABCD$ is a rhombus.

3. A complete solution is given in the answer key at the back of the text.

4. A complete solution is given in the answer key at the back of the text.

Solutions to Chapter 8 Test

1. F The slope of a perpendicular line is $-\frac{1}{2}$.

2. F If C is between A and B, then $AB + BC \neq AC$, but $AC + CB = AB$.

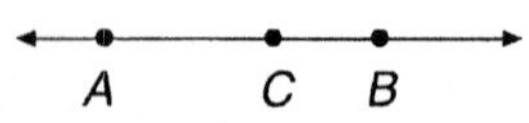

3. T Slope can be calculated using *any* two points on a line.

4. T If the line segments bisect each other, they cut each other in half.

5. T If the slopes of opposite sides are equal *and* the slopes of the diagonals show that the diagonals are perpendicular, then the quadrilateral must be a rhombus.

6. F It may have no solution or infinitely many solutions.

7. T Its rise is zero.

8. F If $ax + by = c$, then $by = -ax + c$, and
$y = -\frac{a}{b}x + \frac{c}{b}$. Therefore, the slope is $-\frac{a}{b}$.

9. F Doubling the coordinates doubles the dimensions and quadruples the area.

Chapter Test

10. T The y-intercept is the constant when the equation is written in slope-intercept form.

11. (a) The midpoint of $\overline{AB}$ is $\left(\frac{3+9}{2}, \frac{-2+4}{2}\right) = (6, 1)$.

    (b) Using the distance formula, you have
    $AB = \sqrt{(9-3)^2 + [4-(-2)]^2} = \sqrt{6^2 + 6^2}$
    $= \sqrt{36+36} = \sqrt{72} = \sqrt{36 \cdot 2} = 6\sqrt{2} \approx 8.5.$

    (c) Using the slope formula, slope of $\overline{AB} = \frac{4-(-2)}{9-3} = \frac{6}{6} = 1.$

    (d) Use the slope from part (c) and point A in the point-slope equation, and you get

$$y - (-2) = 1(x-3)$$
$$y + 2 = x - 3$$
$$y = x - 5$$

12. You can use either slope or the distance formula to determine whether points are collinear. Using slope, find slope of $\overline{AB}$ and slope of $\overline{BC}$. If they are equal, then the points are collinear.

Slope of $\overline{AB} = \frac{232-400}{-16-(-30)} = \frac{-168}{14} = -12$, and

Slope of $\overline{BC} = \frac{-200-232}{20-(-16)} = \frac{-432}{36} = -12.$

Slope of $\overline{AB}$ = Slope of $\overline{BC}$, so A, B and C are collinear. (You do not have to check the slope of $\overline{AC}$ since B is a point common to $\overline{AB}$ and $\overline{BC}$.)

13. A complete solution is given in the answer key at the back of the text.

14. (a) Find the solution to $3x - y = 11$ and $5x + 4y = 24$. Solving the first equation for y yields $y = 3x - 11$. Substituting for y in the second equation gives you

$$5x + 4(3x - 11) = 24$$
$$5x + 12x - 44 = 24$$
$$17x - 44 = 24$$
$$17x = 68$$
$$x = 4.$$

Then $y = 3x - 11 = 3 \cdot 4 - 11 = 1$.

The solution to the system of equations is $(4, 1)$.

Chapter Test

14. (continued)

(b) Solve $y = -x + 5$ and $3x + 3y = 13$. Use $-x + 5$ from the first equation to substitute for y in the second equation.

$$3x + 3(-x + 5) = 13$$
$$3x - 3x + 15 = 13$$
$$15 = 13$$

This is a contradiction so there is no solution to the system of equations. The lines must be parallel.

(c) You have a line and a circle with radius 2 and center (4, 3). If you graph them accurately, you will see that they do not intersect, and thus, there is no solution. An alternate algebraic method would be to solve by substitution as follows.

Solve $5x + 3y = 15$ for x.

(You could also solve for y.)

$$5x = 15 - 3y$$
$$x = 3 - \frac{3}{5}y.$$

Substitute for x in the second equation.

$$(x - 4)^2 + (y - 3)^2 = 4$$
$$\left(3 - \frac{3}{5}y - 4\right)^2 + (y - 3)^2 = 4$$
$$\left(-\frac{3}{5}y - 1\right)^2 + (y - 3)^2 = 4$$
$$\frac{9}{25}y^2 + \frac{6}{5}y + 1 + y^2 - 6y + 9 = 4$$

Multiply by 25 to eliminate the fractions.

$$9y^2 + 30y + 25 + 25y^2 - 150y + 225 = 100$$
$$34y^2 - 120y + 250 = 100$$
$$34y^2 - 120y + 150 = 0$$

Use the quadratic formula to solve this equation, and you will find that there is no real solution. Thus, the circle and the line do not intersect.

15. The perpendicular bisector will go through the midpoint of $\overline{AB}$ and have a slope that is the negative reciprocal of the slope of $\overline{AB}$.

The midpoint of $\overline{AB}$ is $\left(\frac{1 + 5}{2}, \frac{-6 + 2}{2}\right) = (3, -2)$.

The slope of $\overline{AB}$ is $\frac{2 - (-6)}{5 - 1} = \frac{8}{4} = 2$.

Chapter Test

15. (continued)

Since slope of $\overline{AB} = 2$, the slope of the perpendicular bisector is $-\frac{1}{2}$.

Use this slope and the midpoint in the point-slope form.

$$y - (-2) = -\frac{1}{2}(x - 3)$$

$$y + 2 = -\frac{1}{2}x + \frac{3}{2}$$

$$y = -\frac{1}{2}x + \frac{3}{2} - 2$$

$$y = -\frac{1}{2}x + \frac{3}{2} - \frac{4}{2}$$

$$y = -\frac{1}{2}x - \frac{1}{2}$$

or $x + 2y = -1$

16. $\overline{AB}$ is parallel to the x-axis on which $\overline{CD}$ lies. Therefore, B should have the same y-coordinate as A.
Also, $AB = CD = \sqrt{(c - b)^2 + (0 - 0)^2} = c - b$.
The x-coordinate of B equals the x-coordinate of A plus AB or $0 + (c - b) = c - b$. Therefore, the coordinates of B are $(c - b, a)$.

17. A complete solution is given in the answer key at the back of the text.

18. The radius of the circle is the distance between the center and the given point on the circle.
$r = \sqrt{[5 - (-1)]^2 + (0 - 4)^2} = \sqrt{6^2 + (-4)^2} = \sqrt{36 + 16} = \sqrt{52}$
The equation of the circle with center $O\,(-1, 4)$ and radius $\sqrt{52}$ is
$[(x - (-1)]^2 + (y - 4)^2 = (\sqrt{52})^2$, or
$(x + 1)^2 + (y - 4)^2 = 52$

19. Finding the slopes of the sides will help you to determine what type of quadrilateral $ABCD$ is.

$$\text{Slope of } \overline{AB} = \frac{6 - 4}{8 - (-5)} = \frac{2}{13}$$

$$\text{Slope of } \overline{BC} = \frac{0 - 6}{12 - 8} = \frac{-6}{4} = \frac{-3}{2}$$

$$\text{Slope of } \overline{CD} = \frac{-2 - 0}{-1 - 12} = \frac{-2}{-13} = \frac{2}{13}$$

$$\text{Slope of } \overline{DA} = \frac{4 - (-2)}{-5 - (-1)} = \frac{6}{-4} = \frac{-3}{2}$$

Chapter Test

19. (continued)

Thus $\overline{AB} \parallel \overline{CD}$ and $\overline{BC} \parallel \overline{DA}$. The adjacent sides are not perpendicular and a quick sketch or calculation using the distance formula will show that not all sides have the same length. Therefore, $ABCD$ is a parallelogram, but it is neither a rectangle nor a rhombus.

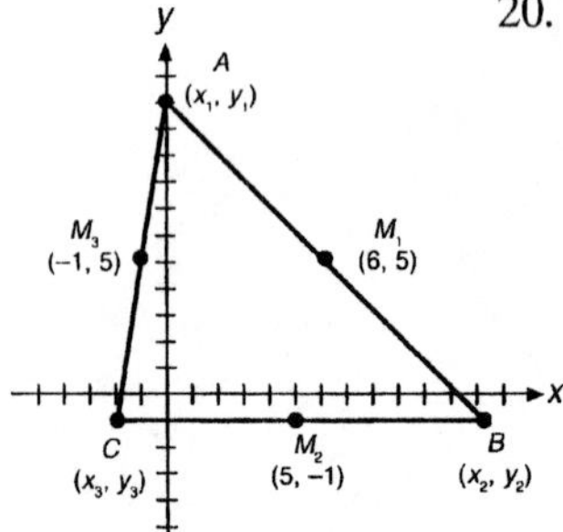

20. Let the coordinates of the vertices of the triangle be (x_1, y_1), (x_2, y_2), and (x_3, y_3).

You know $M_1 = (6, 5) = \left(\frac{x_1 + x_2}{2}, \frac{y_1 + y_2}{2}\right)$, so

$\frac{x_1 + x_2}{2} = 6$, or $x_1 + x_2 = 12$, and

$\frac{y_1 + y_2}{2} = 5$, or $y_1 + y_2 = 10$.

Likewise, $M_2 = (5, -1) = \left(\frac{x_2 + x_3}{2}, \frac{y_2 + y_3}{2}\right)$, so

$\frac{x_2 + x_3}{2} = 5$, or $x_2 + x_3 = 10$, and

$\frac{y_2 + y_3}{2} = -1$, or $y_2 + y_3 = -2$.

Also, $M_3 = (-1, 5) = \left(\frac{x_1 + x_3}{2}, \frac{y_1 + y_3}{2}\right)$, so

$\frac{x_1 + x_3}{2} = -1$, or $x_1 + x_3 = -2$, and

$\frac{y_1 + y_3}{2} = 5$, or $y_1 + y_3 = 10$.

Now you have two systems of three equations in three variables.

(i) $x_1 + x_2 = 12$

$x_2 + x_3 = 10$

$x_1 + x_3 = -2$

(ii) $y_1 + y_2 = 10$

$y_2 + y_3 = -2$

$y_1 + y_3 = 10$

To solve system (i), multiply the second equation by -1 and add it to the first.

$$-1(x_2 + x_3 = 10) \Rightarrow \quad \begin{array}{rrrl} & -x_2 & -x_3 & = -10 \\ +\ x_1 & +\ x_2 & & = 12 \\ \hline x_1 & & -x_3 & = 2 \end{array}$$

Chapter Test

Now add the result to the third equation.

$$\begin{array}{rcl} x_1 - x_3 & = & 2 \\ + \quad x_1 + x_3 & = & -2 \\ \hline 2x_1 & = & 0 \\ x_1 & = & 0 \end{array}$$

Substituting 0 in place of x_1 in the first and third original equations will give you x_2 and x_3.

$$0 + x_2 = 12 \text{ and } 0 + x_3 = -2$$
$$x_2 = 12 \qquad x_3 = -2.$$

Similarly, you can solve for y_1, y_2 and y_3.

$$-1(y_2 + y_3 = -2) \Rightarrow \qquad \begin{array}{rcl} -\, y_2 - y_3 & = & 2 \\ + \; y_1 + y_2 \qquad & = & 10 \\ \hline y_1 \qquad - y_3 & = & 12 \end{array}$$

$$\begin{array}{rcl} \text{and } y_1 - y_3 & = & 12 \\ + \; y_1 + y_3 & = & 10 \\ \hline 2y_1 & = & 22 \\ y_1 & = & 11 \end{array}$$

Substituting 11 for y_1 in the first and third original equations gives you y_2 and y_3.

$$11 + y_2 = 10 \text{ and } 11 + y_3 = 10$$
$$y_2 = -1 \qquad y_3 = -1.$$

Therefore, the three vertices of the triangle are $A(0, 11)$, $B(12, -1)$, and $C(-2, -1)$.

21. A complete solution is given in the answer key at the back of the text.

22. A complete solution is given in the answer key at the back of the text.

23. Refer to the solution to problem #44 in Section 8.2. It might be easier to calculate the slope if you first convert to similar units.

The horizontal distance is $\frac{0.8 \text{ mi}}{1} \times \frac{5280 \text{ ft}}{1 \text{ mi}} = 4224$ ft.

The slope of the road is $\frac{220 \text{ ft}}{4224 \text{ ft}} = \frac{5}{96} \approx 0.052$, so the percent grade is $\approx 5.2\%$.

Chapter Test

24. (a) The wire is just a very tall, skinny cylinder. The volume of a cylinder is $\pi r^2 h$. In this case $r = 0.2 \div 2 = 0.1$ cm. The length x is also the height of the cylinder. If y is the volume, then $y = \pi(0.1)^2 x$ or $y = 0.01\pi x$.

(b) The slope is the coefficient of x, or $0.01\pi = \frac{0.01\pi}{1}$. For every 1 cm change in length (x) there is a 0.01π cm^3 change in volume (y).

(c) Use the fact that mass = density · volume. If y represents mass, then $y = 8.94(0.01\pi)x = 0.0894\pi x$.

(d) The slope is 0.089π. This means that for every 1 cm change in length, there is a 0.0894π gram change in mass.

9

Transformation Geometry

Section 9.1

Tips:

✔ A rotation has a center of rotation about which the rotation is performed. A reflection is always performed with respect to a specified line.

✔ When tracing a polygon to perform an isometry, it is a good idea to mark the vertices first and then use a straightedge to draw the sides.

✔ Remember, a rotation of 30° means 30° *counterclockwise*, while a rotation of −30° is 30° *clockwise*.

Solutions to odd-numbered textbook problems

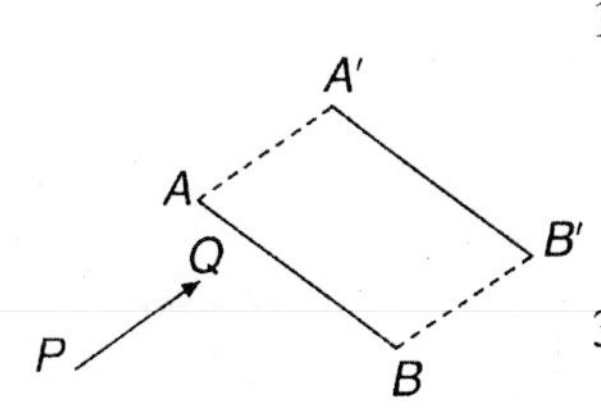

1. The dashed lines represent the path that point A slides along to get to A' and the path from B to B'. $\overline{AA'} \cong \overline{BB'}$ and $\overline{AA'} \parallel \overline{BB'}$ because a translation takes all points the same distance and direction. Thus, $ABB'A'$ is a parallelogram and $\overline{A'B'} \parallel \overline{AB}$. Translations take lines to parallel lines.

3. (a) A complete solution is given in the answer key at the back of the text.

   (b) Many vectors are possible. The head of the vector is 5 units right and 4 units down from the tail of the vector. One such vector is given in the answer key at the back of the text.

5. $A(-1,2)$ and $B(3,-3)$ are given.

   (a) The given vector describes the translation that moves points 1 unit left and 3 units down.

   Therefore, A' has coordinates $(-1-1, 2-3) = (-2,-1)$ and B' has coordinates $(3-1, -3-3) = (2,-6)$.

   (b) The given vector describes the translation that moves points 2 units left and 2 units up.

   Therefore, A' has coordinates $(-1-2, 2+2) = (-3, 4)$ and B' has coordinates $(3-2, -3+2) = (1,-1)$.

   (c) The given vector describes the translation that moves points 2 units right and 4 units up.

   Therefore, A' has coordinates $(-1+2, 2+4) = (1, 6)$ and B' has coordinates $(3+2, -3+4) = (5, 1)$.

   (d) The given vector describes the translation that moves points 3 units right and 2 units down.

   Therefore, A' has coordinates $(-1+3, 2-2) = (2, 0)$ and B' has coordinates $(3+3, -3-2) = (6,-5)$.

Section 9.1

7. Yes, $\Delta A''B''C'' \cong \Delta ABC$.

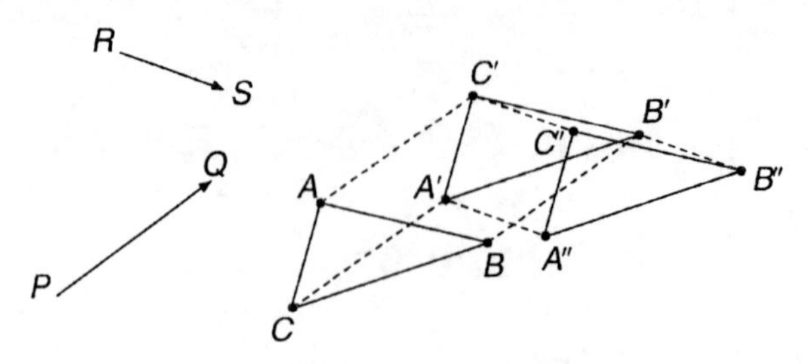

Since each of the translations performed is an isometry, $\Delta ABC \cong \Delta A'B'C'$ and also $\Delta A'B'C' \cong \Delta A''B''C''$. Thus, $\Delta ABC \cong \Delta A''B''C''$ (See Theorem 9.5).

9. Step 1: Construct a line parallel to $\overrightarrow{PQ}$ containing point A. (See Construction 7 of Section 5.5.)

   Step 2: Construct a line parallel to $\overrightarrow{PQ}$ containing point B.

   Step 3: Set compass to radius PQ and place point of compass on A and mark off length PQ. Label this intersection point A'.

   Step 4: Repeat Step 3 placing point of compass on B to locate intersection B'.

   Step 5: Connect points A' and B' to construct $\overline{A'B'}$ the image of $\overline{AB}$ under translation $\overrightarrow{PQ}$.

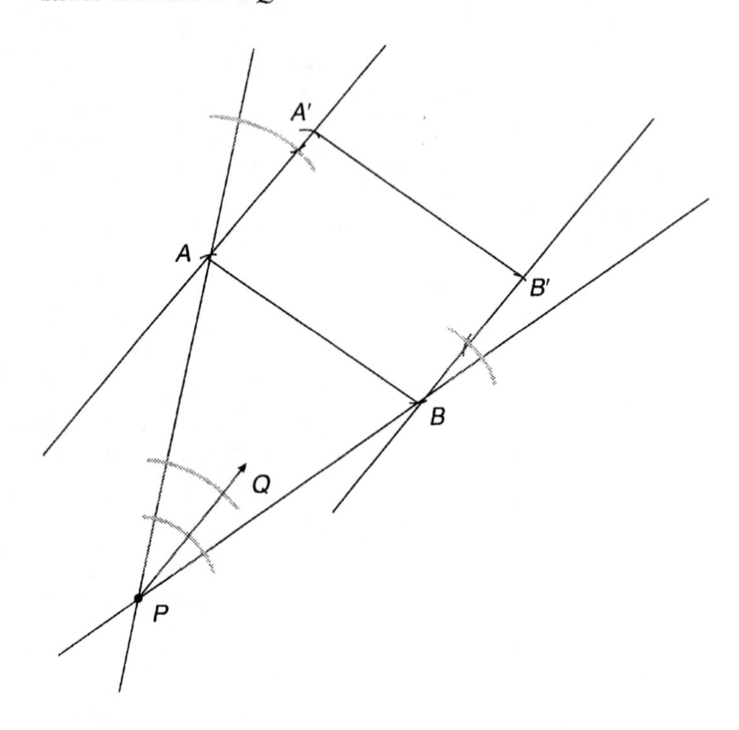

Section 9.1

11. (a) The translation that takes $P(0, 0)$ to $Q(1, 3)$ moves points 1 unit right and 3 units up.

    Therefore, $A' = (-1 + 1, 1 + 3) = (0, 4)$,

    $B' = (3 + 1, 3 + 3) = (4, 6)$, and

    $C' = (4 + 1, -2 + 3) = (5, 1)$.

    (b) The translation that takes $P(0, -1)$ to $Q(4, -2)$ moves points 4 units right and 1 unit down.

    Therefore, $A' = (-1 + 4, 1 - 1) = (3, 0)$,

    $B' = (3 + 4, 3 - 1) = (7, 2)$, and

    $C' = (4 + 4, -2 - 1) = (8, -3)$.

    (c) The translation that takes $P(-1, -2)$ to $Q(-2, 2)$ moves points 1 unit left and 4 units up.

    Therefore, $A' = (-1 - 1, 1 + 4) = (-2, 5)$,

    $B' = (3 - 1, 3 + 4) = (2, 7)$, and

    $C' = (4 - 1, -2 + 4) = (3, 2)$.

13. With $\overline{OP}$ as the initial side, use your protractor to draw a 90° angle, $\angle POT$, in a clockwise direction. Trace O and P and rotate the tracing paper about O through the angle you drew until P is on the terminal side of the angle. Make an imprint of P on $\overline{OP}$ to get P'. A diagram of the solution is shown in the answer key at the back of the text.

15. For each $\angle ABC$, $\overrightarrow{AB}$ is the initial side and $\overrightarrow{BC}$ is the terminal side of the angle.

    (a) $\angle ABC = -60°$ as it is a clockwise angle.

    (b) $\angle ABC = 110°$ as it is a counterclockwise angle.

    (c) $\angle ABC = 180°$ (or $-180°$) A straight angle measures 180° and the directed angle is either positive or negative.

    (d) $\angle ABC = -20°$ as it is a clockwise angle.

17. A complete solution is given in the answer key at the back of the text.

19. A complete solution is given in the answer key at the back of the text.

21. Step 1: Draw line segments $\overline{OA}, \overline{OB}$, and $\overline{OC}$.

    Step 2: Construct circles centered at O with radii $\overline{OA}, \overline{OB}$, and $\overline{OC}$.

    Step 3: With $\overline{OA}$ as initial side, use a protractor to locate point A' on circle of radius $\overline{OA}$ with $\angle AOA' = -60°$.

    Step 4: With $\overline{OB}$ as initial side, use a protractor to locate point B' on circle of radius $\overline{OB}$ with $\angle BOB' = -60°$.

Section 9.1

21. (continued)

Step 5: With $\overline{OC}$ as initial side, use a protractor to locate point C' on circle of radius $\overline{OC}$ with $\angle COC' = -60°$.

Step 6: Draw line segments joining A', B', and C' to form $\Delta A'B'C'$.

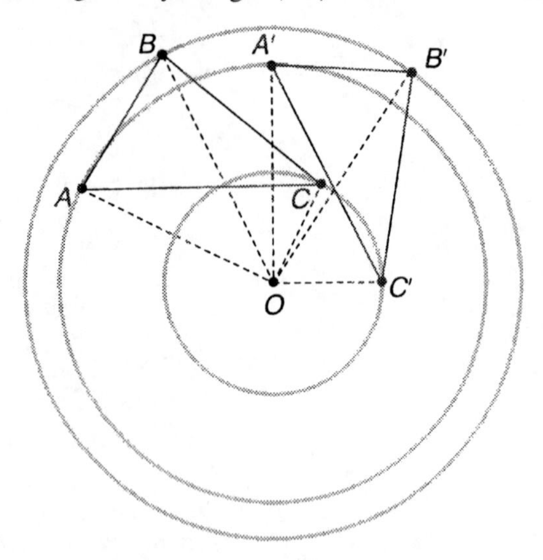

23. Step 1: To find $\Delta A'B'C'$, follow the procedure given in problem #21 using an angle of $-70°$ instead of $-60°$.

Step 2: Use tracing paper (See problem #1) to find the translation image $\Delta A''B''C''$ of $\Delta A'B'C'$ using translation vector $\overrightarrow{PQ}$.

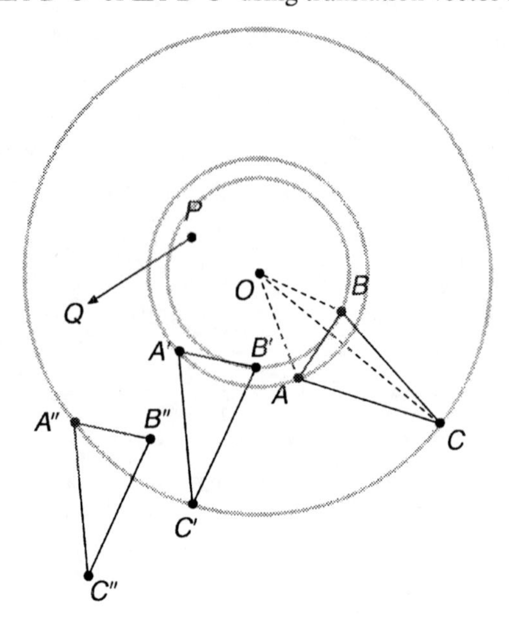

Section 9.1

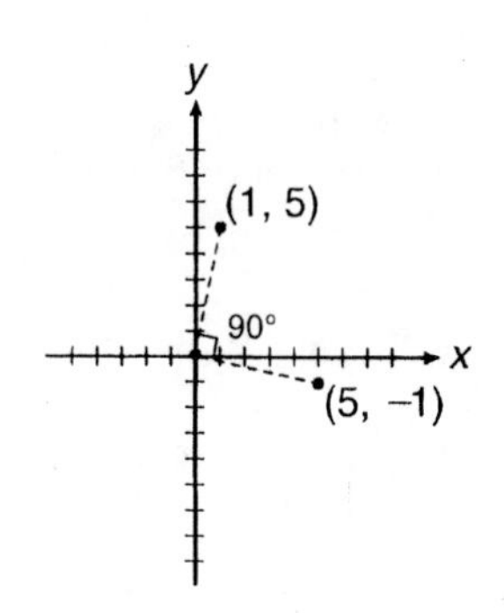

25. Draw a sketch for each point.

	point		image
(a)	$(1, 5)$	$\rightarrow$	$(5, -1)$
(b)	$(-1, 3)$	$\rightarrow$	$(3, 1)$
(c)	$(-2, 4)$	$\rightarrow$	$(4, 2)$
(d)	$(-3, -1)$	$\rightarrow$	$(-1, 3)$
(e)	$(5, -2)$	$\rightarrow$	$(-2, -5)$

Notice that each time the y-coordinate and the x-coordinate switch places and the new y-coordinate changes signs.

(f) $(x, y) \rightarrow (y, -x)$

27. A complete solution is given in the answer key at the back of the text.

29.

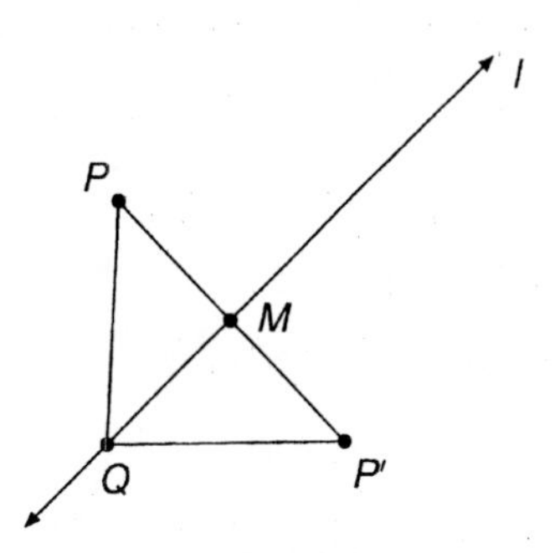

(a) $\Delta PP'Q$ is an isosceles triangle because $PQ = P'Q$.

(b) ΔPMQ is a right triangle because $\overline{PM} \perp l$.(Also, $\overline{PP'} \perp l$.)

31. Step 1: With point of compass on A and radius set to AB, swing an arc on the other side of line l.

Step 2: With point of compass on C and radius set to CB, swing an arc on the other side of line l.

Step 3: Label the intersection of the two arcs B'.

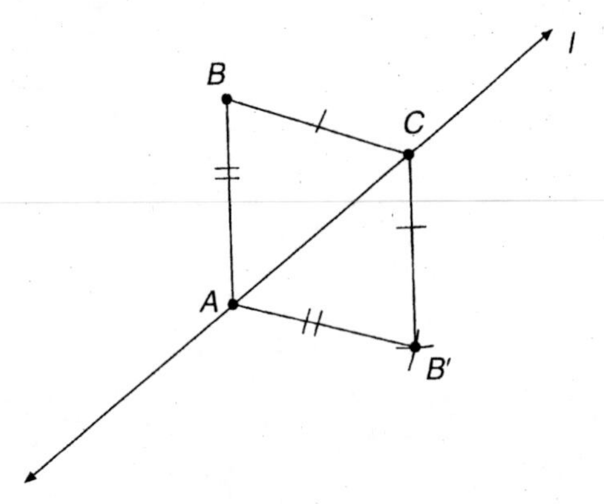

$ABCB'$ is a kite because $CB = CB'$ and $AB = AB'$.

Section 9.1

33. To construct the reflection image of A with respect to line l, do the following. (See diagram in the answer key at the back of the text.)

    Step 1: Trace line l and point A.

    Step 2: Proceed as if you were constructing a perpendicular line from A to line l (See Section 4.4). The intersection of the arcs on the side of l opposite A marks A'.

35. A complete solution is given in the answer key at the back of the text.

37. Step 1: Follow the procedure given in problem #33 to find points A' and B'.

    Step 2: Draw $\overline{A'B'}$.

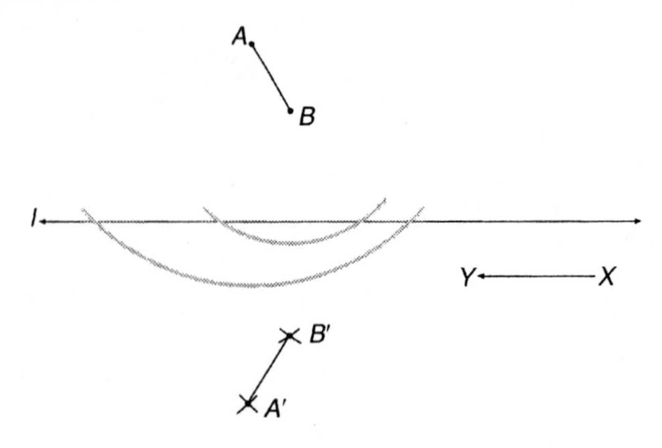

Step 3: Construct a line parallel to $\overrightarrow{XY}$ containing point A'. (See Construction 7 of Section 5.5.)

Step 4: Set compass to radius XY and place point of compass on A' and mark off length XY. Label this intersection point A''.

Step 5: Repeat steps 3 and 4 using point B' and locating intersection point B''.

Step 6: Connect points A'' and B'' to construct $\overline{A''B''}$.

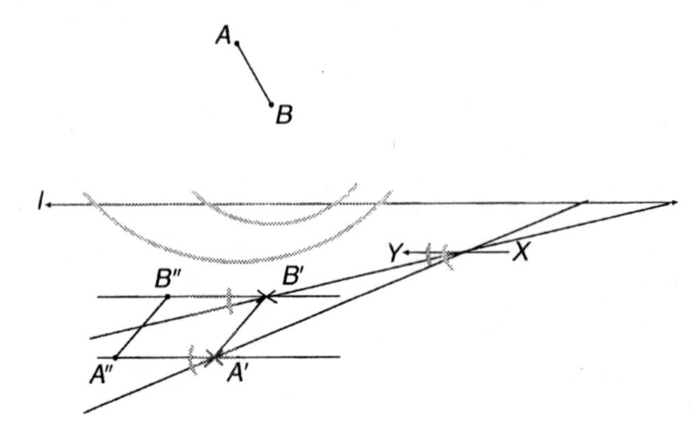

Section 9.1 39.

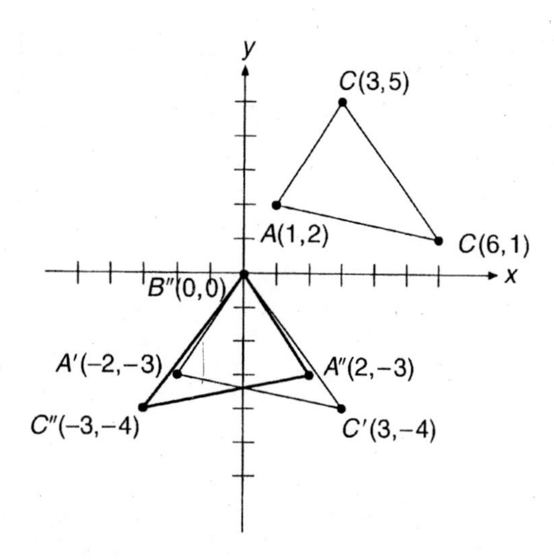

(a) The translation that takes $P(0, 0)$ to $Q(-3, -5)$ moves points 3 units left and 5 units down.

Thus $A' = (1 - 3, 2 - 5) = (-2, -3)$,

$B' = (3 - 3, 5 - 5) = (0, 0)$, and

$C' = (6 - 3, 1 - 5) = (3, -4)$.

Now, reflect $\Delta A'B'C'$ about the y-axis

(b) $A''(2, -3), B''(0, 0)$, and $C''(-3, -4)$

(c) (i) The coordinates of (a, b) under the given translation is $(a - 3, b - 5)$.

(ii) The coordinates of $(a - 3, b - 5)$ after reflecting about the y-axis are $(-(a - 3), b - 5) = (3 - a, b - 5)$ because the x-coordinate of the image is opposite its preimage and y-coordinate of the image is the same as its preimage.

(iii) Therefore, the point (a, b) is mapped to $(3 - a, b - 5)$ under the glide reflection.

41. (a)

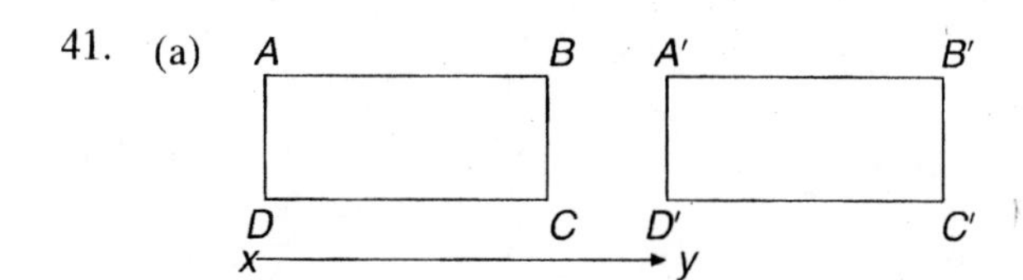

Rectangle $A'B'C'D'$ is the result of the translation from X to Y of rectangle $ABCD$.

(b)

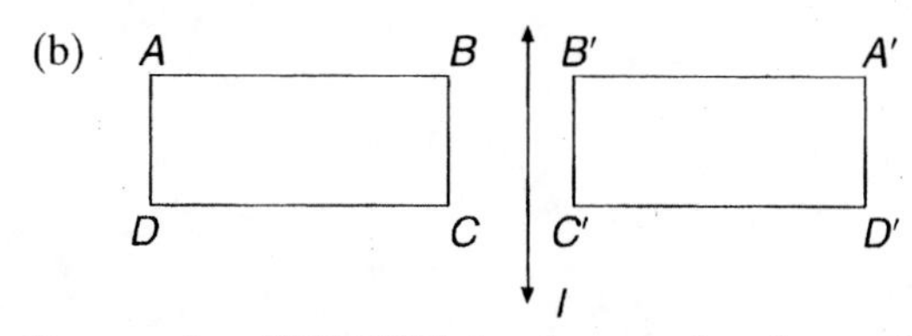

Rectangle $A'B'C'D'$ is the result of a reflection about line l of rectangle $ABCD$.

Section 9.1

43. Find the perpendicular bisector of $\overline{AA'}$, $\overline{BB'}$ or $\overline{CC'}$. (See Construction 6 of Section 4.4.) The perpendicular bisector is the reflection line.

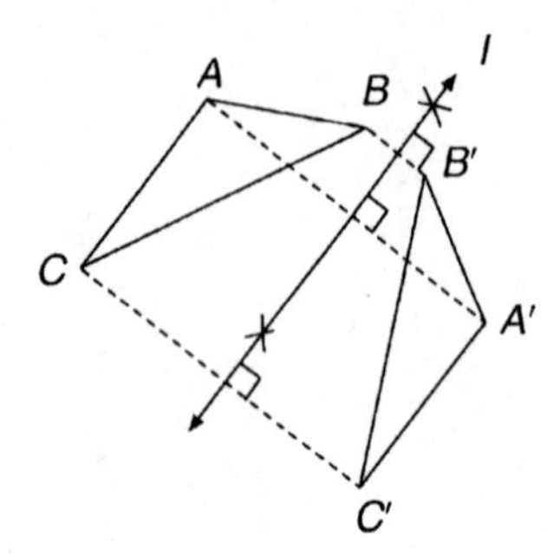

45. (a) Reflect $ABCD$ about line l and translate by the given translation vector to obtain $A''B''C''D''$.

Thus, the isometry is a glide reflection.

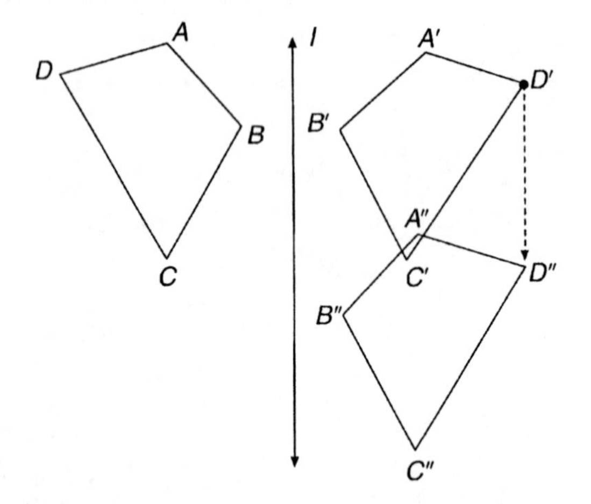

(b) The isometry is a rotation about O with an angle of $-90°$.

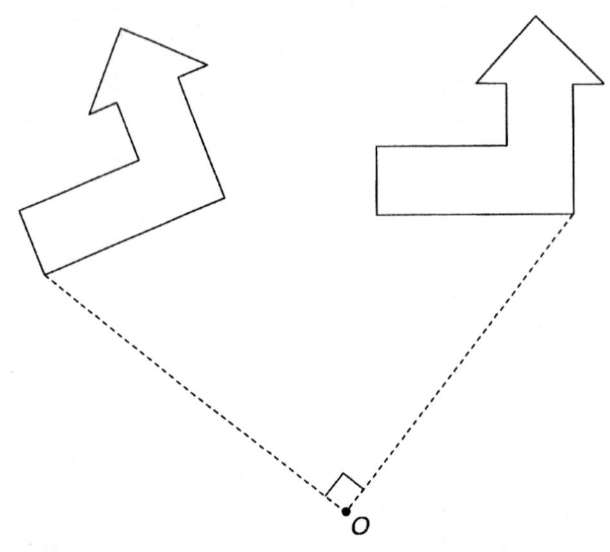

Section 9.1

45. (continued)

(c) The isometry is a translation. The translation vector is given.

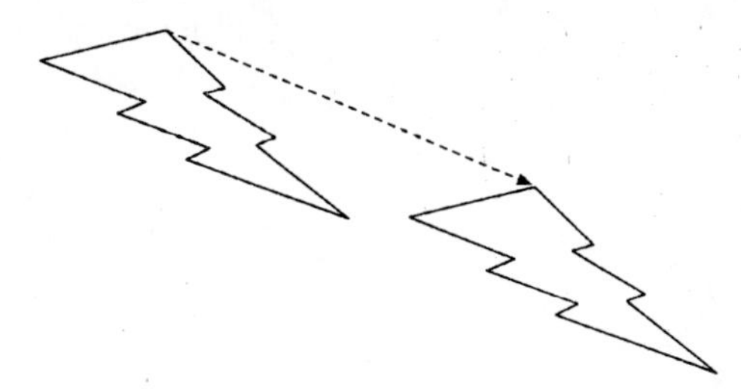

47. A complete solution is given in the answer key at the back of the text.

49. A complete solution is given in the answer key at the back of the text.

51. A complete solution is given in the answer key at the back of the text.

53. Given: $\overline{AB} \perp \overline{CD}$. Let A', B', C', and D' be the images of A, B, C, and D, respectively, under a rotation about point O.

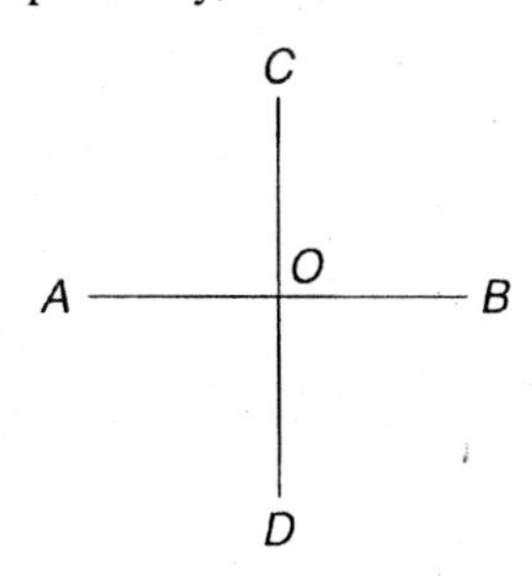

Prove: $\overline{A'B'} \perp \overline{C'D'}$

Proof:

Rotate $\overline{AB}$ and $\overline{CD}$ through $\angle XPT$ to $\overline{A'B'}$ and $\overline{C'D'}$, respectively. Then $\overline{A'B'}$ is the image of $\overline{AB}$, and $\overline{C'D'}$ is the image of $\overline{CD}$.

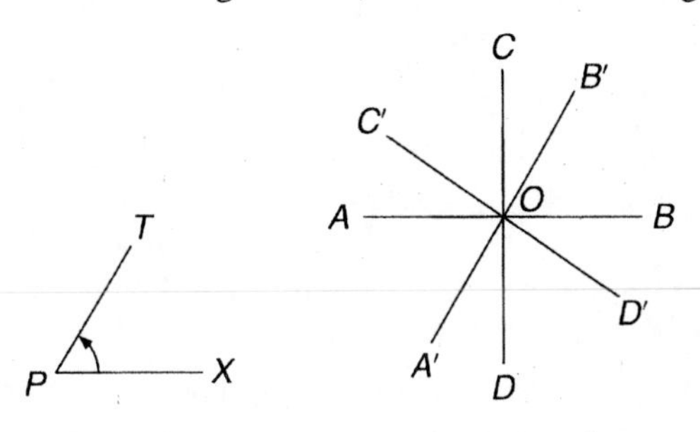

$\angle AOC = 90°$. Since rotations preserve angle measure, $\angle A'O'C' = 90°$ also. Thus, $\overline{A'B'} \perp \overline{C'D'}$ by the definition of perpendicular.

55. A complete solution is given in the answer key at the back of the text.

Section 9.1

57. Given: $\overline{AB} \perp \overline{CD}$

Prove: $\overline{A'B'} \perp \overline{C'D'}$, where A' is the image of A under the reflection with respect to line l.

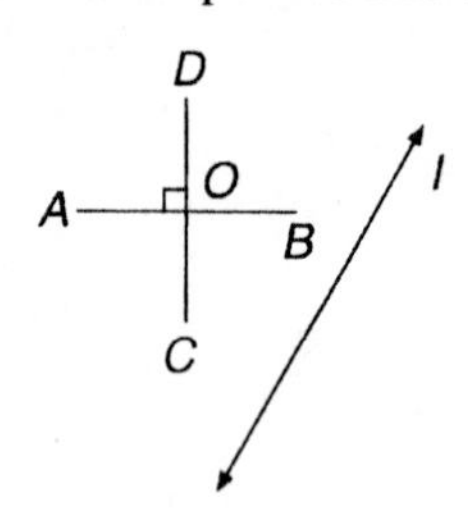

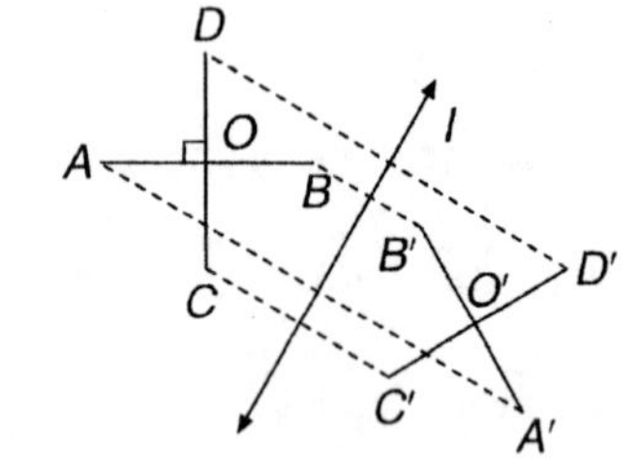

Proof:

Reflect $\overline{AB}$ across line l. Also reflect $\overline{CD}$ across line l. The reflection image of $\angle AOD$ is $\angle A'O'D'$. $\angle AOD \cong \angle A'O'D'$ because reflections preserve angle measure (Theorem 9.3). $\angle AOD = 90°$ by the definition of perpendicular. $\angle A'O'D' = 90°$, and $\overline{A'B'} \perp \overline{C'D'}$.

59. Let $\overline{A''B''}$ be the image of $\overline{AB}$ under the glide reflection consisting of reflection with respect to line l and translation $\overrightarrow{PQ}$.

Reflect $\overline{AB}$ about line l obtaining $\overline{A'B'}$. Since reflections preserve distance, $AB = A'B'$.

Translate $\overline{A'B'}$ by vector $\overrightarrow{PQ}$ to obtain $\overline{A''B''}$.

Since translations preserve distance, $A'B' = A''B''$.

Therefore, by the transitive property $AB = A''B''$ and glide reflections preserve distance.

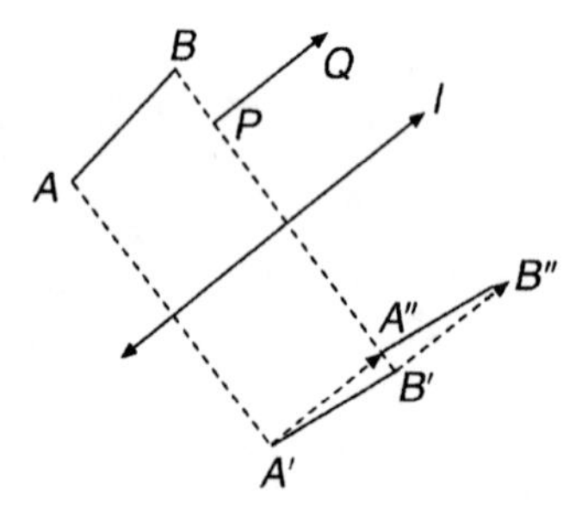

61. Depending upon how you look at these borders, other answers besides those in the answer key are also possible. Focus on one key part of the pattern and imagine what would map it to an identical part somewhere else.

(a) This pattern could be made by a translation or rotation.

(b) This design can be formed by a translation or glide reflection (across a line parallel to the line containing the centers of the flowers) or a rotation about a point not in the design.

(c) This design can be formed by a translation.

(d) Imagine mapping one crescent to another crescent and one dot to another dot. A translation, rotation, reflection, or glide reflection could produce this pattern.

Section 9.1

63. A complete solution is given in the answer key at the back of the text.

65. A complete solution is given in the answer key at the back of the text.

67. (a) A complete solution is given in the answer key at the back of the text.

(b) (i) If player O places an O in the center, then there are two unique moves for player *X*.

The following four moves are equivalent.

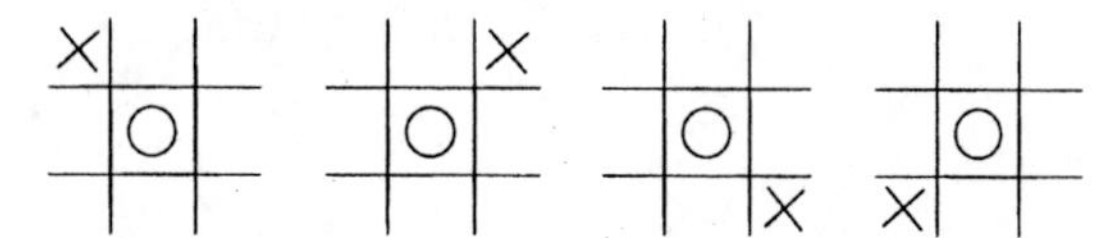

The following four moves are equivalent.

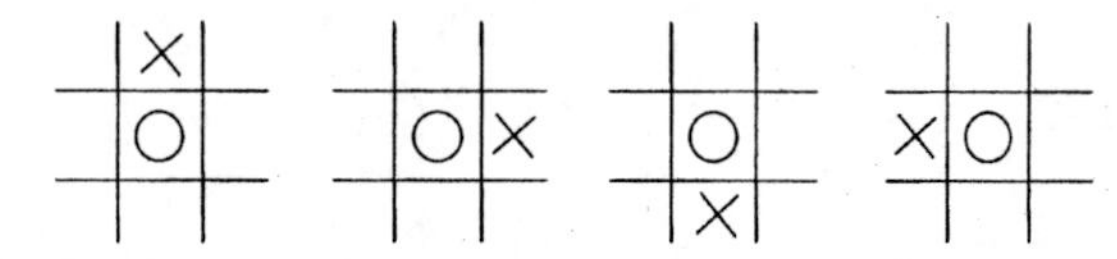

(ii) If player O places an O in a corner, then there are five unique moves for player X.

The following two moves are equivalent.

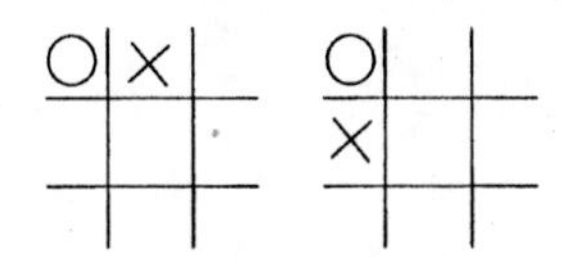

The following two moves are equivalent.

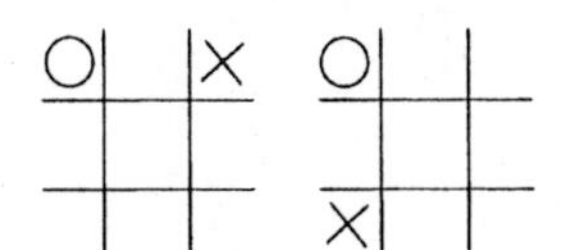

The following two moves are equivalent.

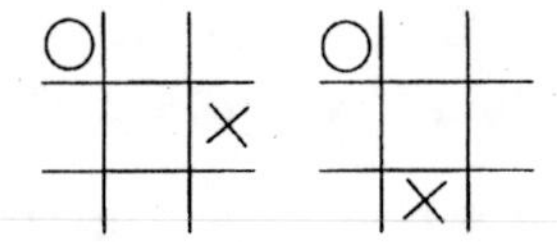

The following are the other unique moves that can be made.

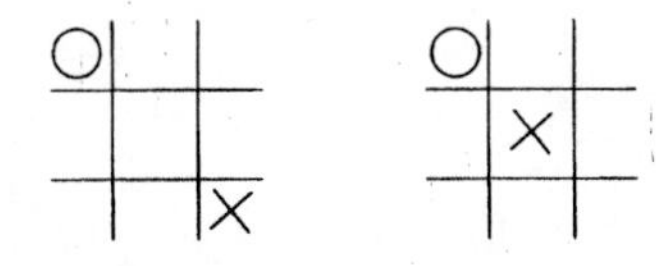

67. (continued)

(b) (iii) If player O places an O in the middle of an outside row or column, then there are five unique moves for player X.

The following two moves are equivalent.

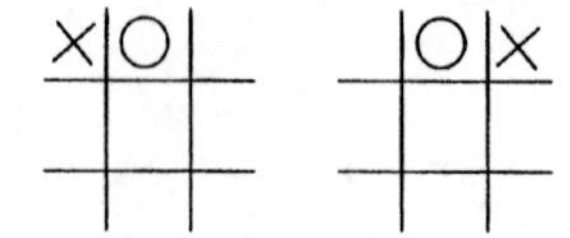

The following two moves are equivalent.

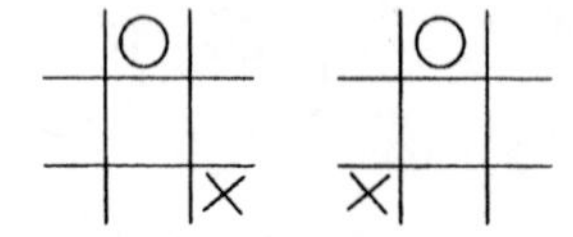

The following two moves are equivalent.

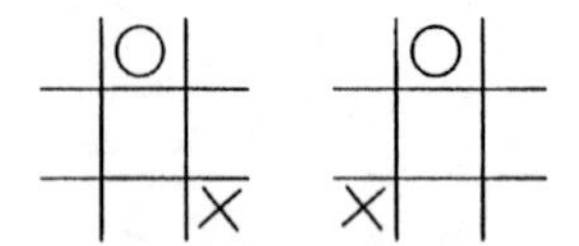

The following are the other unique moves that can be made.

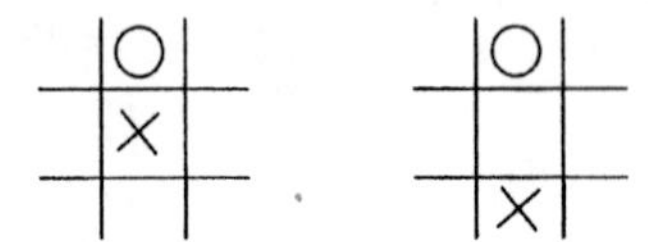

Section 9.2

Tip:

✔ A similitude is a combination of a size transformation with an isometry (such as a translation, rotation, reflection or glide reflection).

Solutions to odd-numbered textbook problems

1. (a) Trace O and A and draw ray $\overrightarrow{OA}$. Measure OA with a ruler. With a scale factor of four, OA' must be four times OA. Mark the point on $\overrightarrow{OA}$ that is a distance of $4(OA)$ from O. That point is A'.

   (b) Follow the directions for part (a) except mark A' at a distance of $3.75(OA)$ from O on $\overrightarrow{OA}$.

   (c) Follow the directions for part (a) except mark A' at a distance of $2\frac{5}{8}(OA)$ from O on $\overrightarrow{OA}$.

Section 9.2

1. (continued)

 (d) Follow the directions for part (a) except mark A' at a distance of $\frac{2}{3}(OA)$ from O on $\overrightarrow{OA}$. Notice that A' will be closer to the point O than A is when the scale factor is less than one.

3. Use a ruler to measure $\overline{OP}$. Mark P' at a distance of $3(OP)$ from O on $\overrightarrow{OP}$. A diagram of the solution is shown in the answer key at the back of the text.

5. (a) Use a ruler to measure OC, which is the radius of the given circle. Since $k = 2$, the radius of the new circle is $2(OC)$. Set the radius of your compass at that length. Placing the point of your compass on O, draw circle C' with the new radius. A diagram of the solution is shown in the answer key at the back of the text.

 (b) Use a ruler to measure OC. Calculate $\frac{1}{2}(OC)$ and set the radius of your compass to that length. With the point of your compass on O, draw circle C' with the new radius. A diagram of the solution is shown in the answer key at the back of the text.

7. (a) Measure $\overline{OA}$. Mark A' at a distance of $2(OA)$ from O on $\overrightarrow{OA}$. Measure $\overline{OB}$. Mark B' at a distance of $2(OB)$ from O on $\overrightarrow{OB}$. Measure $\overline{OC}$. Mark C' at a distance of $2(OC)$ from O on $\overrightarrow{OC}$.

 Connect points A and B, and B and C. A diagram of the solution is shown in the answer key at the back of the text.

 (b) $\angle ABC \cong \angle A'B'C'$ as size transformations preserve angle measure.

9. (a) Since the center of the size transformation is the origin, to find B' and C' you multiply both coordinates of each point by $k = 2$.

$$A' = (2 \cdot 0, 2 \cdot 0) = (0, 0)$$
$$B' = (2 \cdot 4, 2 \cdot 0) = (8, 0)$$
$$C' = (2 \cdot 4, 2 \cdot 3) = (8, 6)$$

 (b) Area of $\Delta A'B'C' = \frac{1}{2}(8)(6) = 24$

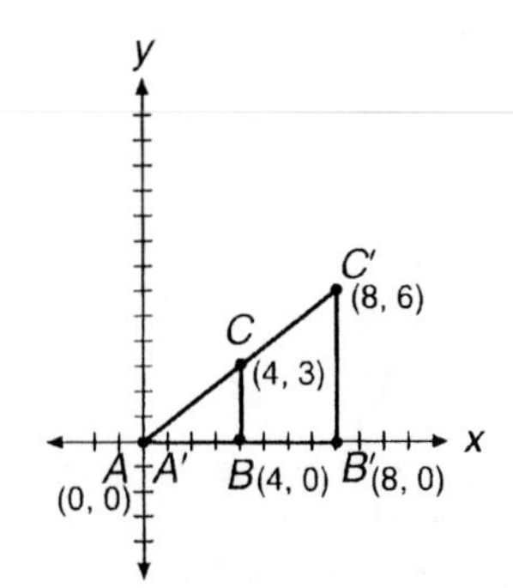

$$\text{Area of } \Delta ABC = \frac{1}{2}(4)(3) = 6$$

$$\frac{\text{Area of } \Delta A'B'C'}{\text{Area of } \Delta ABC} = \frac{24}{6} = 4$$

Notice that the ratio is the square of the scale factor.

Section 9.2

11. Let P be a point on circle O. Then OP is the radius of circle O. Measure $\overline{OP}$. Set compass to radius $3(OP)$ and draw circle centered at O.

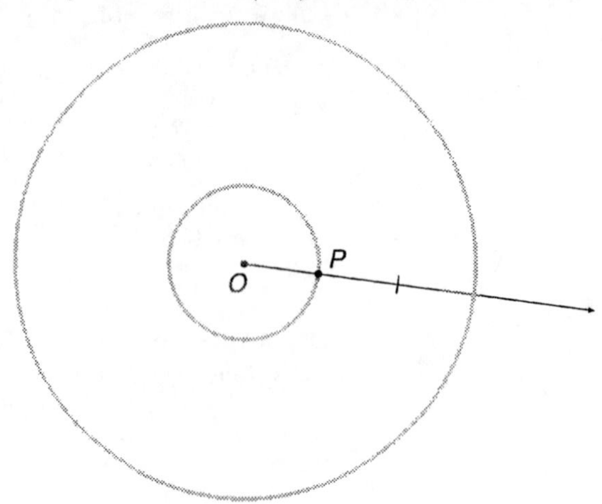

(i) The circumference of the original circle is $2\pi(OP)$.

The circumference of the image circle is $2\pi[3(OP)] = 3[2\pi(OP)]$.

Thus, the circumference of the image circle is three times the circumference of the original circle.

(ii) The area of the original circle is $\pi(OP)^2$.

The area of the image circle is $\pi[3(OP)]^2 = 9[\pi(OP)^2]$.

Thus, the area of the image circle is nine times the area of the original circle.

13. (a) The center of the size transformation is found by extending the lines $\overleftrightarrow{Q'Q}$ and $\overleftrightarrow{P'P}$. Their point of intersection is the center of the size transformation. A diagram of the solution is shown in the answer key at the back of the text.

(b) Label as O the point of intersection of the lines $\overleftrightarrow{P'P}$ and $\overleftrightarrow{Q'Q}$. Then, using a ruler, measure $\overline{Q'O}$ and $\overline{QO}$. The scale factor is the ratio of these two lengths. $\frac{Q'O}{QO} \approx 2$. Similarly, $\frac{P'O}{PO} \approx 2$. Thus, the scale factor is approximately 2.

15. ΔABC and $\Delta A'B'C'$ are shown in the answer key at the back of the text. Each coordinate of A', B' and C' is twice the corresponding coordinate of A, B, or C. Thus, the center of the size transformation is the origin. The center could also be found by finding the intersection of $\overleftrightarrow{AA'}$, $\overleftrightarrow{BB'}$ and $\overleftrightarrow{CC'}$.

Therefore, the center is $(0, 0)$ and the scale factor is 2.

17. (a) The image of $ABCD$ after the size transformation as shown in the answer key is $A'B'C'D'$ and after both the size transformation and the reflection, the image if $A''B''C''D''$.

Section 9.2

17. (continued)

(b) Each of the coordinates of A', B', C', and D' is twice the corresponding coordinate of A, B, C, or D since the center of the transformation is the origin. The reflection across the y-axis changes the sign of the x-coordinate.

(c) If a point has coordinates (a, b), then its image will be $(-2a, 2b)$ under this similitude.

19. (a) The image of $ABCD$ under the rotation as shown in the answer key is $A'B'C'D'$ and after both the size transformation and the rotation, the image is $A''B''C''D''$.

(b) If a point has coordinates (a, b), then under the rotation of 180° about the origin, its image will be $(-a, -b)$.

Hence the coordinates after the rotation are $A'(0, 0)$, $B'(-1, -3)$, $C'(-4, -2)$ and $D'(-5, 0)$.

Now applying the size transformation centered at the origin with scale factor one-half results in each of the coordinates of A'', B'', C'' and D'' being one-half the corresponding coordinate of A', B', C', or D'. That is, $A''(0, 0)$, $B''\left(\frac{1}{2}, -\frac{3}{2}\right)$, $C''(-2, -1)$ and $D''\left(-\frac{5}{2}, 0\right)$.

(c) If a point has coordinates (a, b), then its image will be $\left(-\frac{a}{2}, -\frac{b}{2}\right)$ under this similitude.

21. A similitude is a combination of a size transformation and an isometry.

First find the scale factor k by finding the ratio of the sides of $\Delta A'B'C'$ to the sides of ΔABC.

$$k = \frac{A'B'}{AB} = \frac{9}{2} = 4.5$$

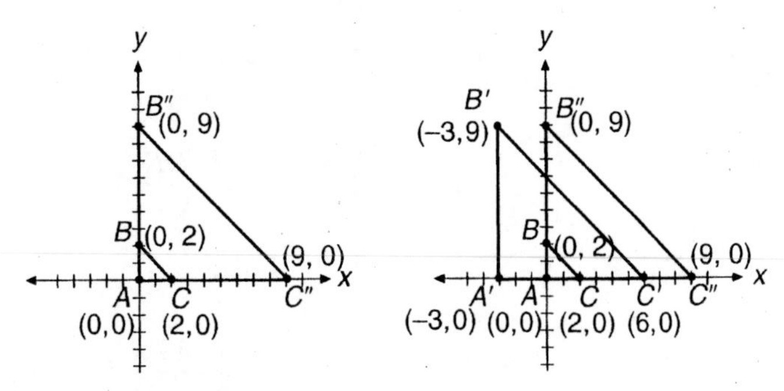

Next, using a size transformation with scale factor $k = 4.5$ and center A, you get $A''(0, 0)$, $B''(0, 9)$, and $C''(9, 0)$. These points give $\Delta A''B''C''$, which is similar to ΔABC. (See diagram at left above.) Now translate $\Delta A''B''C''$ 3 units left and you have $\Delta A'B'C'$. (See diagram at right above.)

23. A complete solution is given in the answer key at the back of the text.

Section 9.2

25. A complete solution is given in the answer key at the back of the text.

27. A complete solution is given in the answer key at the back of the text.

29. Consider the size transformation with center O that takes A to A', B to B', and C to C' with $\overline{AB} \perp \overline{BC}$.

Then $\angle ABC = 90°$ since $\overline{AB} \perp \overline{BC}$.

Hence $\angle A'B'C' = 90°$ since size transformations preserve angle measure.

Therefore, $\overline{A'B'} \perp \overline{B'C'}$. If the measure of an angle is 90° then the sides of the angle are perpendicular.

31. First find the radius of the oil slick 100 square feet in area.

$$\pi r^2 = 100$$

$$r = \sqrt{\frac{100}{\pi}} \text{ ft}$$

Now, let k be the scale factor of the size transformation.

Then $kr = k\sqrt{\frac{100}{\pi}}$ is the radius of the oil slick 450 square feet in area.

Thus, $\pi(kr)^2 = 450$

$$\pi\left(k\sqrt{\frac{100}{\pi}}\right)^2 = 450$$

$$\pi\left(\frac{100k^2}{\pi}\right) = 450$$

$$k^2 = 4.5$$

$$k = \sqrt{4.5} \text{ or } k = -\sqrt{4.5}.$$

Reject $k = -\sqrt{4.5}$ as it does not make sense here.

Hence, the scale factor is $k = \sqrt{4.5} \approx 2.12$.

The center of the size transformation is the oil drum where the oil is leaking.

33. The scale factor is $k = \dfrac{OR}{OQ}$

$$= \frac{OQ + QR}{OQ}$$

$$= \frac{18 + 6}{18}$$

$$= \frac{24}{18}$$

$$= \frac{4}{3}.$$

35. A complete solution is given in the answer key at the back of the text.

Section 9.3

Section 9.3

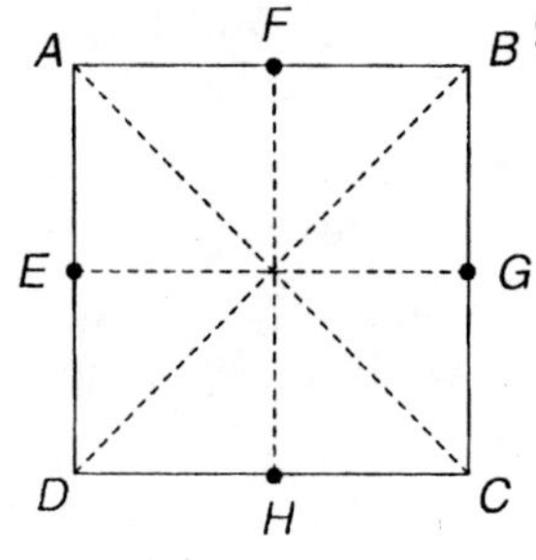

Solutions to odd-numbered textbook problems

1. Draw in the diagonals and connect $\overline{EG}$ and $\overline{FH}$ as shown in the picture. Remember that the diagonals are perpendicular bisectors of each other, as are $\overline{EG}$ and $\overline{FH}$. Reflection with respect to any one of the lines containing a segment $\overline{AC}, \overline{BD}, \overline{FH}$ or $\overline{EG}$ will reflect $ABCD$ onto itself.

3. A complete solution is given in the answer key at the back of the text.

5. (a) Remember that an isometry can be a translation, rotation, reflection or glide reflection. $ABCD$ has reflection symmetry with respect to the lines $\overleftrightarrow{AC}$ and $\overleftrightarrow{BD}$ since when you fold it on either of these lines, the sides match. Thus, $ABCD$ maps onto itself. Similarly, rotations of 180° or 360° around the point of intersection of the diagonals will map $ABCD$ onto itself.

   (b) A complete solution is given in the answer key at the back of the text.

7.

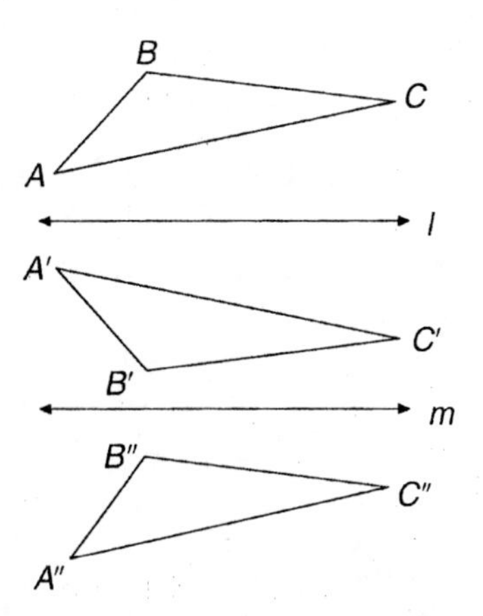

   (a) $\Delta ABC \cong \Delta A''B''C''$ as reflections take triangles to congruent triangles.

   (b) Yes. Both ΔABC and $\Delta A''B''C''$ have a clockwise orientation.

   (c) There are no fixed points.

   (d) A complete solution is given in the answer key at the back of the text.

   (e) A complete solution is given in the answer key at the back of the text.

9. (a) A rotation of 90° about P will map the square onto itself since it will map each side onto the adjacent side and the sides of a square are congruent.

Section 9.3

9. (continued)

(b) Because the diagonals are perpendicular and bisect each other, a reflection with respect to $\overline{AC}$ maps B to D and D to B while keeping A and C fixed. Thus, it maps $ABCD$ onto itself.

(c) The reflection with respect to $\overline{BD}$ maps A to C and C to A while keeping B and D fixed. Thus, it maps $ABCD$ onto itself.

(d) A rotation of 270° maps A to D, B to A, C to B and D to C. Since rotations preserve length and the sides of $ABCD$ are congruent, $ABCD$ is mapped onto itself.

(e) The rotation about P of 90° described in (a) or 270° described in (d) maps each diagonal onto the other. Either rotation can be used to prove that the diagonals are congruent since rotations preserve distance.

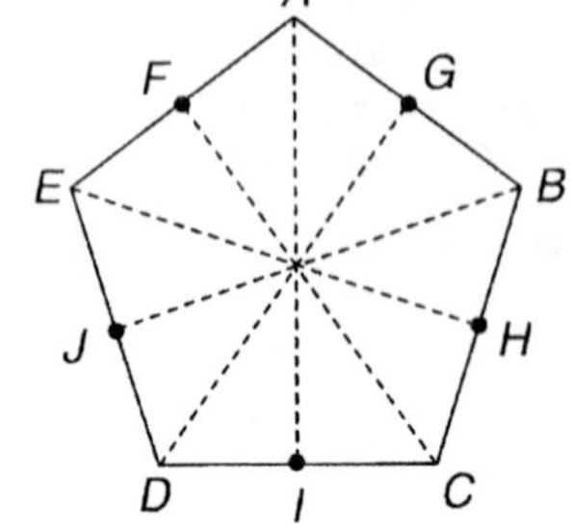

11. (i) The lines of reflection are $\overleftrightarrow{AI}$, $\overleftrightarrow{BJ}$, $\overleftrightarrow{CF}$, $\overleftrightarrow{DG}$, and $\overleftrightarrow{EH}$.

(ii) The angles of rotation are 72° (the central angle) and its multiples. Thus, the angles are 72°, 144°, 216°, and 288° with respect to the center of the pentagon. Note: A rotation of 360° with respect to the center of the pentagon is an identity transformation.

(iii) Rotate pentagon 72°. Then diagonal $\overline{AC}$ rotates to $\overline{EB}$. Thus $\overline{AC} \cong \overline{EB}$. Similarly, $\overline{EB}$ rotates to $\overline{DA}$, and so on.

13. A complete solution is given in the answer key at the back of the text.

15. A complete solution is given in the answer key at the back of the text.

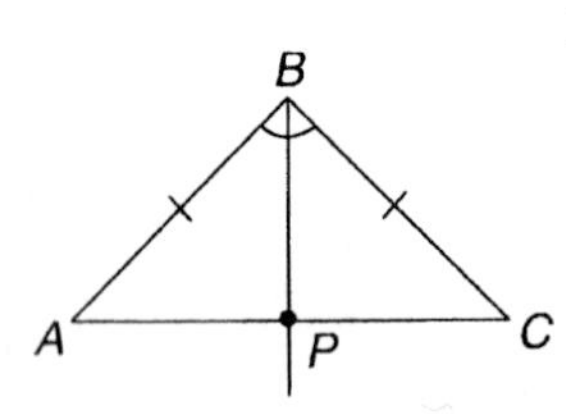

17. Given: B is equidistant from A and C. P is a point on $\overline{AC}$ such that $\overrightarrow{BP}$ bisects $\angle ABC$.

Prove: $\overleftrightarrow{BP}$ is perpendicular bisector of $\overline{AC}$.

Proof:

Reflect $\overline{AB}$ with respect to $\overleftrightarrow{BP}$. Since a reflection preserves angles and distance, $\overline{AB}$ goes to $\overline{CB}$. Therefore, $\Delta ABP \cong \Delta CBP$ by SAS. Then by C.P., you have $\angle APB = \angle CBP$. Since $\angle APB$ and $\angle CPB$ are congruent and supplementary, they must be right angles.

Thus, $\overleftrightarrow{BP}$ is perpendicular to $\overline{AC}$. Since $\overline{AP} = \overline{CP}$ by C.P., $\overleftrightarrow{BP}$ is the perpendicular bisector of $\overline{AC}$.

19. A complete solution is given in the answer key at the back of the text.

21. Given: $ABCD$ is a parallelogram.

Prove: $\overline{AC}$ and $\overline{BD}$ bisect each other.

Section 9.3 21. (continued)

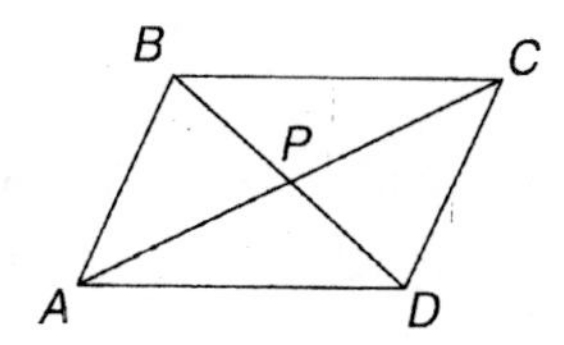

Proof:

Let P be the midpoint of $\overline{AC}$ and rotate $ABCD$ through an angle of 180° about P. Since rotations take parallel lines to parallel lines and A goes to C, then $\overleftrightarrow{AB}$ is mapped onto $\overleftrightarrow{CD}$. Thus B maps onto $\overleftrightarrow{CD}$.

Similarly, since C maps onto A, $\overline{BC}$ maps onto the parallel line $\overleftrightarrow{AD}$. If B is mapped onto $\overleftrightarrow{CD}$ and $\overleftrightarrow{AD}$, then it must be mapped onto the intersection of $\overleftrightarrow{CD}$ and $\overleftrightarrow{AD}$, which is D. Since B goes to D, $\overline{BP}$ must go to $\overline{DP}$. (P remains fixed because it is the center of rotation.) Therefore, $\overline{BP} \cong \overline{DP}$ so P is the midpoint of $\overline{BD}$ as well as the midpoint of $\overline{AC}$. Thus, the diagonals bisect each other.

23. (a) A complete solution is given in the answer key at the back of the text.

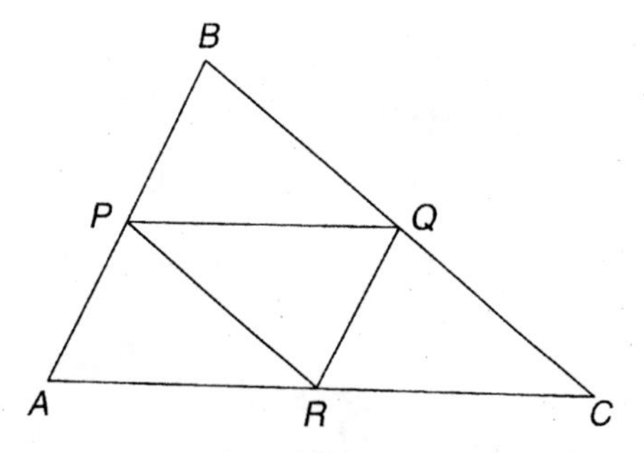

(b) Since P and Q are midpoints of $\overline{BA}$ and $\overline{BC}$, respectively, the size transformation with center B and scale factor 2 takes P to A and Q to C. Therefore, ΔBAC is the image of ΔBPQ under the size transformation with center B and scale factor 2.

(c) Since Q and R are midpoints of $\overline{BC}$ and $\overline{AC}$, respectively, the size transformation with center C and scale factor 2 takes R to A and Q to B. Hence, this size transformation takes ΔQCR to ΔBCA.

25. In each successive picture, parts of the shaded area are translated to other positions. For example, notice that in the second picture a shaded region with the same area as the triangle is translated up to fill the triangle. Still, the total area shaded remains the same. In the third picture, the entire shaded area of the second picture has been translated up. The process continues until the two smaller squares are filled with the shaded parts originating in the largest square. In each transformation, the shaded parts still have the same area as the original shaded region.

Additional Problems

27. One path is shown.
Step 1: Find the reflection of B about line l. Label this point B_1.
Step 2: Find the reflection of B_1, about line m. Label this point B_2.
Step 3: Find the reflection of B_2 about line n. Label this point B_3.
Step 4: To hit ball B after reflecting off three rails, aim A for point B_3.

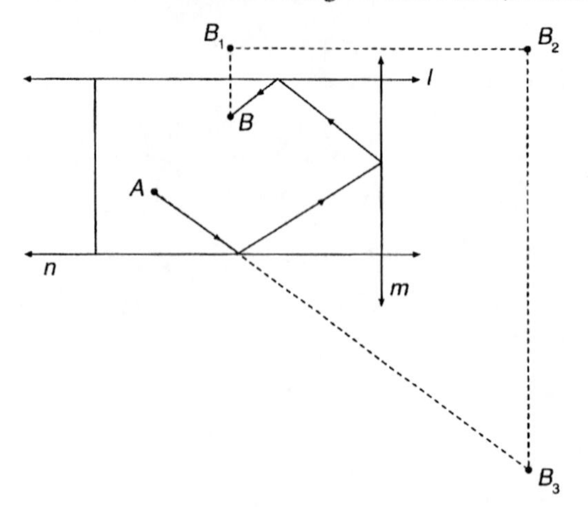

Additional Problems

1. Draw a line of reflection (either horizontal, vertical, or one of the two diagonal lines. Then shade pairs of squares and their images. Here are twelve possible quilts. Many other quilts are possible.

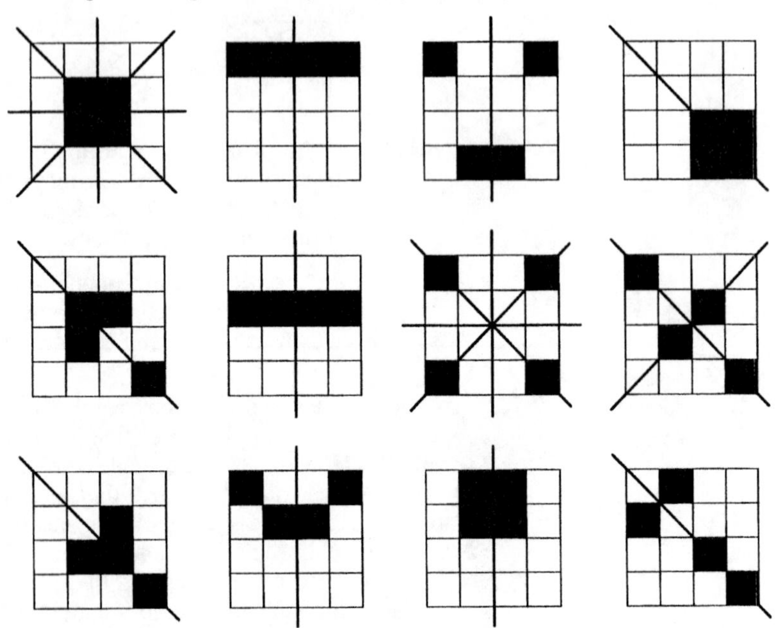

Additional Problems

2. (i) Reflect points $P(a, b)$ and $Q(c, d)$ about the y-axis to obtain points $R(-c, d)$ and $S(-a, b)$. Now, slope of $\overline{PS} = 0$ and slope of $\overline{QR} = 0$, so $\overline{PS} \parallel \overline{QR}$.

$PQ = \sqrt{(a - c)^2 + (b - d)^2}$ and

$SR = \sqrt{[-a - (-c)]^2 + (b - d)^2} = \sqrt{(a - c)^2 + (b - d)^2}$

so, $PQ = SR$ and $\overline{PQ} \cong \overline{SR}$. Thus, $PQRS$ is an isosceles trapezoid.

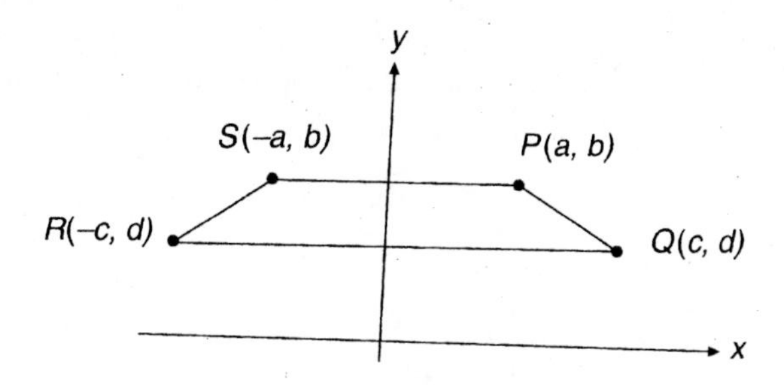

(ii) Reflect points $P(a, b)$ and $Q(c, d)$ about the x-axis to obtain points $R(c, -d)$ and $S(a, -b)$. The slope $\overline{PS}$ is undefined and the slope of $\overline{QR}$ is undefined, so $\overline{PS} \parallel \overline{QR}$.

$PQ = \sqrt{(a - c)^2 + (b - d)^2}$ and

$SR = \sqrt{(a - c)^2 + [-b - (-d)]^2} = \sqrt{(a - c)^2 + (b - d)^2}$

so, $PQ = SR$ and $\overline{PQ} \cong \overline{SR}$. Thus, $PQRS$ is an isosceles trapezoid.

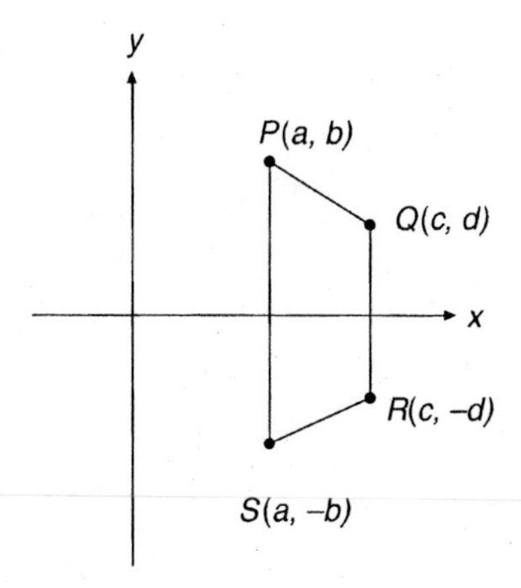

Note: Other answers are possible.

Chapter Review

Solutions to Chapter 9 Review

Section 9.1

1. (a) The image of $ABCD$ after the translation as shown in the answer key is $A'B'C'D'$.

   (b) Since a translation from $P(0, 0)$ to $Q(-1, -2)$ moves points 1 unit left and 2 units down, the image of (x, y) is $(x - 1, y - 2)$.

2. The image of (a, b) under a 90° rotation about the origin is $(-b, a)$. Hence the images of the given points are:

   (a) $(-3, 2)$

   (b) $(-3, -1)$

   (c) $(-4, -1)$

   (d) $(2, -4)$

   (e) $(4, 2)$

   (f) $(-y, x)$.

3.

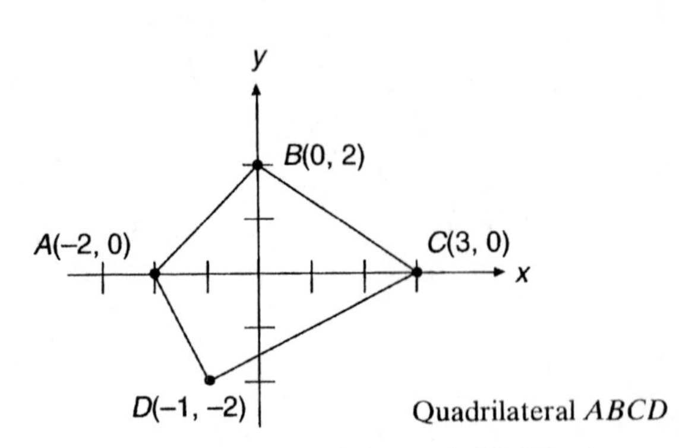

Quadrilateral $ABCD$

   (a) The graph below shows the image of $ABCD$ after a rotation of 180° about the origin. The coordinates of the image $A'B'C'D'$ are $A'(2, 0)$, $B'(0, -2)$, $C'(-3, 0)$, and $D'(1, 2)$.

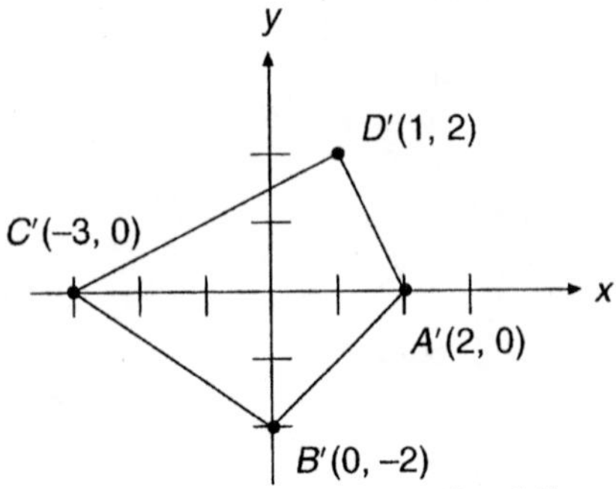

Quadrilateral $A'B'C'D'$

Chapter Review

3. (continued)

   (b) The graph below shows the image of $ABCD$ after a reflection over the x-axis. The coordinates of the image $A''B''C''D''$ are $A''(-2,0)$, $B''(0,-2)$, $C''(3,0)$, and $D''(-1,2)$.

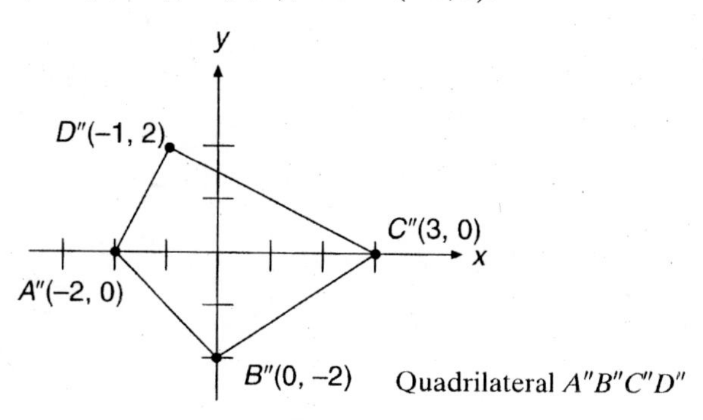

Quadrilateral $A''B''C''D''$

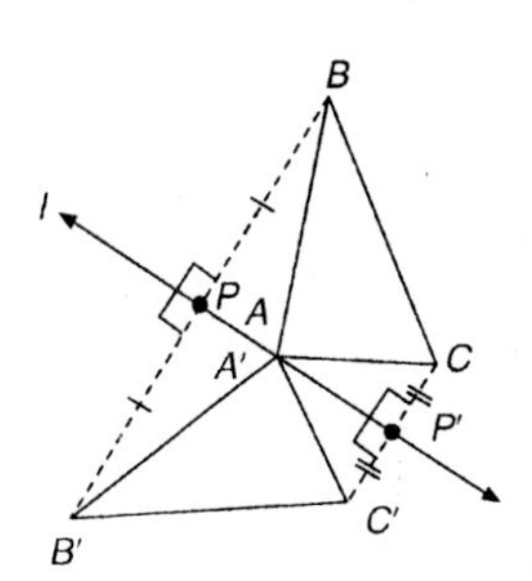

4. Step 1: Find the line perpendicular to l containing B. (See Construction 5 of Section 4.4.) Label P the intersection of l and the perpendicular line.

   Step 2: Find B' on $\overleftrightarrow{PB}$ so that $PB = PB'$.

   Step 3: Find the line perpendicular to l containing C. Label P' the intersection of l and the perpendicular line.

   Step 4: Find C' on $\overleftrightarrow{P'C}$ so that $P'C = P'C'$.

   Step 5: Connect points A, B' and C'.

   Note that points A and A' coincide as point A lies on the line of reflection.

5. (i) A translation from $P(2, 3)$ to $Q(-3, 5)$ moves points 5 units down and 2 units right.

   (ii) A reflection with respect to the y-axis results in the x-coordinate of a point being opposite in sign and the y-coordinate remaining the same.

Point	→	Translation from $P(2,3)$ to $Q(-3,5)$	→	Reflection about y-axis
a) $(6, 10)$	→	$(6-5, 10+2) = (1, 12)$	→	$(-1, 12)$
b) $(-5, 1)$	→	$(-5-5, 1+2) = (0, 3)$	→	$(0, 3)$
c) $(-4, -8)$	→	$(-4-5, -8+2) = (-9, -6)$	→	$(9, -6)$
d) (x, y)	→	$(x-5, y+2)$	→	$(5-x, y+2)$

Section 9.2

1. (a) Measure $\overline{AB}$. Mark B' at a distance of $2(AB)$ on $\overrightarrow{AB}$. Measure $\overline{AC}$. Mark C' at a distance of $2(AC)$ on $\overrightarrow{AC}$.

   Connect A', B' and C'.

   Note: $A = A'$ since A is the center of the size transformation.

Chapter Review

1. (continued)

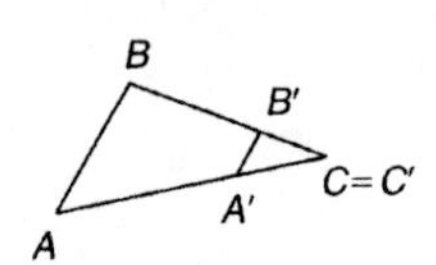

(b) Measure $\overline{CA}$. Mark A' at a distance of $\frac{1}{3}(CA)$ on $\overrightarrow{CA}$. Measure $\overline{CB}$. Mark B' at a distance of $\frac{1}{3}(CB)$ on $\overrightarrow{CB}$. $C' = C$ since C is the center of the size transformation. Connect A', B' and C'.

2. (a) The center, O, of the size transformation is the intersection of $\overleftrightarrow{AA'}, \overleftrightarrow{BB'}, \overleftrightarrow{CC'}$ and $\overleftrightarrow{DD'}$. The scale factor is the ratio

$$\frac{OA'}{OA} = \frac{OB'}{OB} = \frac{OC'}{OC'} = \frac{OD'}{OD} = 3.$$

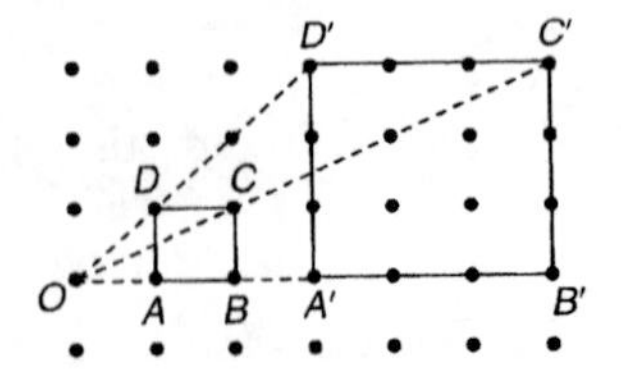

(b) $\frac{A'B'}{AB} = \frac{3}{1}; \frac{B'C'}{BC} = \frac{3}{1}; \frac{C'D'}{CD} = \frac{3}{1}$ and $\frac{D'A'}{DA} = \frac{3}{1}$

Thus each ratio is 3.

(c) $\frac{\text{Area of } A'B'C'D'}{\text{Area of } ABCD} = \frac{9}{1} = 9$

3.

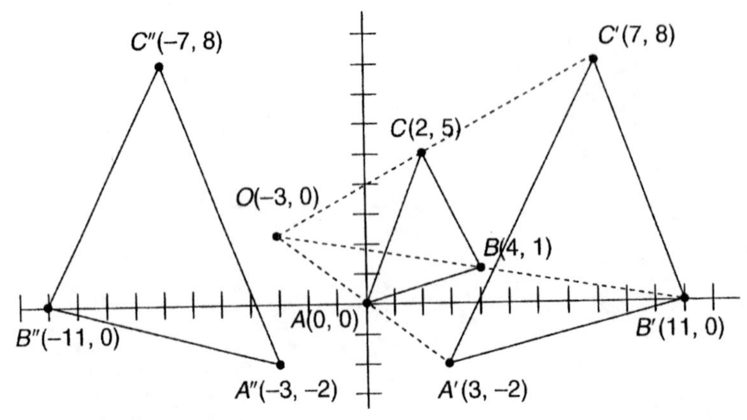

To find $A'(x_1, y_1)$ note that A is the midpoint of $\overline{OA'}$ hence

$O = \frac{-3 + x_1}{2}$ and $O = \frac{2 + y_1}{2}$; thus $A'(x_1, y_1) = (3, -2)$.

To find $B'(x_2, y_2)$, note that B is the midpoint of $\overline{OB'}$, hence

$4 = \frac{-3 + x_2}{2}$ and $1 = \frac{2 + y_2}{2}$; thus $B'(x_2, y_2) = (11, 0)$.

Chapter Review

3. (continued)

To find $C'(x_3, y_3)$, note that C is the midpoint of $\overline{OC'}$, hence

$2 = \dfrac{-3 + x_3}{2}$ and $5 = \dfrac{2 + y_3}{2}$; thus $C'(x_3, y_3) = (7, 8)$.

To find the coordinates of A'', B'' and C'', take the opposite of the x-coordinates of A', B', and C', and leave the y-coordinates the same. Hence, we have

$A''(-3, -2), B''(-11, 0)$, and $C''(-7, 8)$.

$\Delta A''B''C''$ is similar to ΔABC because there is a similitude that takes ΔABC to $\Delta A''B''C''$.

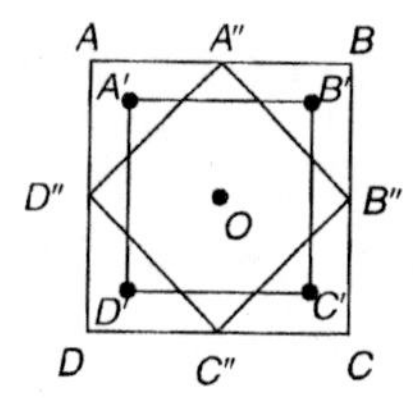

4. Find the size transformation that takes $ABCD$ to $A'B'C'D'$. The center of the size transformation is O, the center of the square. The scale factor is $\dfrac{OA''}{OA}$. ΔAOB is a 45°–45° right triangle with hypotenuse $AB = 4$; thus $OA = 2\sqrt{2}$. Using center O and scale factor $\dfrac{OA''}{OA} = \dfrac{2}{2\sqrt{2}} = \dfrac{1}{\sqrt{2}} = \dfrac{\sqrt{2}}{2}$, we obtain square $A'B'C'D'$. Now, rotate $A'B'C'D'$ about O an angle of $-45°$ to obtain the image of $A''B''C''D''$.

5. Apply a size transformation of $\dfrac{GH}{AB}$ about O, the center of $ABCDEF$ to obtain $A'B'C'D'E'F'$. Then use a translation $\overrightarrow{A'G}$ to take $A'B'C'D'E'F'$ to $GHIJKL$.

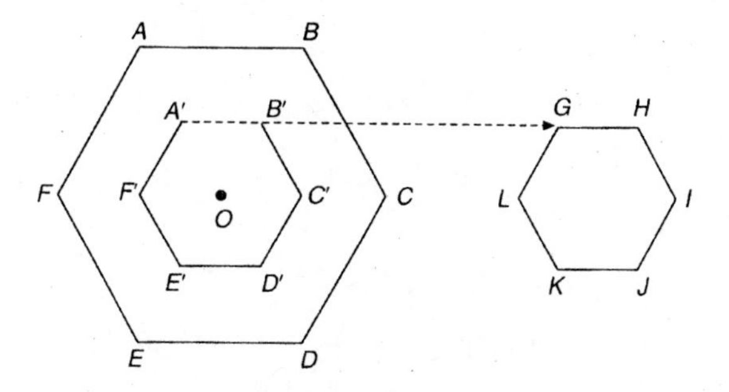

Section 9.3

1. Reflect $ABCD$ about $\overleftrightarrow{EG}$ and $\overleftrightarrow{HF}$ will map $ABCD$ onto itself.

Rotate $ABCD$ about point O 180° will map $ABCD$ onto itself. Note that a rotation of 360° about O will also map $ABCD$ onto itself, as it is an identity transformation.

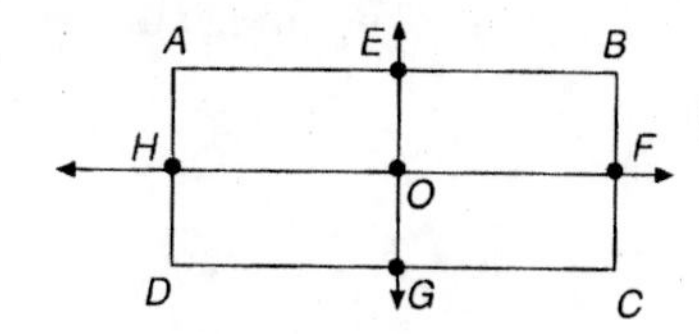

Chapter Test

2. This diagram accompanies the complete solution given in the answer key at the back of the text.

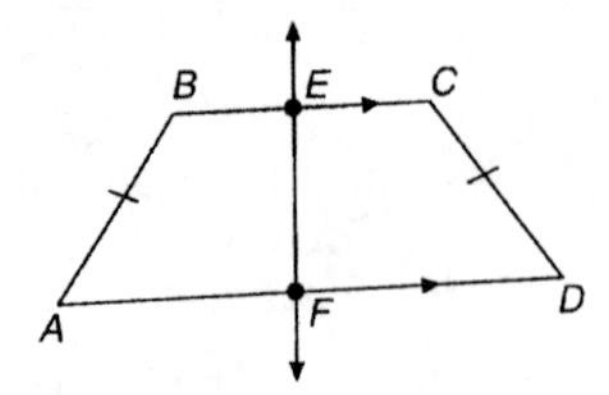

3. A complete solution is given in the answer key at the back of the text.

Solutions to Chapter 9 Test

1. T The center of the rotation is always fixed.

2. T Every point on the line of reflection is fixed.

3. T See Theorem 9.7.

4. F Because a glide reflection involves a reflection, the line segment does not necessarily map onto a parallel line segment.

5. T An isometry is, by definition, a transformation that perserves distance.

6. F Translations and rotations preserve orientation, but a reflection reverses the orientation of a triangle.

7. T The transformation shifts every point left 3 units.

8. T The transformation is a reflection with respect to the y-axis.

9. F The transformation may be a rotation.

10. F There is a similitude, but that may involve a size transformation and an isometry.

11. $\overrightarrow{PQ}$ translates each point to the right $1 - (-2)$ or 3 units and then up $6 - 4$ or 2 units.

    (a) Thus, $A' = (-3 + 3, -1 + 2) = (0, 1)$,

    $B' = (0 + 3, 2 + 2) = (3, 4)$,

    and $C' = (4 + 3, 1 + 2) = (7, 3)$.

    (b) The image of (a, b) is $(a + 3, b + 2)$

Chapter Test

12. Rotate each vertex separately. The dashed lines and the 40° angle for vertex B are shown.

    Step 1: Draw a dashed line from the vertex to O.

    Step 2: Use your protractor to measure a 40° clockwise angle with the dashed line as the initial side.

    Step 3: Measure the distance from the vertex to O. Mark off that distance from O on the terminal side of the angle. Label the point as the image of the vertex.

    Step 4: Use a straightedge to join the new set of vertices to form the rotated image.

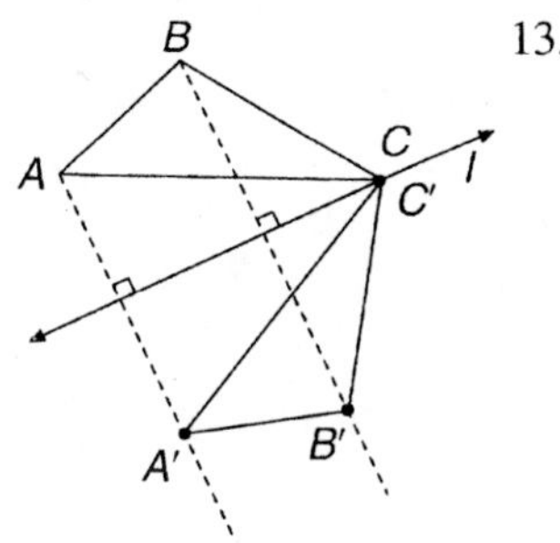

13. Step 1: Construct a perpendicular line from A through line l.

    Step 2: Measure the distance from A to the intersection of the perpendicular and line l. Mark off that distance on the perpendicular line on the other side of line l. Label the point A'.

    Step 3: Repeat Step 1 and Step 2 with point B to find B'.

    Step 4: Since C is on line l, it is a fixed point, and $C = C'$.

    Step 5: Use your straightedge to join A', B', and C' to form $\Delta A'B'C'$.

14. First find $A'B'C'D'$, the image of $ABCD$ under the translation $\overrightarrow{AB}$.

    Step 1: Extend $\overline{AB}$ and use the ruler to measure $\overline{AB}$ and then locate B' on $\overrightarrow{AB}$.

    Step 2: Use the protractor to measure $\angle B'BC$. Then use the protractor to copy $\angle B'BC$ at vertex C and D.

    Step 3: Use the ruler to locate points C' and D'.

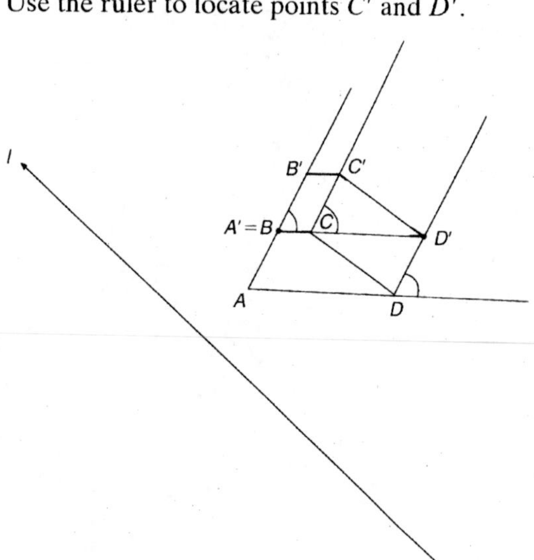

Chapter Test

14. (continued)

Step 4: Use the protractor to find the line perpendicular to l passing through point A'.

Step 5: Repeat Step 4 for points B', C' and D'.

Step 6: Connect points A', B', C' and D'.

Step 7: Use a ruler and measure the distance from A' to line l. Use this distance to locate A''.

Step 8: Repeat Step 6 using points B', C', and D' to locate points B'', C'' and D'' respectively.

Step 9: Connect points A'', B'', C'' and D''.

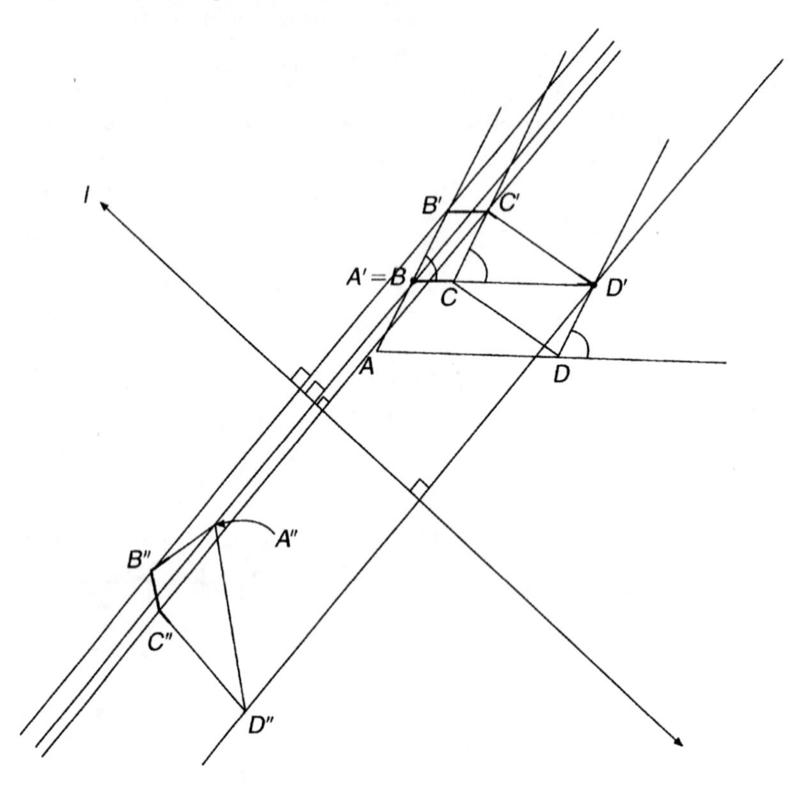

15. Remember that $\angle AOA'$ and $\angle BOB'$ each need to be 270°, and the distances AO and BO should be preserved. (That is, make sure that $AO = A'O$ and $BO = B'O$.) A sketch of the solution is given in the answer key in the text.

16. Don't forget that the distance from each vertex to O must be doubled. For example, $A'O = 2(AO)$. A sketch of the solution is given in the answer key in the text.

Chapter Test

17. (a) See the sketch below.

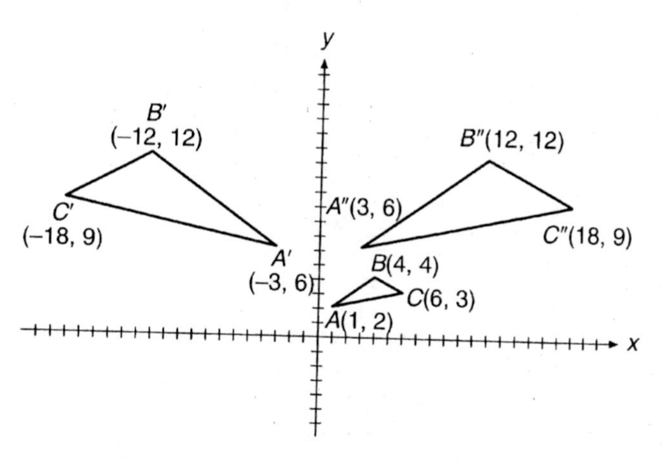

(b) A size transformation centered at the origin with a scale factor of 3 will triple each of the coordinates of the vertices.

$$A = (1,2) \rightarrow (3 \times 1, 3 \times 2) = (3,6) = A''$$
$$B = (4,4) \rightarrow (3 \times 4, 3 \times 4) = (12,12) = B''$$
$$C = (6,3) \rightarrow (3 \times 6, 3 \times 3) = (18,9) = C''$$

To reflect with respect to the y-axis, change the sign of the x-coordinate.

$$A'' = (3,6) \rightarrow (-3,6) = A'$$
$$B'' = (12,12) \rightarrow (-12,12) = B'$$
$$C'' = (18,9) \rightarrow (-18,9) = C'$$

(c) In general, the size transformation and reflection change (a, b) to $(3a, 3b)$ to $(-3a, 3b)$.

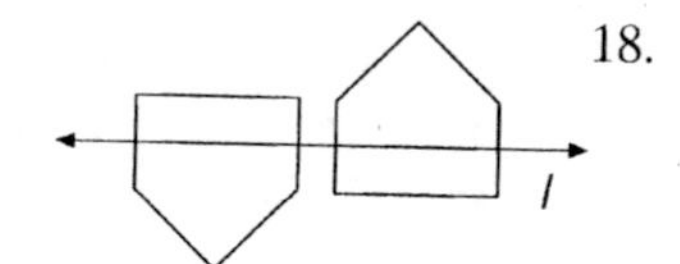

18. (a) The figure has been translated by gliding to the right and then reflected about line 1 which goes through the middle of the rectangular section (See diagram).

(b) Since orientation is preserved this must be a rotation rather than a reflection. To find the center of rotation, join each pair or corresponding vertices with a line segment. Then draw the perpendicular bisector of each line segment.

(c) The top figure has been reflected with respect to a horizontal line halfway between the two figures.

19. A complete solution is given in the answer key at the back of the text.

20. A complete solution is given in the answer key at the back of the text.

21. You could picture a glide reflection as a point A moving along one side of a rectangle to the next corner A''. (See the figure at left below.) Then A'' is reflected across line l to the opposite corner A'. In other words, A'' now moves along the side $\overline{A''A'}$ of the rectangle.

Glide, then reflect

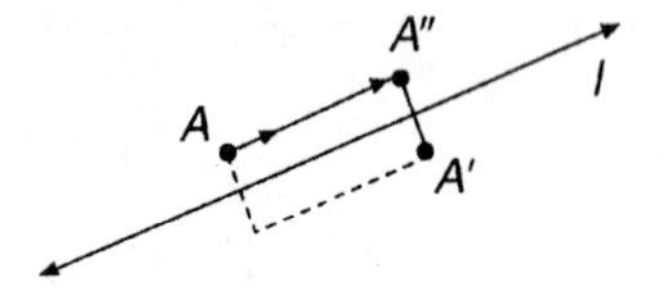

Reflect, then glide

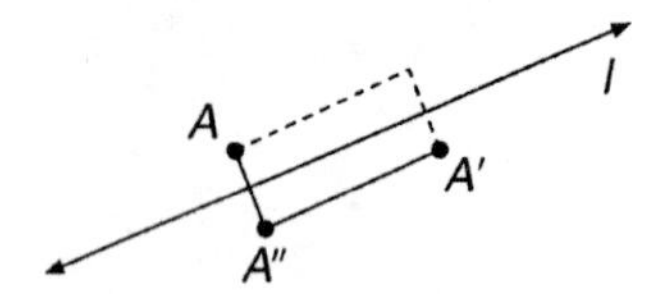

If the reflection were performed first, then A would first be reflected across line l, moving along side $\overline{AA''}$. (See the figure at right above.) Then A'' would move along the side opposite A to A'.

The results are the same.

22. A complete solution is given in the answer key at the back of the text.

23. One solution has been given in the answer key. Others are possible. Remember that the angle at which the ball hits the side has the same measure as the angle that it makes when it bounces off the side.

24. You want to double the area of the logo shown. If each dimension is doubled, the area is quadrupled because the dimensions are multiplied by each other. (See Section 3.1 problem #20, Section 3.2 problem #23 and Section 9.2 problem #9.) In order to double the area, you need a scale factor that when squared gives you 2, i.e., $k^2 = 2$. Therefore, $k = \sqrt{2}$ is the needed scale factor. If she places the lower left-hand corner of the logo at the origin and uses a size transformation centered at the origin with scale $\sqrt{2}$, the area will be doubled.